BANGLADESH: WHOSE IDEAS, WHOSE INTERESTS?

BANGLADESH
WHOSE IDEAS, WHOSE INTERESTS?

Geoffrey D Wood

Practical Action Publishing Ltd
25 Albert Street, Rugby, CV21 2SD, Warwickshire, UK
www.practicalactionpublishing.com

in association with

University Press Ltd
Red Crescent Building
114 Motijheel C/A, PO Box 2611,
Dhaka 100, Bangladesh

© University Press Ltd, 1994

First published in 1994
Transferred to digital printing in 2008

ISBN 10: 1 85339 246 4
ISBN 13 Paperback: 9781853392467
ISBN Library Ebook: 9781780444543
Book DOI: https://doi.org/10.3362/9781780444543

All rights reserved. No part of this publication may be reprinted or reproduced or utilized in any form or by any electronic, mechanical, or other means, now known or hereafter invented, including photocopying and recording, or in any information storage or retrieval system, without the written permission of the publishers.

A catalogue record for this book is available from the British Library.

The authors, contributors and/or editors have asserted their rights under the Copyright Designs and Patents Act 1988 to be identified as authors of their respective contributions.

Since 1974, Practical Action Publishing has published and disseminated books and information in support of international development work throughout the world. Practical Action Publishing is a trading name of Practical Action Publishing Ltd (Company Reg. No. 01159018), the wholly owned publishing company of Practical Action. Practical Action Publishing trades only in support of its parent charity objectives and any profits are covenanted back to Practical Action (Charity Reg. No. 247257, Group VAT Registration No. 880 9924 76).

Reasonable efforts have been made to publish reliable data and information, but the author and publisher cannot assume responsibility for the validity of all materials or for the consequences of their use.

Cover design by Mohiuddin Ahmed

The manufacturer's authorised representative in the EU for product safety is Lightning Source France, 1 Av. Johannes Gutenberg, 78310 Maurepas, France. compliance@lightningsource.fr

Contents

Acknowledgements vii

Introduction: Understanding and Practice   1

PART I
CLASS FORMATION AND UNIVERSAL RURAL DEVELOPMENT MODELS

Chapter 1	Exploitation and the Rural Poor	31
Chapter 2	Rural Class Formation in Bangladesh 1940-80	100
Chapter 3	Rural Development in Bangladesh: Whose Framework?	127
Chapter 4	Women and Gender	152

PART II
LANDLESS PARTICIPATION IN AGRICULTURAL GROWTH

Chapter 5	Rural Employment and Patterns of Agricultural Development	163
Chapter 6	The Rural Poor in Bangladesh: A New Framework?	172
Chapter 7	Provision of Irrigation Services by the Landless: An Approach to Agrarian Reform in Bangladesh	193
Chapter 8	The Social Framework of Rural Exchange in Bangladesh	215
Chapter 9	Agrarian Entrepreneurialism in Bangladesh	233

PART III
RURAL WORKS: DEVELOPMENT OR WELFARE

Chapter 10	Landless Labour Participation and Mobilisation in Rural Works Programmes	259
Chapter 11	Targets Strike Back—Rural Works Claimants in Bangladesh	290

vi *Bangladesh: Whose Ideas, Whose Interests?*

Chapter 12 Rural Infrastructure and Social Relations: The Intensive Rural Works Programme in Bangladesh   309

Chapter 13 Plunder without Danger: Avoiding Responsibility in Rural Works Administration in Bangladesh   325

PART IV
FISH AND POVERTY

Chapter 14 Off the Page and into the Pond: Fish Extension Strategies   341

Chapter 15 Open Water Bodies and Capture Fishery: The Poverty of Policy   384

PART V
STRATEGIES WITH THE POOR

Chapter 16 Government Approaches towards the Rural Poor in Bangladesh   403

Chapter 17 Sirs and Sahibs: Government and Technical Assistance Relations in Rural Development Projects   428

Chapter 18 Philosophies of Economic Change: Three EIG models   475

Chapter 19 NGOs and the Theory of Struggle   484

Chapter 20 Target Groups and the Resource Profile Approach   490

PART VI
GOOD GOVERNANCE AND THE FRANCHISE STATE

Chapter 21 Parallel Rationalities in Service Provision: The General Case of Corruption   519

Chapter 22 States without Citizens: The Problem of the Franchise State   541

Conclusion: Strategic Dilemmas   557

Bibliography   565

Glossary of Non-English Words   577

Acronyms   580

Index   583

Acknowledgements

With a volume of this kind, papers drawn together over nearly 20 years, the list of people to thank could be very large indeed. At the outset, I should apologise to those not named in person but included in institutional references, and also to those whom I have failed to identify as having a special influence over my thinking.

My initial thanks go to Ms. Hazel Wallis, who has undertaken, with tremendous goodwill and patience, the work of assembling the original items (involving scanning mimeo texts and complicated transfers from other word programmes), and editing them with me. Her editorial proposals always seemed to improve the clarity of meaning.

I must also thank The Ford Foundation in Dhaka, Bangladesh for a small grant to prepare the book. The Representative, Raymond Offenheiser, and then Doris Capistrano made constructive inputs to the original proposal as part of their support. The Ford Foundation has to be acknowledged more widely as, in effect, the sole sponsor of my first decade on Bangladesh through the Academy at Comilla, with various ministries and with Proshika. In particular I should mention George Zeidenstein, Bill Fuller, Marty Hanratty, Anthony Bottrall, the late Ken Marshall, Adrienne Germaine and Mr. A. Hannan and his team of support staff. All have displayed endless patience as deadlines were met, programmes changed, meetings organised, and fanciful ideas floated.

The Swedish International Development Authority (SIDA) started to commission inputs from me in the second decade. My first association was with Daniel Asplund, followed rapidly by Christer Holtsberg, Mats Svensson and, later, Bo Sundstrom. Intense discussions about aid and Bangladesh have occurred between us. I must especially thank Matts and Inge, his wife, for incredible forebearance and endless stimulation. They may never forget a New Years eve in Bangkok! Although not a Swede, Jean-Louis Leterme will know why I bow to him in this paragraph.

My formal association with the UK-ODA in Bangladesh only began in the late eighties in the context of the NW Fish Culture project. My particular thanks to Rick Gregory and his wife, Jan, who were my hosts on many occasions in Saidpur. Rick's feel for fish and his basic common-sense led to much creative interaction between us. Paul Francis at SEADD, Bangkok, was a constructive critic, sharpening my act. Fiona Duby has been a wonderfully supportive friend, as has more recently David Clarke.

From my time at Comilla, I must record the friendship of the late Ameerul Huq and his wife and daughters. He guided me through my first, faltering steps in the country. Md. Yahiya assisted me during fieldwork, as did Aroma Goon (Bua) who continues to advise me on most aspects of my life. Our lives are intertwined. Yahiya subsequently introduced me to Proshika, for which I shall ever be grateful. I also met Mahera Khatoun, now with UNICEF, at that time. She remains a valued friend. Although we have not met often, Peter Bertocci is mentioned here for the value of his pioneering anthropology in the region and for providing a set of findings and arguments with which I could engage. While our analyses differed, I have hugely enjoyed our early debates and all too rare meetings.

Many visitors to Bangladesh have reason to thank Abu Abdullah at BIDS. His place in the history of agrarian studies in Bangladesh is assured, and he continues to place the work of upstarts in proper context. Rehman Sobhan has also been generous with his time. Other BIDS fellows, past and present must forgive me for not mentioning further names. At the University of Dhaka, Mosharaff Hossein and B.K. Jahangir have been constant mentors, and I am especially grateful to B.K. for publishing me on a number of occasions in his journal (The Journal of Social Studies). Over the years, there have been many meetings, workshops and seminars in other universities in the country, but I will single out Sattar Mandal from the Agricultural University in Mymensingh and Nurul Islam from the Institute of Appropriate Technology at BUET. Then, of course, there was Md. Yunus at Chittagong who subsequently formed Grameen Bank.

Within GOB, I was privileged to work closely with A.K.M. Obaidullah Khan, especially when he was in Agriculture. He and several other senior officers demonstrated a wide range of interests and erudition, reminding narrow-minded *bideshis* of the depth of Bengali culture. Mr. Anisussaman was another example in both Rural Development and later Agriculture, brimming with ideas and keen to enter debate. Hopefully I gave as good as

Acknowledgements ix

I got on a number of heated, but constructive occasions. Both were in a position to give me the freedom to think the unthinkable and experiment, and both did so. Hasnat Abdul Hye, at various points in his career, has also critically discussed ideas, bringing some of the wilder ones into the realm of reality. In his team, when at BRDB, I should also cite two ex-Comilla colleagues, Mahmudur Rahman and Rezaul Karim. Both continued to be patient and generous as they encountered my PhD students, after me!

I would not feel happy if the following did not appear here as friends, advisers and debaters: Harry Blair, Kiki McCarthy, Eiric Jansen and Bodil Maal, Therese Blanchet, Jim Boyce and Betsy Hartman, Jon and Polly Griffiths, Richard Palmer-Jones (my co-author for *The Water-Sellers*), Bosse Kramsjo (my co-author for *Breaking the Chains*), Don Parker, Wahida Huq, Rokheya Kabir, Sultana Begum, Abed, Salehuddin Ahmed, Shams Mustafa, Albert Malakar, Sven Olaf Storm, Shapan Adnan, Syed Atiqullah, Kerry-Jane Wilson, Mark Goldring, Tanya Rahman, Bruce and Mahtab Currey, Zahin Ahmed, Richard Holloway, Keith Pitman, Linda Hearne, Hugh Brammer, M.A. Hamid, Ro Cole, Farida Akter, Susan Davis, Edward Clay, Bill Harvey, Allison Barret, Ines Zalitis, Shelley Feldman, Stephen Biggs.

The Bath-Bangladesh connection was further established through a 'team' of PhD students during the eighties: Allister McGregor, Sarah White, Marion Glaser and David Lewis. This has been a most stimulating experience, with much cross fertilisation and mutual support. They will recognise themselves in various sections of this book. It was their initiative to establish the European Network of Bangladesh Studies to support postgraduates and other students of Bangladesh resident (permanently or temporarily) in Europe. Mark Ellison at Bath is the secretary, and other leading members include long-standing friends like Kirsten Westergaard and Willem van Schendel.

I have left some names until the end of this lengthy 'acknowledgements'. They occupy a special place in the evolution of my thinking. Now the director of Gono Shahajo Sangstha (GSS), Mahmud Hasan and I met first in the UK when he was finishing a PhD and working for the BBC World Service. That was when we first started to argue, as any two leftists will. He honoured his London agenda on his return to Bangladesh, though, first, he analysed other NGOs as the basis for creating GSS. His manuscript was ahead of its time and should have been published widely. I kept in touch with him and GSS as it evolved, though not always agreeing. It was an honour to be selected to lead the evaluation team which examined GSS in February 1993, and still gratifying to be arguing about things that really matter.

Mahmud continues to tease me, gently, about my strong association with Proshika Manobik Unnayan Kendra. While there is not much specifically about Proshika in this volume, its experience informs much of it. This association began in 1979, and has been intimate ever since. Among the many colleagues and friends throughout Proshika, there are some special thanks due. Qasi Faruque Ahmed, the Executive Director, has exceeded any imaginable principle of generosity, patience and insight. Together with other colleagues we have created ideas through words and practice... I leave the reader to guess the division of labour! My times with Proshika always leave me feeling guilty that I have taken, not given. I have been privileged with widespread fieldwork opportunities and access to very private institutional moments which have enriched my appreciation of the development process. The most sensitive discussion and criticism has always been encouraged. As a result, Proshika remains at the frontier of development initiative. Faruque is strongly supported by his colleagues. Mabhub Karim, now the Director of Proshika's Institute for Policy Analysis and Advocacy, has been equally significant in the evolution of strategy, and always keen to deal with fundamentals despite his responsibilities in building up training capacity and the centre at Koitta. His grasp of domestic politics and the political economy of the country in the broader international and historical context has been shared on numerous occasions. Other colleagues enhance this picture: David Biswas, Khaze Alam, Shahabuddin, Assad, Serajuddin, Santosh and many, many others at all levels. Many nights of discussion and song, long days and nights in area development centres and village meetings with group members. Then Shahnewaz and his wife, Eva, who also works for Proshika, occupy a special place in these acknowledgements. Shahnewaz and I have worked together on a number of assignments, especially with SIDA. He also came to Bath for an MSc, with Eva joining later. He represents a freedom of spirit, an independence of thinking, a fearless disregard for personal consequences which displays the best Bengali tradition. Between them, they have educated me—though they are not responsible for the inadequacies of what follows, those are all down to me.

Finally Angela, my wife, and Kai and Marsha. All have been to Bangladesh, with Kai and Marsha displayed as babies to the village which saw Angela pregnant. Angela interrupted her own career to participate in the first fieldwork in Comilla, shameless exploitation on my part. From the seventies, they have endured countless absences, often coinciding with the worst weather here as the winter bird flew East. Others can judge from the pages which follow whether the extent of family disruption has been worth it. But I use this occasion to thank them publicly for their patience and forebearance.

Throughout this volume, it is clear that my most important teachers have been the many men and women, especially the poorer ones, whom I have encountered across the villages of Bangladesh. Some of them, appropriately disguised, appear as characters in what follows. It is to them that I dedicate this volume.

Geoffrey D Wood
Bath
September 1993

Introduction

Understanding and Practice

This book brings together a selected and edited collection of my various writings on Bangladesh since I started working on problems of rural poverty and development in 1974. Some of these pieces have been published in various places in and outside Bangladesh, others have appeared in the form of reports for different agencies, and still others have not appeared in any other medium. This introduction opens with a discussion of the relationship between social science and the development policy process, with references to various stages of my own work but also commenting on the wider connection between research and development aid. It proceeds with a presentation of the rationale for the various sections which follow and the 'intellectual' conditions under which items were written. Finally, it outlines an overall argument which considers options for: an agrarian future which acknowledges urbanisation and the continuing fragmentation of farms; the interaction between agricultural involution and intrusion of agricultural capital; the viability of alternative organic strategies; and the consequences of market imperfections and weak states for the organisational landscape in development policy and implementation.

The rationale for the volume has been prompted by the continuing (and therefore, of course, flattering) requests for various pieces, often accompanied by a complaint that they can no longer be found or that they are inaccessible in some agency office. Such an arbitrary 'testing' of the market may not, in itself, be a sufficient condition for such a volume. It is important that it contains new, unpublished thinking alongside the history of ideas deriving from earlier pieces, i.e. that it be considered relevant to the present and future of Bangladesh not just a record of the past. I should, however, admit to a certain desire to demonstrate some of the prophetic aspects of earlier pieces.

It is always difficult to claim original insight for oneself, but whether illusory or not, I claim some originality for providing an analysis of:

- the class characteristics of exploitation among the small farming communities of Comilla;
- the inappropriate transfer of the Comilla model to other parts of Bangladesh;
- the regional variations in the rural political economy, distinguishing between the mechanisms of rents and usury;
- the significance of landlessness as a major policy issue;
- the opportunities, as opposed to the constraints, for employment and rural business activity arising out of the process of farm fragmentation and the related futility of pursuing redistributive land reform as a poverty measure ('the disappearance of the Bangladesh farm');
- the ways in which the landless rural poor can be supported to capture some of these opportunities, especially via the provision of irrigation services;
- the limits of target-group strategies in poverty alleviation, especially in the context of rural works programmes (which prompted more theoretical work on 'labelling in development policy');
- the prospects for improving rural workers' rights through 'labour contracting societies' to overcome official corruption and exploitative forms of labour mobilisation;
- the need to understand people in terms of their total portfolio of survival options rather than a single, indicative label like 'landless' or 'fish-farmer';
- (with R. Gregory, I. Golder and D. Lewis) the significance of trading systems in fish pond culture and the use of traders as extension agents;
- the limits of 'entrepreneurial' panaceas to enhance rural incomes and the need to consider other employment models;
- and the implications of official rent-seeking in understanding local state-society interaction.

Even if only some of these claims were accepted by others as valid, the main purpose of this 'list' is to offer an explanation of the themes tackled in this volume, which reflect my package of ideas and interests over the last 2 decades.

Class Formation and Universal Rural Development Models

Since I have often been accused of being over-theoretical, and my writing at times too obtuse for widespread comprehension, it may come as a surprise that I have always considered my research and writing efforts in Bangladesh to have a direct policy intent. This may be the policy of governments (where deliberate assumptions have been made by me about the benevolence of the state or at least that the ideas of sensible people within government will prevail) or the policy of poor people and the organisations directly supporting their interests.

My first entry to Bangladesh was via government through the Bangladesh Academy of Rural Development at Comilla, an institution under the ultimate control of the Ministry of Local Government, Rural Development and Cooperatives. The work at Comilla, which culminated in Exploitation and the Rural Poor, was essentially based on the brief to understand why cooperatives did not deliver equality of benefits to their members, and additionally the consequences for those excluded from membership by virtue of their landlessness, or the weakness of other connections to local mini-elites. Later work in the seventies was located in the Planning Cell of the Ministry of Agriculture. Basically I followed the Secretary, A.Z.M. Obaidullah Khan, from the Rural Development Ministry into the Agriculture Ministry. Apart from channelling academic work across the country on agrarian change into the Ministry's strategic thinking, I started to prepare papers directly on rural development policy.

Two main themes arose from these papers: first, a critique of attempts to establish a universal model or framework for rural institutions and patterns of investment across the country; and second, in the context of World Bank support for the privatisation of an expanded agricultural sector, an argument for the direct inclusion of the landless in agricultural production via the ownership of other productive assets such as irrigation.

Landless Participation in Agricultural Growth

It must be noted that such 'ideas' did not just come out of the blue. They were rooted in a prior understanding of agrarian change and class formation in which minifundist farming systems and intensive local competition for resources undermined the prospect of farmers managing lumpy technologies like deep tube wells for themselves. At the same time, the multi-dimensional relationships of dependency within the village did not constitute facilitating conditions for individual landless families to enter new agricultural service markets. Solidarity was essential: group formation among landless workers was a prerequisite for success in these localised, interlocked markets. Thus these ideas for direct landless participation in agriculture (similar arguments were made for rural works) looked as though they would be stillborn, since there was little evidence of radical political party interest in such an agenda.

It was at this stage of 'great idea, shame about the practice' that earlier connections with some new, small, almost clandestine NGOs became relevant. The venality of successive governments, which also remained conveniently attached to the image of rural Bangladesh as a society of homogenous peasantry, and the inability of radical parties to devise a meaningful rural programme beyond cliches about land reform, had prompted the growth of a number of developmental, poverty-focused NGOs. Among these NGOs, Proshika Manobik Unnayan Kendra (hereafter just Proshika) developed one of the strongest traditions of organising groups of landless men and women based upon an unashamedly class analysis of the rural political economy. My assistant for the Comilla fieldwork, Md. Yahiya, had subsequently joined the Comilla wing of Proshika and he had introduced me to its small staff on my return to Bangladesh in 1978. Later, I met the Dhaka Proshika staff, and started to be an informal resource person on 'theory' from 1979 onwards while working on Ministry assignments.

Thus in 1980 it was possible, with Ministry support, to merge the ideas in 'The Rural Poor in Bangladesh: A New Framework' (written initially as a critique of World Bank plans for commercialisation of agriculture in the 2nd 5YP) with Proshika's interests in deploying landless group solidarity to enable them to capture productive assets in the countryside. The Ford Foundation, which had supported all my efforts to date, funded a project design and the costs of establishing a programme within Proshika's portfolio of employment and income generation projects (EIG). The results of a decade of irrigation experimentation on these lines has been published separately in The Water-Sellers (Wood and Palmer-Jones 1992), so that only an early article describing the process and its merits are included in Part II of this volume: 'Provision of Irrigation Services by the Landless: An Approach to Agrarian Reform in Bangladesh' (Chapter 7).

While pursuing the landless irrigation strategy, much else was learnt both about the pattern of rural transactions, processes of farm fragmentation and the effects of new capital upon the minifundist agrarian system. This prompted related writing of my own, as well as being the basis of securing grants for PhD students to work on aspects of this process (especially Glaser and Lewis). The first draft of the paper on 'Agrarian Entrepreneurialism...', presented as the keynote address to the Bengal Studies Association Conference in the USA in 1988, particularly explored the proposition that farmers were forced to buy services from 'agricultural companies' operating in the different command areas to which their plots were distributed, and that the interlocked character of these markets worked against the interests of the smaller, less well networked farmers, with the result that many small farms would in effect disappear as integrally managed entities. Glaser's work indicated that small farmers could avoid this outcome through reverse, seasonal and cost-share leasing; Lewis's work drew attention to ability of brokerage networks to mediate lumpy technologies like tractors into small-scale situations. Both of these findings modified my pessimistic prognosis for the viability of the small farmer. More recently, the experimentation through Proshika with organic cereal farming may have more profound implications for the way we should understand the future of the agrarian system in Bangladesh.

It is early days yet, but it is important to speculate on the significance of a new movement in farming and technology and to connect this to other potential features of the social landscape. Taking the technological issue first, the current findings from organic agriculture reveal the possibility of average size rural families being able to subsist on much smaller acreages through integrating the annual cycle of crop rotation more comprehensively for composting purposes and including in that cycle products from homesteads. Initial experiments among Proshika members reveal yields at par or even above those of high-tech, chemically dependent agriculture, and rates of return far in excess, with farmers avoiding expensive dependency on markets and adverse terms of trade on chemical and mechanical inputs (especially in the policy context of the withdrawal of subsidies on many of these input commodities).

If these findings are sustained under more rigorous testing and are seen to be widespread, then Bangladesh may be able to contemplate a viable strategy for sustainable agricultural development which is not import dependent, a drain on scarce foreign exchange, or environmentally damaging. However the difference between yields and rates of return to farmers is significant. The obvious question concerns the generation of agricultural surplus to feed the growing cities, i.e. gross yields are important, especially as the dependency ratio on agriculture declines. The productivity of labour and land has to increase. While some of these arguments are not yet resolved, particularly for example in the possibility of decline in the productivity of high-tech agriculture with declining fertiliser-yield coefficients, it is worth speculating on the social landscape itself.

The issue may be captured in the rather clumsy and unfashionable term 'rurbanisation'. Will urbanisation in Bangladesh be concentrated in the primate cities, as has happened elsewhere in the sub-continent and other parts of the developing world, or will it take the form of a more widely dispersed expansion of rural, market towns and decentralised administrative centres. In terms of theory, nothing very original here, with 'growth pole' patterns of development. Admittedly impressionistic at this stage, but both my casual observations as well as those of Bengali colleagues who have travelled consistently throughout Bangladesh over the last 20 years reveal substantial growth in these rural towns. These constitute significant markets for local and regional produce, agricultural as well as horticultural. We must expect these patterns of 'rurbanisation' to have an impact upon cropping patterns and land use. Whereas there has been a general trend towards increasing the gross cropped acreage under cereals, displacing the traditional cash crop of jute as well as vegetables and pulses, we might now expect a resurgence of local markets for vegetables and pulses, offering price incentives to change cropping patterns towards these crops, with potentially positive implications for rural people's diets as well. New market systems and transportation will develop accordingly. Land may be used in a more sustainable way, with natural but commercial use of nitrogen-fixing crops and mutually reinforcing rotations. Since Proshika has such a wide basis of activity in most regions of the country, it will be possible both to experiment with and research these possibilities over the next few years through its newly formed Institute for Development Policy Analysis and Advocacy, and to disseminate findings more widely into the national policy process.

Rural Works: Development or Welfare

Alongside a decade of praxis linking poverty alleviation directly to agriculture and agricultural growth, I was also able to contribute to other policy sectors where opportunities for the poor to participate more fully were being explored. I had indicated, rather fancifully, in 'The Rural Poor in Bangladesh: A New Framework' that landless rural labour should be able to own the product of their own rural works labour and charge rent to the beneficiaries of such infrastructure and thereby gain long term incomes rather than short term 'relief' wages. The rent might be justified by the continuing function of the infrastructure (e.g. a flood control embankment, an irrigation canal, a feeder road) as well as the continuing cost of maintenance, especially earthworks, damaged by regular monsoon conditions. I even suggested that sluice gates could be installed as part of a flood control embankment structure, a gate which could be raised to demonstrate the value of the service of providing protection from flood water to those farmers whose land was protected but who were reluctant to pay.

While this latter suggestion was dismissed as the musings of an irresponsible *bideshi*, aspects of the general argument were taken up by SIDA in conjunction with the Local Government division of the MLGRDC in the form of the Intensive Rural Works Programme (IRWP), which later became the present Rural Employment Sector Programme (RESP). The explicit purpose of this programme was to use the vehicle of necessary infrastructure building to ensure better wages and improved longer term rights for rural labour, without going as far as some of my suggestions. When invited by SIDA to look at the problems of rural labour participation after a couple of years of the programme, the opportunity for a new research direction was provided. This provides the basis for Part III of this volume, exploring themes of: labour participation; the relation between engineering and social science inputs to project design and implementation; of course the issues of corruption when the local state is involved in the disbursement of inherently leaky resources with little fear of accountability under virtual monopoly conditions ('Plunder Without Danger'); and the exploration of improving voice among workers to secure their rights and overcome corruption.

With the current controversy about the Flood Action Plan, in which engineering solutions and many related vested interests (from local to international) seem to prevail, this agenda now has a renewed contemporary relevance and these connections are made, especially in Chapter 12. For readers not so familiar with Bangladesh, the Flood Action Plan has been a multi-donor reaction to the floods of 1987 and 1988, proposing a huge programme of investment in physical infrastructure to control water flow. Such a strategy is regarded by many critics as potentially disastrous not just for the environment in the long term but for the environment and rural people in the short term. It seems to represent a clear case of engineering solutions taking precedence, leaving social analysis to fight a rearguard action either to oppose or to modify physical proposals. There is much evidence that where the large number of studies have incorporated social science analysis, critical findings have been filtered out in the higher level decision-making.

Not only does this process reveal the power of various consultancy and contractor interests and likely government and aid agency collusion (with tied interests to promote), but it has also raised important questions about the intellectual autonomy of social scientists who have been drawn into collaboration, both local and international. Indeed, such has been the scale of social science consultancy opportunities in the various pre-studies that a half-generation of Bengali social science talent has been diverted from the analysis of fundamental processes in the society to underpin development policy making in the future. There is an irony in the observation that many of the external aid agencies recruiting such local talent trained that talent in the first place to aid basic, independent research and policy development in the society. If these comments appear intemperate in tone, it is because this Flood Action Plan process is so symptomatic of a wider, more endemic reproduction of ignorance.

There is much knowledge in Bangladesh about the detailed conditions within which poor, rural people have to arrange their survival. Alongside this is other knowledge about the ecology and options for harmonious human interaction with it. Social science, especially in the anthropological tradition, is especially equipped to access this knowledge and participate in the wider policy process. However, with development in Bangladesh so dominated by the aid agency discourse (short-termism, rapid turn-around projects, quick and minimum rates of return, unrealistic application of the privatisation agenda of performance indicators, 'modernisation' assumptions of a commercialised, capitalised economy enjoying sufficient comparative advantage in some production sectors to deliver growth to the rest of the economy) and corresponding funds to ensure the application of this discourse, Bengali intellectuals are frequent victims of this cultural imperialism. Their own government prefers the messages emanating from the aid discourse than from any independent, local tradition of thought. The university, research institute and local consultancy sector has been entirely seduced in this process and now cannot deliver much reliable, independent, alternative thinking that might reflect the special circumstances of the Bangladesh political and cultural economy. There are a few shining lights in this gloom, who might be embarrassed to be singled out by name in a *bideshi* volume! Certainly it is this gloom which has prompted NGOs like Proshika to realise that it must take steps to create the institutional conditions for independent, critical enquiry about the development process, which hopefully will constitute a bolt hole for some of these shining lights if they are prepared to turn their back on rich international pickings for the sake of the longer game.

It is in the context of these issues, highlighted by the Flood Action Plan, that the title of this volume is satirically offered. At the same time, the title embodies other ironies and ambiguities, some to be revealed at later points, others to be left to the imagination of the reader.

Fish and Poverty

Having worked with cooperative specialists, agronomists, irrigation engineers and rural works engineers, it was time to engage with biologists, in particular aquaculturalists. For me, a UK citizen, this coincided with my first work with the UK-ODA in Bangladesh in the late eighties. I had declined an earlier offer to become involved in the notorious UK-ODA funded Neemgacchi fish farm, fishpond project which had run into trouble on social development issues, arising out of the 'technological fix' point of departure for the project. I was not comfortable with a brief to salvage the social aspects of a project, so far advanced in conception and implementation.

The UK-ODA funded Dinajpur Fish Project at Parbatipur was a sequel to Neemgacchi, which ignored critical evaluations of the Neemgacchi project and adopted a similar strategy of seeking to stimulate private sector fish pond culture through the construction of a large-scale public sector Fish Seed Multiplication Farm (FSMF) to provide cultured hatchlings and nursery reared fingerlings to a region, assumed to be both deficit in wild fish stocks (correct) and without cultured supplies or private sector nursery ponds (incorrect). The design of the project flew in the face of a consensus among aquaculture experts in the country that public sector FSMFs do not work against any indicator of efficiency, and completely overlooked the poverty-focus which had become a key criterion of UK-ODA policy in the country. However, despite the design, it was embryonic enough for me to accept an assignment to secure a poverty-focus for the project by working with its extension staff and the ODA TCO (Rick Gregory).

This work, which stretched over 2 years, is reflected in Part IV of this volume, especially Chapter 14. During this time, I was also asked to reflect on the poverty and distributional implications of the emerging strategies for open water, capture fishery (to be funded by a consortium of donors led by the World Bank, but including the UK-ODA and others). This work appears as Chapter 15, and to some extent has underpinned subsequent work at the Centre for Development Studies at Bath on these issues in Bangladesh, Thailand and Indonesia.

The work on the fishpond culture project reflected the close association between biology and sociology, a joint fascination for detailed processes. Fish culture represents particular excitement for the social scientist. The swift growth of young fish from artificial breeding to their introduction into large ponds for mature growth, as well as subsequent catching and marketing, requires a rapid sequence of transactions to disperse the stock into appropriate rearing conditions. These transactions operate in complex market structures, segmented by cultural restrictions and overlaid by credit lubrication. No fish culture system can operate without this network of transactions and no policy, especially a poverty-focused one can be devised without comprehending this network in context. Once Gregory and I had identified the significance of this, we persuaded colleagues in the project and in the ODA that further studies were needed. We devised the TOR for a series of supporting studies which were carried out during 1989 into early 1990 on: fingerling trading; foodfish marketing; credit; quick and dirty appraisal techniques on the physical and social aspects of foodfish ponds; women's participation in fish culture; and an assessment of NGO interest in poverty-focused fish culture strategies. Colleagues from CDS, Bath and the Bangladesh Centre for Advanced Studies (BCAS) were involved under my supervision, with Gregory. The report 'Off the Page and Into the Pond...' summarised the findings and analysis of these studies and certain themes have been selected for this volume, where my own input in the development of the ideas was the heaviest.

The point of departure for both the fish culture project and the open water, capture fisheries strategy has been the decline of wild fish stocks due to over-catching and population pressure, along with recognition of the importance of fish protein, especially in poor people's diets. The additional issues are the recognition of the inland fish sector as an opportunity to control productive assets and rates of return on trading activities, and as a generator of employment. The concern, as always, is the distribution of these assets and trading and labour opportunities. At this point, the findings for fishpond culture and capture fisheries vary, and the picture becomes very complicated.

With fishpond culture, trading and rearing activities appear to attract the highest rates of return. This is partly because there are no 'fish farmers', as assumed by many of these projects. Fishpond holders rarely manage their ponds as if they were a commercial production enterprise. They either own them as individuals, or more usually jointly as part of an undivided inheritance in a multiple-inheritance system. If individual, their product is seen more as a bonus and anyway such ponds tend to have multiple use both by the family, and sometimes by the community (bathing, washing clothes, even drinking, duck rearing etc. in addition to rearing foodfish). If jointly, the principal problem is cooperation between nuclear households in the commercial exploitation of the asset where disputes over division of costs and returns are paramount even without a formal production strategy.

It has been an interesting project to convince our aquaculture colleagues of these 'social' issues. Just as the agronomist and agricultural economists of the past (and sometimes the present) have constructed the image of the farmer as a rational investor of new technology to maximise yields and returns (not always compatible objectives) so the aquaculturists rely upon the construction of the fish farmer as the 'target group' for enhancing the production and supply of cultured fish into a local society. In truth, householders who own ponds, individually or with their brothers and cousins, are casual purchasers of fingerlings and make little efforts to 'manage' their ponds. The purchased fingerlings, often of the wrong size and species for the conditions, are thrown into the pond and with a low survival rate as the predators (snake-head wild fish, snakes and other reptiles, even mammals, as well as people of course) get to work. This casual end-user situation can nevertheless be consistent with a buoyant demand for fingerlings—indeed low survival rates keep up the demand for fingerlings to stock! Thus, despite 'commercial' inefficiency at the point of rearing fingerlings into mature fish of edible size and weight, the production and trading of hatchlings and fingerlings (often referred to as 'seed') can remain highly profitable, which explains both the numbers of operators in these stages but also the level of competition and the interlocked, imperfect character of the markets.

As in any trading system, there is a hierarchy of operators involving the conventional distinctions between wholesalers and retailers. The dimensions to be considered are: scale of operation (excuse the pun), degree of personal labour effort, turnover, risk and credit with interest rates further affecting the distribution of the real rates of return to different sets of participants. The complex, and necessarily lagged, credit transactions partly reflect the distribution of risk and are partly responsible for that distribution. Large-scale wholesalers may dominate in price terms the producer supplier of hatchlings, but pass on their risks of transportation to the small scale traders who shoulder-load *patils* into the villagers by insisting on loan repayments at interest, regardless of success in selling to nursery pond holders or foodfish pondholders. The 'seed' is always at greater risk as the transportation sequence progresses. Thus smaller traders ('fry traders') invest more of their own labour, operate at low turnover (often no more than daily labour rates in agriculture), undertake greater risks with increasingly 'stressed' stock, which affects their bargaining strength in end sales, and are unable to operate in competitive credit markets, with their wholesale suppliers also functioning as their creditors. So, although there are operators gaining profits at these stages of the fish culture system, the distribution of these profits is highly skewed in favour of large wholesale traders, thus undermining any poverty-focus in terms of labour, trading or asset holding opportunities.

At the other end of the process, there is a real danger that cultured foodfish are not consumed by those for whom fish protein is most important, namely the urban and rural poor. To the extent that these technological approaches to the expansion of fish production dominate the type, size and weight of fish produced, then the most typical outcome is the production of fish for consumption which the poor can neither afford to buy nor produce for themselves (if they lack access to ponds). Under such conditions of producing large carp for increasingly urban middle class markets, the poverty-focus for such projects is lost. The technical focus has been on species which flourish in large pond situations (the major carps), and which require capital for corresponding pond investment and management. The poor are unable to benefit either as producers or as consumers. These were the pessimistic conclusions from 'Off the Page and into the Pond', and the analytic basis for proposing a number of possible poverty-focused directions for a continued fish culture strategy in the NW of Bangladesh, but a model which might be applied subsequently elsewhere.

Some of these ideas have been pursued, but many others have been stifled by the combined interests of the technological-fix oriented Directorate of Fisheries and its overseas advisors. It is worth reflecting on the extent to which the institutional homes of the advisors have also educated and trained their counterparts in the Directorate, developing a reinforcing mind-set unable to comprehend the social context within which such technical-fix initiatives have to work. Little has been learnt by donors and counterparts alike. The clash of analytic paradigms was a major feature of this experience, in which the donors especially made several attempts to return to a comfortable story by sending in other aquaculturists to refute our findings. They simply could not accept that the original assumptions of their project designers were entirely misplaced—namely that a private sector in fish culture (especially nursery pond rearing of fingerlings) needed to be stimulated by a public sector FSMF which would be physically dominated by (guess what?) nursery ponds. That a sociologist could refute this assumption empirically within a few hours of the first reconnoitre field trip was far too great a challenge.

The collusion between the donors and the Directorate of Fisheries went beyond sharing an intellectual paradigm provided by the donors in the first place. Both institutions had a direct interest in the 'hardware' of an FSMF construction and subsequent management. It represented financial 'leakage' for the Directorate and steady consultancy work for those hired by the donors (to design and subsequently technically 'advise' the project). In short, we witnessed corruption and intellectual dishonesty. The latter continued with a re-writing of the story, to imply that a later introduction of Participatory Rural Appraisal (PRA) techniques to determine wealth rankings in the village somehow would alter the entire orientation of the project towards the poor without actually altering much of its basic content (namely the FSMF continues to be the main project, with a slight scaling down of later phases of constructing accommodation and some re-definition of the function of ponds within the site). The theme of 'Whose Ideas, Whose Interests' seems to fit this case admirably.

The open-water capture fishery story matches the previous one for 'the poverty of policy'. I have recently encountered the anger of the present Third Fisheries project team who had been kept unaware of this paper, criticising the original joint ODA-World Bank appraisal. As it appears here, there is a technical opening to the paper since it consists initially of a close textual examination of this appraisal document but then it moves into the implications of the project objectives and design for equity and poverty alleviation. The criticised document represents a litany of misplaced assumptions and internal contradictions: the now familiar problem of identifying real, poor fishermen in contrast to businessmen and entrepreneurs, dominating others in interlocked transactions; the problem of inappropriate species for harvesting by small-scale fishermen; the costs of rearing, transporting and stocking fingerlings of sufficient size to reach optimal survival rates in predated water systems; the contradiction between the objective of cost-recovery through leasing and ensuring access for the poor; the problems of common-property resource management with competing communities around large water bodies. The paper is included in this volume as evidence of the availability of critical thinking in the face of disbelief by the technical-fix lobby, again confirmed by other aquaculturists engaged for further reviews by a UK-ODA unhappy with this message. One can only speculate why the paper was withheld from those subsequently engaged actually to implement the project and to secure where possible a poverty-focus within ODA's overall rhetoric for its aid.

Strategies with the Poor

These 3 parts of the book, addressing poverty-focused development in the context of the three sectors of agriculture, rural works, and fishing, have all entailed an institutional sub-text concerned both with government and NGO relations with people, but also the strategic content of rural development policy. On various occasions from the mid-eighties, these institutional issues have been explicitly analysed in the context of competing prescriptions about organisational style as well as the potential clashes between these prescriptions and the problems of organisational culture in Bangladesh. There is a triangular problem between government, donors and NGOs which operates at the levels of style, culture and strategic models. Thus the background to the 'Strategies with the Poor' papers (Part V) partly concerns the rise of NGOs and the increasing adoption of them and their approaches by donors, often as a basis for criticism of government.

Although I have a personal record of close work with one of the leading development NGOs in the country, Proshika, and certainly share, through other writing, critiques of the state in both a class sense as well as the inappropriateness of its rationality in relation to the needs of its clients, I am not blind to the fact that the recent discourse about development administration in Bangladesh has been dominated by the NGO panacea, without considering broader questions of national level coverage, statutory responsibility, accountability, democratisation and human rights. Some of these issues are pursued more theoretically in the final section on 'Good Governance and the Franchise State'(Part VI).

The 'Strategies with the Poor' papers are derived significantly from work with SIDA and intense debates about models and institutions, and in particular the virtues or otherwise of the Bangladesh Rural Development Board as a vehicle of creative rural development. The SIDA interest arose from the earlier critiques of its IRWP as a vehicle for a broad EIG agenda and the preparation of a sequel project which would have 2 wings: a continuation of infrastructure building with enhanced employment rights; and the introduction of a Production and Employment Programme (PEP) implemented through the Rural Poor Programme of the BRDB. During this preparation, it was important to assess the government's own activity, commitment and capacity to work in this way. The paper 'Government

Approaches to the Rural Poor' was written for this purpose, and had as its theme the government's own record of innovation since the Comilla model. In particular, it included some primary fieldwork on the RPP record with 'its' landless groups (BSS and MBSS), as well as some description of the history of institutional innovation through special rural development programmes. The case studies of landless groups have been retained here, since there still remains little empirical material of this kind and because the summary descriptions of the cases lead into an analytic conclusion of the RPP experience, dominated by an argument of 'bureaucratic unrealism'.

The dilemma for me was the necessity to criticise the BRDB and its RPP, while remaining wary of the NGO route to universal coverage and statutory responsibility to work with the poor. This dilemma remains in Bangladesh. There are those, even within the aid community, who entertain a model of total 'rural development' coverage via a proliferation of NGOs arguing that they have the potential to substitute for the state rather than functioning as watchdogs and experimental innovators. They base this model on the relative advantages of existing NGOs, with their size and sector-specific functions. There is much hype about the potential of the handful of large NGOs to take over mainline service functions for the society, like education and primary health care. This is donor fantasy, a product of the small aid-client community which aid officials inhabit. There is little knowledge of the problems of scaling-up within these organisations. They can not easily be separated from the prevailing organisational culture if they are to survive or pursue their specific agendas. Without NGOs as a panacea, as the convenient alternative to dealing with the state, and without any faith in imperfect, interlocked markets to deliver basic welfare (wages, wage goods, services etc.) to the poor, then one has to turn back to the state and the democratic process. Again, these issues are pursued more deliberately in Part VI. However, these arguments were the basis for a further piece of work with SIDA, examining its support for the PEP within RPP/BRDB in the paper 'Sirs and Sahibs' prepared, with supporting fieldwork in Faridpur, Madaripur and Kurigram, in December 1988.

I made some enemies with this paper among the expatriate project staff who formally occupied the technical assistance roles in relation to counterpart staff within the RPP/BRDB. My criticisms of the TA project staff (expatriate and Bangladeshi) were based not on their personal dedication and performance, but on their domination of the implementation of the project (mobilising the landless and supporting them with credit and training for EIG projects—just like NGOs). With inflated and non-replicable resources, they had constructed an enclave project albeit relatively successful in achieving the immediate goals of mobilisation and credit disbursal. However, there was little development content in relation to the counterpart agency (the RPP/BRDB) since the TA staff saw the RPP/BRDB not as the main objective of their efforts but at best as assistants and at worst as a hindrance. It was a surprise to the TA staff that it was precisely the weaknesses of the RPP/BRDB staff in pursuing participatory rural development which justified the presence of TA staff in the first place. The 'Sirs and Sahibs' title was used as a dramatic reference to this theme which had many, more detailed facets:

- inequalities of pay, conditions and transport between TA and counterpart staff, even of the same nationality;
- a certain arrogance of ex-NGO Bangladeshi TA staff in their attitudes towards counterpart staff;
- hierarchies and one-way flows of command and information both between TA and counterpart organisations, and within their respective cadres, especially in relation to group organisers in the field, who should have been the source of information and ideas rather than the passive recipients and implementers of others;
- supplicant, deferential 'beneficiaries' with little role in defining themselves as targets let alone their needs or rights in relation to the project staff.

This paper includes a section which was, in effect, a manifesto of good rural development practice. There is little doubt for me that this paper fits in well with the overall theme of 'whose ideas, whose interests' for the book.

When we turn to the content rather than institutional form or style of poverty-focused rural development, there is a danger that the discourse has settled only upon the general formula of entrepreneurialism and groups. This, of course, was the basis for the landless irrigation programme. Indeed the lengthy technical report, from which The Water-Sellers was extracted, was originally titled 'Social Entrepreneurialism'. While this approach has not yet been exhausted, we should be alert to its limitations as a general formula. I am always struck by the inconsistencies within expatriate arguments which advocate the small-business solution for illiterate, trade-saturated rural Bangladesh with severely constrained effective demand for new products. The same expatriates would be sceptical of the validity of such an approach in their own, better resourced countries. Often this inconsistency has its parallel in a strong emphasis upon development (preferably self-sustaining) and an antipathy to relief, while extensive welfare budgets are defended tenaciously at home. Although aware that the two sets of countries cannot be so simply compared, especially in terms of revenue per head available for public expenditure, the central issue of employment generation outside of small-scale, short term local projects or through landless group business activity requires some fresh thinking.

Employment generation in Bangladesh obviously has to be considered in the context of the country's position in the global economy, and where if any its comparative advantage lies. The ideas presented to SIDA (as part of a review of its aid policy to the country) were contained in a paper, entitled 'Making a Little Go a Long Way'. The economic problem for Bangladesh is a weak trading position internationally, combined with a weak domestic market in labour-intensive sectors outside subsistence agriculture. Gas and garments will not do the trick, though they are an important contribution to foreign exchange earnings. This was the point of departure for offering some speculation on employment generation strategies across 'natural resource sectors'.

Alongside the entrepreneurial model, therefore, private sector wage employment and public sector, 'Keynesian' strategies are considered by each natural resource sector. Such a matrix is offered as a guiding framework for an employment generation discourse which recognises the dangers of entrepreneurial, informal sector panaceas, and is realistic about the skewed distribution of human capital and other social and cultural resources necessary to survive successfully as an individual or even small group in the market place.

This reference to human capital and the other resources as necessary conditions for the entrepreneurial option also apply to some extent in a more conventional employment context (labour rights) as well as in relation to official projects. In their search for political safety, many NGOs have moved towards service delivery functions on behalf of the state or donor sponsors. In the process, much of the earlier rhetoric about struggle has been left behind. Of course, some of that rhetoric was always overplayed and never seriously meant on the road to respectability. It is a cause for some regret that this agenda has receded for agencies like BRAC and Grameen Bank. Defining and treating people as clients is not the same as training people to become successful clients, itself a different activity from traditional conceptions of the revolution.

There has recently been an interesting exchange between OXFAM and the UK Charity Commissioners on references to struggle by groups in India, supported by OXFAM. Why, asks the Right wing in the UK, should OXFAM enjoy tax concessionary status in order to participate in the political struggles of other countries against (it must be 'against') democratically elected governments. Such a reaction is, of course, to misread the situation in many ways—a revelation of dogmatic ignorance. Struggle is a deeply rhetorical utterance, a clarion call, an emotive term of mobilisation—but it has not referred exclusively to revolution for sometime now. To any sane person, even to the self-appointed guardians of UK ethics about charities, poverty is politics. To anyone prepared to think beyond the odd charitable gesture, poverty is rooted in the distribution of the means of production and, where the state controls significant productive assets, the distribution of the means of administration as well. Poverty must, therefore, be about power—in both micro and macro situations. It is true that some NGOs may have entertained the hope of a revolution by the dispossessed, but organised landless workers—the macro agenda. But there have not been many examples of success by this route, and even these have degenerated into Animal Farm. Struggle is the micro agenda, these days—made anodyne in the notion of participation. In the context of the cultural and social hierarchies prevailing in the subcontinent, small victories have to be celebrated.

The recently published Breaking the Chains (with B. Kramsjo) explores the significance of these small victories. The extract reprinted here (Chapter 19, Part V) draws upon the classic distinctions between protest and revolution, and acknowledges the significance of accumulated protest activity in the gaining of strength by the poor to participate systematically both in the capturing of resources and the allocation of public sector ones. This is a realistic agenda, dealing with people's actually existing consciousness (to steal from Bahro) rather than presuming it en route to grander ambitions. This may be dismissed by the newly re-emerging 'radical' political parties and the occasional NGO camp follower as revisionism, but there can be no place for Leftist adventurism (in contrast to Leftist realism) in Bangladesh, which amounts to gambling with the lives of those who live it closer to the edge than almost anywhere else in the world. Furthermore, it is not just a question of preparing the groundwork for some mythical final shove. Where life expectancy remains in the late forties, programmes have to have meaning within people's shorter lifespans rather beyond. In this sense, it is important that people experience regular, albeit minor, victories over the forces that oppress them: a better lease deal; wage improvement; respect for contribution to community decisions; honour for poor women; respect for managing complicated resources (like mechanised irrigation). Poor people gain an immediate dignity in this process. It is in this sense that breaking the chains is the first step. However such an agenda has to be connected to demographic and economic trends, if achievements are to become plateaux for the future rather than peaks never to be re-climbed. In particular, as NGOs with these realistic agendas move into the second generation, the issue of sustainability at the level of consciousness, motivation and commitment will become paramount. Less a question of scaling up, more a question of keeping the flame of struggle alive through descendant generations as a condition of a structural transformation, based on the principle of equity. It is no accident that more NGOs are turning to education and literacy as key ingredients in their next decade of work. In developmental terms, it is now time to pay more attention to the children.

While the next chapter (20, Part V) does not explicitly focus upon the themes of education and literacy, it pursues the theme of struggle through a concentration upon the development of human capital among poor people along with other key sets of social and cultural resources to improve the effectiveness of their action in the wider society. The paper originally appeared as Annex 1 in the evaluation of Gono Shahajo Sangstha (GSS), which I led in February 1993. The basic argument is that GSS's activities can be represented as 'state of the art' poverty alleviation in the sense of engaging with each of the 4 sets of resource profiles which constitute people's survival and accumulation portfolio. More specifically, though important for this volume, the annex argues that GSS should recognise the interdependence of these resources (material, human, social and cultural) and seek to focus all its programmes in any area of operation to exploit this interdependence. For GSS, the implication of this advice entailed withdrawal from areas where the prospects for applying the full package were currently weak. For our purposes here, it highlights the utility of the 'resource profile' approach in analysing programme priorities, and indicates within this framework the dilemma of service delivery versus mobilisation, indeed the meaning of struggle.

The 'resource profile' approach has been coined at CDS, Bath to bring together ideas about portfolios (people as stories rather than cases, as in the Fish Culture studies and elsewhere), exchange entitlements (Sen), vulnerability and survival strategies under acute stress (Swift and de Waal) and was deployed in 'Going It Alone: Female-Headed Households in Bangladesh' (D. Lewis *et al*—including myself and McGregor) and in McGregor and Lewis's report on 'Poverty and Social Development in Albania' (both are now CDS Occasional Papers).

The GSS programme has, until recently, opted strongly for an emphasis upon the human, social and cultural dimensions of a person's portfolio, relegating the implications of the material dimension to 'revisionist service delivery'. Apart from its dilemma of the extent to which it should follow the 'credit and training support for EIGs' model, thereby abandoning earlier principles, GSS's strategy (including its Legal Aid and Legal Education programme, Annex 2 of the Evaluation Report, also by me but not included here) focusses upon the issue of 'struggle' as discussed above. Chapter 20 therefore becomes, in effect, an extension of the previous chapter, but drawing on GSS experience rather than Proshika's. In particular, it discusses the contested notion of empowerment and the problems of measurement.

With organisations requesting large-scale funds from donors to pursue 'empowerment' programmes, funding agencies these days have to provide a 'value for money' justification to their domestic constituencies (whether taxpayers or voluntary donors) via the use of performance indicators. This paper explores the utility of both activity-input indicators and social action output indicators, and the need for sensitivity in the use of qualitative assessments. The issue of performance indicators, and the attempts by agencies to produce numerical indices, reflects a preoccupation with spurious objectivity by officials who are oriented in their own training and imagination towards the validity of large, numerical data-sets whether in economics or sociology. This issue is part of the wider problem of establishing the validity of other forms of knowledge (i.e. from anthropology) in the policy process. I have recently encountered precisely this problem in assisting a study in Bulgaria on risk groups and poverty in the economic transition, where only large-scale quantitative surveys have been regarded as valid forms of knowledge.

Good Governance and the Franchise State

In international development discourse, it is now clear that we have entered the era of 'good governance': the new conditionality moving for the first time away from economic targets or management criteria. There is much hypocrisy embodied in the western preoccupation with this theme, and much room for debate about concepts and meanings. In the West (or the North, or OECD countries, or....?), there is a presumption of a successfully working democratic process with strong accountability between state and people, removing the prospect of dictatorial oppressive governments and underpinning, therefore, the protection of fundamental human rights. In the West, it is often difficult to recognise this description of ourselves and our political systems. I can be accused (I often am) of arguing out of perspective; that by comparison with many societies in Asia, Africa and Latin America, the democratic political systems and strong civil society of the West do deliver accountability and human rights. In response, while I can concede some of this comparison, I maintain that our systems are fragile in many respects and when tested quickly resort to the curtailment of people's rights and the suppression of truth. I would, however, go further and argue that via processes of labelling and ideological manipulation (both aspects of our political culture) unwary citizens in the West are managed more subtly but no less completely, i.e. I share the Habermas, Marcuse, Frankfurt School position. Furthermore, it is noticeable under recent conditions of higher unemployment how quiescent European citizens are. In short, I am suggesting that 'good governance' represents a revival of ethnocentric, modernising ideology, attempting to apply the myths of one society as reality in another.

The complexity does not stop with this observation. Part of the claim for 'good governance' is participation, a wider involvement of citizenry in managing their own affairs, taking responsibility for them and so on. For many in the UK, such a theme verges on a sick joke. The idea that hospital trusts or locally managed schools empower consumers with choice is frankly laughable as budgets are constrained and increasingly earmarked. Local government in the UK has been systematically undermined since 1979 by the very forces which are now placing 'good governance' on the agenda of other societies with' much weaker traditions of local government in their political culture. If a highly literate, IT-saturated society cannot manage successful voice, how ethical is it to expect, indeed insist, on better performance elsewhere? However, I often fantasise about group mobilisation among the poor and dispossessed in the UK—perhaps good governance is more possible elsewhere than in those countries which purport to be the keepers of the discourse!

In the context of Bangladesh, there are two further, specific aspects of the 'good governance' issue which form the basis of Part VI of this volume: corruption and the franchising of state responsibilities to NGOs. These constitute Chapters 21 and 22 respectively.

Corruption has featured strongly in the good governance rhetoric and possibly represents the greatest area of western hypocrisy. The discussion offered here was first presented to a conference on state and non-state service provision at the Centre for Development Research in Copenhagen in 1992. I had been asked to do a stocktaking exercise of themes in development administration concerning state-client relations. Alongside this request had been an invitation from SIDA friends in Dhaka to offer some ideas about the corruption issue. This prompted me to merge the two agendas. My interest in corruption as a generic process of state-society relations under certain conditions was stimulated by my fieldwork in North Bihar, India during 1990-91, leading to propositions about marketisation of the local state replete with imperfections and interlocking. The analysis of micro-encounters between client and state has to be set in a wider framework of conditions. Using the 'exit, voice and loyalty' arguments of Hirschman which are connected to notions of monopoly of supply and, for this discussion, the balance between state and non-state (market, not just NGO) provision, I have offered a continuum of official rent-seeking conditions.

The neo-liberal thinkers of the Right have long argued that such conditions can be simply removed by the privatisation of public services. However this proposition relies upon the naive or wilfully ideological assumption of perfect competition removing the scope for rent-seeking. This assumption is far from reality in most societies, and certainly in rural South Asia. It is in this sense that transactions within the so-called marketplace resemble those between state and client, hence 'parallel rationalities'. Furthermore, since the state (usually in a local form) remains a significant actor in the marketplace anyway (with target allocations and subsidies), the cultures of these two theoretically contrasted systems of allocation have actually merged, empirically. This seems to be an important way of trying to understand this 'frontier' in ˙ rural development administration, and it certainly renders good governance conditionality as hopelessly unrealistic and insensitive to actually existing processes, and therefore unethical as a donor posture.

The second specific good governance issue concerns the relationship between government and NGOs, mediated to a considerable extent by the ideological prescriptions of donors. Within the community of donors and their immediate clients, it is easy to over-estimate the significance of NGO development and service activity. The debate, therefore, about the state franchising its responsibilities to NGOs is about a trend which needs to be examined carefully before it takes further root in the society. There are now several large NGOs in Bangladesh in effect tendering with government and donors for the franchise to take over major services in the society: primary education; adult literacy; primary health; rural banking. Since these and other NGOs are also involved in other sectors (agriculture, pisciculture, forestry, horticulture, rural works, veterinary, crop storage, sericulture) through institution building, training, extension and credit as well as monitoring and performance evaluation, there is no reason to suppose that this 'tendering' will not continue with the support of donors who are increasingly turning their back on the inefficiency and corruption of government. NGOs look like a better bet for spending the revenues of hard pressed western taxpayers.

Many questions can be raised about whether NGOs enjoy a relative advantage over government in the provision of services (especially rural ones), and whether these advantages continue as NGOs expand the scale of their operations. Such questions would be central if NGOs were to attempt universal, nation-wide coverage in any of these sectors. Some of these questions arise from the implications of organisational size, and the extent to which procedures have to become increasingly bureaucratised to cope with the maintenance of universal standards and services, and whether in this process the relative advantage of flexibility and adaptability (associated with small organisational size) is lost. There are further points, also, about organisational culture. Small, ideologically and ethically motivated organisations in the early years of their formation may be able to insulate themselves from the prevailing patron-client norms of the surrounding hierarchical and authoritarian culture with a strong deference to status and leadership. Such organisations are, almost by definition, insignificant in the sense of not challenging the norms of the society in their objectives for it. However as they become larger, both more significant in the realm of ideas and policy as well as actual implementation and allocation of scarce resources and as employers of the educated middle classes, so that insulation is harder to preserve.

However, despite the importance of these issues, they are not the main ones addressed in Chapter 22 which focuses upon the problem of governance and the state in this context of the NGO franchise option: hence my coining of the term 'The Franchise State'. The question is posed: to what extent do citizens of the state lose basic political rights if the delivery of universal services and entitlements is entrusted to other bodies which would at best only be accountable to the state rather than directly to those with service entitlements? Can the state devolve responsibility for implementation without losing control over policy (since practice is policy) and therefore losing responsibility for upholding the rights of its citizens? If the answer to the first question is 'yes' and to the second 'no', then we have states without citizens. The irony, of course, is that the international advocates of the process of democratisation are the same ones who wish to privatise the functions of the state, thus rendering the purpose of democracy toothless and meaningless.

A response to this line of argument is that 'governance' is not just about government. I accept this point, but if the suppliers of essential universal services and the guarantors of basic rights and entitlements are dispersed throughout the society into other agencies, then such fragmentation of responsibility is hardly compatible with effective control by a corresponding fragmentation of voice through the creation of specialist constituencies. In a society like Bangladesh, let alone elsewhere like the UK, it is difficult enough to mobilise the population into localised, meaningful groups without then re-ordering them into specialist constituences of specific service agencies. The rhetoric of governance is disingenuous if it cannot realistically be given good effect, rather like the community care controversy in the UK.

The Strategic Agenda

How closely do the themes addressed in this volume hang together? Is there an overall pattern and structure to the arguments made even though they cover a range of topics from the analysis of agrarian systems to state-society encounters, explored through various sectors as well as in the choices of institutional form and style? The reader will not fail to notice strands of: impatience with those insufficiently patient; irritation with those who shoot too quickly from the hip; scepticism of the fashionable and trendy; ridicule of the panaceas of the newly arrived and soon to depart; and suspicion of vested interests among Bangladeshis and donors alike. But the reader needs also to ask why Bangladesh continues to be such a stimulating experience to development theory and practitioners, and to those more generally concerned to celebrate the human spirit.

It is becoming a cliche to the sensitive lay consumers of television images in the West to express admiration for Bengali resilience in the face of floods and cyclones. That new cliche is not unimportant; it is a step forward from vacuous sympathy towards a respect for other people's ability to cope with objectively awful conditions. It is a more positive image. Face to face, in country, this respect has to intensify, but only alongside some recognition of the relationships within the political economy which account for the skewed distribution of accumulation and survival options. Perhaps a distinctiveness of Bangladesh is a sense of room for manouevre. Even the class and gender relations are not so extreme and negative in effect as in other parts of the subcontinent. Many years ago, in an exchange with our so-called progressive Minister of Overseas Development, the late Judith Hart, she justified in correspondence the use of aid to fund intelligence gathering equipment and training in Bangladesh on the grounds that stability was a precondition of development. Such an odd stance for a radical Labour minister. On the contrary, the persistence of instability, in which no regime can be sure of its social support base, obliges those in power to retreat from extreme positions, having tested their strength, into concessionary behaviour. As a result, the door is not closed on experimentation, nor is it closed on the practices and ambitions for widespread mobilisation of the poor in the society through NGO activity. Periodically attempts are made to control the activities of NGOs, labelling them as essentially political organisations, but so far these attempts have been successfully fought back. Partly the explanation for this lies in the intervention of donors, who act as advocates for the NGOs. But this is too simple. The state in Bangladesh (as also elsewhere) is not a monolith, reflecting a single ideology and set of interests. It consists of many fractions and factions, some of whom recognise that security for their own lineages can only be achieved by supporting NGO efforts with the rural and urban poor (especially urban!). If neither the government nor the market can deliver the minimum level of politically controllable economic returns to the poor, then support for the NGOs is the only option left. It is precisely the fragility of the state which establishes the conditions for promoting a sense of struggle and hope among the poor. To be continuously among all this in Bangladesh over the last two decades has indeed been a privilege.

This time period has enabled me to be a witness to other processes as well. Bangladesh remains an agrarian society for the immediate future, despite the rise in absolute urban numbers. The discourse about this agrarian system has been entirely transformed from the mid-seventies preoccupation with ex-landlords, moneylenders and small peasants (as a mix of owners and sharecroppers) leading to an unreconstructed advocacy of land reform as the radical agenda. The discovery of pervasive, structural and increasing landlessness along with significant regional variations in agrarian structure required a fundamental shift in the Bengali self-image of a society of small, subsistence peasants living in relative harmony with each other and the environment; undermined, thereby, the populist appeal of the nationalist party of liberation (the Awami League); left radical parties stranded ideologically; caused 2 decades of search among substitute populist organisations (Jatiya, BNP etc.) for a post-Comilla ideology and strategy relevant to the new conditions while at the same time not challenging the combined interests of the shifting coalitions at the centre; and may possibly provide the social underpinning for the expansion of Islamic fundamentalism as a cohering movement of the dispossessed and the frustrated.

To me, the lack of grounded speculation about the emerging structure of agrarian social relations is worrying. Apart from some pockets among the

NGOs and radical parties re-working their positions, there is an almost total absence of strategic thinking. I hope that this volume contributes to this process and entices others into debate about some of the fundamental structural trends in the society. To put myself on the line therefore, this is what I think is happening to agriculture.

- Farms are continuing to fragment in response to the dual processes of population pressure and low rates of rural surplus labour absorption into other sectors of the economy.
- Other potentially labour intensive parts of the economy cannot expand sufficiently because of limited comparative advantage for Bangladesh in any of its current sectors of activity, and because of weak domestic demand.
- The fragmentation of farms occurs in both senses of a smaller size of overall holdings per household and the division of dispersed plots into smaller units to spread risk at the time of inheritance.
- The intrusion of high-tech agricultural capital into this picture of agricultural involution increases the problems of smaller farmers in accessing such technology and inputs—markets are interlocked against them, technology may be lumpy, maintenance of successful networks is essential, transaction costs of self-governing cooperatives are too high, state services are too unreliable and also characterised by rent-seeking.
- Such agrarian conditions create increasing landlessness on the one hand (without a corresponding polarisation in ownership) and non-viable farmers on the other, who are increasingly dependent on purchasing services from others to continue production.
- Where the landless are unable to enter these new agricultural service business opportunities, they experience increasing seasonal underemployment and will continue to migrate seasonally or permanently unless the scale of public and private sector rural investment is increased and focused upon the sustainable use of natural resource endowments as well as upon products with potential domestic demand.
- Models of urbanisation need to include 'rurbanisation', i.e. the growth of small, market towns partly reflecting patterns of step migration, partly government investment in its own decentralised administration.
- Such rurbanisation may constitute a pattern of localised demand for rural products (handicrafts and horticultural as well as agricultural)

which may assist a change in cropping and land use patterns, and may assist the transformation of farming from high-tech, chemical based towards organic options.
- Farmers may be attracted by the prospect of higher rates of return on various forms of organic production methods, though the incentives to maximise yields and therefore commercial surpluses is less clear or proven at this stage of knowledge.
- Organic methods of cultivation will less water-demanding per unit of output and therefore more sustainable and less demanding on the environment in the longer term.
- The jury is still out on whether organic methods will be more or less labour intensive, though seasonal fluctuations are likely to be levelled out.
- New patterns of localised demand resulting from rurbanisation should impact upon employment and investment opportunities in transportation, storage, processing and trading.

The implications of this scenario for agrarian relations and rural-urban relations are quite profound. There are also implications for the required direction of investment internally, as well as a reduction in the dependence upon scarce foreign exchange if the imported content of agricultural inputs is reduced. This would alter sets of commercial opportunities, driving out some old sets of actors and introducing new ones, who might be strategically supported, as part of a poverty-focused agenda. Throughout all of this, there is an implicit lack of faith in market mechanisms or government to orchestrate this process of change, and a recognition of the principle that people need to be 'rich in people' (i.e. be well placed in networks) to operate successfully and that this will occur through a combination of private networking as at present and NGO institutional support for those defined as targets because they are 'poor in people'.

PART I

CLASS FORMATION AND UNIVERSAL RURAL DEVELOPMENT MODELS

Chapter 1

Exploitation and the Rural Poor

Introduction

The Political Process—A Research Note

At the outset this chapter intends to establish its own brief—derived from the problematics raised in the general theoretical literature concerning the analysis of the peasantry in the disciplines of political sociology, Marxist political economy, and the transition to socialism. It is asserted, in effect, that these problematics are of high relevance to the development of Bangladesh, and that their analysis should be pursued in a public manner. In this way, even where our conclusions are critical of the role of certain classes and are unpalatable to them, they should play some part in the debate about the radical path for Bangladesh. More specifically, the research focus discussed in this chapter is addressed to the sheer lack of knowledge about the rural social formation. It starts from the assumption that:

> "there will be no long-term economic growth, no general increase in the well-being of the people, and no movement towards a just and equal society in Bangladesh without a great deal of objective study of its people, society and economy. At present our ignorance is profound." (Nicholas 1973, p. 26)

An understanding of the rural power structure is especially relevant to any plan about rural development. This is particularly the case in Bangladesh where the scarcity of administrative capacity excludes from consideration an authoritarian, well-centralised alternative such as that adopted by Stalin in the post-NEP period. Therefore the specialist professionals and political leaders have no alternative but to devise a strategy which recognises the strength and weaknesses of the rural social formation, regardless of whether the objectives of growth or equality are uppermost. However, planners are further constrained for no strategy will progress beyond the drawing board unless it is one which strikes the majority of the peasantry as legitimate and worthwhile. Legitimate, in the sense that it commands their support and is seen as being consistent with their culture. Worthwhile in the sense that there is a realistic prospect of any plans and targets not being subverted by a minority of the peasantry which sees the possibility of private accumulation at the expense of the public. A genuine socialistic content to the rural development programme is therefore not only desirable, but a prerequisite for success. Genuine in the sense of identifying the loopholes which leak public funds to private pockets, with stern sanctions against those who persist in obstructing the transition by employing the traditional methods of domination—in the market, usufructuary loans, manipulating the relationships of access to public resources and to justice, and terror. But there will be no genuine programmes until these dominant features of the contemporary social formation are understood: understood, not just in general terms supported by flimsy estimates and projection, but in detail—in terms of behavioural relationships between different classes and groups: relationships which can rarely be articulated in the form of quantities, but which nevertheless pervade the social structure of the countryside and constitute a real constraint (or possibly a resource) to the overturning of its exploitative characteristics.

If, as everyone seems to agree, the co-operative strategy and other programmes are being sabotaged for reasons in addition to administrative inefficiency and corruption, then it seems obvious that the social context into which the co-operatives were introduced has to be examined in depth. Why is it that co-operation between individuals and groups even within the same objective class is difficult to sustain? Was it ever realistic to assume that small-scale co-operation is a state of nature in conditions of adversity? Should we consider in this context the propositions derived from studies of peasant societies elsewhere about 'amoral familism,' 'zero-sum games,' and 'the image of the limited Good'? (Banfield, 1958; Foster, 1967). We certainly need to know where the boundaries of co-operation have been traditionally set—at the *'paribar'* (*'chula'*) the *'bari'*, the *'reyai'*, or the *'samaj'*? Under what conditions and for what purposes do these different social units become operative units of cohesion and solidarity? There are possibilities enough for division and conflict by just referring to these institutions, but interwoven among them are other relationships and transactions based on class distinctions; the need for credit; differential access to bureaucratic institutions; relation to the market; and a range of other variables. With such a collection of structures competing for loyalty, programmes based on formal institutions for co-operation must be treated with great caution. Perhaps the central issue is being avoided and co-operation should be overtly political from the outset—the creation of a social class of the poor peasantry.

It would be misleading though to focus research specifically around the co-operatives since it is tempting to frame one's conclusions in terms of shortfalls from the objective, a failure of goals which anyhow may have been unrealistic or inappropriate in the first place. This kind of analysis constitutes an exercise in judgement of behaviour against a set of criteria introduced from outside the culture of the community. This is evaluation, and occupies a justifiably important place in the work at the Bangladesh Academy for Rural Development (BARD). However this proposal is concerned with system, and its description in a manner which both reflects the reality of the peasant's daily existence, and is at the same time suggestive of strategies for institutional experimentation. In this way it has to be more comprehensive than a piece of evaluation, and essentially exploratory. Only by this method can we determine: which features have been responsible for low rate of investment and capital formation; the preference for money lending at high rates of interest rather than productive investment; the enduring pattern of differential access to bureaucratically allocated resources etc. as well as the problems of the co-operatives.

The Question of Integration: A Discussion of Prevailing Theory

There is an important theoretical and therefore methodological issue which must be confronted at the outset. Much of the conventional analysis of the rural political process treats the rural community as a set of isolated social structures, traditionally independent of other rural communities and of other national level institutions. This has encouraged a conceptualisation of the development and nation-building process which assumes a dichotomy between the traditional and modern sectors in the society. It therefore views the process as one of incorporating the traditional structures and institutions into modern ones, thereby destroying the discrete social entities that characterise the rural social formation. As a corollary of this perspective, rural institutions represent 'obstacles' in the path of development. In the literature on political development this corresponds to Pye's 'crisis of integration' (Pye, 1966).

Underlying this approach is an elite-mass formulation, with the elite (or part of it) as a modernising force, engaged in the building of a new nation which demands the integration of the society to the point where national institutions (the political party, the bureaucracy, the judicial process, the Planning Commission) become the primary focus of loyalty (instead of the family, the caste, the region, a religious community, linguistic group), and which become therefore legitimate. It is a stimulus-response model of political and social change, where the 'modernising' elite both sets the goals and evaluates the extent of their achievement. But the goals and the evaluation derive from the assumption that rural political and social structures are discrete and encapsulated most significantly in the village. Thus for 'rural political structure' substitute 'the traditional village'. This is the language both of the national elites and the social scientists who have indulged them in this fantasy.

This theoretical perspective is considered to be misguided in several respects. Firstly, since it represents an ethnocentric western ideological formulation of the political system it may be countered by an alternative ideology of greater relevance to Bangladesh which allows the possibility of rural class organisation and peasant initiative in any transformation to socialism. With the fragile nature of the State, the inability of the elite to command independent resources and the limitations of administrative capacity, development programmes which do not depend upon the organisational resources of the peasants themselves cannot be envisaged. Indeed it has been BARD's position to articulate this ideology over many years. The elite-mass models which have pervaded descriptions of Bengali political life must go. They contain no relevant solutions and they are inaccurate.

Secondly, the basis of this inaccuracy is an historical approach to social analysis, which has instead depended upon imported concepts. For example, the structure known as the political party system performs, according to Almond 1960, the function of interest articulation and aggregation in the political system of Western pluralist-democracies. Thus when institutions, labelled as political parties, are identified in other societies it is assumed that because of the label similar functions to that of their western counterparts are being performed, or should be performed by them. There is the assumption that the political system itself is essentially the same, despite some imperfections (or 'crises') which have to be overcome. Such analysis is historical since it totally ignores the evolution of particular groups and class formations in the society, their relationships to the imperial power and later on their position in the nation-state. The picture of the political process which thus emerges reveals a network of relationships which has traditionally placed the peasantry in a mode of production of wider significance than the village: the quasi-feudal system of *zamindari*; the relationship between subsistence and cash crops; the market institutions of distribution and exchange; a hierarchical structure of money lending and credit; and only finally the manipulation of this network by the classes which dominate it to obtain a formalistic legitimisation of their class power through the establishment of political parties.

Thirdly, it follows from this that the peasantry cannot be treated as an undifferentiated mass with no institutions of political, economic and social organisation beyond the boundary of the village. Therefore the elite concept is seriously weakened in favour of a class one, (elites exist in the context of an undifferentiated mass; a ruling class however, dominates other classes) and it is the interplay of these class forces rather than elite-mass confrontations which is the focus of this research. To underline this point, the peasantry is not composed of discrete social units, which are co-terminous with the village, but is composed of sets of institutions which extend critically beyond the village to include the cities and ultimately the nature of the state. Nowhere is this more true than in Bangladesh. A study of the rural political process cannot ignore these dimensions and nor can strategies for development and transformation which recognise the historical role of the peasantry in the agrarian societies of the twentieth century (Wolf 1969). It is in this way, for example, that village based schemes for development may be inappropriate—especially in South Asia where '*baris*' and '*caste-tola*' may under certain conditions (e.g. low incidence of hired labour) be very independent of the 'village' but closely interrelated to '*baris*' of other villages (through kin relations), and to classes of rich peasants/ landowners/ landlords/ moneylenders who dwell outside the village, perhaps in small market towns or larger cities. Co-operatives involving co-operation between '*baris*' in the same village would appear very strange with such an analysis.

Before considering these extended features of the peasant social formation, it is nevertheless important to refer to the treatment of village social structure in the tradition of political anthropology. This work is relevant to our purpose in two respects particularly: the notion of encapsulation, and the analysis of factional systems of political interaction. Both of these approaches are derived from the theoretical tradition of which we are critical—the first explicitly, the second implicitly. Nevertheless they represent important part truths and should be harnessed to the perspective followed here, thereby enriching it.

The model of encapsulation was presented by the British social anthropologist, F.G. Bailey, based upon the analysis of village social structure and the wider political process in Orissa State, India (Bailey 1963). It postulates a significant cleavage in the contemporary peasant societies of modernising nations at the boundary of the village. The diffuse, undifferentiated structure of social relations at the village level is contrasted with the structure of social interaction existing at other levels elsewhere in the society. The process of modernisation is expressed as the blurring of the cleavage. Bailey distinguishes between encapsulating political structures (B) and encapsulated political structures (A), thus:

> "Structure B disposes of greater political resources than structure A, [structure B] is large scale where structure A is small scale, and [structure B] tends to be made up of specialised roles while the roles of structure A tend to be undifferentiated..." (Bailey 1969, P. 147)

Broadly stated, the central hypothesis in this model maintains that structural tension between A and B (village institutions and those of the modernising elite) would be resolved in the dissolution of the encapsulated structure (A) (that of the village) into the encapsulating structure (B) (that of the nation-state), and that this process would reflect a higher degree of overall structural differentiation consistent with the political and economic systems of an integrated nation-state. This formulation represents the cornerstones of modernisation theory: peasant communities are regarded as discrete social entities responding to an interference in their value-system by State and institutions in a manner which inevitably leads to an overall pattern of increased role differentiation.

The general theory from which this model derives has been criticised above, but one specific comment should be added here. The model assumes the existence at one end of the tradition-modernity continuum of differentiated sets of encapsulating structures which are the institutional characteristics of the modern state. These are the structures—it is argued—which disturb the (structural functional) equilibrium of the encapsulated structures by absorbing them into the new equilibrium of the modern state. However this approach begs the question of how the encapsulating structures came into existence in the first place. It adheres to a stimulus-response mechanism of change without explaining the process whereby the dichotomous situation emerged.

In a country numerically dominated by its peasantry, the State cannot be regarded as a fixed or even prior entity in the process of agrarian change.

Structural change among the peasantry must inevitably alter the complexion of the state. If resources are distributed among the peasantry as part of the process of 'intervention' or 'penetration' by the State, then it is important to understand not only the impact of the distribution but also why the State was associated with one pattern of distribution and not another.

The institutions of the State are themselves the result of political and economic forces, so the rulers of the State are likely to defer to those forces in their allocative decisions. In this way the institutions of the State cannot be opposed to the forces from which they derive, and these forces exist in the village as well as elsewhere. It is therefore important analytically to distinguish between the State opposed to sections of people in a village or society, and the State as a reflection of the forces which differentiate the village into sections or classes. Seen in this way the stimulus-response dichotomy between State and village confuses the conflicts of interest between the regime (representatives of the class which dominates state apparatus) and large sections of the weaker peasantry with the fallacious notion of the State as being conceptually independent of the political and economic forces which constitute it.

However although Bailey does not include this vital dimension to his analysis, his treatment of the relationships between the village and the outside world through this model of encapsulation does provide us with a useful tool of analysis with which to examine these wider extensions of the rural social formation. His notion of the 'broker' is a particularly good example (Bailey 1963, pp 58-63). To the extent that a formalistic contrast exists between the normative characteristics of the nation-state institutions and some of the features of the rural social structure, then the precise nature of the interaction between them is very important to those involved in the frontline of extension activities for rural development. If one discovers for example that the internal structure of the effective unit for co-operation (whether it be a *bari*, *reyai*, *samaj*, or groups of *baris* from different villages but related to one another by kinship) consists for the most part of diffuse rather than specific role-sets, while the structure of the rural administration is formally made up of specific compartmentalised role-sets, the ideal type which rarely exists in South Asia. Then what is the structure of communication between them? The 'broker' role refers to institutions, or personalities, which have credibility in each of these contradictory structures. The market is a good example, where traders have traditionally mediated between the physically isolated community and the outside world. But with the peasant society of Bangladesh the situation is more complex, where the variables are values and social structure rather than the simple ones of proximity. Perhaps the broker function is performed by educated sons or by wage-labourers who seek employment in urban areas, or by people who have been in one of the national services, or groups of migrants and peddlers, or by the local landlord who has always 'fixed things' with the authorities, or shopkeepers, or money-lenders, and so on. The problem is essentially one of access, communication, understanding and trust. And where the rich act as brokers for the poor, then there are the additional dimensions of patronage and dependence.

The point about interacting with strange institutions is: can they be trusted? If he is not familiar with their roles how can the peasant predict the likelihood of one outcome or another? How can he trust promises made to him if he has no personal knowledge of the official's reliability, nor any sanctions at his command if things go wrong? Thus the peasant must have a mechanism of interaction which he can trust—a procedure or person who is credible and legitimate both for him and for the rural administration.

Hart 1971, in his excellent paper "The Village and Development Administration" (pp 32-90) pursues a similar theme taking up a distinction between 'hinged' and 'linear fields' (p 77) as a way of conceptualising the interrelations between peasant, social institutions and government. Basing his discussion on Redfield's notion of 'fields' of social relations (Redfield 1969). Hart suggests that a hinged field is where values are not shared between parties to the transaction, where they are not members of the same reference group ('them and us', rather than 'we') and where there is no "continuum of cognitive meanings expressing and maintaining communication upon the subject matter of the transactions." In this situation there is a need for a broker, or a series of brokers in the field. By contrast, the linear field is where values, identities and communications are shared. With these concepts, Hart suggests the following proposition: "Development will be facilitated other things being equal, by channelling a transaction through a linear rather than a hinged field."

This is obviously a very important proposition for us to consider, and indeed Hart offers as an example of this two experiments undertaken at Comilla: the distribution of contraceptives through existing retail merchants in the villages; and the use of *imams* to teach both pre-primary children and illiterate adults. However it is my premise that we cannot simply rest content at that proposition, since we have to understand the nature of the 'linear field' and identify precisely which social relations and social units represent 'linear fields' and which do not. To do this, we have to weigh the significance of other variables like power, class differentiation, access, factional structures of political competition and kinship. In this way the concept of the 'linear field' can be made more empirically precise, with operational conclusions drawn about appropriate 'brokerage' experiments.

At this point we turn to the second pervasive theme in the political anthropological literature on South Asia the analysis of village political structure in terms of factions. Ralph Nicholas' work in West Bengal has been pioneering in this respect. In his paper on the role of social science research in Bangladesh, he concludes:

> "Villages, which appear to be the most 'natural' of human communities and the most obvious basis for co-operative development organisation, frequently prove to be cockpits of bitter struggle, factionalism and the very opposite of co-operation: no-one has an effective remedy, but it is clear that rural development is severely hampered by the prevalence of such conflict. Social scientific research cannot solve any of these problems but by identifying them and dispelling some of the general ignorance that prevails about them, it may provide policy makers with an initial basis for action." (Nicholas 1973, p.26)

Furthermore since the problem of factionalism is often regarded as an ongoing dynamic one, in which divisions in the village are actually created as the result of development programmes (the introduction of new institutions such as co-operatives, and the influx of new resources such as seed, fertiliser and tubewells), it is vital to review some aspects of theory and extract an appropriate tool of analysis for ourselves.

The term 'faction' is used to describe a pervasive form of peasant political interaction, and detailed accounts of peasant societies (especially in India) are replete with examples of 'factionalism' (Nicholas 1963; Banton 1965; Swartz et al 1966; Singer & Cohn 1968). The central point about this mode of politics is that factional alignments cut across class alignments. Thus the analysis of factions to a large extent constitutes an examination of the manner in which factions are recruited and led. The political cleavages in peasant societies are often vertical cleavages which run across class lines, rather than horizontal ones of class conflict. Faction leaders then are typically landlords or rich peasants, or more rarely just manipulating political entrepreneurs (brokers in effect), who organise political groups out of followers who are either economically dependent (labourers, sharecroppers or debtors) or who are obligated as a result of past favours (perhaps the broker has arranged a government licence or a loan). The relationships between leader and followers may be diverse, that is different principles of recruitment may be applied, and transactional in the sense that a single interest constitutes the bond. Bailey distinguishes between 'core' and 'following', in that the 'core' represents an inner circle of allies with multiple relationships to the leader, usually involving close kinship links

(Bailey 1969, p.49). Only the core represents a permanent, corporate group, while the rest of the faction is essentially impermanent. Alavi, 1973 summarises the significance of this internal structure:

> "An important aspect of factional conflict is that rival factions are, in general, structurally similar namely that they represent similar configurations of social groups, although that is by no means always the case. Where that is so, the faction model describes a segmental rather than class conflict. Such conflicts, therefore do not have an ideological expression, because rival factions or faction leaders fight for control over resources, power and status as available within the existing framework of society rather than for changes in the social structure."

It is important for our purposes to distinguish between the concept of faction as an organising concept which we can employ as a tool of analysis, and the theoretical propositions and generalisations about the factional mode of politics in peasant societies. The latter tend to be part of the general tradition (of which we are critical) which treats villages as discrete social entities, often ignoring the significance of the extended nature of these vertical cleavages throughout the rural social formation by confining their analysis to the village. The concept is interesting however as a method of identifying precise aspects of the rural political process. It is useful in *describing* actual political alignments—a necessary step before proceeding to the *explanation* of the alignments which we observe. In this way, the concept in certain circumstances will provide us with a procedure for stating the initial problem—a map of political and social interaction, which must then be analysed in terms of wider regularities in the exercise of power, deriving from the mode of production.

Thus even where a preponderance of factional alignments characterised by vertical cleavages existed in the villages, such a pattern may be successfully incorporated into a class analysis of the rural society. For example, the themes not only of exploitation but also of class solidarity and horizontal alliances may be pursued so that the divisiveness of factional politics in the villages constitutes the counterpart of class formation elsewhere in the system, for example, rich peasants, petty landlords and rural entrepreneurs). A hypothesis on these lines might then be stated in the form:

The prevalence of factional politics in the villages (characterised by vertical cleavages) hinders the emergence of class solidarity among poor peasants and landless labourers and thereby contributes to the strength and solidarity of the dominant classes both in the countryside and among the entrepreneurial and commercial groups in the towns.

Two further interrelated points about factions are germane to our purpose. First, the relationship between factional patterns of political interaction and change; and secondly, the significance of changes in recruitment patterns to political groups.

It is often argued that development activities themselves stimulate factional divisions within the village as different 'patrons' compete with each other for new resources in order to retain the loyalty of their followers and remain as a credible leader in their eyes. Factions are therefore often regarded as disruptive of a prior social order which was (ideally) characterised by consensus and unity. Much of the thinking about co-operatives is based on this premise. However factions are typically found in peasant societies which have not really been subjected to rapid social change, for example in South Asia most of which, despite the claims, has not yet been touched by the 'Green Revolution'. On the contrary therefore, factional patterns of competition have long been expressive of the political order of peasant societies where patron-client relationships have prevailed (e.g. *jajmani* in Hindu caste society, or some relationships between peasant and labourer in Bangladesh) and where resources have been essentially static. Under these conditions life is a zero-sum game where it is more appropriate to compete for control over existing, known resources and organise to protect one's possession (for example exclusive rights to the labour and service of certain groups at fixed, traditional prices), than to organise for the generation and secure possession of new resources (productive investment in high-yield producing technology). However under conditions of induced change like the 'Green Revolution', the increased concentration on productive investment by those peasants who initially have adequate resources, leads to considerations of efficiency, cost, profitability as well as a secure environment for accumulation. The tendency here then is to sacrifice the patron-client relationships inherent in the prior mode of production where fixed proportions of the harvest were distributed between landlord, peasant and dependents, relationships which were often the basis of factional recruitment. For the peasant innovator, the concern is to enter into new alliances which will jointly provide a secure environment for his productive investments. This has to be a class alliance which extends into the towns and administration, for although the rich peasant may be subordinate in such a network, this is the location of the objective interest of his class. For example, if the labourers of the village (under these conditions of rapid change) perceived their relative disadvantage and organised to demand an increase in the wage rate, then the rich peasant must have the contacts to obtain a tractor, or other forms of labour substituting mechanisation. With these considerations, it will be extremely interesting to compare the organisation of political interaction in Kotwali Thana with an area as yet relatively untouched by development activity to obtain an idea of the conditions in which factional conflicts give way to cleavages of class. The analytical tools provided by the factional model enable us to contrast the respective structure of relationships with some precision but only if research is conducted at the level of empirical depth indicated in the scenario just outlined.

The second issue (recruitment) is a methodological refinement of the first. As has now been emphasised several times, the political and social structures of those who reside in the countryside are complex in that they cannot simply be treated as discrete entities and placed somewhere on the tradition-modernity continuum and sent on a predictable path of change. It is likely then that one observes a variety of political groupings from the *reyai* (or factions within it as in one case which I have already come across) to membership of political parties. However regardless of the formal labels of these groups (for example, a political party which claims certain ideological leanings), the significance of a group and explanations of its formation are to be found by identifying the principles of recruitment to the group. Where the principles of recruitment within one group are diverse, transactional and based on single interest relationships with the leader then this *group* is probably small-scale, non-ideological, with a fluctuating membership and therefore is unlikely to constitute an enduring force for structural change. In addition, and very importantly for any programme of institutional experimentation, it is difficult to harness such groups to development programmes and convert them into co-operative units since no single incentive to co-operate will appeal to all members of the group. And yet, for the same reason—this diversity of allegiance these groups cannot easily be broken down. They are difficult to compete against. As a corollary where the co-operative as formally constituted does not have the same internal structure of diverse relationship between leader and follower, it cannot then perform the patronclient functions for its members. *In this situation one of three possibilities might follow: either the co-operative adopts the internal structural characteristics of the faction and competes with other factions outside the co-operative; or the co-operative itself becomes divided into factions, probably reflecting the factional alignments which exist in other spheres outside the co-operative or the co-operative is simply by-passed as a relevant institution in which to compete for resources.* The third possibility is an extreme version of the second. We might consider these possibilities in the nature of alternative hypotheses about the relationship between co-operatives and factions.

Alternatively, where alignments are based on a single principle of recruitment-ideology, lineage (the genealogy of a lineage is often 'ideological' where solidarity between families is explained by referring to 'common' ancestors who are often in fact not related), economic status—these groups are more exclusive in their membership, since the rules for inclusion are more rigid, and so the commitment to the precise purpose of their membership is high. Generally there is also the recognition that the individual and the group interests are intimately connected, so the groups are more stable and permanent in their membership. This may be the ideal-typical model of a political party, but few parties in South Asia actually recruit and maintain their membership in this way. But where one observes such alignments, or their emergence, in the countryside, then significant changes are occurring in the organisational and structural characteristics of the political process. It is indicative of horizontal cleavages in which the dominant mode of production is purposively either being attacked or defended, rather than simply being expressed and thereby reinforced. It represents a shift in the structure of power, and will itself be the outcome of change in the mode of production in which vertical relationships of interdependence are giving way to polarisation, structural antagonism, and relationships of exploitation and dependence (as opposed to interdependence).

Thus while the identification of recruitment patterns opens up for us a Pandora's box, at least it does allow us to look inside and speculate on the dialectical interrelationship between programmes of change and the rural political process. *Thus where group formations emerge on the basis of a single principle of recruitment, we may be forced to the conclusion that it is the programmes of change themselves which are producing not a factional mode of interaction, but a class struggle involving horizontal cleavages within the village.* Such a struggle involves those with a better start to accumulate, organising over a wide area to protect themselves against those who are progressively excluded from reaping the benefits of the new technology. And where the weaker peasantry by virtue of their position in the mode of production (Wolf 1969, pp 289-291) are unable to organise and develop their class solidarity on a complementary scale to the rich peasantry, how will a co-operative help them? *In the co-operative they might remain weak, even if rich peasants are excluded, precisely because of their subordinate position in other structures of dependence which constitute the greater part of their lives.*

We have been trying to illustrate here how this sort of analysis enables us to grapple with the problem: to what extent can political and social structures of inequality in the countryside be manipulated through the institution of the co-operative within the overall strategy of IRDP? Or alternatively, are we expecting too much of the co-operative; in what ways can it be reinforced to defend it against subversion? Or finally is the co-operative too feeble an institution with which to contain the class struggle, or at least to provide the poor with a fighting chance?

Class Characteristics of the Rural Social Formation

Having discussed aspects of theory which might enrich our methodology, we proceed by referring to some recent analyses of the political economy of the peasantry in Bangladesh to be found in the detailed village-level empirical work of Bertocci (1970; 1972 pp 28-52) and in the general, sometimes speculative, paper by Abdullah and Nations (1974). This will help us to define more precisely the empirical concentration of our work, and the significance of the hypotheses which are to be pursued.

Perhaps it should be emphasised here that an examination of the rural power structure in the context of new agricultural technology has to be concerned not only with the impact of the technology and new resources on the political dominance upon the policy formulation process about development itself. That is, a study of the rural power structure must be related to the total system, so that its conclusions contribute to an understanding of the process of development both in the making of decisions and the effect of their implementation. Only thus can it realistically guide a programme of institutional experimentation.

A critical element in such an analysis must then be to trace the interconnections between the national institutions of power and the way power is exercised in the countryside. It seems to be generally acknowledged for example that the richer peasants of rural East Bengal are crucial in mobilising the large electoral majority obtained by the Awami League in December 1970 (Ayub, Md 1971). Ayub argued that this class of peasants had direct linkages to urban areas, as many of its members had become part of the newly-urbanised intelligentsia, small businessmen and student groups, which provided the political base of Bengali nationalism and therefore the most important social base of the Awami League. Naturally, these links would remain two-way for this class between town and country. In the absence of further evidence, this analysis cannot be pushed much further, but it does establish an important hypothesis concerning the relationship between the party and the dominant surplus farmer in the countryside (that is the relatively rich peasants). Where this relationship is intimate and strong, it highlights the kinds of constraints which surround the policy-making activities of the ruling party in the area of agricultural and rural development, constraints which will modify good intentions at the implementation stage.

At the same time such hypotheses have to be pursued in a specific manner so that detailed behavioural relationships are understood. We must go beyond the generalisations and platitudes of radical social scientists and pessimistic officials and if necessary beyond the obsession for data which is statistically respectable (for example counting heads of those who occupy official positions) but often meaningless and unreliable, to the point where the process of power is observed and its structures identified. It is in this sense that Bertocci's work is invaluable and perhaps unique for Bangladesh.

We must begin by discussing the social and political significance of possession of the principal means of production: Land. Immediately the picture is complicated for Bangladesh in that the extent of a family's landholding gives no clear universal indication of that family's social and political status as it might do in the Gangetic Plain of India, for example. Yet ownership of land constitutes an important variable in the determination of a family's status in any particular community. It is evident even allowing for landholding data which fails to aggregate to the level of the cultivating household unit (*chula* or *paribar*), that variation in landholding size is very narrow in Bangladesh (compared with North India). It would seem though that there are important regional variations in Bangladesh, with the centre Dhaka-Comilla belt displaying much more homogeneity than elsewhere. Perhaps areas like Dinajpur and Khulna are at the other extreme. However this narrow range of landholding size is often cited as evidence of the classless nature of the Bengali peasantry and is of course a particularly important theme in the ideology of the new state.

Bertocci's research suggests that we need to be much more cautious in the interpretation of this low variation in size of holding. Abdullah and Nations (1974) point out that policy-makers may have been misled in not attributing enough power to wealthier sections of the peasantry itself. The emphasis that the rural economy of Bangladesh is characterised principally by a peasant mode of production rather than for example, a feudal or quasi-feudal mode. Thus family farms are operated predominantly by family labour with a low incidence of sharecropping and use of landless agricultural labour. Therefore "it is surplus farmers who dominate the rural economy with the help of large landowners" (Abdullah and Nations 1974, p.9) rather than exclusively landlord classes relying solely on rent or interest, or manipulation of the market. Rich peasants—the surplus farmers—are in a position to diversify into these activities, especially moneylending, but importantly they remain as cultivators and are therefore included in the subsidised programmes of agricultural development. Abdullah and Nations, (p.10) conclude this section: "... while the large owners of land are exploitative and non-productive their position in the rural economy is secondary to that of the rich peasants whose influence lies principally in the *relative* control of land." (my emphasis). Such a conclusion might well have to be modified when regional variations are considered, but there seems little doubt that it remains broadly true for Bangladesh at this time.

Bertocci, continuing this theme, concluded from his close observations of Hajipur and Tinpara (pseudonyms for two villages studied in Kotwali Thana of Comilla District) that:

"*Absolute smallness in farm size should not obscure the importance of small differences in landownership, and associated economic activities associated with them as these reflect clear variations in class, status, life style and power.*" (Bertocci 1972, p.37). This conclusion should be regarded as a central text for our study: the key hypothesis to be explored and examined in detail.

It is worthwhile summarising the analysis which led Bertocci to this conclusion. In this population, about two-thirds of the farms (the landholding of a cultivating family unit) are less than two acres in size, and two acres is regarded as the amount of land needed by an average Bangladesh rural family of 5.5 to maintain a consistent subsistence level of income. (Bertocci 1972, p.36 Footnote 3; 1970, pp. 60-61) "Those with substantially more than 2 acres are on the edge of surplus, and peasants with 4 acres and above are, in this overall minifundist context, surplus farmers." (Bertocci 1972, p.36 Footnote 3). Half of the subsistence farmers in his population were deeply in debt and faced certain downward economic mobility. "By contrast the surplus farmer, who is likely to own upwards of 4 acres, is well able to withstand the vicissitudes of most years and to be able in fact to count regularly on a small crop surplus which is the lending of money or rice to his less fortunate neighbours. Thus, a surplus farm holding allows its owner typically to augment his income by lending activities, taking the land of his creditors in mortgage (which is by custom usufructuary). In the literature on Bangladesh agriculture and rural society, the term *mahajan* (moneylender) refers more often than not to the surplus farmers". The broader conclusion of Abdullah and Nations is certainly underlined in this case study. However, a flexible analysis of any particular situation and the significance of the precise variation in landholding size is required since a 6 acre family may be dominant in one village but not in another where 10 acre families exist. In this way landholding size constitutes a variable but not a precise indicator of status and political strength. The overall message is clear enough.

Furthermore competition for land itself generates important empirical data for our research. Since land is such a vital but scarce resource in Bangladesh with prices rocketing, its possession constitutes an important objective of economic and political activity. Firstly, there are the economic and social processes by which families gain and lose land (purchasing, mortgaging, and multiple inheritance); secondly, the way in which families and groups obtain and defend titles to plots for example, through inheritance and bribery; and thirdly, the inevitable disputes which arise over boundaries and the structure of their resolution. This 'absolute smallness in farm size' will render relative differences of even a minute scale, quite significant—both as a source of disputes and as a political variable in their resolution. Although these disputes may occur between individuals or groups in the village, often their pursuit will critically involve the patrons of the respective parties to the quarrel, access relationships to politicians, administration and the Courts, and a variety of methods designed to ruin the opponent without simultaneously ruining oneself. Clues to the operation of the political structure obtained in this way can then be tested for their significance in the overall social formation.

The relationship between landholding, status and class is complex and should be pursued in our work with precision and caution. Although Bertocci's data confirmed "an overall association of landownership with relative social rank" (Bertocci 1972, p.41), he nevertheless considered that it was important to maintain a conceptual and empirical distinction between status and class. The villagers themselves discriminate between *ucho-bangsho* (high status lineage) *madhya-bangsho* (middle status lineage) and *nicho-bangsho* (low status lineage) families (Bertocci 1972, p.4). However, not all those families with traditionally high status titles are wealthy peasant households, since within the same *bari* patrilineal extended family there can be considerable variation in landholding (the *paribar* or *chula*). It is likely that there will be a continually changing balance of strength between *paribars* in a *bari*, and perhaps sometimes between *baris* as families divide holdings amongst sons and the demographic balance of the nuclear family shifts (Chayanov 1966; Kerblay 1971). In this respect, Bertocci has distinguished in his thesis between *sardari* lineages (having traditional high status) and non-*sardari* lineages which have acquired some economic strength and have thereby become politically important and in some cases dominant through moneylending activities and other forms of patronage. From this he attributes a great deal of fluidity to the system in which different lineages rise to power as rich peasants, and where they are not *sardari* then sometimes acquiring status by marrying into *sardari* lineages. He refers to this process as a 'cyclical kulakism'. However, he seems to overstate this circular mobility (marginal/ subsistence peasant lineages are anyhow virtually excluded) and he certainly recognises the tendency for property and status to coincide.

It is a very important focus to take up this analysis, especially to contrast a situation which has been heavily involved in the new technology with an area which has been relatively untouched. The hypothesis here would be that *the tendency for status (especially sardari) and landownership to coincide would increase, and the extent of circular mobility decrease under conditions of new opportunies for increasing agricultural productivity which thereby also provided greater scope for investment in non-agricultural activity* like moneylending, manipulation of the market, primary processing plants. In other words, the hypothesis suggests that a stable rich peasant class will emerge under conditions which reduce the vulnerability of families in that class to the weakening process of fragmentation through inheritance as other non-agricultural sources of income become available as a result of initial higher productivity of holdings. This hypothesis is in contrast to Bertocci's own speculations (Bertocci 1972, p.47), which do not seem to take account of either the possibility for richer peasants to prey on the misfortunes of others by acquiring their land, or the role of the new technology in stimulating other forms of rural (but non-agricultural) economic activity like moneylending itself which hitherto plays a prominent part in his analysis. Nor does he take account of the enhanced capacity to establish sons in other forms of employment. Furthermore, he appears over-optimistic about the potential for subsistence-holding families to improve economically "especially if they are able to engage successfully in lending activities, in particular the taking of land in mortgage." Surely subsistence level households could not compete with families already well-established in these activities? The economic interrelation characterised by exploitation *between* the two classes in the village would prevent such competition in the first place.

The dimension of status becomes important to such an analysis of class formation since it can function to legitimise and reinforce class differentiation by providing familiar and socially acceptable institutions through

which it can be represented and the internal cohesion of classes maintained. But the function of status is complex. Bertocci found that:
"the titled lineages married others of like status at an average rate of 42.3% of their total marriages recorded. By contrast, non-titled lineages show evidence of marriage into titled lineages at an average rate of 13.4% of their total recorded marriages." (Bertocci 1972, p.46). Furthermore, 50% of the surplus farmers in his population are members of titled lineages. From this data one can point to a definite tendency for status (in the form of membership of a titled lineage) to be coterminous with class, although it is not yet exclusively so. Interestingly, there seems to be a tendency for "the assignment of status to follow the rise and fall of various families" (Bertocci 1972, p.48) where a distinction is made between *ashol* (original or real) and *dak* (so-called) status. This seems to correspond to the process of 'Sanskritisation' which has been identified in many of the studies in India.

This data also points to the function of marriage in uniting families along horizontally stratified lines. But for the analysis of the relationship between marriage patterns and class formation in the countryside a further refinement is required—the identification of spatial variations between differing economic categories. We might hypothesise for example that *higher lineages who were also richer peasants would be able to sustain kinship links of a greater distance than the poorer families of their village*, although, of course, local strategic marriages would also be made for alliance purposes. The importance of such extended kinship links for class consciousness and solidarity is self-evident—a familiar network of relationships through which specific, particular interests can be related to universal sets of interests which can be perceived to be held in common. In Hindu society where village exogamy is practised quite strictly along with caste endogamy, such a proposition is relatively easy to handle, but in a Muslim society where the rules are less overtly institutionalised and therefore less clear (to me, at any rate) the research must be done carefully.

However this horizontal solidarity function has to be qualified. In his discussion of the dominant *sardari* lineages, to which other lineages (*baris*) in the *para* give allegiance, Bertocci identifies a vertical solidarity function for some of the marriages between *sardari* lineages and other lineages. By trying to establish the extent of intermarriage between *sardari* lineages within the *samaj*, Bertocci was concerned to identify the role of kinship in maintaining the social and political order of the area covered by the *samaj* through increasing the solidarity of the *sardari* lineages which dominated the *samaj*. He found that the rate of intermarriage between the 5 *sardari* lineages in the same *samaj* was 11.4% of all their marriages. But he also found that the rate of intermarriage between the 16 non-*sardari* lineages and the *sardari* lineages in the *samaj* was 8.7% of all recorded marriages over several generations (Bertocci 1972, pp. 49-50). Thus the *samaj* was to some extent unified both horizontally and vertically through intermarriage, functioning presumably to constrain conflict between *sardari* lineages while also strengthening the vertical ties of dominance between *sardari* lineages and those of lower status (and usually of poorer class) providing *sardars* with an additional lever to keep their own followers from disturbing the balance of power through their own conflicts. This principle of access of non-*sardari* lineages to *sardari* families by marriage can also function as a device by which the economic strength of a non-*sardari* lineage can be socially recognised and corresponding status (*dak*) acquired. Thus we can supplement the earlier hypothesis: *Where the process of class differentiation is stimulated by new opportunities for the accumulation of wealth, rich peasants who come neither from titled lineage, nor from sardari lineages can be incorporated into the traditional institutions of local power by acquiring status through intermarriage with lineages of traditionally high status. In this way the dimension of class and status will became mutually reinforcing.*

Bondokgram—the Minifundist Case

Introduction

The following description and analysis of Bondokgram represents an early stage in our understanding of political and economic power in the villages of Bangladesh. The description and analysis presented here is itself incomplete in two senses. First, the fieldwork period allowed to us was very short, so that we could not reasonably expect to obtain a full picture of the village. Second, we have generated much more data than can be presented or analysed here.

In particular, the omissions in this text are: a detailed class and kinship analysis of KSS (farmers' co-operatives) membership: a more extended analysis of mortgage and lending transactions, involving a typology of arrangements; class and kinship analysis of capital flows through lending and land transfers; greater examination of the class differentiation and domestic cycle issue; a closer analysis of the structural significance of the disputes in the village and the way in which they were resolved.

There are many further omissions, such as: a more thorough analysis of actual water distribution; more accurate data on yields, costs of production and so on. This sort of data is required for a rigorous conclusion about the social relations of production in the village. Furthermore, we have not dwelt upon the linkages through Union, Thana and district-level political institutions and activities. However, my purpose has been to provide a qualitative picture of the class structure in the countryside rather than the operations of these higher levels in the political system. These levels are an expression of class forces, but will require further research. We have gone to the people first, as a basis for understanding the superstructure.

The identity of the village and its inhabitants have of course been protected in writing up the case study.

Summary Description of Village

Location

Since the Kotwali Thana of Comilla District has been so intensively researched and is so easily and readily accessible to visits by academics and officialdom, the inhabitants of Bondokgram are also entitled to some vagueness about the location of their village. It lies at some distance north of the Gumti river, and on the northern side of the river in this area no four-wheel mechanised forms of transport are available. The village is therefore reached on foot. Bullock carts are a rare sight, but rickshaws can occasionally be seen carrying the women of richer peasant families or visitors from the town. Bicycles are frequently used by minor officials and some of the service-holders commuting daily from the village to the town. Sons of richer peasants attending schools or college in the town also regularly use bicycles, while others have to walk, thus losing a few hours potential studying time every day. Two motor bikes are often seen, one belonging to the Chairman of the Union Parishad who lives in the village adjacent to Bondokgram, the other belonging to the member of the Union Parishad, who resides in the East Para of Bondokgram.

The relative proximity of the village to the town affects its economic life in several ways. The richer peasant families are able to diversify their economic activity in ways which include the resources and employment opportunities of the town. Thus several families have business interests in the town—medicine, stationery and electrical stores; and some family members are employed in minor government positions as clerks, typists etc. Furthermore, this relative proximity has facilitated the education of sons from the richer families especially so that employment further afield lies within their grasp for example in the police service, but the army, air-force and the teaching profession are also represented. This accessibility to the town has also extended some opportunities to the poorer peasants for engaging themselves in petty trading activity (usually of rice, and vegetables in season), where the produce is purchased either from the village itself, or from other villages and nearby *bazars*, and then carried on shoulders to the towns. The return on such a day's work rarely exceeds Tk. 10.00 and often falls below that, being equivalent therefore to little over a *seer* of rice at current prices. Despite its prevalence, petty trading is generally regarded by the poor as a last resort when other opportunities for an income are unavailable.

The village was described to us by one of our richer informants—Boshrat Ali—as "Mujib's Paradise", and indeed our analysis may at times appear unreasonable as Bondokgram must be one of the most fortunate villages in Bangladesh. Apart from its accessibility to the town, the village is distinguished by the absence of flooding during and after the monsoon. And yet, there are low land areas, particularly in the North-East of the village which are regularly inundated but which allow a late *aman* crop, and which even retain sufficient moisture in some of the lowest stretches as the water recedes to allow a natural *boro* rice crop to be cultivated. Of course information about yields was regularly prefaced by tales of unprecedented flood damage, but our indiscreet friend assured us that fortunes were being made out of the *aus* and *aman* crops as the yields were as high as ever, and the prices even higher. Most of the *nal* land is double cropped—*aus* and *aman*; a small proportion in the North-East is naturally irrigated during *boro*, and areas of the village in the North lie within the command of two deep water tube-wells—including land owned by some of the residents from West and North Paras. At present, the Southern half of the village and the East Para have little scope for *boro* rice cultivation, although East Para with its own KSS expects to receive its own tube-well shortly. Thus although there is scope for improvement in irrigation facilities, the village itself must be among the most favoured in Bangladesh with respect to climate, topography, irrigation facilities, accessibility, and of course its general situation of being in the 'command area' of the Comilla Academy for Rural Development.

These characteristics become an important feature of the methodology for this preliminary study, since this village to some extent represents the outcome of applying the 'Comilla approach', which has remained (perhaps by default) the inspiration for contemporary rural development strategy in Bangladesh. By selecting such a village we have purposively chosen a community which exists under conditions whose replication has been pursued elsewhere in the country. For the purposes of radical criticism, we

Exploitation and the Rural Poor 53

have taken a village where our arguments—according to the ideals and objectives of the 'Comilla approach' ought to be the weakest. But if we can show behaviour and a structure of the political economy which are unpalatable and inconsistent with the socialist objectives of leaders and critics alike in Bangladesh, then the implications for present institutional solutions for rural development are far-reaching.

Population

Our figures for the population of the village are approximations since our fieldwork time did not allow us to do a complete census. The national census—itself an approximation unfortunately—also included the adjacent village, although such an aggregation does not represent the social reality of an integrated community consisting of both villages. In this area at least, the national census appears to have been based on the memory of one or two informants in each village rather than on a thoroughgoing household survey. As a result, we have found the census tale to be unreliable; even the number of *paribars* identified is inaccurate. From our own incomplete investigations we are able to present in table I.

TABLE I

DISTRIBUTION OF *PARIBARS* WITHIN THE VILLAGE

No.	No. of Para	No. of Baris	%	Paribars	%	Estimate of Population*
1.	East	16	49	89	39	623
2.	Middle	5	15	36	16	252
3.	North	3	9	37	16	259
4.	South	3	9	21	9	147
5.	West	6	18	45	20	315
Total		33	100	228	100	1596

The village consists of 228 *paribars* distributed between 33 baris and 5 *paras*.
* We completed a preliminary random survey of 76 out of 228 *paribars*, with a total population of 544. Thus the average size of each paribar from this sample is 7.1. For computing the overall population estimate, we have therefore employed a multiple of 7. By this calculation, the census has undercounted for Bondokgram and the adjacent village by a factor of about 15%. We know that the number of *paribars* for the two villages is greater than that shown in the census—303 as against 290—so that our total estimated figures for the two villages is 2121, compared with the census figure of 1835.

The village is thus one of the largest in this area, reflecting our concern to work in a village which is large enough to contain a fuller range of relationships than might be found in a 'typical', smaller village—a true microcosm in effect.

Landholding

From the initial random survey, the following distribution of the *paribars* by land owned and effective landholding is found. Land owned by a *paribar* is not necessarily all cultivated by that *paribar*, so that we have adopted a more complex indicator of a peasant's real landholding position by calculating for each *paribar* the land from which an effective and regular income through cultivation is obtained. Thus for example, an acre of land sharecropped by a family where the crop is divided 50:50 between owner and sharecropper has been included in the overall effective landholding figure as 50 decimals. Similar adjustments have been made for mortgaging arrangements.

The complete formula for 'effective landholding' is therefore:
- − land owned and cultivated; + half land sharecropped out
- + half land mortgaged out but cultivated by the household
- − land mortgaged out and not cultivated by the household
- + half land sharecropped in by the household
- + land received in mortgage and cultivated
- + half land received in mortgage but not cultivated by the recipient household.

TABLE II

LANDHOLDING BY SIZES AND PERCENTAGE

Acres	Land owned		Effective Landholding	
	Number	% age	Number	% age
0	14	18.4	12	15.8
0.01-0.99	18	23.7	18	23.7
1.00-2.49	25	32.8	31	40.9
2.5 +	19	25.1	15	19.6
Total	76	100.00	76	100.00

With the aggregated figures, the comparison does not reveal a great variation between the two columns, with a tendency to reduce the frequency of the extremes and enlarge the middle range. Individually, however there are some interesting movements of certain *paribars* from one land size to another. Throughout the discussion which follows, the effective landholding column is used in most of the tabulations, except where the identification of land transactions demands otherwise, and in the following table. Here the implications for inheritance patterns, division of holdings and fragmentation is shown by presenting the distribution of sons.

TABLE III

NUMBER OF SONS BY LANDOWNERSHIP

Land owned (acres)	No. of sons per *Paribar*								Total Sons	Total *Paribars*	Average Sons per *Paribar*
	0|	1|	2|	3|	4|	5|	6|	7|			
0	4	3	5	1	1	-	-	-	20	14	1.4
0.01-0.99	1	5	5	3	2	2	-	-	42	18	2.3
1.00-2.49	4	3	8	8	2	-	-	-	51	25	2.0
2.5+	3	5	3	4	2	1	-	1	43	19	5.0
Total	12	16	21	16	7	3	0	1	156	76	Average 2.0

Perhaps the main conclusion to be drawn from these figures is that, of those who own some land (62 out of 76), 41 *paribars* (or 66%) will have to divide their holdings between 2 or more sons. The distribution of the average only warns us that richer families are not yet controlling the size of their families as a way of maintaining their economic position through generations, although evidence shown below indicates that they are certainly responding to this problem by diversifying. This factor of size of household is repeated in the following table, when all the *paribar* members are included.

TABLE IV

HOUSEHOLD SIZE BY EFFECTIVE LANDHOLDING

Effective Landholding Size (acres)	Number of *Paribar* Members						No. of *Paribars* Total
	0-2	3-5	6-8	9-12	13-16	17-21	
0	1	4	7	-	-	-	12
0.01-0.99	2	6	9	1	-	-	18
1.00-2.49	1	6	15	9	-	-	31
2.5+	-	-	6	5	3	1	15
Total *Paribars*	4	16	37	15	3	1	76

Main Forms of Economic Activity

Some people in Bond(m obtain their income from a variety of sources in addition to cultivat_ _ _ _ There are two main items of interest to us here: the employment position of the landless and the poor peasant; and the pattern of diversification away from agriculture among different classes of peasant. The analysis of power structure depends critically on the nature of the dependency of the poor on the rich in terms of actually providing employment, arranging employment elsewhere, or providing loaned capital to engage in petty trading activities. At the same time, the extent to which the richer peasant class is able to diversify into other secure forms of non agricultural activity is vital to the preservation and indeed reinforcement of the strength of that class. We hypothesised earlier on the political process in rural Bangladesh, *that a stable, rich peasant class will emerge under conditions which reduce the vulnerability of families in that class to the weakening process of fragmentation through inheritance as other non-agricultural sources of income become available as a result of initial higher productivity of holdings.* We will pursue this analysis, and its significance later, confining ourselves here to a description of the main types of economic activity found amongst the peasants of this village. We begin by presenting a summary table (Table V) of the relationship between cultivation and other occupations by effective landholding category for the *paribars* and their members in the village. This table is derived from our initial random sample survey of *paribars*.

TABLE V

RELATIONSHIP BETWEEN CULTIVATION AND OTHER OCCUPATIONS

Effective Landholding Size (acres)	Occupational Groups					Total by land-holding	% by land-holding
	A	B	C	D	E		
0	1	-	-	-	11	12	15.79
0.01-0.99	8	9	1	-	-	18	23.68
1.00-2.49	17	7	4	3	-	31	40.79
2.5 +	6	3	4	2	-	15	19.74
Total (by Occupation)	32	19	9	5	11	76	100.00

(Occupational Groups: A = Cultivation only
B = Cultivation + one other activity
C = Cultivation + two other activities
D = Cultivation + three or more other activities
E = Other sources only)

Diversification into Off-Holding Employment

The first point to note is that about 16% of the *paribars* in the village are landless, by our definition of effective landholding, with about 20% regularly enjoying the product of land in excess of 2.5 acres. In this '2.5+' category: 60% of *paribars* have at least one member in an off-holding occupation; and 40% have at least 2 members in off-holding occupations. The single largest category in the village is the group holding between 1.00-2.49 acres, representing about 41% of the total. Within this category, 45.2% of the *paribars* have at least one member in an offholding occupation (compared with 60% for the 2.5+ category); and 22.6% have at least 2 members employed outside the cultivation of the *paribar's* holding.

From these figures, the hypothesis about rich peasant diversification is not refuted and would seem to warrant a more widespread analysis of the phenomenon. However, by considering the 'less than one acre' category, where 55.5% of *paribars* have at least one member in off-holding employment it is clear that the concept of diversification has to be clarified and distinctions made between different types of occupation. Furthermore from our data, a high proportion of the *paribar* members in the 'less than one acre' category with one off-holding occupation in addition to own cultivation are the *paribar* heads themselves. This reflects a position of underemployment on the land, and the off-holding employment is very often irregular—for example agricultural labouring or petty trading. Some of these cause reservations about specifying the type of off-holding employment characteristics of the landless. One further note of caution from this table is that the figure of 42.1% for those *paribars* engaged only in cultivation of own holding may be extraordinarily low for the majority of villages in Bangladesh. But comparisons between villages of varying locations (both in a regional and urban proximity sense) is absolutely essential for an understanding of the pattern of rural employment.

In proceeding to a breakdown of this off-holding employment pattern in the village, it might be useful to present first some data on the distribution of educational qualifications by effective landholding category. The table VI refers to *paribar* heads only.

It is immediately apparent that the level of education for the *paribar* heads is low, surprisingly so when the favourable location of the village is considered. Only those with holdings above 25 acres are significantly represented beyond class 8, and then only at the SSC level (Secondary Matriculation). The proportion of those with an educational level below class 5 steadily declines through the landholding categories from 83.3%-72.2%;

64.5%-40%; but the figure is high. The figure for the landless—while expected—is of course a very depressing indicator of the dependency of this class, where the *paribar* heads themselves are obliged to search for employment throughout the entire year either within agriculture or outside. The lack of education reduces the possibility of entering regular employment, and so they are continually obliged to resolve the problem of access to employment.

TABLE VI

LANDHOLDING SIZE AND EDUCATIONAL QUALIFICATIONS.

Size of Land-holding (acres)	Education Level of Household Head						Total
	Nil	Class 4	Class 5-8	SSC	HSC	BA	
0	6	4	4	1	-	-	12
0.01-0.99	10	3	4	1	-	-	18
1.00-2.49	13	7	8	1	2	-	31
2.5+	3	3	5	3	1	-	15
TOTAL	32	17	18	8	3	0	76
% of Total	42.1	22.6	23.6	7.8	9.9	0	100

The impact of this distribution of educational qualifications on the pattern of off-holding occupations for individual *paribar* members can be clearly seen from the table VII.

TABLE VII

PERCENTAGE DISTRIBUTION OF INDIVIDUAL OCCUPATIONS (OFF-HOLDING) BY EFFECTIVE LANDHOLDING. (FROM 44 *PARIBARS.*)

Effective Landholding	Business	Service	Profession	Petty Trading	Agric. Labour	Peons etc.
0	-	8%	-	25%	75%	25%
0.01-0.9	-	12%	-	37.5%	-	31.25%
1.00-2.49	50%	52%	20%	12.5%	25%	31.25%
2.5+	50%	28%	80%	25%	-	22.5%

Note: Service = government clerks, officials, military service and skilled service occupations like electrician. Profession =Teachers, Doctors. Petty Trading = distinguished from business by absence of fixed capital, scale of operation, and seasonal variations. Peons includes industrial workers, Artisans and non-agricultural labour.

The main division here is between those with less than one acre, and those with holdings over one acre. The most secure and usually most remunerative occupations—business, service and professions—are almost solely confined to those *paribars* with holdings over one acre. The reverse does not hold with equal emphasis for the petty trading, agricultural labouring, and the other manual jobs. It is interesting to note that in the manual job category where there is often a higher element of security (e.g. peons, textile workers or artisans, including the carpenters of the village), those in *paribars* with effective holdings in excess of one acre are more highly represented (43.75%).

If we return to the unit of the *paribar* from the individual, we find that the groups of off-holding occupations business/ service/ profession and petty trading/ agricultural labouring/ other manual jobs—correlate inversely with the size of effective landholding. By coincidence, 24 *paribars* out of 76 in the sample (31.6%) are involved in each group of off-holding occupations, and a few *paribars* are involved in both groups.

TABLE VIII

PERCENTAGE OF *PARIBARS* ENGAGED IN TWO GROUPS OF OFF-HOLDING OCCUPATIONS BY EFFECTIVE LANDHOLDINGS.

Effective Landholding	Business/Service Profession	Petty Trading/ Agric. Labour/ Other Manual Work
0	8.3%	37.5%
0.01-0.99	12.5%	37.2%
1.00-2.49	50.0%	20.8%
2.5+	29.2%	12.5%

Again it has to be noted that, despite the correlation, the 1.00-2.49 acre paribar category is still well represented in the second column. This is perhaps some indication of the tenuous nature of their security on the land and a response to the problem of multiple-inheritance and fragmentation. However this conclusion has to be modified, by recognising that the 1.00-2.49 acre category is the largest in the sample (40.79%). Furthermore, 54.8% of the category is exclusively involved in the cultivation of own holdings whereas the corresponding figure for *paribars* with holdings over 2.5 acres is 40%.

From this general description of the occupational structure, some qualitative description of the occupational types is relevant to our analysis. However, we will not present an orthodox description of cultivation practice, since there are numerous comprehensive accounts for this area. We will focus only on those issues which relate to our overall theme.

Cultivation

Despite the location of the village the amount of land irrigated by tube-well for *boro* rice is now very low having declined over the last few years. East Bondokgram KSS has yet to acquire a tube-well, therefore we have to consider only Bondokgram KSS and Bondokgram-Dhaninagar KSS. Bondokgram KSS has been operating a tube-well since 1970. According to the records, in 1970-71, 100 acres were irrigated; 70 acres in 1971-72; and 60 acres in 1972-73. During 1973-74 and up to the present, the tube-well remained idle. There are several explanations for this, one's preferred explanation depending on the position occupied in the system. Thus faculty at the Academy will emphasise factionalism within the society; the A.C.F. understandably points out the high incidence of defaulting payments on loans; the villagers refer to organisational inefficiency of the 'Government' (the more informed will cite the BADC); and politicians will remind us of the post-liberation problems of the country in obtaining essential supplies of oil, fuel and spare-parts. The truth is most likely a combination of these observations.

The Bondokgram-Dhaninagar KSS has been operating two tube-wells since 1966, one of which has land owned by Bondokgram residents from North Para under its command. These two tube-wells have irrigated on average 50 acres each per year, but again the coverage per tube-well has been sharply declining from year to year so that the current acreage irrigated is some way below 50. The only significant addition to the quantity of land available for *boro* rice cultivation is that low land area on the edge of the canal, one-third of a mile North-East from the new Haji Bari of East Para. There is some irrigation by tank, representing an only an insignificant amount.

Thus at present the extent of *boro* rice cultivation in the village is very low, probably representing no more than 10% of the land cultivated under *aus* and *aman*. However it has been higher in the past, affecting particularly some of the peasants from North and West Para. But we could not conclude that a triple cropping annual system with *boro* rice had been responsible for the wealth of the stronger *baris* in these *paras*, since our historical data on the families indicates that their superiority stretches further back than this.

When the tubewells were introduced these families were of course in a position to take advantage of the new resource, but their adoption of the innovation seems to have been an outcome of their economic superiority rather than the cause of it. Importantly, the continuation of this economic superiority does not apparently depend upon the regular achievement of surpluses of rice derived from *boro* cultivation since alternative secure sources of income besides cultivation are already available to this class.

A further observation should be made about the impact of *boro* rice cultivation on the annual cropping cycle. There is some evidence from the village to suggest that much of the *boro* cultivation functions to displace an *aus* crop from the same land-so that the double-cropping pattern is altered but only a small proportion of the land under *boro* is actually triple-cropped. Our data is not rigorous in this respect and the observation for Bondokgram must remain impressionistic. However, more specific data for the *thana* has been collected by others at the Academy which support this observation. In addition to this, it has to be noted that the yields from *boro* rice in practice often do not exceed those obtained from the cultivation of an *aus* crop, since the appropriate conditions for realising the full potential of HYV *boro*—water and adequate supplies of fertiliser—are rare at this time. Thus the involvement in *boro* rice cultivation is not necessarily an indicator of a net addition to a peasant's annual surplus. With these reservations, we present the following figures based on the smaller, stratified sample of village households.

TABLE IX

NUMBER OF *PARIBARS* ENGAGED IN *BORO* RICE CULTIVATION WITH LAND OWNED, AND EFFECTIVE LANDHOLDING

Serial No.	Name of *Paribar* Head	Land owned Acres	Effective Landholding
1.	Abdul Mannaff	4.0	3.2
2.	Boshat Ali	4.0	2.8
3.	Muklesur Rahman	0.33	0.33
4.	Sirajul Islam	2.5	3.3
5.	Rusmat Ali	2.8	2.5
6.	Askar Ahmed	7.2	5.6
7.	Junab Mia	3.1	3.1

(Total in stratified sample = 23
Total in stratified sample with land owned and cultivated = 21)

Although the total is small, as is the stratified sample, the message from these figures is clear enough. With the exception of Muklesur Rahman, all of those from the sample engaged in some *boro* rice cultivation come from the richest class of peasant in the village. If we were to extrapolate from this data, then 33.3% of the landholding peasants in the village are engaged to some extent in *boro* rice cultivation—although the amounts of land are small, *aus* on the same land is often displaced; and the yields of HYV *boro* fall a long way short of their potential. If we were to extrapolate further, about 90% of those cultivating *boro* rice have effective landholdings above 2.5 acres—i.e. a category of peasant which might normally expect to produce a surplus of foodgrain in addition to family consumption requirements with *aus* and *aman* alone, without the help of *boro*.

The issue of fertiliser for cultivation is obviously a critical one, and its overall scarcity apparent to all. In this situation, data on the use of fertiliser can be a vital index of power and influence in the village. Certainly one of major points of controversy between different groups of villagers concerns the way in which the supply of fertiliser is manipulated and controlled, and the varying rates of successful access to fertiliser by different classes or factions among the villagers. However, it is precisely this intense competition for fertiliser in a situation of extreme scarcity which makes the data on usage very unreliable partly because fertiliser is obtained sporadically and in small amounts so that recall is difficult, and partly because the method by which it was obtained involved some form of corrupt practice. The importance of fertiliser is certainly not lost on the peasant and the problem is very definitely not one of extension but of access. Most of the cultivators would like to use fertiliser on all their paddy crops where there is a risk of the crop being damaged or destroyed through flooding, untimely cyclones, or lack of water (i.e. in rain-fed *boro*). In this last year, most of the cultivators would have applied more fertiliser where they did use it, if they had been able to obtain more. A few were satisfied, as they had sufficient power and access to obtain adequate supplies. We intend to illustrate these points in detail later, referring to the structure of allocation, and leaving quantitative analysis for further investigation.

The village as a whole has a shortage of labour, which does not however mean that the landless of the village find regular employment on the holdings of the richer peasants. The village landless have to compete with a regular influx of migrant labourers from elsewhere in Comilla District—particularly the flood-affected areas. As agricultural labour will be discussed in the following section, we need only to note here that the peak demand for labour by peasant cultivators is during the period of *aus* harvesting and *aman* transplanting, with two lower peaks at *aus* transplanting and *aman* harvesting. The healthy cultivator with no other form of employment regularly engages himself in most operations, especially ploughing, hiring labourers where the timing of an operation is critical. Other landholders who are perhaps older; engaged in other activities; or simply richer, employ labourers as a substitute for their own energy or time—sometimes supervising themselves, or appointing a son or relative to the task.

The disposal of marketable surplus of foodgrain and vegetables involves the larger cultivator in a transaction with village or migrant petty traders. They move from house to house buying odd amounts of paddy, collecting to an average maximum of 1 *maund* 10 *seers* before carrying the load to their own houses, converting it to rice and then carrying the same weight either to a local *haat*, or to the town. It appears to be cheaper for a cultivator with a large surplus of paddy to sell to a petty trader rather than engage labourers to convert paddy to rice. Peasants with a smaller surplus may sometimes do their own conversion to rice (involving the women of household) and transporting it to the market. There exists therefore the possibility of a range of combinations of methods by which the crop is marketed, depending upon the economic position and status of the cultivator; and the size of his surplus. The additional dimension of procurement features later in this analysis.

Agricultural Labour

The attitude of the major employers of labour in the village was very frank, and displayed an acute appreciation of the harsh reality of the laws of supply and demand. The issue for us is the employment position of the landless and the consequent nature of their dependence upon their economic superiors in the village. Our main conclusion for the village is that, despite a general shortage of labour in the village, the position of the Bondokgram landless in need of employment was weakened by the entry of outsiders into the local labour market.

These migrant labourers come from other *thanas* in Comilla District—Chandina, Burichang, Debidwar, Muradnagar, Daudkandi, 'Laksam and Barura, and most frequently from Chandina and Burichang. These are areas regularly affected by flooding. The labourers are often landowners whose crops have either been damaged, or whose transplanting and harvesting operations occur at different times, or who are unable to crop at such an intensity as in Kotwali Thana, or whose average yields are lower even though they may own more than their employer (rare). Often, of course there is a combination of these reasons. 1974 was a good year for Bondokgram and its employers of labour, since the flooding had forced a higher number of peasants from flood-affected areas on to the labour market, keeping the price of labour low. Thus one informant—Sardar Askar Ahmed—admitted that he and others could have afforded to pay Tk. 7 plus three meals for *aman* transplanting and *aus* harvesting during 1974, since the price of rice was so high; but they were able to pay Tk. 4/5 plus three meals as "the wage-rate is a commodity fluctuating like chillie, it depends on supply."

In general outsiders are referred to by the local villagers as daily labourers. Another informant—Boshat Ali (the indiscreet one)—remarked: "Local labourers get distracted by the demands of their own family. But outside labourers sleep here, and as they have no house here (i.e. no food), so they have to work even if the rate is low." Several informants said that the landless from the village were idle, and that "they go in for petty business instead of working." It seems likely that the local landless and the poor peasants are forced into these other insecure activities with narrow and fluctuating profit margins, (for selling rice, on average Tk. 10 a load—i.e. per day, which includes the labour of conversion to rice. A rare and lucky day might bring in Tk. 30, but more often even losses can be sustained. Out of this 'profit' interest repayments on loans have to be paid, since the capital required for the trading is usually loaned). They are 'forced' precisely because the migrant competitors are prepared to accept low rates for a short period of time to cover their own subsistence since others in their family are also migrants and their womenfolk and children have remained to subsist from the crop which has been harvested. But the landless of Bondokgram have no such means of providing for their dependants.

The outside labourers moreover are not given wages daily. They work for different people in Bondokgram—a few days here, a few days there— and then come to collect the total amount of cash owed to them when they decide to leave the village. One of the functions of this mode of payment is of course to regulate the behaviour of these strangers staying in the village. The main variation in the rate during the year is that only two meals are offered for *aman* transplanting, plus Tk. 4/5, "since the day is shorter." For *aus* transplanting the rate is usually highest—Tk. 5/6 plus three meals is common, although there are instances of the rate going as high as Tk. 8 plus three meals. This is not the highest peak in demand for labour, but the supply is much less at this time since the flooding does not usually affect the homes of those migrant labourers until later in the year. The opportunities for Bondokgram landless are higher at this time. In previous years, Bondokgram has also been attractive to migrants in search of work during the winter, where the introduction of HYV *boro* rice creates another season of labour demands. The decline in acreage under *boro* has affected them adversely too.

This phenomenon of migration is not the immediate concern of this study except insofar as it functions to depress the local wage rate and further weaken the position of the landless in the village. But the general issue of the relationship between patterns of labour migration and other variables such as flooding, the extent of HYV cultivation and levels of cropping intensity is an important one sociologically and politically, and should not be overlooked.

Political and Social Organisation

The Nature of Politics

The five *paras* of Bondokgram have strong *baris* with interconnections between them. The *paras* are significant as distinct social entities; and there are acknowledged *sardars* within each *para* and for the village as a whole. Other institutions exist, like the Union Parishad, co-ops, dealerships, Youth associations, the school and procurement operations. The following will describe them further, identifying their significance for our analysis of the structure of power.

The aspect of our lives which we call 'political' refers to two themes. How the social order of our community is maintained at a minimum acceptable level for those who dominate in that community. How, as a corollary, the competition for scarce resources can be regulated to achieve a distribution of resources which reflects the reality of social and class power and thereby remains consistent with the maintenance of social order and security. These social processes are maintained through a variety of forms— traditional lineage status; religious sanction; the syndrome of economic superiority and dependence; the social structure of formal education (teacher-pupil); the internal structure of a *bari* or *paribar* (domination based on criteria of age, sex, education, earning power); and a general category of purposive transactions between individuals for long or short-term mutual benefit: leader-follower, friends of equal status, clients and brokers. The list can be extended, but the point is made that there exists a complex network of relationships through which the social fabric is maintained. These relationships are inter-dependent, and represent different levels of social organisation referring to different aspects of social behaviour. The following discussion is organised around two themes, the structure within which domination and exploitation occur; and the significance of this domination for the future of the system as a whole.

Since we conceive of the problems of rural development in a materialist perspective, we consider the analysis of the social relations of production and exchange to be the primary focus in discussions about domination and power. The poor in most societies have not the leisure to consider the more refined aspects of domination and freedom; and in Bangladesh where such leisure exists for the poor it is enforced and implies an absence of material welfare. In short, such leisure means they are dying. Accordingly, we are concerned with identifying the reality which lies behind the myth of a liberated people. If it is painful to read, it is because the reality is a vicious one, not because we select only those features which attract moral disapproval.

The units of social cohesion within which authority was recognised by villagers as legitimate are perceived differently by the inhabitants of Bondokgram. Different groups have different social horizons—for some the *bari* is the primary point of reference; for others the *para*; for a few the village as a whole; and for some 'territorial' analysis has surrendered to class analysis, although still within limited territorial horizons. The terminology with which different social institutions were described also varied between informants. This variation in perceptions perhaps suggests that if there was ever previously a consensus among the village people about the description of their social system, this has been disturbed by an 'untidy' reality in which the notion of a *para* or village as an integral community (a societal isolate) is undermined by the differential relationships between classes of the peasantry and other classes in the country, and the State itself.

The social groupings in the village appear to exist alongside a structure of pervasive class domination which often crosses the boundaries between these territorially-based social entities as well as occurring within them. The rich peasants of one *para* (or sphere within it) on the one hand do not have to restrict their economic transactions to that *para* (or sphere), but as a corollary neither do they have a monopoly of economic transactions within their *para*. We will have to analyse our data further to compare precisely the levels of intra-*para* economic interaction with that between.

However, this picture alters when the 'purer' forms of political authority are examined. Then there seems to be a definite hierarchy of authority extending from the *paribar* head, through the *bari* and the *para* (with perhaps an intervening level here referring spheres of influence within the *para*), to the village as a whole. Beyond the village the picture is not so clear and several alternatives appear.

The Bari Heads

Apart from the direction of his own family's affairs, the political role of the *paribar* head is one of loyalty to the *bari*, and its recognised head, if it has one. Such loyalty from the members of the *bari* is essential if the *bari* is to enjoy any respect or leadership position in the village. Such respect among the village is important for the *bari*'s leaders to represent successfully the interests of *bari* members—whether it be in marriage negotiations, the resolution of disputes, the allocation of fertiliser or relief, the location of a tube-well, and the distribution of water etc.

The internal solidarity of the *bari* is also important in this respect. A *bari* whose members have the reputation of quarelling among themselves— like Sardar Askar Ahmed's *bari* in South Para during the 1940s—is unable to command any political authority in the village, leaving it vulnerable to ostracism or sabotage. However, since internal conflict and competition between *bari* members is prevalent in most of the *baris* of the village, certain levels are tolerated and are not regarded as being prejudicial to the status of the *bari*. Thus loyalty to the *bari*'s leaders often co-exists with a series of long-running disputes and relationships of exploitation between *paribars*— where the *bari*'s leaders recognise they are powerless to intervene. Indeed they are frequently involved in such relationships themselves. However, perhaps more importantly, to support one *paribar* against another risks splitting the *bari* into factions, thus weakening the position of the *bari* as a whole in relation to outsiders. In this way, the *paribar* heads are obliged to accept minimum levels of conflict, competition and exploitation among themselves where protection from another source may not be forthcoming.

Underlying, therefore, a facade of internal *bari* solidarity, a bitter struggle for survival and advantage occurs. Our data contains several examples of brothers borrowing from and mortgaging out land to brothers under extortionate conditions, of brother forced to sell to brother, as well as disputes over the division of the inherited holding. Under these conditions of great pressure on the land, the bond of kinship has been displaced by considerations of market price for credit and 'assistance'. The 'zero-sum game' aspect of inter-*paribar* relationships is modified only by the recognition that other resources (e.g. water or fertiliser) can only be obtained if the public status of the *bari* as a whole is maintained. This explains the gap between words and deeds—the presented image of harmony betrayed by land transactions and 100% interest rates often repayable only in kind between *bari* members (no compensation effects for the declining real value of inflated currency during inflation.)

For Bondokgram, these general observations have to be modified in the case of East Para where most of the *baris* recognise strong kinship links between themselves. Since the *baris* by strong genealogical ideology do not exist independently of each other, so individual *bari* status is dependent not only on the behaviour and internal characteristics of the *bari* but also upon its social status historically ascribed. Most of the *baris* claim membership of the same patrilineal descent group, although there is nevertheless a social ranking between them according to the intensity of the '*Haji*' connection. In this way the boundaries between the *baris* are more diffuse, and the status of individual *baris* consequently less important in determining the flow of resources to them, or as a principle of recruitment of its members to wider leadership status. Thus while this larger kinship group contributes towards the identity of East Para as a distinct social grouping (with two additions— Abdul Rahim (Member) Bari of Middle Para, and Sardar Askar Ahmed's Bari of South Para, both immediately adjacent) with a higher degree of internal solidarity and cohesion than other *paras*, the wider kinship group has also facilitated a freer interplay of class forces within its own boundaries, with richer *paribars* preying on members of their own *baris* according to the market-rate for transactions irrespective of any sense of *bari*-territory. Perhaps more importantly, as a result, this had generated a greater sense of class identity, and an acceptance of class interests as a more powerful criterion for behaviour than factional or *bari* interests. It is within this context that its new co-operative functions, though it will be interesting to note whether factional conflict emerges after the tube-well has been introduced, (that is after the advantages of class unity have been achieved, when the allocation of water *within* the class becomes the issue).

Most *baris* have an acknowledged head, but the small *baris* are often indistinguishable from larger neighbours and so their heads are not generally recognised as leaders in the village. Some of the larger, more influential *baris* may in addition to their acknowledged head have other leaders. Where the head of a *bari* has recently died, it may be sometime before a successor emerges. Where the *bari* head is old or infirm, his status is a formal one with other prominent *bari*-members attending village meetings and representing the *bari*'s interests. We could not identify all the *bari* leaders in the village, but of the 22 whose leadership seemed to be generally recognised and about whom we have some data (not necessarily from the sample)—16 have effective holdings in the 2.5+ acre category, and six have between 1.00-2.49 acres. From this we might conclude at least that *bari* leaders do not usually come from the landless and poor peasant class.

Besides a position of economic security (if not strength), age is obviously an important qualification and usually precedes a succession through the line of the eldest son. Where several brothers or cousin brothers are of similar age, then the eldest may simple be regarded as *primus inter pares*, unless he has clearly been superseded by another on further criteria of economic position, education, perhaps a high-status off-holding occupation, and a history of wise and operational decisions.

The Para Leaders

The criterion of age is reduced in significance at higher levels in the leadership hierarchy. Thus one informant—Abdul Mannaf—who was clearly a respected leader of his *bari* (new Haji Bari), although not a *sardar*, suggested that to be a *sardar* "a man should have a sense of justice and the ability to dispense it." But he also added that a *sardar* should come from a strong *gosti* (lineage). By this he meant a numerically large *bari*, with respected social status and extensive economic influence in the village. In this way, such a *bari* leader in the role of *sardar* would obtain respect for his decisions since his *bari*—between its members—would have a wide range of sanctions at its command. Thus, if a party to a dispute was ordered to make redress in some way, failure to do so might incur retaliation against himself or one of his kinsmen in the form of: physical assault; stricter conditions for obtaining loans and other forms of assistance; denial of resources such as water, fertiliser, relief, a licence (if the *sardar* or one of his *bari* members is well-connected at the *thana* level); being 'slack' about grazing cattle near his plots; and by virtue of the *bari* strength, encouraging others to ostracise him.

Apart from Middle Para, there is at least one acknowledged *sardar* from each *para*. In some *paras*, East especially, more than one is recognised although not all of them have an effective *sardari* status beyond the *para*, and may indeed have only a smaller 'sphere of influence' within the *para*. These 'minor' *sardars* constitute the occasional intervening level between the *bari* and the *para*. Where the *para* did not possess an obvious leader, the different *sardars* (of minor and of village-level status) might represent different factional groups within the *para*, instead of being ranked in a hierarchy.

In East Para, partly because of its size, the political structure is very complex. Three men were most frequently referred to as *sardars*—Ansar Ali, Riasat Ali and Siddiqur Rahman—all of whom in this local context may be regarded as rich peasants. Among these, Ansar Ali was the eldest, the most respected, and came from the 'strongest' *bari*. However, his advancing years and the death of his only surviving son has reduced his *sardari* activity, thereby increasing the importance of the other two leaders. Ansar Ali's overall position remains strong enough to offset any factional tendencies between all three of the *sardars*, although it is also clear that each of the *sardars* have their own sphere of influence within the *para*. The internal structure of these spheres of influence is interesting—Ansar Ali's depends upon his leadership of a large, economically strong *bari*, so that kinship links represent an important source of his power. Riasat Ali, the head of a much smaller *bari*, is engaged in diverse, one-dimensional transactions with dependants (money-lending, mortgaging, exchanging favours for obligations etc.). Siddiqur Rahman, coming from the 'policemen's' *bari* is head of his *bari* simply because elder brothers are serving in the police outside the area. His own economic position is secure, but he is not involved in extensive transactional relationships with dependants. His influence is based more strictly on the principle of territorial proximity to his own *bari*. Yet none of these *sardars* may be regarded as 'minor', for their leadership is not restricted to these immediate spheres but extends through to the level of the village. These spheres are not then rigid social groups with distinct boundaries, united in support of one leader. They are looser entities, and able also to acknowledge other leaders—recognising the reality of their strength and the utility of their patronage. In short, the ties cross-cut between leaders and spheres outside their immediate one. This functions to reduce factional conflict.

Part of the explanation for this structure of political relations can undoubtedly be found in the pervasive acceptance by richer peasants of market principles in their economic transactions with other peasants in the village (see for example the discussion on agricultural labour, above). The very openness of these transactions—not only with respect to labour, but in land and debt transactions too—and the advantage to rich peasants of disengaging themselves from the obligations which accompany the more traditional forms of patron-client relationship, preclude the emergence of exclusive political and social groups tied to one leader. In this way, vertical cleavages between such groups are weak, and horizontal cleavages between emergent classes much stronger.

However, we have introduced the notion of a 'minor' *sardar*, to deal with a qualification to the impression given above. Ali Mia, the head of old Haji Bari, is also referred to by some as a *sardar*, and thinks of himself as such. But the reference is not widespread. His *bari* is of course *ucho-*

bangsha along with the members of new Haji Bari, and yet neither of these *ucho-bangsha baris* have produced acknowledged leaders in the village. Individually they are respected and indeed others intermarry with them to 'purchase' their *bangsha* status. These two *Haji baris* are very conscious of their *bangsha* status and have been antagonistic towards the *sardars* of the *para*. Thus Dr. Abdus Samad's son from the new Haji Bari alleged that the *sardars* "all take bribes from both sides to a dispute". Ansar Ali and Raisat Ali were also bitterly criticised for arrogantly allowing their goats to graze all over other people's plots. Furthermore, there have been disputes between Riasat Ali and the Haji Baris. It would seem that Ali Mia's sphere is more exclusive, based largely on the *ucho-bangsha* connection with little influence outside it. His claim to the title of *sardar* probably reflects a desire on the part of the Haji Baris to regain some of their political status, lost when Reajuddin (the long-serving Member of the Union Council) and Nazir (his murdered nephew, who was Member for a short time), died. However, although Ali Mia may be restricted to this narrow sphere, other *sardars* are not, so that *ucho-bangsha* criterion of group cohesion is waning. Perhaps 'minor' *sardars* of this type are a transitional phenomenon in the shift from a vertical to horizontal cleavage pattern.

The *sardari* structure of the other *paras* is not so complicated, and has been partially described in the survey of the *paras*. Middle Para has its Member, who is not regarded as a *sardar*, but who *de facto* functions as such for many members of the Middle Para. The three *sardars* from East Para are most recognised, particularly in the Member's own *bari*, while the relationships of the other members of Middle Para tend to be mediated through the Member or one of his brothers. In North Para, the late Kala Mia was regarded as a *sardar*; but now the remaining *sardar* is Sirajul Islam, whom we described earlier. As Sirajul Islam himself is old, his brother Bazlur Rahman is slowly taking over many of his functions. Although Dr. Sadu Mia is not recognised as a *sardar*, his position at the head of a large, wealthy *bari* gives him a leadership position in addition to the status carried as a doctor. South Para, much smaller, is dominated by *sardar* Askar Ahmed, his position undisputed.

In West Para, two *sardars* are generally recognised—Minnat Ali and Abdur Rashid. Minnat Ali is the head of the strongest and richest *bari* in the *para*—his son is a Member; and his grandfather's brother's son's son—Panjat Ali—who is also in the *bari* is occasionally referred to as a *sardar*. Panjat Ali has a narrow sphere of influence and should be regarded as 'minor' *sardar*. Abdur Rashid is one of the richest peasants in the village; was a Member of the Union Council; a headmaster; and is now a master in the

Bondokgram Primary School. Minnat Ali's position is based to a large extent on his leadership of a 'strong' *bari*, and he is the most dominant leader of the *para*. By contrast, Abdur Rashid's political status is based more on his personal reputation outside the *para* as well as within it. He does not dominate a narrow support base, and is to some extent independent of one. Therefore, he does not compete with Minnat Ali for a sphere of influence within the *para*. The reality of Minnat Ali's domination in the *para* is accepted grudgingly, as many stories exist about the involvement of his bari members in corrupt practice—for example with the Bondokgram KSS and fertiliser distribution.

These leaders are integrated into a political structure for the village as a whole by the criteria of recruitment (described above); the manner of their recruitment; and the functions they perform at different institutional levels in the village.

Recruitment of Traditional Leaders

The way in which a peasant becomes accepted as a leader provides an important clue to the overall leadership structure. The main characteristics which *sardars* have in common (as described above and in the earlier discussion of the *paras*) are: they are heads of *baris*, which are usually 'strong'; have a reputation for making decisions which maintain and reinforce the social order; and they are rich peasants. But not all people with those qualifications become *sardars*. Therefore, the critical question for us is whether *sardars* are accepted and afforded legitimacy for their status by other members of the dominant class from which they come, or whether acceptance by the subordinate classes of the village or *para* is a precondition? The question is important, since *sardars* who are primarily accountable to their own class define social order and acceptable competition by reference to the interests of their own class, and can be removed by that class if they fail to maintain and reinforce these interests. Any radical content to their leadership is therefore fantasy. Dependence on their own class also acts as a constraint to factional division, since class rather than factional support is responsible for the maintenance of leadership status. Disputes may occur within the class, but usually only where the class is not in competition with other classes for a resource e.g. under certain conditions, tube-well water or even other forms of irrigation like canals.

From popular commentary, this recruitment process involved an institutionalised form of acceptance through the mass feeding of the village *paribar* heads. Attendance by other *sardars* indicated that the claim to *sardari* status had been accepted by them. To feed in this way implies the economic capacity to afford it, and therefore the acceptance comes from other leaders who must themselves also be rich peasants. Of course the provider of all this food should already have acquired some of the other qualities of leadership, otherwise his claim could not be taken seriously at all. This "feeding of the village" has disappeared from Bondokgram, and according to informants was never a strong tradition. Thus the process of acceptance is more informal, and less public. *Sardar* informants will say that acceptance depends upon the quality of decisions, but that becomes a self-fulfilling prophesy if the *sardar* comes from a 'strong' *bari* and his decisions can be enforced. But perhaps, more importantly the decisions made by a *sardar* (or 'suggestions' made by an aspirant *sardar*) have to be of a nature which the 'strong' *bari* is prepared to enforce if necessary. It is unlikely then that decisions will conflict with the interests of the 'strong' *bari*, and those who dominate within it.

Leadership Succession

With these considerations, the element of succession can be introduced, where a *sardar* in the evening of his life begins to nominate a younger member of his *bari* to take on some of his functions (mainly attending meetings at first). This nominee will already have gained some reputation for being interested in village affairs, concerned with the maintenance of ethical conduct, and free from public criticism for faults and previous misdemeanors. The nominee may not always be the first son, but a younger brother, another son, a nephew or other close relative. A nominee with kinship connections to more than one *sardar* would clearly be in a stronger position.

One example of this process is the case of Ansar Ali—the senior *sardar* in the village from East Para. His father had been a *sardar*. When the father was old, he sent Ansar Ali to the meetings in the village, and Ansar Ali was encouraged to be a *sardar*. In addition, his father's cousin's brother was Reajuddin from the new Haji Bari who was a member of the Union Council. When this Reajuddin made a pilgrimage to Mecca just before the formation of Pakistan, he nominated Ansar Ali to stand in his place. Ansar Ali thus became a Member for five years. Now, Ansar Ali is an old man; his brothers too are old, and he has no son. At first he remarked, "After me, Allah will give power to someone." However one member of his *bari*—Rafiqul Islam—was being 'groomed' for office. He is the manager of the

East Bondokgram KSS, and he has obviously devoted much time to mobilising support for it within the *para*. Furthermore, he is diligent in reminding the ACF to give the KSS high priority in allocating a tube-well, and in securing fertiliser dealership licences from the Chairman of the Union for the KSS. Ansar Ali named Rafiqul Islam as his likely successor, and increasingly depends on him for advice.

Another example is the case of Sirajul Islam from North Para—whose father was a *sardar*, and who is now passing on his duties and office to his younger brother Bazlur Rahman. Perhaps the one obvious exception is the case of *sardar* Askar Ahmed from South Para, whose leadership position is a reflection of his superior economic status, thus indicating that the rule of succession is not an exclusive one and that *sardari* status can be accepted when the strength of the client makes it difficult to do otherwise. This exception therefore functions to reinforce the analysis of a class basis to the leadership structure.

It is mainly through their conflict-resolution activities that these leaders articulate the political structure of the village and therefore maintain the system of social order. Occasionally they may act on some general issue in the village, jointly requesting the assistance of the Union Parishad and its Chairman. One such example was over the provision of watchmen for the village, and another over the school (to be discussed later). But otherwise their leadership functions do not extend to the organisation of the village for development, and indeed the *sardars* are regarded as rather reactionary in this respect by the younger men of the village. In addition to conflict resolution, *sardars* have traditionally been involved in the sensitive issue of feeding at marriages and other occasions. They are consulted by the family concerned about who should be invited. The importance of this problem must not be underestimated. The unity and solidarity of a group can be disturbed and undermined if the correct formula is not found so that the powerful are not excluded where to do so would constitute an insult. But since the prospect of having to feed an entire *para* (or even just its *paribar* heads) is impossible for many families, a delicate balance of guests has to be achieved with strategic invitations thinly spread outside the immediate kinship group. It is not unknown for the *sardars* themselves to make the invitations on the family's behalf so that the family is relieved of the burden of having to exclude people. The social order functions of these decisions about feeding are significant, as the activity of feeding in most societies represents an important social gesture and statement of group identification.

The inclusive group for feeding is known in the village as the *reyai*, and for Bondokgram this unit coincides for the most part with each *para*. (I have been in a village where there has been more than one *reyai* to a *para*, each *reyai* headed by a *sardar* having some of the appearance of function). Disputes within these *reyais* are in the first instance discussed by the respective *sardars*, and so the *reyai* becomes a reference group of political authority. However, as we noted in our description of the *paras*, the structure is not this neat and has to be modified since the Member Bari of Abdur Rahim in Middle Para, and Sardar Askar Ahmed's Bari of South Para are often included by our informants in the political territory of East Para.

As a result of this pattern, there is some ambiguity in moving to and describing the wider institutional context to which these *reyais* or *paras* are related. In its simplest form, "there is one *reyai* for each *para*, and five *reyais* constitute the *samaj*". However this informant, Tiku Mia, from South Para, has not been accepted by the dominant peasants of East Para in the way that Sardar Askar Ahmed has. Tiku Mia's Bari is quite hostile to East Para, and there has also been a disagreement between East and South Para youth organisations (see below). However Sardar Sirajul Islam from North Para also informed us that there was one *reyai* for each *para*, adding the complication that the *samaj* consisted of the five *paras* of Bondokgram and the one *para* of Dhaninagar, although no-one else includes Dhaninagar. But we have also been informed by Muklesur Rahman from Abdur Rahim's (Member) Bari that they "are in the same *samaj* as East Para *for settling disputes*", and he refers to the same group in different contexts as a '*reyai*'. From this information, we conclude that *reyai* refers to the inter-dining group; that because of the social significance of inter-dining for group identification, the *reyai* is also more indirectly used to describe a territorial unit with an integrated structure of political authority; that this looser use of the term *reyai* may include neighbours who do not normally inter-dine; that the term '*samaj*' is most usefully and commonly applied to the institutional arena in which these *reyais* are represented by their *sardars*; and that the term *samaj* refers exclusively to a political grouping. A further term, perhaps less common, was used in the village, referring to the exclusively political aspect of the reyai as a '*mel-darbar*' (public court). It was also applied to the level of *samaj*.

Most disputes were settled at the level of the *reyai* in *mel-darbar*; where parties to a dispute call one or more *sardars* from the *reyai* to hear their complaints and defences. Disputes are discussed outside the *reyai* if more than one *reyai* has been involved in the incident, or if one or both of the parties remain dissatisfied with a proposed solution and seek further discussion or support. *Sardars* at the *reyai* level are frequently called to settle disputes within *baris* and even *paribars*—for example, brothers or other close relatives in conflict over an inheritance of land; division of the holding; or some transaction (often involving land) which began initially as an informal arrangement with subsequent disagreement over rates of interest, foreclosures and confiscation of property, and whether land was sold or mortgaged and so on. Such disputes over land occur between other people as well as close relatives of course. Few disputes are discussed at the level of the *samaj*, and according to our *sardar* informants no dispute in Bondokgram has required intervention from *sardars* outside the village elsewhere in the Union. However, the Chairman of the Union Parishad has his *bari* in Dhaninagar (although he livers mainly in Comilla), and has become involved sometimes in the affairs of Bondokgram both through the two Members, and the *sardars*.

Challenge to Traditional Authority

We were only given one clear example of challenges to the authority of the *sardars*, but it is structurally important and has wider implications for the exercise of political control in the village. The dispute was between Jahed Mia (of the 'policemen's *bari*') and Khorshed Alam from Sardar Ansar Ali's Bari. Jahed Mia and Khorshed Alam are cousin brothers. The dispute was over land which Jahed Mia purchased from his father's sister, who is also Khorshed Alam's paternal aunt. Jahed Mia paid Tk. 3000 three years ago, and demands receipt of the land. Khorshed Alam has retained the land (on behalf of his aunt) displaying a false document saying that only Tk. 1000 has been paid and that a further Tk. 2000 must be paid before the land can be released to Jahed Mia. The case was taken to the *sardars* of the *reyai* and they decided in Jahed Mia's favour, that the land should be transferred to him. Khorshed Alam—despite his membership of Sardar Ansar Ali's Bari— has ignored this decision, although it has been restated on many occasions. Khorshed Alam is individually a powerful man in the village. He lends money and receives land in mortgage (including some from Jahed Mia, which he also refuses to return, although the money has been offered to release it). Jahed Mia is not without friends, and is supported here by the *sardars*, but to no avail. Neither he nor his closest supporter—Sardar Siddiqur Rahman from the 'policemen's *bari*'—think that he will now be able to obtain the land. Jahed Mia is reluctant to go to court, as the decision usually goes in favour of the person holding the land and verbal agreements are difficult to prove. The analysis of this challenge to authority was suggested by Jahed Mia and Siddiqur Rahman themselves. They argue that the extraordinary increases in the value of land, partly of course the result of increasing pressure of population on the land, stimulates an intense competition to accumulate land by any means possible at a price below the market rate. The phenomenon of mortgaging can be explained in this way, and is discussed below. What concerns us here is the impact of this intense competition upon the structure of social and political behaviour. The high value of land encourages land-hungry peasants with power to ignore the ethical code established in the village, in which verbal agreements are adhered to as a grave matter of honour. I myself was sharply reminded several times by poor peasants that they would never break their word and claim back land which they had mortgaged out unless they had the money to release it. The *sardars*—their position and role—epitomise this ethical code, and it is of course their duty to maintain and defend it. However, in their defence of such a system, which depends heavily on trust and the ability of *sardars* to insist upon and maintain the value of trust, *sardars*—and therefore the political structure which they embody—become highly vulnerable to successful challenges from the more unscrupulous of their own class. Furthermore, the *sardars* themselves, as rich peasants depending on their economic strength for their political power, have two options. They may suppress these unethical methods of enhancing economic strength in order to defend their own positions, or they must adopt similar methods. This latter course thereby undermines still further the principles of accepted behaviour—trust and honour—which constitute the ideological basis of the political structure. The activities of some of the *sardars* are already close to this second alternative. The lending and mortgaging activities of Sardar Askar Ahmed are an example. There are also a considerable number of allegations that *sardars* take bribes and are not impartial.

Traditional Leaders Co-opted into New System of Patronage

The next addition to the discussion concerns the involvement of the *sardars* in relief distribution. For the first time, in November 1974, the Chairman and Members of the Union Parishad asked the *sardars* to participate in actually disposing of the relief commodities. Nine *sardars* from Bondokgram were nominated. The timing of this invitation coincided with the arrest of several local black marketeers and Members from neighbouring Unions on charges of corruption. The distribution of relief notoriously provides an opportunity for corrupt behaviour. However, the lists of those to receive relief goods had already been compiled by the Members, so that although the *sardars* were asked to distribute accordingly, they had not been involved in the selection. The lists were an object of great suspicion in the village, as the criteria by which a name appeared was never made public. The function of this invitation to the *sardars* seems to have been a desire on the part of the Union politicians (or those at other levels) to deflect some of the unpopularity which was accruing to them, by using local leaders to 'mediate' their patronage to the village. In this way the *sardars* became exposed to the antagonism of those who were disappointed, since it appeared that they, rather than the Union politicians, were responsible for the selection. It is too early to conclude, but one possible outcome of this involvement might be to undermine the *sardars'* position in the political structure of the village by alienating them from their narrower support base. This in turn would prepare the way for a more thorough assertion of political authority by the Members, who at present are very unpopular in the village and compete with *sardars* for loyalty and support on very unfavourable terms.

Concluding Analysis: Class Differentiation and Power

This section is preoccupied with five overlapping, interrelated themes: the nature and direction of class differentiation; the pattern of cumulative control over the principal means of production—land; relationships of dependence and exploitation; unequal access to new opportunities and resources; and the patterns of diversification of economic activity, such as labouring, sharecropping, petty trading, business and service, which critically affect whether the peasantry is described in terms of 'cyclical kulakism' or in terms of families remaining as stable members of persistent classes. Even for heuristic purposes it is not possible to file our remarks exclusively under each of these headings.

First, it is quite clear from the landholding data (and confirmed by other studies in the Comilla area) that the socio-economic structure of the peasantry here is essentially minifundist. The range of landholding is very narrow, and no household in the village owns more than 7.2 acres of cultivable land. (See Table II, 'effective landholding' column).

This situation is of course in great contrast to other parts of Bangladesh —Rangpur, Kustia for example—where the range of landholding is much more extended. The issue is not just ownership (or control) of land; but the implications of landholding size for one's role as an employer of labour. The richer peasant in Bondokgram does employ labour—not usually on a permanent basis, but seasonally, mostly employing migrant workers. Thus the relationships between rich and poor within the village are not primarily those of employer-employee, centred around the direct extraction of surplus value from labour. This is in contrast to the other areas cited above, where the richer peasant in the village is a significant employer of labourers originating from the village itself. The relationships between rich and poor in that situation are very direct economic dependence mirrored by political subordination; a rich man's labourers are also his political retainers in the quasi-feudal conditions (particularly where the dimension of sharecropping is added to the equation). There is then the problem of labelling—does the rich peasant in Comilla have much in common with the rich peasant in Rangpur in the context of his objective class position? Likewise the rural poor? There is not only the problem of generalising about rural classes in Bangladesh, but also the problem of identifying common interests, states of consciousness, alliance and therefore potential for mobilisation of any sort.

In Bondokgram it is of course important to differentiate between households in terms of ownership over land, but the nature of the dependency or exploitation of the poor does not occur by and large through the activity of labour. In this regard it is necessary to make distinction between ownership and control over land, which immediately introduces the issue of sharecropping, mortgaging and money-lending. The table X describes the sharecropping and mortgaging characteristics for the village.

TABLE X

PROPORTION OF *PARIBARS*: SHARECROPPING OUT, SHARECROPPING IN, MORTGAGING OUT, BY LANDOWNERSHIP CATEGORY

Effective Landholding	*PARIBARS*						
	Total *Paribars*	Sharecrop out		Sharecrop in		Mortgage out	
		YES	NO	YES	NO	YES	NO
0	14	-	14	3	11	-	14
0.01-0.9	18	1	17	8	10	5	12
1.00-2.49	25	2	23	6	19	13	12
2.5 +	19	6	13	-	19	8	11
Total	**76**	**9**	**67**	**17**	**59**	**27**	**49**

Sharecropping

From this table we can see that the incidence of sharecropping in Bondokgram is not very high. This is not surprising for such a minifundist pattern of landholding, and certainly over the years the amount of land available for sharecropping has declined. There are the usual caveats about peasants being reluctant to admit to sharecropping out (no problem usually about sharecropping in), but our cross-checking and re-interviews confirm that informants have on the whole been honest. There is considerable competition among the poorer peasants to obtain land for sharecropping by offering terms to the owner. The conjuncture between desperation and a minimum level of resources to engage in cultivation (bullocks etc.) seems to settle at the 0.01-0.99 category of landowner, where 44.5% in that category sharecrop in some land. We can conclude for the purposes of this discussion that sharecropping does not significantly affect the pattern of control over land in the village and does not impinge greatly upon our analysis of its class structure. However, this again highlights an important variation between the minifundist character of the Dhaka-Comilla belt and areas elsewhere in Bangladesh, where sharecropping is a much more significant element in the delineation of classes, and the relationships between them.

Mortgaging

For Bondokgram, the issue of control revolves around mortgaging and moneylending transactions. First, it should be pointed out that although there is not much variation between the landholding and effective landholding tables in aggregate terms, quite a number of households have shifted between categories as a result of the second calculation. This indicates the number of households whose effective control over land has to be modified where their sharecropping and mortgaging transactions are taken into account. Second, out of the moneylending transactions which were identified, 58.5% of these transactions involved accompanying arrangements for mortgaging pieces of land at the outset of the negotiations. In addition, there are cases of moneylending, where mortgage 'clauses' are included at a later stage, usually as a response to non-payment of interest (let alone the principal). Third, of the households which own land in the village, 43.5% have some land mortgaged out, which seriously disturbs the pattern of asset control and income distribution as indicated by simple landowning data.

The significance of these mortgages cannot be overstated in this minifundist context, since they are likely to be a much more important part of the class structure and power distribution story in this area than elsewhere. The cumulative impact of these mortgage transactions function is to increase and stabilise the gap between the richer and poorer peasants, thus offsetting the cyclical mobility patterns too often associated with minifundist situations. Whereas the process of land acquisition can occur through sales and purchases in more polarised situations, the market price for land in densely populated areas where most of the holdings are below ten acres is often beyond the reach even of the richest peasant households in that area. Thus the mortgage transaction becomes the principal means whereby land ultimately changes hands, and it is this relation which truly characterises the antagonistic relations between rich and poor classes and differentiates between them, rather than labour or subordinate and inferior tenancies.

It is difficult to present a typical scenario for these transactions, since in detail there are many variations. However, a critical part of the process is the conversion of inflated interest rates on loans which have accumulated over a year or two into an equivalent land value based on the going market price for land. Since the interest rates on credit are inflated beyond all reasonable hope for repayment, this amounts to an acquisition of land by the lender below the market price if he is prepared to wait for the interest to accumulate and run the minimum risk (to his prospects for acquisition, not to his loaned-out capital) of repayment and foreclosure. In many cases it is not even a question of waiting where the lender has use of the land unit, either the debt is redeemed or he formally acquires it. This process reflects a situation in which the demand for credit is in excess of the demand for land at the respective rates of supply.

Thus mortgaging not only underlines the need for a distinction between ownership and control, it also represents the principal device by which land is transferred between households in the village (and occasionally to urban-based entrepreneurs). It need hardly be added that this process of acquisition is both a reflection and expression of power, as well as functioning to reinforce the economic and political dominance of those who are steadily acquiring land in this way. However, it is possible that this conclusion cannot stand alone, but requires support; that this process of land acquisition and control can be contained within the cyclical mobility thesis and may not necessarily lead towards a greater polarisation of the class structure. That is to say, the dynamic aspect of this process of acquisition is not regarded as structurally significant since it is offset by the demographic cycle of large families, which are naturally in a stronger position to accumulate land until the sons inherit and divide the holding. This process, it is argued, acts as a leveller. Bertocci, 1972 drew this conclusion from his work in the Comilla area. It is best to quote his position:

"...it is unlikely that a family can maintain superior wealth over a long period of time without some difficulty... over time, unless land is consistently accumulated, a given lineage taken collectively becomes vulnerable to the inexorable problems of agriculture in a monsoon environment, in that as its property is progressively divided into smaller and smaller shares, the size of its individual segments/ holdings progressively diminishes and renders individual members of the lineage each less capable of maintaining amounts of land sufficient to ensure adequate production. Hence over time, unless accumulation of land is kept up a lineage's collective wealth stands to be dissipated." (p.47).

In response to this conclusion, we had hypothesised that the extent of this circular mobility (political as well as economic) would decrease under conditions of new opportunities for increasing agricultural productivity, since such opportunities would thereby provide greater scope for investment in non-agricultural activities—like moneylending, but also forms of off-holding employment. Thus our reply to Bertocci is two-fold. First that land is consistently being accumulated; and second the opportunities provided both by superior landholding and contact with development agencies facilitates diversification of economic activity into off-holding sources of income. In other words, we now argue from the Bondokgram case that: a stable rich peasant class will emerge under conditions which reduce the vulnerability of families in that class to the weakening process of fragmentation through inheritance as other non-agricultural sources of income become available as a result (partially) of initial higher productivity of holdings.

Diversification

Bertocci does not seem to take account of either the possibility for richer peasants to prey on the misfortunes of others by acquiring land, or the role of the new technology in stimulating others forms of rural (but non-agricultural) economic activity like moneylending. Nor does he take account of the enhanced capacity thereby gained to establish sons in other forms of employment (often with education as a prior condition). Furthermore, he appears over-optimistic about the potential for subsistence-holding families to improve economically especially if they are able to engage successfully in lending activities, in particular the taking of land in mortgage. Surely—as we have discovered—subsistence level could not compete with families already established in these activities? Also the economic (and political) interrelation characterised by exploitation between the two classes in the village would prevent such competition in the first place.

It is obviously important to distinguish between different types of off-holding employment not only in terms of quantity of income, but also in relation to its security, its ability to keep pace with inflation, and other benefits which might accrue to a strategically useful occupation. From Table V we can see that the extent of diversification is higher among those *paribars* in the '2.5+ acre' effective landholding category. From Table VII, we can also conclude that the members of this '2.5+ acre' *paribar* category are well represented in business and professional activities.

However, it is also clear that the 1.00-2.49 acre category too is well represented in business and especially service, but the service category does include some semi-skilled activities which account for some of this representation. The data as it appears here is not as convincing as it should be to support our propositions above. However, higher up the landholding range this diversification becomes more evident with most of the *sardari* and *ucho-bangsha baris* and the two 'Member' *baris* heavily engaged in off-holding economic activities. It is this kind of observation which reveals the importance of looking behind quantitative data for its qualitative aspects, since the *baris* are strategically very significant in the analysis of political and economic structures in the village.

Economic and Political Access

The issue of diversification extends beyond additional income and its structural impact. If a family is involved in business, professional activities or certain types of service occupation (such as a clerk at the Collectorate, the District Court or the TCCA), then these provide connections and access to the various resources and items of patronage which are being distributed. Even the power to 'lose a file' can be an important asset, and one which is convertible into hard cash, a loan, or a fertiliser permit. Regular employment in the town naturally makes it easier to petition offices and lobby officials: and since education and literacy are highly correlated with business and professional activity anyway, the access of this class to the resources and patronage of the state is further enhanced. Such access does of course strengthen the position of the family within the village—lending it superiority in a variety of transactions in the courts, getting land registered, obtaining credit and fertiliser at subsidised rates, getting licences to trade and so on.

Off-holding employment reduces the vulnerability of the family to the inexorable problems of agriculture in a monsoon environment through the provision of alternative sources of income, and functions therefore to insulate the family from the necessity of borrowing with its concomitant downward spiral effect. In this way, off-holding employment (of the right kind) can actually facilitate innovation in agriculture, since capital for agricultural investment and access to information, credit and inputs has been acquired. This extends the willingness to invest in high-yielding varieties of seed on a greater proportion of the holding. The set of preconditions for innovation in HYVs should not be overlooked, although it constitutes a reversal of the more expected direction of cause and effect.

In addition to these features reinforcing the position of the class within the village, there are wider implications. The diversifying class becomes intimate with urban-based entrepreneurs. The links between rich peasants and petty bourgeois are often those between father and son, younger and elder brother. As for Bondokgram at least, this relationship is reflected in the recruitment of political leaders from the village and the structure of political power between the state and the countryside. Both of the Members (of the Union Parishad) are from rich peasant families in the village which have enjoyed considerable traditional political influence in village affairs. However they are both now successful entrepreneurs in Comilla town, and live there rather than in the village. They have become agents of the state, and it is difficult to see how a political class which achieves its position through the subjection of others in the village can possibly become the champion of these dependants. Their activities in distributing relief and fertiliser is adequate testimony of their disregard for the poor, whose poverty is a precondition for their own success and power.

Impact of 'Market' Behaviour on Traditional Social Order

The continuous and intensifying struggle for control over the means of production in the village—not only land, but water and other inputs—has repercussions throughout the peasant social system. It is apparent now that the traditional structures of political and social order embodied in the *sardari* system and kinship relationships are being undermined. It is no longer possible for example (was it ever?) to identify all the members of one *bari* with the same class. We described earlier the exploitation and class relations which can occur between brothers and other members even within the same *bari*. Under these present conditions of population pressure on the land, disputes over division of holdings are frequent. Instead of formal cooperation, the bond of kinship has been displaced by considerations of the market price, for example, the prevailing the interest rate to extend credit to a kinsman for purchasing and rearing a bullock.

There is now a pervasive acceptance by richer peasants of market principles in their economic transactions. These transactions are very open with respect to labour, land and moneylending. Migrant labourers and local labourers are hired at rates which consciously take account of the supply and demand situation for labour. This openness is very convenient for the richer peasants, for they are disengaging themselves from the obligations which accompany the more traditional forms of patron-client relationship, under conditions of increasing landlessness where the market price alone will ensure the availability of labour. This principle of open recruitment, together with the increasing phenomenon of class differentiation within *baris* (and certainly *gostis*) inhibits the traditional formation of exclusive political and social groups tied to one leader.

Thus groups consisting of leaders and followers or dependants are now less stable, and opposition between such groups is declining as a feature of conflict in the villages. In this way, the vertical cleavages which describe the opposition between such groups dominated by different leaders (rich patrons) in the village are becoming much weaker as the character of the group formation changes. Certainly in terms of perceptions of the poorer peasants in the village, the horizontal cleavages between emerging classes are becoming much stronger. The representatives of dominant classes in any society will of course always deny such a proposition.

The Undermining of Trust

It is interesting to note that the element of trust, a belief in the 'word' of another, has usually characterised the style of transactions between peasants in a village. The traditional institutions of authority in the village embody these principles of trust. Besides, everybody knows most of everybody's business, so that to renege on a deal is to incur the moral disapproval and disaffection of others in the village, making it difficult to turn to them for assistance at a later date. Thus the element of trust has been a vital component in restraining conflict within the village, constituting the essence of a minimum ethical code of conduct. To be a *sardar* not only required membership of a 'strong' *bari*, but also trust in one's word; for only then could the decisions which resolved conflicts (i.e. justice) be respected and adhered to. However this ingredient of the social order has gone sour, so the competition to accumulate land and other resources increasingly subverts the ethical code (e.g. disagreements over whether a piece of land was mortgaged or sold, refusal to release deeds when the loans are paid off, refusal to allow foreclosures, reference to false documents and witnesses) and the structure of political authority which embodies it. Even *Sardars* are now tempted to exchange the benefits of a respected leadership position for those based upon the more naked reality of economic power. The strength of their lineages and therefore their own status is undermined unless they adopt the new rules for capital accumulation. The result of course is an increasing dependency upon the political order provided by the state as a substitute for the social and political order of the village.

Interdependence of the State and Village Politics

The authority of the state is impersonal, ordered as it is around capitalist market relations; it is maintained by more universal principles of recruitment to the political positions of authority—usually centred around a position of economic superiority as the means of obtaining political support with consequent favourable access to the state as the method of obtaining political patronage. A combination of market forces and access represents the new political order.

The two members of Bondokgram are good examples of this process, and they increasingly confront and undermine the traditional structure of authority and leadership in the village. New resources are brought into the political arena, as well as the economic sphere: the power to distribute or withhold relief; issue licences to deal in scarce commodities (fertiliser, food, salt, kerosene); obtain favourable court decisions; acquire fuel and service for pump sets; have loan defaults ignored in the co-operative; and so on. Such power establishes one's authority in the village over other matters particularly where disputes over the title and status of land are concerned.

To ensure their re-election and a continuation of their leadership, members still depend upon the support of their own *baris* and their capacity to maintain a minimum loyalty of potential competitors from other *baris* (however pragmatic such loyalty might be). As a result there is little scope for benevolent leadership, where access and power might be used in favour of the poor. The trade-offs have to made with one's own class. Potential competitors have to be 'squared'.

Thus the 'true colours' of the state are revealed not by the pronouncements of national leaders whose inclinations are to transcend the issue of class, but through these mechanics of recruitment to political positions in the countryside—the roles through which the state interacts with the people, the ongoing *practice* of politics in the fields and the components. It is here that the state acquires its class character, and from where—through the different levels of the political system—policies at the national level for rural development become constrained and biased: if not in word, then certainly in deed and implementation.

The Methodology of Class Analysis

The discussion so far has concentrated upon class formation, class cohesion and linkages outside the village by referring to the behavioural characteristics of the economically and politically dominant peasants in the village. It should be clear by now that our definition of 'richer peasant' is based as much on these behavioural characteristics of domination as it is on the more conventional, quantitative indicators which emphasise the amount of land held, levels of income, the production of a surplus and so on. Since we are concerned with the question of consciousness among different classes and their respective capacities to act as a class in the interests of that class (whether the interests be those of exploitation by the rich, or revolutionary mobilisation among the poor), we emphasise in this section the interactions between classes rather than merely differentiating between them on the basis of quantative scales—whether they have more or less of this and that.

Particularly in village situations, the activity of one group or class is a critical part of the explanation of the behaviour of other groups or classes. The poor are poor because of the rich, and the socio-political characteristics of each class are a function of the exploitative relations between rich and poor. If the poor peasantry is a fragmented, incoherent class, this is not just a function of poverty, but a function of poverty *plus* the social relations of production which are responsible for that poverty. Therefore the behavioural characteristics of the poorer peasantry, that is to say their position in the social relations of production are likewise part of the characteristics of the poorer peasant class. Here regional variations must again be emphasised. As the rich peasant in the Comilla-Dhaka belt is different from the rich peasant in Rangpur, so the 'poor peasant class' is a variable concept according to the structure of social relations of production obtaining in the particular situation. The minifundist large landholder variation affects the nature of the poor peasant class, as it does that of the rich peasant.

The Impact of Migrant Labourers on the Landless

We must ask whether the landless have benefited in any way from the Comilla HYV experiment in a non-flooded area? We must also remember that conditions could hardly be more favourable than in Bondokgram—indeed the demand for labour in Bondokgram is substantial when compared with flooded neighbours outside Kotwali Thana. Yet we noted above that relations between the rich and poor of the village were not primarily centred around the issue of labour. We concluded earlier that in Bondokgram, despite a general shortage of labour, the position of the village landless in need of employment is weakened by the entry of outsiders into the local labour market. there is no need to repeat that analysis here, except to point out that the presence of these migrant labourers force the landless into insecure, non-agricultural economic activities.

The effect of the migrant labourers on the employment pattern of Bondokgram was to confine the number of adult males (from households effectively holding less than one acre) who were regularly dependent upon agricultural labouring to approximately 15%. Non-agricultural manual work (industrial workers in local textile mills), peons, artisan occupations and petty trading accounted for the employment of about 70% of adult males from *paribars* with an effective landholding of less than one acre. However, the economic independence of these groups from others in the village was not thereby greatly enhanced, for in addition to other loan transactions which might exist, petty traders require loans from patrons (sufficient at least to purchase a load, which is 1 *maund* 10 kg of grain) and the other occupations require patronage, access from those who are more influential in the village, and could arrange jobs outside the village. A poor man does not have to work directly for a rich man to be dependent on him. In this situation, by financing petty traders, the wealthier peasants indirectly participate in and control the process of marketing the surplus—their own, and the smaller ones of others.

There is a further regional variation to note here. The relative proximity of Bondokgram to the town of Comilla enlarges the possibilities for landless and near-landless to engage in these non-agricultural occupations, but where villages are more remote these opportunities are reduced and the dependence of the poor on the rich more absolute and desperate. However, despite Bondokgram's favoured situation, we can now see that the poor peasant is becoming steadily more alienated from his land through usufructuary mortgage, and the landless are denied opportunities to work as agricultural wage labourers by their migrant rivals. Naturally as a result the employment structure of the poor becomes fragmented and inhibits the potential for class solidarity, perceptions of mutual interest and so on. In this connection, the migrant labourers are very significant and we return to them.

Migrant Labourers and Class Consciousness

No-one concerned with rural development should overlook this issue of the relationship between patterns of labour migration and other variables such as flooding, the extent of HYV cultivation and levels of cropping intensity. There are important socio-political conclusions to be drawn. Here is a class of landless and poor peasants which is increasingly and continually being forced to be geographically mobile during parts of the year. Everywhere they go, they have the effect of depressing the local wage rates—making themselves popular with employers but unpopular with others competing for employment. However, it is this enforced mobility which reduces their independence on *individuals* although it transfers the relationship to one of dependence on a *class* of peasant employer.

Under minifundist conditions, our data suggests that peasant employers have a realistic appreciation of market forces. As we observed above, they are disengaging themselves from traditional patron-client relationships (the multiple, undifferentiated role structure within which a loyal and regular labour supply could be secured) as their demand for labour decreases—either in relation to supply, or partly as a result of the decreasing size of the farm through inheritance. As a result of this process labour becomes untied, it is 'freed'. Where opportunities do exist for diversification into off-holding employment among richer peasants the demand for non-family agricultural labour may be sustained. But where such opportunities do not exist among poorer peasants and in very remote areas, then the overall demand for this 'freed' and mobile labour will fall as holdings decline to a size where they can only absorb family labour.

In other societies where a rural labour force has become 'liberated' (for example England during the enclosure movements), the period has also been associated with an accumulation of capital which has provided the basis for capitalist industrial development and the absorption of this labour force. The same cannot be said of *this* process of 'liberating' the rural labour force in Bangladesh, to the extent that minifundist conditions obtain. It is now associated with an accumulation of bourgeois capital, but with the increased division of holdings (the very opposite of enclosures, with their subsequent repercussions for the industrial sector—wool manufacture, production of machinery and so on). The class potential of such a migrant population of 'liberated' labour is explosive, and represents a key element in the mobilisation of the rural poor into a class conscious of itself as such, for it becomes the transmitter of information and discontent and of course recruits eventually the traditionally dependent labour force which it has itself displaced.

Poor Peasants Lack Fertiliser

Having considered the position of the landless and near-landless, we can now turn to the poor peasant cultivators. From Table V, our data indicates that 54.8% of the 1.00-2.49 acre effective landholding category *paribars* derive their income solely from cultivation. This group is mainly at subsistence level, rarely obtaining a surplus. Over half of the group (see Table X) has an average of 29.4% of its land mortgaged out.

TABLE XI

DISTRIBUTION OF PROPORTION OF OWNED LAND WHICH IS MORTGAGE OUT

Ownership Category (acres)	Ave. proportion of mortgaged out land by *Paribars**
0.01-0.99	37.3%
1.00-2.49	29.4%
2.5+	14.5%

*Includes only those *paribars* with mortgaged out land

From Table X, only 24% of this 1.00-2.49 acre group sharecrops land in. The economic position of this group is obviously very precarious, leaving an increased intensity of cropping with higher yields as the only means of improvement. The opportunities for this group to accumulate land or even extend their sharecropping activities are very slim. However, our data also suggests that despite the village's location in the 'command' area of the Comilla Programme this group has not greatly benefited.

Although the poorer peasants have some disadvantages in obtaining tube-well water—usually as a result of the initial decision about the location of the boring, they are mainly hampered as a *weaker economic class* in obtaining fertiliser, which is not just applied for *boro* rice cultivation, but in the other seasons as well. This is an important observation, for in this minifundist context land is not really distributed around the village in any manner related to class. The location of a boring may be biased in favour of the lands of a strong *bari* or two, but richer as well as poor peasants may also have their land excluded from the command area of the well.

Fertiliser is obviously a critical resource and its overall scarcity apparent to all. Data on the use of fertiliser is a vital index of power and influence in the village. One of the major points of controversy between different groups of villagers concerns the way in which the supply of fertiliser is manipulated and controlled and the differential access to fertiliser by classes or factions within the village. The importance of fertiliser is not lost on any class of peasant and the problem is very definitely not one of extension but one of access, circumscribed as it is by socio-political variables. The fertiliser dealerships in Bondokgram are either run by the cooperative dominated by a strong *bari*, or are directly in the hands of another strong *bari*; or the members are involved and they are primarily concerned to square potential competitors for their positions, rather than effect an egalitarian distribution. Most of our poor peasant respondents referred to their inability to obtain adequate supplies of fertiliser, alleging at the same time that the available supplies had been cornered by the rich and powerful families in the village.

Class Characteristics of HYV Innovation

Only 33% of landholders are engaged in *boro* rice cultivation, and 90% cultivating *boro* rice have effective landholdings above 2.5 acres. This is a category of peasant which might normally expect to produce a surplus of foodgrain in addition to family consumption requirements with *aus* and *aman* alone, without the help of *boro* (assuming a constant family size of around six). It is difficult to see (although our data is not rigorous in this respect) how more than 10% of the land cultivated under *aus* and *aman* was also cultivated for rice in the 1973-73 *boro* season. The extent of *boro* rice cultivation has been higher in the past, as the decline in area irrigated by the tubewells would suggest.

Several explanations are given for this decline: factionalism within the societies; high incidence of defaulting payments on loans; organisational inefficiency of BADC and other government institutions; and post-liberation problems in obtaining supplies of fuel and spare parts for pump sets. Even recognising that there has been a decline, we could not conclude that a triple cropping annual system with *boro* rice *had* been responsible for the advancement of any group in the village. Our historical data on the richer families indicates that their superiority stretches back further than this. When the tubewells were introduced, the richer families were in a position to take advantage of the new resources, but their adoption of the innovation seems to have been an outcome of their economic superiority rather than a cause of it. Furthermore, there is a tendency to innovate on only a small proportion of the holding; and often *aus* is displaced so that double cropping persists.

Implications for Rural Development Strategy in Bangladesh

In this final section we attempt to set out briefly some of the connections between our analyses and the problems of formulating a socialist rural development policy at the national level. It is for this reason that we have been so concerned not to exclude the nature of the state from our discussions. The relationship between different classes in the countryside and the state has to occupy the central place in any serious thinking about rural development strategy. An analysis of policy alternatives themselves can tell us something about this relationship between the state and society.

Alternative Strategies

Perhaps for Bangladesh there are four broad alternative sets of rural development objectives and methods—each appealing to a different combination of classes, groups and interests. The different objectives refer to the varying intentions to maintain or change the dominant mode of production. The methods (or process) will vary between state-sponsored, self-reliance or revolutionary peasant organisation according to the philosophy of those combinations. Radical objectives are normally associated with a radical process of changes. Whichever alternative is dominant depends upon historical circumstances and the balance of class forces.

First, there would be the policy of pursuing agricultural development within the prevailing dominant mode of production, which itself might vary for different parts of Bangladesh, quasi-feudal in some places, Chayanov's 'peasant economy' in others, or quasi-capitalist. The issue here is best regarded as one of intensity of governmental penetration in the countryside. Until the late sixties (post Ayub) the main rural development activities of the state in East Pakistan were confined to the Rural Works Programme (RWP), the Thana Irrigation Project (TIP) and the Union Multi-Purpose Co-operative Society (UMPCS) in the districts outside Comilla. Until the expansion of the Comilla Programme of KSSs, the extent of formal government involvement in the provision of new agricultural inputs in the province as a whole was slight, with little emphasis on a class of poor or middle peasants as a specific and exclusive target group.

Second, there is the equation of rural development with agricultural growth and modernisation. There is a strong emphasis here upon the introduction of new technology in agriculture with only a secondary concern for the nature of distribution among different classes of peasant. The assumptions here are often apolitical, involving a preoccupation with technical aspects of production (the optimum ratio between labour inputs, fertiliser dose, regulated water supply, timing of operations) and a belief that other problems are resolved by increases in output—increases which are dependent upon technological rather than political variables. Although claiming neutrality with respect to class, this technological approach to rural development is often associated with an attempt to alter the socio-economic formation. Even where such interference is not intended, the differential access of rich and poor peasants to new resources usually implies that a technological definition of modernisation amounts to a policy of betting on the strong.

This brings us to the third and fourth sets of objectives, where the technological opportunities are formally incorporated into political and economic definitions of rural modernisation. There is, however, an ideological distinction with corresponding institutional variations. One refers to the transformation of pre-capitalist modes of production in the context of an expansion of the capitalist mode of production, changing peasants into farmers, involving higher levels of investment, crop specialisation, profit-maximisation as a motive, incorporation into the capitalist market, employers of wage labour and so on. Again this is the concern to transform both the pre-capitalist mode of production *and* the capitalist mode of production in agriculture into a mode where the means of production are publicly owned.

It is within these broad sets of alternative objectives that the various strategies for rural development must be understood in the past and in the future. In this way we obtain a fundamental framework in which to locate policies and their organisational implications. In so doing it is also necessary to distinguish between words and deeds. By applying such a frame of reference, a phrase like 'the socialist strata to development through co-operatives' becomes a contradiction if it is clear that the cooperative is the organisational form which corresponds to the third rather than the fourth set of objectives. It is clear to us that if contemporary pronouncements about a radical, socialist path to development are to be given their logical expression in policy and commitment then we are concerned with the fourth set of objectives: the transformation of both pre-capitalist and capitalist modes of production into a mode where the means of production are publicly owned. The debate concerns the capacity of the state apparatus to implement such a policy and the appropriate institutional forms to be applied. However, an explanation of this capacity depends precisely upon the way in which the state has until now been involved in pursuing one or more of the other sets of objectives, and whether the political logic which lay behind those pursuits remains too strong to permit any genuine shift to the fourth set of rural development objectives. [The pronouncements have changed, some of the personnel have changed, but has the nature of the state changed?] To what extent does the class structure in Bangladesh, which has evolved since 1947 and given the state its particular complexion, now impede the socialist transformation?

Regional Variations

From our studies of minifundist conditions we have become alerted to the importance of regional variations in socio-economic structure for the problem of devising strategies for rural development (whether by the state or among revolutionary movements). These variations—historically and ecologically determined—are significant at the level of the mode of production and are therefore relevant to any discussion of attempts by the state to intervene at that level. The precise regional characteristics of a mode of production will circumscribe the role and capacity of the state to engage in development activity in that region. We must stress that our activities have been concentrated in the Comilla-Dhaka belt, so they are representative only of the minifundist socio-economic structures which typify that area. At the same time, we have established a basis for speculative comparison between this and other parts of Bangladesh. In this way, we hope to contribute to the planning of similar research in other areas.

These regional variations concern rural development strategies whichever of the alternative sets of objectives are preferred, but in different ways. If the fourth set of objectives is considered—the transformation of pre-capitalist and capitalist modes of production in the context of radical upheaval—then the issue of consciousness and dependence is essential. Without the direct participation of the rural poor—those who are in a subordinate position in their respective modes of production (tied labourer or indebted poor peasant)—with class consciousness as its precondition, this fourth set of objectives is unlikely to be achieved. A small benevolent elite at the centre committed to socialism cannot alone simply subordinate the politically and economically powerful classes; it depends on the mobilisation of the rural poor themselves. But the structural position of the rural poor affects the nature and extent of their consciousness. A high proportion of 'tied' landless labourers and sharecroppers directly dependent for their livelihood upon the favour of big landlords may provide fertile ground for the development of a class consciousness but at the same time offer little scope for collective action—either in the form of taking control over resources such as land, and inputs, or in the sense of joint productive enterprises such as co-operative farming and marketing schemes. In this situation it would also be necessary to discover whether lineage differences run along class lines or cross-cut them (are members of a *bari* all in one class or distributed between classes?), although we would hypothesise that they would tend to follow the same cleavages as class.

However, the minifundist situation differs: the range of landholding is very narrow, making small differences in holding very significant in terms of class; the structure of dependency tends to be indirect through usufructuary mortgage and other forms of indebtedness. The extent of competition within *baris* (between brothers, for example) over land, and the existence of dependency relationships within *baris* indicate that the dimension of class frequently cross-cuts that of kin. One would expect the development of a class-based consciousness to be inhibited by these factors. Furthermore, in this situation it may be particularly difficult for policy makers to isolate a category of poor peasants as a target group for co-operative farming activities, special subsidies on inputs, or consciousness mobilisation, precisely because of the complex, intimate and multi-dimensional nature of the relationships between rich and poor peasants in a minifundist context. If the relationships of exploitation are relatively simple i.e. one-dimensional, devoid of kinship ties—then it is relatively easier to manipulate or sever that relationship.

Regional Variations Prevent Universal Strategies

There is a distinction however between dealing with the existing mode of production in order to reform it and dealing with it as a necessary part of the task of radically reforming it. Where reform is envisaged—and anything short of a pervasive socialist ownership of land is regarded as reformist— then variations in the mode of production will determine quite directly what can be achieved. It is, therefore, illogical and unrealistic to attempt to implement a universal reform when the structures to be reformed are not universal. The land reform legislation in Bangladesh (and the same is true for India), is a case in point, having no impact whatsoever on the problems of land distribution in minifundist areas. On the other hand, co-operatives of the Comilla type may work better and be more appropriate in the minifundist situation, although the politico-economic problems have been recounted at length even for the Comilla area by many of the BARD members themselves. When transplanted to other areas, where the gulf between rich and poor peasant is greater, the problems of political inequality within the co-operative are correspondingly more significant. Unless the ingredient of socialist ownership of land is added to new village co-operative schemes, the same criticisms of universal policies of reform still apply. It seems to us that the inequalities in the ownership and control of land will always threaten to subvert the more purely political arrangements to ensure grassroots democracy. This is probably recognised, but what is important is that the extent of subversion will vary as a result of the regional differences in the dominant mode of production in peasant agriculture. What may be possible in a minifundist area like Comilla is not likely to achieve the same degree of success in an area where a rich peasant/landlord has 50 acres or so, monopolises access and patronage, and has a monopoly influence over the price of labour.

However, this problem of regional variation is not removed simply by incorporating the socialist ownership of the means of production as an exploit objective. The coercive powers of the state are not sufficient to implement such a programme of radical transformation over the heads of those classes which are dominant by virtue of their private ownership and control over land. Thus the dominant mode of production in an area still has to be confronted even when the fourth set of objectives is contemplated. Whichever group of radicals is involved, whether elements of the state or revolutionary forces in the countryside, universal strategies for a radical transformation involving rural mobilisation and the raising of consciousness among the poor are also inappropriate. At its most extreme, the class of large landlords can be eliminated to achieve relatively profound effects if there is no significant stratum of middle peasants ready to take their place. (The strategy of the Chinese Communist Party during the revolution, especially after the Long March, was to involve middle peasants in the elimination of landlord classes thus committing them to new institutions and a rejection of the old order). However, such a strategy would be very inappropriate in the minifundist areas of Bangladesh, with the additional Bengali-Islamic complication of large *baris*. The class to be eliminated is more diffuse—behaviourally its position depends upon a variety of relationships rather than an absolute quantitative distinction referring to land. Compared with areas of larger landlords, in the minifundist situation richer peasants are more numerous so that the ownership or control of the majority of village land is not confined to a tiny group.

The Poor Need Higher Forms of Organisation than the Rich

There are further problems for radical policy formulation in addition to those of regional variation in socio-economic structure. The first refers again to the issue of group or class consciousness and mobilisation—a major component in current thinking about rural development in Bangladesh. Observers are frequently confused by the existence of factional competition between richer peasants of the same class into thinking that these richer peasants are so divided among themselves as to prevent economic behaviour which is objectively in their class interest. These observers assume that the richer peasants are concerned to maintain a traditional structure of clients and dependents in the village—a secure band of followers—but on the contrary, these richer peasants are increasingly following 'market' forms of behaviour in their transactions with poorer people in the village. That is to say, the richer peasants are behaving according to economic principles which are currently running in their favour (in the payment of labour, usufructuary mortgage even with kin). The competition and conflict between richer peasants is by comparison insignificant and does not function to weaken the class differentiation revealed by this 'market' behaviour. The factors of cohesion and solidarity are not therefore so important in determining the objective nature of their class behaviour.

The position for the poorer peasant, near-landless and landless, is different. To improve their economic and political prospects, they are really bound to oppose and overturn the dominant mode of production and exchange which is not running in their favour. An ambition to improve their relative position within the dominant mode of production (higher wages, less interest, more land) is probably equally difficult to achieve. However, for both strategies, the essential ingredient is organisation, the preconditions for which are consciousness, cohesion and solidarity, in that order. If you are in a subordinate oppressed position in a mode of production, it is logically impossible to behave in favour of your objective class interest *as an individual* since you have to change the status quo rather than merely adhere to it. In this situation, the poor require class-based organisation, while the rich can do without it until such time as they positively have to defend the status quo. The poor peasantry and the landless find class consciousness and organisation particularly difficult precisely because of their poverty and subordinate position.

First, they depend on their patrons who are employers, or landlords, or creditors, and who have the freedom and strength to take their patronage elsewhere at the going market rate. Where simple survival is involved, the risk of the unknown (the untested 'unity' of one's own class) is too great to contemplate. Second, to survive they have to try and keep alive by seeking employment, working hard on their own plots, begging, trying to obtain loans, looking for the best bargains in the market and so on. This type of behaviour tends to isolate them from each other (even within the same village)—these are all independent, lonely activities. Third, at the end of the day (food or no food) they are exhausted—angry perhaps, but too tired to act on it. They are unlikely to attend a meeting which might have to be clandestine (on the edge of a village) or in another village altogether. Fourth, their low level of literacy and education makes communication very difficult, prevents the undermining of traditional loyalties to the patrons in the village, and the sense of community which is the rich peasant socio-political order. Fifth, with some important exceptions to be noted presently, they are less mobile both in terms of leisure time and financial capacity. It is difficult to envisage membership of a class of poor throughout the country when you have never travelled more than eight miles in your life. All these constraints somehow have to be overcome as part of a strategy to confront the mode of production in order to change it.

The Poor Should Organise Outside the Village

Perhaps this picture is not entirely hopeless—there are some exceptions to poor people's lack of mobility. As petty traders, they move from village to village in the course of their work, meeting others of similar status. Also there are migrant workers and they are very significant as has been noted earlier. Finally, we noticed an interesting correlation between poverty and the frequency of visits to the local *haat* (market). If you are very poor, you cannot make large purchases; you live from day to day, hand to mouth. Except in periods of peak employment like transplanting where there is less time and higher income, the poorer people of Bandakgram and Mahajanpur were visiting a market three or four times a week. The *haat* therefore represents an important meeting place for poor peasants and the landless from different villages. Their weak position in the mode of production obliges them to be frequently mobile in this way, while at the same time an opportunity for organisation and mobilisation is provided. The factors of communication and mobility which are necessary for this are already part of a way of life for the poor. This pattern of movement involving the *haat* seems to offer greater scope for mobilisation of the rural poor, than units of organisation confined to the village where there are no inherent features of the poor's behaviour to facilitate their solidarity and counteract the superior power of the larger landholders.

Our analysis therefore indicates that the *socio-economic relationships within the village are not conducive to poor peasant and landless political organisation within the village itself.* Here, the process of class differentiation—encouraged by the introduction of new agricultural technologies—maintains the poor in an increasingly weak economic and political condition. By concentrating units of political and administrative organisation at the village level without altering the competitive equipment of the oppressed groups (representation alone will not do as shown by the failures of Community Development and Panchayati Raj in India), the probable outcome is a consolidation of the power of those who are already dominant in the village economy. There is always a danger with total village programmes (through co-operatives etc.) that the theme 'village community' is re-emphasised. But class differentiation is not thereby swept under the carpet. Instead, the notion of the community would be defined by the powerful in the village as embodying the behaviour and values of the status quo. The system of Panchayati Raj in India, where there are quotas for committee membership in addition to the 'democratic' process of election, has been unable to circumvent the power of the richer landlords and peasants in the village. On the contrary, it has provided an additional statutory framework within which they can consolidate their class superiority. The poor can be stronger when they are not confined to institutional arenas where they confront only their patrons and landlords. In wider arenas extending beyond the boundaries of the village, these confrontations can be more impersonal and the solidarity of the poor (once generated) becomes less vulnerable to subversion. But as a precondition for radicalisation through institutions at the village level, it is worth recalling Barrington Moore's 1969 remarks referring to India: "the villages cannot be democratised without democratising property relations".

Perhaps there are many in Bangladesh who accept the validity of this proposition, but the problem is that legislation at the centre regarding abolition of private property in agricultural land will have no effect in the countryside unless the poor are mobilised in support of such an objective. It is difficult to mobilise them without the abolition of private property in land as an explicit purpose. There are no intermediate proposals of reform which are exciting enough to stimulate the poor into action. On the other hand, if the abolition of private property in agricultural land is announced as a specific objective, is there a danger of provoking the mobilisation of the richer peasant class (perhaps in alliance with the petty bourgeois of the towns) in defence of the status quo? This is the current dilemma for the state in Bangladesh. But the possibility of achieving the fourth set of objectives in rural development recedes the longer it is postponed, for in the meantime the richer peasant classes consolidate their position anyway.

Chapter 2

Rural Class Formation in Bangladesh, 1940-1980

Introduction

Rural class formation in Bangladesh is not a straightforward and uniform process of polarisation between expanding capitalist farmers and poor peasants/tenants who are being steadily pushed off their land. There is certainly a rise in the incidence of landlessness but it cannot simply be attributed to rural capitalist expansion or the rapid increase in population, although the latter is a significant variable (Arthur & McNicholl, 1978). This chapter argues that during the last four decades of political upheaval and development efforts in East Bengal other uses for rural capital have been more attractive than its investment in transforming and expanding the scale of agricultural production. This implies that the description of rural Bangladesh as merely consisting of millions of small peasants, destined to become uniformly pauperised unless aid is forthcoming, is also not accurate. At any one time, it is of course possible to identify a high proportion of small peasant households, but their condition is not stable. The ways in which they decline varies in different regions of Bangladesh, revealing distinctive patterns of rural class relations. Each variation has sometimes been generalised to apply to the whole of Bangladesh either through the dogmas of political analysis, or because a researcher or commentator has extrapolated from studies in a particular area. The purpose of this chapter is to see whether these distinctive patterns can be contained within a single framework which does justice to the common as well as varied experiences within Bangladesh. In doing so it will be necessary in this introduction to consider: the weaknesses of previous accounts; the use of class as a tool of analysis in this context; opposing analyses of class relations; a framework for analysis; and the role of unproductive capital in the processes of accumulation and polarisation.

Weaknesses of Previous Accounts of Agrarian Structure in Bangladesh

Until the mid-seventies, the accounts of rural and more specifically agrarian social relations in Bangladesh could be criticised in three ways: for being ahistorical; for their commitment to homogeneity; and for their methodological bias in using Comilla District in East Bangladesh as the basis for generalisations.

First, any discussion of class formation in Bangladesh cannot be separated either from the general history of class relations in East Bengal or from the particular conditions under which the independent state emerged: the Partition of Bengal into West Bengal and East Pakistan in 1947; the economic and cultural subordination of most classes in East Pakistan to the interests of a combination of propertied classes in West Pakistan represented mainly through a military and bureaucratic leadership (Alavi, 1972; Nations, 1971; Sengupta, 1971); and the participation of India (with USSR support) in the "liberation" of Bangladesh in 1971 (Lifschultz, 1979). During these events, the connection should not be overlooked between emerging nationalist ideology and the way the agrarian structure of East Bengal was presented "as a peasant economy based on small family farms operated primarily with family labour" (Abdullah & Nations, 1974; Bose, 1973). This became the language of nationalist struggle for an autonomous Bangladesh against the "landed aristocrats" of the dominant Western wing of Pakistan. It was also consistent with an earlier picture within East Bengal of Hindu *zamindars* (landlords) dominating Muslim *raiyats* (permanent occupancy tenants) and *bargadars* (sharecroppers) before partition in 1947.

Secondly, most of the attempts to generalise relied on aggregate data collected in the 1960s about the distribution of landholding and tenure which emphasised the uniformity of landholding patterns and disguised important regional variations with related ecological and historical specificities (Khan, 1977). It is interesting that even the Bureau of Statistics (Government of Bangladesh) Land Occupancy Survey 1977-1978 (financed and led by US AID) conducted its analysis almost entirely in aggregate terms, never presenting a breakdown on a district or some other regional basis.

Thirdly, the content of these generalisations had been biased by a concentration of research in the densely populated and fertile region of Comilla, based on agro-economic surveys inducted by the Bangladesh Academy for Rural Development (BARD) (Blair, 1974; van Schendel, 1976). The Comilla findings formed the basis of extrapolations for Bangladesh as a whole, providing an image of an homogeneous agrarian structure described in terms of smallholding farmers.

These weaknesses gave rise to two standard but misleading generalisations about the absence of a substantial class of landlords and landholders and the low level of absolute average size of holding. As a result, the Comilla co-operative programme from the early sixties was predicated on the ideological assumption that class division within the "peasant economy" was structurally insignificant. This programme was later *extended* in 1973 throughout rural Bangladesh in the form of the Integrated Rural Development Programme (IRDP), and today still constitutes the principal "instrument" of rural development policy in Bangladesh despite modifications, distortions and takeovers (Wood, 1980a & 1980b). Furthermore, the IRDP itself is a crucial ideological apparatus of the state, representing and reproducing a view of class merely by persisting in a form of intervention based on "farmers cooperatives" (*Krishi Samabaya Samitis*). An investigation of the process of rural class formation in Bangladesh is therefore also an important act of political criticism.

Use of "Class" in the Analysis of Rural Bangladesh

There is dispute over whether it is possible to employ a concept of social class at all in describing the dynamics of rural Bangladesh. Thus, when some pattern of rural differentiation has been recognised, class analysis has been rejected because the differentiation was not regarded as structurally significant. For example, even in the Comilla area a distinction was accepted between "surplus" peasants on the one hand and subsistence and below subsistence peasants on the other, based on minor variations in landholding (Bertocci, 1972). But such a distinction was not credited with long term dynamic implications for changes in the pattern of landholding distribution, since surplus peasants can only temporarily extend their control over land. An important aspect of this proposition is that surplus peasants (the rural elite) do not constitute a class since they cannot easily reproduce their economic position in the society. The prevailing system of multiple inheritance requires a division of the holding between sons and ensures a process of cyclical mobility, or indeed the "cyclical Kulakism" of Peter Bertocci (1972). Differentiation in this perspective is analysed as stratification, i.e., categories of wealth, income and status through which families move up and down. I have elsewhere refuted this analysis of "cyclical Kulakism" on the basis of evidence that surplus peasants are able to reproduce and even expand the conditions of their own class situation at the expense of other classes (Wood, 1976).

In a subsequent paper, Bertocci regards class as a blunt instrument of analysis, arguing that it fails to express the "structural fragmentation" of rural social relations and as a consequence is a bad guide to the formation of consciousness which ought to be implicit in the notion of class. But Bertocci's understanding of structural fragmentation involves a Weberian distinction between class and status, a distinction which enables him to be concerned about the failure in Bangladesh and elsewhere in the subcontinent "to correlate economic indicators of class with non-economic, culturally significant markers of relative prestige or social rank." (Bertocci, 1977). Thus, for Bertocci, and the mode of analysis he represents, structural fragmentation is a system of status frameworks (Bailey, 1963), which allows for the possibility of individual households occupying asymmetrical positions in different hierarchies—economic, lineage, ritual, education. Fluctuating economic fortunes of the individual peasant family through multiple inheritance need not therefore affect the family's ability to reproduce and maintain its lineage status, or maintain other indicators of social rank. Thus class expressed in terms of economic indicators is not regarded as a determining principle of rank, and is not therefore the basis of consciousness or solidarity. Indeed the patron-client relation is identified as the main form of social solidarity at the village level, with a consequent fragmentation of loyalties and attachments on a vertical basis undermining the potentiality for class solidarities on horizontal, class lines. These forms of solidarity competing with class are further reinforced by the religious ideologies of the Bengali Muslim involving notions of brotherhood. It is this analysis which enables Bertocci to make use (out of context) of the problematic distinction between 'classes in themselves' and 'classes for themselves' employing both Marx and Messaros (1972) in his support:

> "One might cite a variety of complexities in the rural political economy which hinder the transformation of 'peasant classes' thus far identified in focal power structure and stratification research from 'classes in themselves' to 'classes for themselves'." (Bertocci, 1977).

There are two problems with Bertocci's logic here. First, his own Weberian view of class contained in his understanding of structural fragmentation outlined above precludes the necessity of occupying a class position and one's consciousness being primarily determined by it; and secondly, he assumes that Marxist class analysis requires that objective conditions must always involve a corresponding subjective awareness of these conditions, and that the absence of such awareness is evidence of false consciousness.

The notion of class which is employed in such approaches is really superfluous, since it does not arise out of a primary concern with characterising the exploitative, contradictory and therefore dynamic aspects of the social relations of production and exchange in which both rural-economic activity and forms of rural culture and political relations are located. The Marxist concept of class is distinctive in that it cannot be defined by reference to internal attributes. It is a relational concept. A class only exists in relation to another class either through struggle and opposition or through domination and bondage with it. Collective action involving a collective consciousness is only one (advanced) form of such struggle and opposition. Marx never insisted that collective action or consciousness was a necessary part of the definition of class—only that such collective consciousness of common exploitation was a distinguishing feature of the proletariat under conditions of advanced capitalism. To demand otherwise would make nonsense of any references by Marx to feudal, landed, yeoman, peasant, serf *classes* (Marx, 1964). Marx made use of the distinction between "classes in themselves" and "classes for themselves" in *The Eighteenth Brumaire* to compare the peasantry in 1850 to the situation of the Paris proletariat in 1848. This was a precise, historically-located comparison referring to the relationship between the social conditions of production and the development of a collective consciousness. It was certainly not a denial of class struggle, merely a statement of its necessary form under the conditions of transformation in mid-nineteenth century France. Once consciousness is accepted as a contingent rather than necessary expression of class in this relational sense, then it becomes a concept for explaining structural fragmentation instead of being regarded as an alternative social form. Vertical solidarities of the patron-client type become the form of class relations under the conditions of agrarian underdevelopment in Bangladesh throughout this period. In this way, the analysis can go beyond the level of appearances and provide an account not only of where different categories of people are located in a system but how they are located in terms of these features of structural fragmentation.

Opposing Analyses of Class Relations in Bangladesh

The use of class as a tool of analysis for rural Bangladesh is not entirely resolved, however, by designating it as a relational concept indicative of a wider set of social conditions and relationships. The essence of this relational concept is the particular relation between capital, labour and accumulation with implications for the reproduction of transformation of the class relation. In Bangladesh, among the handful of Marxist-oriented academics and political party theorists the interpretation of this particular relation between capital, labour and accumulation was divided into two principal opposing accounts.

Some groups, like the Marxist-Leninist Party of East Bengal (Maniauzzaman, 1975) described this relation as being feudal or semi-feudal. In particular, they focused on the relations of exploitation such as sharecropping and bonded labour, and pointed especially to the classes of landlords which remain significant in the North and West regions of the country (Banerji, 1972). Elsewhere, they maintained, sharecropping is still seen as an important component of many of the small farms, even in the densely populated Dhaka-Comilla region including Noakhali.

Others, however, such as the *Jatiyo Samajtantrik Dal* (The Revolutionary Socialist Party) argue that the rural economy is becoming pervasively capitalist (Lifschultz, 1979). Historically, they cite the impact of dominant capital from the Western wing of Pakistan in commercialising agriculture through the development of jute production, and indeed view the liberation struggle of 1971 as the "progressive" assertion of infant East Bengal capital against the foreign capital of Pakistan's Western wing. They emphasise evidence on the polarisation of landholding. Finally, they refer to the ways in which agricultural programmes have become big-farmer oriented both through the extension of the Comilla co-operative strategy into other, more differentiated regions, and through the more recent privatisation and commercialisation strategies of the Second Five Year Plan encouraged by the World Bank and US AID.

A Framework for Analysis

How can such contradictory analyses be reconciled without simply dismissing them as ill-informed dogmatisms? As with any analysis in terms of structural fragmentation, the problem is that relationships are being abstracted out of context and asked to bear the whole weight of theory. It is then not surprising that variations in the relations of exploitation will be found throughout Bangladesh—landlordism here, commercial market transactions there—but conclusions about the nature of the agrarian social system cannot be achieved by identifying which one is numerically dominant. The task is to account for these variations in the relations of exploitation within a consistent framework which will at the same time reflect the uniform aspects of Bangladesh's recent history from 1947 and which combines a recognition of the historical specificity of Bangladesh with a sensitivity towards its internal regional variations. This requires for Bangladesh an appreciation of the role of unproductive capital in the processes of accumulation and class formation since the opportunities for any class to develop capital as a production relation in agriculture were severely constrained. In summary, with the colonial domination of East Bengal by West Pakistan involving restrictions on Bengali capital accumulation, the formal destruction of the *zamindari* landlord-tenant system did not bring about a direct development of capitalist relations of production in the countryside. Instead other forms of unproductive rural capital expanded, involving a proliferation of rental income based on insecure leasing arrangements between rich peasant/petty landlords and sharecroppers with little production input by the owner, the development of usurious moneylending connected in some areas to land mortgaging, and an increase in petty commodity trading activity (Rahman, 1977; Westergaard, 1978; Wood, 1976; Hossain, 1980). Those forms of capital characterised the relations between different classes of Muslims after the Hindu landlords and moneylenders departed.

While the general expansion of unproductive rural relations reflects the uniform experience of Bangladesh as a colony of West Pakistan, significant regional variations in the relations between classes can also be gleaned from available data. For example, the regions in the North and the West of the country are characterised by a greater differentiation in landholding, larger landlords, more sharecroppers (at least of the conventional, dependent type) and higher landlessness. (BARC, 1978b). This pattern is associated both with specific historical conditions and a lower density of rural population than in the Eastern regions of the country. Thus, while a general picture of unproductive (or "antediluvian") capital prevails, the precise forms of that capital and its effects vary—more petty leasing in the North and West, more usury and petty commodity exchange in the East. Because these regional sub-forms of unproductive capital involve the appearance of different relations of exploitation they encouraged the different accounts of class relations noted above and also provoked Bertocci's structural fragmentation thesis. But this chapter is proposing that such sub-forms of unproductive capital are as the outcomes of regional conjunctures the various class expressions of Bangladesh's uniform history, and they determine the precise circumstances of structural fragmentation, i.e., the specific expressions of dependency, conflict and struggle.

Polarisation and Accumulation: The Role of Unproductive Capital

The final question to be considered in this introduction is the role of such unproductive, antediluvian capital in transforming Bangladesh rural society. The historical role of such capital has been hotly debated for the Indian subcontinent and elsewhere (Patnaik, 1972; etc.). While Marx considered accumulations of monetary wealth derived from trade and usury as an essential part of the development of capitalism out of feudalism in Europe, such accumulations "belong to the pre-history of bourgeois economy" and they do not automatically produce capitalist development—otherwise "...ancient Rome, Byzantium etc. would have ended their history with free labour and capital." (Marx, 1964).

In Bangladesh, such monetary wealth (including land rents) has not been invested in the development of productive agricultural capital and has not entailed a transformation of the labour process itself through increasing the productivity of labour and therefore its relative surplus value (Marx, 1970). Instead, this pattern of wealth stimulates the attempt among tenants and labourers to achieve a rise in the level of absolute surplus-value of their labour, where they try to work longer or harder without any capital-assisted increase in their productivity to meet their debt, rental and purchase obligations (Ghose, 1979).

This process is circular and cumulative. Under the conditions of a rapidly increasing labour force due to population growth, the opportunities for tenants to rent more land or labourers to find the necessary additional employment are continually undermined. Their efforts to do so have the function of displacing other labour and weakening other tenants. Furthermore, interest rates, rents and prices continually outpace either wages or incomes from distress sales of subsistence commodities which have to be repurchased at higher prices elsewhere in the year. In this process tenants and poor peasants are steadily alienated from their land, being obliged to sell off portions to meet debts and consumption requirements or losing it through mortgage arrangements which are impossible to redeem. Such land is acquired either for multiple inheritance purposes or for recirculation in rental markets. The explanation for the failure to invest in expanded agricultural production can thus be found within this form of consequent polarisation whereby the rise in landlessness and rural unemployment reduces the level of effective demand for domestic agricultural output at market prices (de Vylder & Asplund, 1979; Wood, 1980). In this process, there has been a stimulus toward the acquisition of non-agricultural assets, although incomes thus derived are often 'reinvested' in land acquisition and rural moneylending (Wood, 1976; Adnan et al., 1978).

It is crucially important to recognise that the development of these relations, even under the "minifundist" agrarian conditions of the Dhaka-Comilla region, pre-dates the government and aid agency strategies of attempting to intensify the level of capital in agricultural production through credit and subsidies. As a result these strategies, which dominate the Government's Second Five Year Plan and the contemporary thinking of the World Bank and US AID, have only a limited impact (so far at least) upon the creation of the capitalist production relations advocated by these agencies.

Topography, Population, and Regional Variations

Bangladesh primarily consists of a low-lying fertile delta region, which is the flood plain of two rivers: Ganges and Brahmaputra. Together with other rivers, these two bring down silt and deposit it on the plain—both enhancing soil fertility and extending the delta into the Bay of Bengal. These 'new' land formations are known as *char* lands. The annual combination of fertility replacement, flooding and warm humid climate provides good conditions for unirrigated rice growing.

At the same time, these conditions have contributed to the emergence of one of the most densely populated rural regions of the world: now approximately 89 million—an all-Bangladesh average of nearly 1500 persons per square mile of cultivated land with the average for the East-Central area (the Dhaka Comilla belt) ranging from 2245 to 2643. The population growth rate is also one of the world's highest at around three percent (World Bank, 1981) under conditions where private gain conflicts with social cost (Arthur & McNicholl, 1978). The ramifications of population growth throughout the agrarian structure are pervasive: larger families living off the same size or reduced size of holding; fragmentation of holdings; dispossession of infrasubsistence peasants; decline in availability of land for sharecropping; increase in the number of landless; enhanced value of children as incomeearners and providers of old-age security; depressive effect on wages; undermining of patron-client dependency relations which offered minimal security to labour, as a reserve army of labour contributes to the commercialisation of labour relations.

An estimated 91 percent of the population live in rural areas, and approximately 80 percent are considered to be directly dependent on agriculture for their income. The ownership of land has become increasingly concentrated. Eleven percent of rural households own more than 52 percent of all land, while over 30 percent own no land at all; they and those owning less than half an acre constitute about 48 percent of the rural population. This situation is accompanied by high rural underemployment and a decline in real agricultural wages. The per capita income was reported in 1980 as equivalent to U.S. $90, but if the concentration of incomes in urban areas is taken into account then the average *rural* income would be considerably lower. It has been estimated that 62 percent of rural people receive only enough income to satisfy 90 percent or less of the necessary daily intake of calories. By the end of the 1970s the production of major agricultural crops was only 7.3 percent higher than 1970 (GOB, 1977; etc.).

The land is flat and low with few parts of the country being more than 50 feet above sea level. The only significant elevations are the Sylhet hills in the North-East and the Chittagong hills in the South-East. Rainfall varies from 50 inches a year in the North-West to up to 200 inches in the North-Eastern region of Sylhet, below Assam. The rivers flood from late July until mid-September, during which the main traditional unirrigated ricecrop is transplanted (*Aman*). The pattern of flooding is complex and brings out the ecological heterogeneity of the apparently homogeneous delta due to minor variations in elevation and proximity to flooding rivers. Some of this variation has been represented in a *thana* based map reporting susceptibility to famine based on 10 criteria: flooding; growth; population pressure; food deficit; lack of employment opportunities; low crop yields; poor land transport; river erosion; incidence of cyclones; and maldistribution of agricultural inputs (GOB 1976). The areas south and west of the rivers Ganges, Padma, and Lower Meghna are the worst affected, but almost all *thanas* bordering the major rivers of the Ganges, Teesta, Dharla and Brahmaputra are at risk. However, there are regional patterns too which reflect significant differences in soil fertility, population density, historical circumstances, or incidence of owner cultivators versus tenancy and wage labour. In Comilla, where soil fertility and population density is high, owner-cultivators predominate (34.6 percent); landlessness there is the lowest in Bangladesh (26.1 percent); the proportion of sharecroppers is smallest at 13 percent; and the smallest (but one) proportion of big farmers (more than 5 acres) is there (6.3 percent) (BARC, 1978). A similar though less extreme pattern holds for other districts in the Dhaka-Comilla belt: Noakhali, Dhaka and Tangail. Dinajpur, in the North-West, by contrast has 31.1 percent landless; 25.6 percent sharecroppers; 28.5 percent owner cultivators; and 14.9 percent big farmers. Here and in other Northern and Western Districts such as Khulna, Kushthia, Rajshahi and Rangpur soil fertility is poorest and population densities are far lower. .

It would be unwise to attach too much significance to these patterns because the BARC data does not arise from a rigorous sample, but the variations noted do conform to other pictures of regional variations gleaned from village studies (Arens & van Beurden, 1977; etc.). In the context of these variations it is important to note that in Comilla Kotwali *thana* (the most heavily researched area during the sixties and the basis of the cooperative model which has been replicated all over Bangladesh through IRDP), owner-cultivators represented 74.5 percent of the cultivating population (i.e. excluding landless) and the Comilla District mean was 73.7 percent (BARC, 1978) The equivalent figure for Dinajpur is 41 percent which certainly belies the 'peasant economy' description for all of Bangladesh.

Aman rice (referred to earlier) represents approximately 60 percent of the annual rice crop. But cropping patterns vary too by regions and within them, mainly on account of elevation but also climate. The other cropping seasons are *aus* and *boro*. *Aus* (25 percent of total rice production) is primarily a broadcast crop, dependent upon the caprice of early monsoon rains, and harvested just before the main flooding in July. *Boro* refers to the dry winter season which lasts from late November through to June; in this season only those with some form of irrigation can plant *boro* (rice) which is harvested around April. The same applies to the rapidly expanding winter wheat crop, although less water is required. The planting of *boro* has traditionally been more prevalent in the Southern and Eastern regions of the country where the winter climate is warmer and there is a higher proportion of low-lying land which retains some moisture. State intervention in agriculture, with international assistance, has in the past focused primarily on the expansion of the *boro* season (i.e. the Comilla programme, etc.) through the provision of irrigation and other inputs. This is still the case but with the critical addition of wheat to the programme which allows the North-West to be included. The area under wheat has doubled between 1975-80, and the output in 1980-81 exceeded 1975 levels by five times to reach approximately one million tons.

The main cash crop is jute. During the sixties the acreage under jute was always higher than for *boro* rice, though never more than twice as high. Acreage under jute reached its peak in 1970 and since then has been overtaken by *boro* rice (Bose, 1973). Jute was an important export earner for Pakistan as a whole and its production by peasants in East Bengal provided the basis of their exploitation by the bourgeoisie in West Pakistan (see below). In Bangladesh jute has provided about three-quarters of the country's export earnings and remains (precariously because of competition from synthetic substitutes) a vital crop in the economy as a whole.

Landholding and Tenancy Before 1950

Under British rule the province of Bengal was a 'Permanently Settled' area where most of the land was divided among *zamindars*—a hereditary class of tax farmers and 'super landlords' taken over by the British from the declining Mughal empire during the eighteenth century. The cultivator with permanent occupancy rights was a *raiyat* and might in turn rent out all or part of his land to under-*raiyats* or sharecroppers *(bargadars)*. Between the *zamindar* and the *raiyat*, intermediaries or tenure-holders proliferated as *zamindars* sub-divided the rent-collecting functions. Apart from the State, then, the classes variously connected with the land were: *zamindars*, tenure-holders, *raiyats*, under-*raiyats*, *bargadars* and landless labourers. *Bargadars* and landless labourers were the main classes of labour on the land with *raiyats* occupying the structural position of aggregating the surplus from sub-tenants and labour and transmitting it upwards in the form of rent (Wood, 1973)) The pervasiveness of this system, and therefore its structural importance for the social formation after 1950, is demonstrated by noting that in prepartition East Bengal (1938-40), out of a total of 28.8 million acres, approximately 18 million acres were held by *raiyats*, and a further two million acres held directly by under-*raiyats*. i.e. approximately 70 percent. This *zamindar* system was finally abolished in 1950 by the East Bengal State Acquisition Act, although the partition of India in 1947 had already undermined some of its features.

The Report of the Land Revenue Commission, Bengal (1940) reveals regional variations within this overall *zamindari* structure of rural social relations which undermine the assumptions of homogeneity based on national level averages which comprise the more recently published data of landholdings. The Report gives a breakdown for the 14 districts of what was then East Bengal, and which along with Sylhet and the Chittagong Hill Tracts were subsequently included East Pakistan. From the sample data (11314 families) the overall percentage distribution in 1940 is shown in Table 1.

TABLE 1

PERCENTAGE OF FAMILY HOLDING[a]

Av. holding per family (acres)	< 2	2-3	3-4	4-5	5-10	10+
4.02	45.8	11.0	9.3	7.5	16.1	7.7

[a] Unfortunately the distribution of land involved is not available.

The district breakdown of this data (see Table 2) shows that the districts in the North and West of East Bengal were distinguished from elsewhere in the province by greater proportions of sharecroppers and landless and that classes of *raiyats* and tenants were more differentiated than elsewhere by variations in the size of landholding in the Rajshahi Division (Dinajpur, Rangpur, Bogra Rajshahi and Pabna) with the exception of Pabna district and the addition of Jessore district.

The districts in the North and West (Dinajpur, Rangpur, Jessore, Rajshahi) fall significantly below the 'less than 2 acre' average, with percentages ranging from 24.2 to 31.8 percent and *rise* significantly above the 'more than 10 acres' average with percentages ranging from 15 to 11.2 percent. Bogra district which is also in the North-West falls below the 'less than 2 acres' average with 34.5 percent, although it falls below the 'more than 10 acres' average with 7.1 percent. There is a similar distribution for Mymensingh district, although it drops out of the grouping of North-Western Districts when the incidence of sharecropping and wage labour is considered (see sections 2 and 3 of Table 2).

The total percentage of families living mainly either as sharecroppers or agricultural labourers in 1938-9 was 31 percent. The districts which exceeded this average were, in order of magnitude: Khulna, Pabna, Faridpur, Bogra, Dinajpur, Rajshahi and Rangpur. All of these districts are situated on the Western side of the North-South river complex (Brahmaputra, Jamuna, Ganges, Padma, lower Meghna). Thus the landless population ranged between 32-41 percent in the North-Western region of the province to be contrasted to Comilla with 5.9 percent, Jessore with 12.1 percent and Chittagong 13.4 percent at the other end of the scale. To these proportions of landless tenants and labourers must be added a proportion of small peasants with insufficient holdings who were obliged to enter sharecropping arrangements or sell their labour. Where there existed a high proportion of peasants with holdings of less than 2 acres, the demand for additional sharecropping or labour opportunities would be high but the availability would be low.

TABLE 2

SUMMARY TABLE SHOWING

1. Distribution of Families by size of landholding 1938-39
2. Distribution of land cultivated by family member (FM), sharecroppers (Sc) and labourers (Lab)
3. Percentage of families living mainly as sharecroppers (Sc) and labourers (Lab).

Districts	1 % age with			2 % age of land cultivated by			3 % age families living as	
	<2 acres	>10 acres	Av. acres	FM	Sc	Lab	Sc	Lab
Dinajpur	24.2	15.0	6.38	72.0	14.5	13.6	13.8	23.5
Rangpur	24.6	11.2	6.67	72.2	22.8	5.1	19.1	12.9
Jessore	28.5	13.6	4.78	71.3	22.1	6.6	4.2	7.9
Rajshahi	31.8	14.6	5.52	81.0	15.0	3.9	13.0	23.8
Mymensingh	34.1	6.5	3.86	78.2	10.3	11.5	7.3	20.0
Bogra	34.5	7.1	4.28	80.8	16.0	3.2	13.4	25.6
Khulna	55.6	7.6	4.78	47.7	50.2	2.1	13.2	41.9
Chittagong	60.3	4.3	2.45	78.1	11.9	10.0	1.8	11.6
Bakerganj (Barisal)	61.8	1.2	2.17	55.3	44.7	-	7.1	23.6
Dhaka	62.4	3.5	2.13	60.9	22.9	16.2	6.9	22.2
Tippera (Comilla)	63.9	2.9	2.22	70.6	12.4	17.0	1.1	4.8
Pabna	64.1	2.4	2.39	77.6	19.4	3.0	26.1	15.2
Noakhali	65.3	2.8	2.41	70.5	16.8	12.7	2.2	17.9
Faridpur	81.5	0.6	1.63	80.5	11.4	8.1	23.8	15.1
Total	45.9	7.7	4.02	73.4	18.6	8.0	12.4	18.6

Sources: Report of the Land Revenue Commission (Bengal: Government of Bengal, 1940), Vol. II.
1. Table VIII(b), p. 115
2. Table VIII(e), pp 118-119
3. Table VIII(d), p. 117

From Table 2 there do appear to be some inconsistencies between the amount of land and the number of families involved in sharecropping and wage-labour relationships. The 'labour' columns for Bakerganj and Khulna stand out particularly; as does the sharecropping column for Bakerganj, although this may be explained by the higher proponion of land held by tenureholders who would have cultivated through *bargadars* rather than directly with their families or with labourers. Furthermore, it is interesting that Faridpur had such a high proportion of sharecroppers (23.8 percent) with its distribution of landholding, confirming that sharecropping occurs even where the absolute average size of holding is low. There is more recent evidence for this from some studies in Comilla, although sharecropping is frequently connected to indebtedness and land mortgage deals.

The sharecropping columns for Jessore, Dhaka and to some extent Khulna suggest that some sharecroppers might have relatively large holdings which should remind us that the sharecropper is not always the supplicant in the exchange. However, these figures do not of course reveal what proportion of the sharecropped land is cultivated by *raiyats-cum-bargadars* (i.e. not living 'mainly as sharecroppers'). This may explain the low numbers of families living mainly as sharecroppers or agricultural labourers in Jessore. The relatively high amount of land which was available for sharecropping in Dhaka district perhaps reflects its status as an urban center (although only a small provincial center at that time) with associated absenteeism of landholders involved in various forms of urban employment.

Together with the data on the larger holdings of *raiyats* in the Rajshahi Division, there are descriptions of Muslim *jotedars* (petty landlords/ rich *raiyats*) in Dinajpur in the 1942 Dinajpur Survey and Settlement report (Bell, 1942). This lends weight to the proposition that there were Muslim 'surplus' farmers in certain areas of East Bengal before and during the emergence of Pakistan, as against the more familiar notion that these classes developed all over East Bengal solely as a result of West Pakistan colonial policy or in response to the opportunities created by Comilla type agricultural development strategies.

The explanation for this regional variation in the extent of differentiation among the *raiyat* classes must in part refer to the historical specificity of the 'border' districts during the 1940s. These districts were more exposed to communal tension which developed as a prelude to the Muslim League's campaign for a separate Islamic State. This communal factor restrained the excesses of (Hindu) *zamindar* oppression over Muslim tenants (e.g. 'illegal' charges in addition to statutory rent), thus retaining more surplus at the level of *raiyat* for accumulation and consolidation of holdings. Also, as the rural population in border districts was more evenly distributed between Hindus and Muslims than in the predominantly Muslim districts further East in the province, the impact of inter-caste relations on the pattern of landholding and tenancy was greater. And after 1947, the departure of Hindu *raiyats* as well as *zamindars* and tenureholders created more 'illegal' opportunities than further East for the remaining rich Muslim peasants to appropriate the land vacated by the Hindu *raiyats* as well as landlords. As a result, in this region, which in any case received the greater proportion of new immigrants, the newcomers became sharecroppers and landless labourers rather than *raiyats* more frequently than elsewhere.

Although the data are not available there is every reason to believe that these regional variations in differentiation increased after 1947 and certainly after 1950 when the *zamindari* estates were formally abolished. The East Bengal Acquisition Act legalised this *de facto* appropriation of land by classes of Muslim *jotedars* in the 'border' districts. As part of this historical specificity, the variations in ecology and the subsequent variations in the density of the population will further contribute to the development of rural social relations along different paths in Bangladesh. The high density of population in the Dhaka-Comilla belt is related historically to the fertility of the soil and favourable climatic conditions for growing rice during the winter. This together with the absence of the 'border' characteristics noted above must account to some extent for the long history of lower absolute average size of holding in the Eastern region, and a much greater concentration of holdings at the lower end of the range.

The evidence on *jotedars* is important for several reasons. First, it indicates a structure of rural class relations very different from the Comilla model which has been generally applied—i.e. a homogeneous structure of undifferentiated, smallholding peasants. Second, it undermines related assumptions about the nature of power in East Bengal which has traditionally been analysed solely in terms of the aspirations of an urban petty-bourgeoisie, leading an undifferentiated mass of the peasantry against the common oppressor from West Pakistan—e.g. the language movements and the rise of Bengal nationalism (Alavi, 1972; etc.). Third it focuses our attention on the role of these classes in transforming the entire social formation both through their support for the Muslim League and the quest for an independent Pakistan, and through their subsequent influence over agrarian legislation and state policy in agriculture. The Muslim League's Policy of Partition had the advantage of removing the stratum of Hindu landlords from their stultifying influence over the aspirations of the Muslim *jotedars*. At the same time as in the neighboring states in India, the attack by *jotedars* on feudal rural relations had to be contained at the point where the conditions for their own appropriation of land and surplus had been fulfilled.

Class Relations and the East Pakistan Colony 1947-1971

The political significance of this class of Muslim *jotedars* in the North and West of Bangladesh can only be considered in conjunction with that of the Bengali petty-bourgeoisie consisting of small traders, shopkeepers, professionals, teachers, junior officials and clerks in the provincial services. In a country where the entire urban population was never more than 10 percent, this class had close kinship and other connections with the economically stronger classes of *raiyats* and *jotedars* in the countryside. It is not difficult, therefore, to see the relationship between the frustrated aspirations of a Bengali petty-bourgeoisie whose language (and thereby career prospects) was threatened and those of the larger Muslim farmers and peasants whose own development was restricted by West Pakistani colonial policy. In this way, the language movements could expand into broader Bengali nationalism and lay the foundation for the creation of a reformist nationalist party (The Awami League) This successfully combined the respective concerns of the large rural petty commodity producers and the urban petty-bourgeoisie with the intention of reproducing West Pakistani capitalism in the East through a constitutional bourgeois-democratic transformation. This would involve associated career prospects for the Bengali intelligentsia in the private and public sectors. A combination of populist rhetoric and rural vote banks (Ayoob, 1971) steadily increased support among the poorer classes of Bengali peasantry, culminating in the League's overwhelming victories of 1970 and 1973.

Despite the contemporary significance of these two classes—*jotedars* and urban petty bourgeoisie—and the alliance between them, they remained economically weak from 1947 to 1971 as a result of East Bengal's status as a colony of West Pakistan. This has been well documented elsewhere by Sengupta, 1971 and Nations, 1971. The Muslim League and the Movement for Pakistan had been led essentially by classes outside East Bengal—landlords from Punjab, Sind and Central India; trading communities from Gujarat and Bombay which formed the basis of the Pakistani industrial capitalist class; and westernised Muslim elites from similar regions who constituted a professional middle class and who had staffed the bureaucratic and military apparatuses of the colonial state. East Bengal had been dominated by Hindu traders and officials (as well as landlords) as part of the Calcutta hinterland. Thus Muslims of East Bengal were historically disadvantaged at the time of Pakistan's formation in 1947. This facilitated their colonial exploitation by the western region up to 1971, and accounts for the restricted and underdeveloped nature of a bourgeois class indigenous to East Bengal. East Bengal had a virtual world monopoly in jute production, but the foreign exchange earned by the export of this peasant-produced cash crop was redirected by the military-bureaucratic oligarchy in West Pakistan on behalf of the development of its indigenous, industrial capitalist class (Alavi, 1972).

The compulsory purchase of West Pakistani manufactures by the East at inflated monopoly prices led to a further outflow of capital which might have been deployed locally. Further aspects of this colonialism indirectly restricted the development of Bengali classes. West Pakistani capital and its entrepreneurs were significantly present and in control of East Bengal industry, with Bengali Muslims owning less than 2.9 percent of private industrial assets. Thus, much of the wealth created in East Bengal industry was repatriated to the West in the form of salaries, dividends, interests and profits; and the related opportunity for multiplier effects from indigenous capital accumulation was also lost. Add to the picture the evidence from the First and Second Five Year Plans, in particular, that the greater proportion of funds for development was allocated to the western province and the failure of an East Bengali bourgeoisie to emerge is hardly surprising. The dominant classes in West Pakistan were engaged in an exercise of primitive capital accumulation.

The development of social relations of production in Bangladesh and the pattern of rural class formation cannot be isolated from these colonial effects. The principal contradiction was that the appropriation of East Bengal's rural surplus by the emerging capitalist classes of the West (accompanied by an unwillingness to re-invest in East Bengal's agriculture) jeopardised the reproduction of the social relations required to produce that surplus by denying the capital for investment in: intensifying food production, and diversification out of agriculture into industrial production. In short, the effect was to disarticulate the East Bengal economy—stimulating food crises, vulnerability to disasters, landlessness and mass under-employment.

During the 1950s agricultural output virtually stagnated (Bose, 1973). For example, rice production increased at 0.7 percent per annum compared with a population growth of nearly three percent per annum. Even the production of jute declined in the 1950s, picking up in the 1960s only because of a large increase in the acreage, since yields were declining. This evidence on jute is consistent with the tendency for landholding to polarise, with land being transferred out of the hands of small, mainly subsistence and rice-cultivating peasants. The squeeze on agriculture, both through adverse terms of trade and the low level of state investment, restricted the capacity of richer peasants to expand their control directly over productive assets. However, this squeeze, together with increases in population, increased the vulnerability of the poorest peasants. Without the dubious advantages of a *zamindari* system to rely on, their urgent requirements for credit, for land to sharecrop, or to sell their labour could only be met by the richer, surplus peasants—at a cost.

Landholding Patterns, 1960-1977

Data on landholding cannot reveal all these relationships, both because of the level of generality, and because the data are based upon reported legal titles to property which are unreliable and misleading guides to the real relations of appropriation. But with these caveats, are the data on landholding during the 60s and 70s consistent with the arguments presented here? At the very least the data are not consistent with the development of capitalist agriculture on a wide scale despite state intervention to that purpose since the late 60s. The data reflect the increasing pressure on the land throughout the period through showing a dramatic rise in the landless, a decline in the mean size of owned farms from 3.5 in 1960 to 2.0 acres in 1977, of owner-cum-tenant farms from 4.3 acres to 3.0 acres, of tenant farms from 2.4 acres to 1.5 acres, and an increase in households owning up to 5 acres from 78 percent in 1960 to 93 percent in 1977, representing a shift from 43 percent of total cultivated area to 58 percent in this category (GOB, 1977; GOB, 1960). Other sets of comparisons are marginally affected by the different units of classification in the surveys involving 5.00-7.5 acres and 7.5-12.5 acres in the 1960 and 1968 surveys, and a series of one acres units (5.00-6.00 etc.) in the 1977 survey (GOB, 1977). In the 'middle' range of 5.00-12.5 acres there was a decline in the number of holdings from 18.6 percent of total in 1960 to 14.4 percent in 1968 with a corresponding drop in acreage from 38.4 to 33.2 percent (GOB, 1960). The 1977 survey reported a dramatic decline in the number of holdings in the 5.00-12.00 acre category to 5.46 percent involving an acreage of 26.2 percent of total. With the same marginally affected problem of comparison the number of holdings above 12.5 acres fell from 3.5 percent in 1960 to

2.6 percent in 1968 and continued to fall for an above 12.00 acres category in 1977 to 1.2 percent. This picture of continuous fall is not repeated, however, when acreage is considered: 18.9 percent in 1960 to 15.4 percent in 1968, but holding up at 15.5 percent (for above 12.00 acres) in 1977. This pattern is reflected in the figures for average size of holding per household in this richer category: 19.2 percent in 1960, 18.6 percent in 1968, and 19.4 percent in 1977.

The dynamic process indicated by these figures is complex. The dramatic weakening of the 'middle' range (5 to 12 or 12.5 acres) is not the result of a simple process of polarisation since it is accompanied by a more or less constant average size of holding in the above 12.5 or 12 acre category throughout two decades rather than an increase which would reveal a process of concentration and perhaps consolidation of land by richer peasants and landlords. At the same time the numbers of households in such a category has shrunk to 1.2 percent, suggesting that the pressures for fragmentation of holdings among sons may still await those families who have not diversified through their sons into income generating activities unrelated to the direct or indirect exploitation of land and the labour expended on it. Of course there is some concentration of landholding among parts of this class which distinguishes those who are vulnerable to multiple inheritance from those who are not. Furthermore, during a period when population has risen by 35 percent from 1960 to 1974, such aggregation dangerously ignores regional variations which along with population growth provides the explanation of how a smaller proportion holding above 12.00 acres maintains both the acreage under the class's control and the average size of holding for the households in the class.

On the basis of these figures combined with the arguments outlined earlier in the paper, the 'regional' hypothesis would maintain that the Dhaka-Comilla belt was mainly responsible for the collapse of the 'middle' range of landholding and the expansion of the below 5 acre 'minifundist' category; whereas the 'North-West' represented the picture of some concentration accompanied by loss of holdings by others at all levels resulting in increases both in landlessness and tenancy (the latter rising nationally from 18 percent of total holdings in 1960 to 23 percent in 1977), as well as contributing to the general picture of redistribution downwards. In this way it is not correct to argue that all categories of landholdings are fragmenting steadily, and reducing all classes to the lowest common denominator. A small but significant class of *jotedars* remains, and when the range of landholding to be considered is widened to conform to the government's classification of 'big farmers' with more than 5 acres, the regional concentration of *jotedars* and 'big farmers' is clearly associated with the Northern and Western districts, displaying a continuity with earlier patterns (BARC, 1978).

Processes of Class Formation

Such a picture gleaned from the only available though imperfect macro-data does provide some background to the numerous variations in the real relations of appropriation reported by different micro-studies. The tendency to make extravagant claims of nationwide relevance from such studies has produced sterile exchanges between protagonists advancing their particular thesis. But it should now be possible to recognise where there are common threads to the process of class formation, mediated through precise historical, ecological and local economic variables, and where room for debate still exists over the interpretation of local data. A further problem of analysis has been the unweighted relativism shown by some commentators who, as Rudra has done in a discussion on class formation in India, have successfully identified an extensive list of relationships without commenting upon their significance. Thus Bertocci attempts to refute a class analysis based upon landlord-tenant relations by pointing out that poor peasants also rent out their land; likewise that poor peasants acquire temporary control over mortgaged land by lending money in order to establish viable farms in land-scarce situations; and finally, that poor peasants employ both casual and permanent wage labour. However, it is unsafe to conclude that:

> "those facts would seem to imply myriad, dyadic, owner-tenant, creditor-debtor, patronclient relationships within as well as across the lines of 'peasant classes' as may be conceptually delimited with reference to landholding size and/or tenure type alone." (Bertocci, 1977, p.9).

Such lists are never an adequate substitute for analysis, especially when access to and control over land is such a crucial variable as in Bangladesh. The fact that so many are landless, and many more (70-80 percent) assumed to be living below established poverty lines (Salimullah & Islam, 1977) does not mean that the scramble for tiny pieces of land is insignificant. On the contrary its value is greatly enhanced both in monetary, economic and status terms. This was a mistake made in Comilla where the range of landholding was very narrow, by those who assumed that the absence of a starkly unequal pattern of landholding was evidence of class harmony. But assuming constant family size, the difference of a few decimals in area under such conditions can make all the difference between survival and successful diversification, and a family's collapse and disintegration through a relentless process of land dispossession. (Wood, 1976).

Apart from the issue of the relative significance of relationships, there is also the question of the meaning or interpretation to be given. For example, to return to the case of poor peasants taking land in mortgage—does this amount to a refutation of mortgage as a class relationship of structural significance? Nadigram in Mymensingh District provides some interesting case material (Arn, 1978, p.11). The most common land transaction is *girbi* (mortgaging), where a landowner in need of money transfers the right of use of one or more fields to the highest bidder.

This system functions as a 'banking-system' for landowners who give out land on *girbi* when they need money for other expenditures, mainly to buy land, or for consumption needs during a difficult period (sickness for example) or to pay for the higher education of a son. The general trend within this kind of land transaction is that the richer and middle peasants give out more land on *girbi*, and the small, poor and landless peasants take in more land. For these last categories the *girbi* system is one way of acquiring land to cultivate under circumstances where land for sale is rather scarce and expensive. When land is for sale they can seldom compete with the richer peasants who will give out some of their lands on *girbi* in order to be able to buy.

Under these conditions where a market for land exists, the mortgage relationships assist richer and middle peasants in obtaining the money to enter that market and accumulate land for the family's use in this generation or the next. Although poor and landless peasants are technically the moneylenders in this situation, they are unable to charge interest, the land has to be returned within twelve years and it is usually redeemed earlier (Arn, 1978). These arrangements reveal the structural weakness of the poor peasants, which has to be attributed to their lack of control over land and a desperation for it which encourages such temporary solutions that divert their own savings away from longterm accumulation. It should be remembered that it is cash hungry peasants who are providing land for this market in the first place.

In Comilla, my own study in Bondokogram (1976) reported a high incidence of mortgaging *(bondok)* where, in the absence of a market for land (its increased value in the context of the Comilla programme placed it beyond the reach of open/market purchase by locally richer but still 'minifundist' peasants)—the acquisition of land in mortgage from cash-hungry poor peasants—with accompanying high rates of interest resulting in permanent transfers of the plot after a few years—constituted the main process of land accumulation. The richer peasants were in a strong position to impose these high rates of interest, since they controlled through their land the labour opportunities for peasants who needed to sell their labour. Along with greater opportunities for diversification among rich or peasant families this process contributed to offsetting the 'cyclical kulakism,' whereby families were supposed to decline through multiple inheritance, allowing others to replace them in the hierarchy (Bertocci, 1972). Thus, two 'directions' of mortgaging may appear to have opposing implications for class formation, but only if arbitrarily isolated from other features of the local agrarian system. Taken in context, both reveal forms of dependency consistent with the terms of exchange and patterns of landholding which characterise their respective loyalties. It is in this sense that Peter Bertocci's notion of *structural fragmentation is useful*—not to deny class situations, but to lead us all to specify them more clearly. The forms of struggle that are implied vary crucially between Nadigram and Bondokgram; for example, a campaign against moneylending and high interest rates would make little sense to poor peasants in Nadigram.

This example of interpreting the meaning of relationships beyond the level of appearances applies with equal force to tenancy. The evidence from Jhagrapur (Kushthia District) reveals several 'class-directions' of sharecropping relationship (Arens & van Beurden, 1977). Almost 40 percent of all families in Jhagrapur were reported as sharecropping some land, but of all the land which is sharecropped only one-third was cultivated by small peasants (Arens & van Beurden, 1977). Since they had no draught animals or sources of credit. Fifty percent of sharecropped land was cultivated by sixty percent of the middle-peasant families. Fourteen percent of the sharecropping families in Jhagrapur were from rich peasant classes for whom the relationship was not a necessity. The pattern of subletting did not only involve *rich* landlords, since the authors found that 15 percent of all small peasants sublet some or all of their land. To this extent, Bertocci's caveats about the 'directions' of such relationships are justified—but is the content of the relationship the same in these different cases? Small peasants who lease out their land did so from a position of weakness—the plots were too far away to be personally cultivated and supervised, the small peasant was a widow (40 percent of the land involved); the small peasant has no cash for renting a plough or animals and/or was disabled (25 percent of the land involved). Rich landlords who lease out land for sharecropping conform to the more familiar pattern: they receive a 50 percent share; employ the surplus in construction, land acquisition and moneylending; and fail to make any investment either in the inputs on the sharecropped land or on agricultural innovations on their own land (Arens & van Beurden, 1977).

There are other micro-studies on Bangladesh which substantiate such observations (Pray & Abdullah, 1980). The problem is whether they can be reconciled within an overall thesis, or must be left theoretically (and politically) undigested. An examination of landholding statistics might be suggestive but in themselves they can neither specify the process of class formation nor, in their aggregate form, reveal the structural fragmentation contained in the micro-data.

The common thread both historically and spatially can be found in the issue of rural capital formation—of central concern both to those attempting to characterise the mode of production in Bangladesh and to policy-makers. While Westergaard gives much explanatory significance to usury as a constraint to the development of productive rural capital, the work of Atiqur Rahman qualifies this perspective on the basis of survey material both from Comilla and Phulpur Thana in Mymensingh District in 1975. According to the BARC (1978) Survey, Comilla Kotwali Thana had 26.6 percent landless, 12.1 percent sharecroppers, 54.6 percent owner-cultivators and 6.7 percent 'big farmers'; while the respective figures for Phulpur were 41.7 percent, 23.2 percent, 26.6 percent and 8.5 percent. Rahman's main observation for both areas was that landlords did not have significant credit relations with tenants and that usury income to landlords was much smaller than that derived from rental income so that increases in rental shares were more significant than raising interest rates in appropriating the surplus from labour. Moreover, there was little participation by landlords in input-sharing with their tenants, and when tenants did innovate, the share arrangement was replaced by a fixed rent which left the tenant bearing all the risk. Credit from other non-institutional sources—rich peasants, relatives, friends—was not always linked to land mortgage arrangements, but it was observed that the "...rate of return on moneylending in unorganised rural credit market was, in many cases, higher than returns from productive investment." (Rahman, 1980).

In addition to the motive of lending in order to appropriate mortgaged land, Rahman suggested three other reasons: to ensure the reproduction of debtors' labour, to ensure the supply of labour to creditors when demanded and to maintain cohesion among the creditors' local political factions. On the basis of the landholding variation between Comilla and Phulpur, one would expect Rahman's landlord-tenant explanation of constraints to rural capital formation to have more relevance to Phulpur and the labour-securing motives for lending to apply more closely to Comilla. If so, then

Rahman's findings would constitute a microcosm of the general thesis concerning regional variations in the forms of antediluvian rural capital in Bangladesh.

Mahabub Hossain reinforces this theme about the depressed rate of rural capital formation in his argument for land redistribution in Bangladesh. He maintains that although the marginal rate of saving is higher for the large landowner, the savings are not employed for productive purposes.

> "In the Bangladesh case a common finding is that a large portion of the surplus of the rich basically finances the deficit of the poor, through land purchases and sales and provision of consumption loans. Also the better off families use a portion of their surplus for conspicuous consumption like expenditure on social ceremonies and construction of houses, and for investment in trade and business." (Hossain, 1980).

This description by Hossain hints additionally at a crucial feature of class relations in Bangladesh. With the rapid rise in landless rural population and the relentless process of pauperisation for many other poor peasant classes either through usury, changes in tenancy arrangements, the availability of land to lease and distress sales of goods and land, the obvious question to ask is why rural Bangladesh does not burst asunder in organised revolution or chaotic anarchy. The structural fragmentation of class relations perhaps accounts for the absence of revolutionary struggle since localised conditions have generated localised and narrow forms of class struggle around cropshare and wage disputes. That is, the form of immigration is not universal. But why not chaos? The post-liberation State in Bangladesh has not been effective in maintaining rural law and order either in its repressive or ideological forms; and it has certainly been unable to guarantee the rights to property and work of vulnerable classes. Indeed there have been times when the State had no contribution at all to make toward securing peace; under a rural petty bourgeois order—such as the period after the floods of 1974 through the subsequent famine to Sheikh Mujibur Rahman's death in August 1975.

Such rural stability as exists under these conditions of differentiation has been maintained through *vertical, segmentary* alliances in the form of extended families and lineages in which the poor households are pauperised through unrepayable loans but finally cushioned against destitution. The typical result is servile bondage for some household members (mostly children and adult females) while others (adult males) must migrate on a long and short-term basis in search for employment at wage levels sufficient for their own subsistence but not for reproduction. Vertical kinship relationships of this sort are often described as factions and their existence cited to deny the existence of class relations. But such dependency relationships in their different expressions all over Bangladesh are precise evidence of pre-capitalist, localised relations of real appropriation which, until recently, at least have provided a means for responding to the imperative of securing a supply of labour during peak periods of demand through localised institutions of either debt or tenancy. While it would be wrong to deny an element of compassion among family members of different class positions, it is equally unwise to become trapped in the ideology of 'cultural solidarity' in Bangladeshi rural life. Bengali Muslim peasants are not falsely conscious but some anthropologists, with their fanciful notions of a moral community, are! The participation of landlords and rich peasant moneylenders in the maintenance of rural order is a pragmatic, structural rather than ethical imperative.

Conclusion

By considering the pre-liberation history of Bangladesh and drawing attention to its extended experience of imperialism up to 1971 (at least) this chapter has argued that in general, rural class relations have been characterised by the use of capital in the sphere of exchange rather than production which has effectively raised the level of absolute surplus value accruing to certain classes, but which is not currently contributing to the generation of productive agricultural capital and the real subsumption of labour under capital. Secondly, it has argued that this failure to develop productive agricultural capital resulted in a form of polarisation during the sixties and seventies which involved a dramatic rise both in the numbers of landless and the levels of rural unemployment, thereby creating a problem of effective demand. That process also entailed a decline of middle peasant categories where the marginal rate of saving is highest and most likely to be diverted into capital formation.

Thirdly, the precise forms of this pre-capitalist capital (antediluvian) have been structured by a host of particular and regional conditions, too numerous to discuss separately here, but including soil fertility, structure and elevation, flood vulnerability, drainage and waterlogging, rainfall, temperature, population density and composition, cropping systems, patters of livestock ownership, and migrant settlers on marginal land. These historical and other variables have been responsible for a fragmentation in class relations typified by regional variations in the incidence of formal tenancy, the use of non-family labour, productivity, land values, nature of indebtedness, patterns of migration, scale and direction of diversification from agriculture, family size and lineage structures, and receptivity to various forms of technology having different capital-labour ratios.

The picture which emerges, though not fully documented here, is that the North and the West have more landlords, sharecroppers and landless labourers than the East; that landlords may have better access to inputs than owner-cultivators but cultivate intensively on small proportions of their land; that sharecroppers and owner-cultivators are being slowly dispossessed of their land, or in the case of sharecroppers, are being forced to cultivate higher proportions of marginal, less productive land in their holdings. The holdings of poorer classes are fragmenting and they are increasingly mortgaged, eventually transferred or sold as those classes descend the downward spiral. This process is accompanied by some concentration and polarisation as this land is being transferred to other classes which are either rising or managing to offset the effects of fragmenting their holdings for multiple inheritance through this land acquisition.

Chapter 3

Rural Development in Bangladesh: Whose Framework?

Introduction

The purpose of this chapter is to incite discussion rather than make claims to a comprehensive review of rural development experience in Bangladesh. It was prepared in response to an invitation by the Secretary of Agriculture (GOB) in February, 1979, to write and circulate in Bangladesh an unrestrained critique of rural development policy and modes of intervention in Bangladesh, including the role of some external agencies. This paper was revised and presented to a conference on Bangladesh at the Institute of Development Studies, Sussex in July 1979, at which the Secretary of Agriculture and other officials of the Government of Bangladesh were present, along with officials from the UK-ODA, staff members of OBRD, and academics from a variety of UK and Bangladeshi institutions. (As a sequel, I was asked to return to Bangladesh at the end of 1979 to review the second Five Year Plan, and this paper appears below, Part II, Chapter 6). I was aware, when writing the paper, that it may give offence either through criticism which is too harsh or because many of the smaller scale "success" stories have been overlooked.

There is no scarcity of experimentation in Bangladesh with so many NGOs, both local and expatriate, at work. However, in defence, what follows is intended to be highly selective precisely in order to dismantle (or at least cast doubt upon) the notions of framework, model, formula, and strategy when examining or participating in rural development. Of course, to be 'anti-framework' reveals an implicit commitment to some other formula—but then one cannot entirely leave the stage without deserting the plot.

This is clearly an appropriate time (1979) to be commenting on rural development strategy in Bangladesh with the current formulation of the next FYP being accompanied by a wariness on the part of some senior official towards the various proposals for strengthening or discarding the IRDP framework. Where the issue becomes whether to replicate Comilla or not, the key question must be to what extent 'frameworks' are of limited regional value, emerging only after a period of trial and error, of action research, of evaluation, and above all of interaction between people and planners. To insist upon applying an external formula to a locality becomes sterile imposition if it is either based on false assumptions or imitates an untypical experiment. This is the main criticism of rural development thinking in Bangladesh.

There is a sense in which this discussion could be regarded as a luxury, since it does not address the crisis which I am not alone in believing is quickly returning to Bangladesh. The late monsoon this year (1979) marks the end of a favourable sequence of weather and harvests, and coincides with a further period of world recession which will require emergency responses possibly beyond the current regime's capacity to manage. The appearance of stability in the last 18 months has hardly been sufficient to legitimise the regime.

However, the differential vulnerability of classes, sectors and regions of Bangladesh to these events must be partly explained by shortcomings of contemporary rural development efforts, along with the more familiar and still valid explanations of colonial exploitation before and after 1947, class differentiation and greed, ecological and climatic disadvantages, and interrelated population growth. Presumably these shortcomings, if they continue, will increasingly account for future disasters as they represent failure to control these processes. But it would be misleading to refer to various shortcomings as if they shared a common conceptual status, all capable of being resolved by knowledge alone. Rural development is *par excellence* a social process rather than a technical one—its content and form are the outcome of relations between groups and classes internationally as well as nationally, and only in that way are criteria established for second order choices between institutional and technical alternatives. Although this discussion mainly concentrates on the second category of shortcomings, it does so in the knowledge that they are overdetermined by issues of political economy and political struggle; and that in societies like Bangladesh rural development activity is a central feature of struggle since for the majority the distribution of the most significant resources is at stake.

At the outset, therefore, it is important to recognise that most discussions of rural development (and this one may be no exception) find it necessary to suspend reality by assuming the existence of a benign state facing mainly problems of finance, manpower, technology and knowledge. This assumption is a necessity for anyone involved in deploying public resources for growth and development—whether they are policy makers or village level workers. However, despite our personal hopes, such assumptions are rarely the truth in any society—government and popular leaders (indeed they may be populist) are inevitably constrained by the interests of the dominant class, which critically includes the bureaucracy when other avenues of employment and status are few. There is nothing to be gained by denying this reality, euphemistically referred to as 'absence of political will' and it does not only apply to Bangladesh. First, we have witnessed the continuous search by successive cliques of petty bourgeois leaders for a coherent populist ideology which will capture mass imagination through the provision of attractive formulas in rural development. Secondly, there is the overwhelming presence of international/bilateral and voluntary aid/ technical assistance agencies waiting in the wings for the stage to be constructed and the script finished. Furthermore, the relation between these two processes is significant, since the inevitable populism establishes the preconditions for the way in which different parts of the state apparatus have become captured in thought and deed by permutations of external agencies. This is particularly evident in the policy and executive area of rural development and agriculture, and symptomatic of the internal contradictions within the Bangladesh state.

It is not mere Marxist preoccupation or 'Third Worldism' to develop these themes of dependency and internal disarticulation in the context of this discussion. Where the survival of 'state' classes (including elements of the bureaucracy) has become inextricably linked to the maintenance of different forms of aid, then internal fractional struggles interact with and indeed thrive upon the external presentation of ideology, knowledge (research, evaluations, concepts, theories, methodologies) and schemes. It is not a question of susceptibility to external propositions, but of interdependence, where external agencies establish a territory of clients within the state apparatus and *vice versa*. The creation of models and slogans is a reciprocal process of mutual accommodation inspired by the imperatives of survival, institutionalisation and the consolidation of personal fortunes by national and international civil servants.

In South Asia, and in Bangladesh in particular, the IBRD has been very public in its voyage of thematic discovery from small farmer to marginal farmer, basic needs, the landless, area development, target groups and so on. The commitment to area development as revealed in the IBRD's RD-I (and more or less adopted intact by DANIDA, ADB, NITA and now possibly UK-ODA) strengthened the IRDP interests while deserting the 'true' spirit of the TCCA/KSS framework. These 'rent-a-*thana*' strategies have contributed to the BADC's hegemony over input distribution through bureaucratic allocations and subsidies. Ironically, the concentration of resources (in particular, scarce administrative and extension capacity) on a small proportion of the IRDP *thanas* has seriously weakened efforts elsewhere where indifferent experience is cited as a general criticism of the framework, and encourages even the initiators of RD-I and co. to look now more favourably upon 'commercial', open-market solutions to rural production and distribution. It is certainly true that US AID is increasingly committed to production-oriented, commercial strategies within the Ministry of Agriculture (such as open fertiliser dealing), while at the same time domestic foodgrain prices are depressed by the efforts of US food aid.

There is no sense in being surprised or indignant about the contradiction and paradoxes of government rural policy. For any society, especially those with substantial state involvement in the economy, the view of a monolithic, coherent state apparatus should be discarded. The management of late monopoly capitalism anywhere involves contradictory responses which are reproduced within the state apparatus itself to produce a chaotic, pluralist, even anarchic relationship between its constituent parts (Hirsch, 1978). In an extreme case like Bangladesh where these contradictions are reproduced within the highly significant internal and bilateral donor agencies as well, the lack of coherence in policy is even more understandable.

It is a wonder then that two main trends in policy can actually be discerned. First, a commercial, production growth strategy, based on individual farmers responding to prices for inputs as well as produce with a gradual phasing out of input subsidies, operated primarily through the Ministry of Agriculture. Critically associated with the trend and not the responsibility of the Ministry of Agriculture is a 'target-group—basic needs' element, centred around the provision of rural employment for those classes and groups who have no future in direct cultivation. Within this trend, land reform is more about legitimising existing processes of consolidation and concentration than redistribution. A related aspect of this trend is the recent introduction, with IBRD support, of the T&V (Training and Visits) extension methods, which indicates a further shift from the earlier Comilla-derived integrated development formula of TTDCs and TCCAs. This will be discussed in more detail later. The second trend is even more familiar in Bangladesh, and refers to efforts by Akhter Hameed Khan (Khan, 1979) to lobby for a revival of this earlier integrated strategy using TTDC's local council, co-operatives; and the principle of *thana* planning for: the improvement of drainage and embankments, securing supplies of additional surface water, and increased utilisation of ground water. Alongside these three elements, four other areas of activity are outlined—the generation of equity capital guaranteed supplies of chemical fertilisers, pesticides and seeds; training of co-op managers and model farmers instead of extension agents, with the posting of specialists at the *thana* level (in direct contrast to the T&V system); and the creation of processing and marketing units owned and operated co-operatively. In this reassertion of the original Comilla theme, Khan is arguing that the indifferent experience within IRDP since 1973 is not so much a criticism of the model (the framework), as evidence that it was mainly the KSS part of the IRDP formula that was attempted in the 250 *thanas*, and that it was introduced mechanically without regard for the spirit of the original idea. He argues further that in this programme co-operatives have been misconceived as targets for subsidy and relaxed credit disbursement, whereas they were intended to be fully viable commercial institutions with the TCCAs behaving according to normal banking criteria.

Neither of these approaches involves the principle of local participation in the formulation of rural development policy, although both contain elements of participation for implementation but more in the sense of rural mobilisation on behalf of projects and schemes already designed. The attraction of a framework, *any* framework, for a central planner is that it provides a uniformity of activity and procedure consistent with the norms of managing a large functional Ministry—practices, supervision and accountability can be standardised, in short: central control established. *Which* framework then becomes of secondary importance. The dilemma which some senior officials have expressed in Bangladesh is whether a process of participation in localised and policy and planning activities will lead to a myriad of local philosophies and institutions which cannot be coordinated centrally, let alone into a coherent policy. Furthermore, a critical dimension of rural diversification is the different forms of class struggle that obtain. The ethical dilemma of 'whose framework?' is certainly not just resolved by regionalising policy-making if the pre-capitalist relations between classes render the majority effectively unfree to engage in a genuine articulation of their interests. However, if class struggles were decentralised in this way, the purpose of struggle might become more meaningful to those involved and would certainly realign the pattern of influences on policy formulation.

At the same time it is clear that variations in regional conditions are significant, and that rural/agricultural development thinking has suffered through neglecting the heterogeneity of Bangladesh in terms of climate, hydrology, soil structure and the historical patterns of interaction between different rural classes and the land. Thus even if the overall description of the pattern of influences on the Bangladesh state is accepted as accounting for the absence of radical rural development programmes in the country, it can nevertheless be recognised that considerable scope exists within the current configuration of rural and urban classes for the inclusion of the regional dimension in discussions about frameworks. The issue, however, is by what process local characteristics are discovered and responses to them devised. It is this question which takes the discussion beyond problems of content to problems of method; indeed it substitutes method for content as the major way for outsiders (whether of the same nationality or not) to assist the rural poor. The focus on method is a meta-framework, the precise virtue of which is to avoid prior specification of projects and the like for 'other' people.

The final part of this chapter will return to this theme. However, to support some of the assertions made above, a summary follows of some of the problems and criticisms which are generally associated with the contemporary strategies (of the late seventies). It is intended to serve as a useful checklist of previous experience as well as a basis for pursuing arguments about diversity, methods, participation and the limits of government intervention. Since it consists of a survey of critical literature, the emphasis is upon the presentation of arguments rather than an attempt to reconcile any differences between them.

Review of Problems and Current Strategies

Population

To begin there is the general context of population growth under conditions where private gain conflicts with social cost (Arthur & McNicholl 1978, pp 46-59). Furthermore, the proposition exists, although inadequately tested at this time, that in some areas where the HYV package in rice has been extensively adopted this is associated with relatively high population growth rates between 1961-1974 (Arthur & McNicholl 1978, p.32). If this is true, the implications are alarming; certainly the proposition requires further examination. (BIDS poverty study, careful use of next census data?) The ramifications of population growth throughout the agrarian structure are pervasive: increasing families living off a constant size of holding; fragmentation of holding; dispossession of infrasubsistence peasants; decline in availability of land for sharecropping; increase in the number of landless; enhanced value of children as income-earners and providers of old-age security; depressive effect on wages; and the undermining of patron-client dependency relations which at least offered minimal security to labour, as surplus labour contributes to the commercialisation of labour relations.

Inputs, Institutions, and Class Differentiation

Significant class differentiation occurs despite the predominance of small farms where 79% of the rural households (excluding the landless) own less than 4 acres, representing 48% of the total net area (Bureau of Statistics Land Occupancy Survey 1977). Local councils and co-operatives are captured by the locally dominant classes; the segmentary lineage structure of the village forms the basis of factional struggles preoccupied with division of scarce resources rather than the production of goods; factions prevent collaborative efforts through co-operatives to the benefit of richer factional leaders and at the expense of the poor; factions composed of vertical patron-client dependency relations diffuse class relations; the domination of co-operatives by the rural rich means control over tubewells (their location and management), better access to inputs and credit, and indeed the private re-lending of credit at higher interest rates; and money lending yields higher returns than productive investment. As a result of some of these processes, the introduction of HYV foodgrain packages is responsible for greater rural class differentiation.

Higher Rice Production: Irrigated or Rainfed?

Critically, the diffusion of HYV technology cannot proceed without the availability of pumps, borings, seed, fertiliser—and the provision of these items remains a key constraint involving both import policy, domestic investment decisions in assembly, manufacturing, raw materials extraction and processing plant, and of course the efficacy of BADC domination over distribution. According to the Bureau of Statistics 1977, by the end of *aman* early 1976, 15% of the gross rice cropped area was under high-yielding varieties. Although there have been rapid increases in the use of LLPs, tubewells and fertiliser, these have started from a low base so that approximately 6.5% of cultivated area is irrigated by LLPs and 1.2% by tubewells. Total cropped area as a proportion of net cultivated area moved from 127% in the early 60s to nearly 150% in 1976 (Bureau of Statistics 1977). The inescapable conclusion from this is that rice production has not been greatly affected by irrigation and use of other inputs, and that higher production still depends on good rainfall for the traditional *aman* crop.

Significance of Co-Operative Membership

More especially in the context of IRDP for 250 *thanas* with a related expansion of TCCAs and KSSs, several studies conclude that the differences in yield between members and non-members for the same crop are not significant as long as non-members have access to irrigation; and that KSS membership *per se* is less important than the widespread availability of HYV package of inputs (Blair 1978). Even assuming an early success in Comilla with the rapid expansion of IRDP the KSSs in particular and TCCAs have become dominated by surplus farmers who monopolise institutional credit capture irrigation and are the largest loan defaulters (Islam 1978; Khan 1971; Stroberg 1977).

Subsidies

There seems to be an important relationship between credit discipline and the maintenance of subsidies. The rapid expansion of the TCCA/KSS system was accompanied by a sharp decline in credit discipline among the landholders especially (the proportion of loan outstanding from rice production in the RD-I *thanas* in 1977 was 79.7% at the beginning of March 1978—Sinali Bank) and KSS savings were correspondingly low. As a result, the programme became heavily dependent upon high input subsidies, thus undermining the ideology of self-supporting co-operatives which originally lay behind the scheme. (Blair 1974). But the picture on subsidies has to be set against overall government investment in agriculture. Fertiliser subsidies are at 48% (BADC 1977 10:11), capital costs of irrigation have always been completely subsidised and recurring costs subsidy for LLPs is 77%, for DTWs, 88%. In 1976/77 fertiliser subsidies represented 17% of total expenditure on agricultural development (CPRB 1978 p.82 and Table 13) but there was a declining proportion of public sector development expenditure (34% in 1973/4 to 29% in 1977/8) devoted to agriculture, rural institutions and water resources (CPRB 1978 Table 1.8). Even less attention has been given to the appropriate technology aspects of agriculture which might be expected both to assist the poorer farmers and provide more employment. Manual and low cost methods of irrigation receive no subsidies at all, and Edwards argues that they are in effect taxed because of import duties on raw materials (Edwards *et al* 1978).

Landlessness

Increasing attention by observers and some officials to the plight of the landless has stimulated criticisms about the extent of their exclusion from IRDP and other public expenditure programmes. We hardly need reminding of the grim statistics—the proportion of households in rural Bangladesh with only homestead land or none at all in 1978 was 43.47% and a further 31.29% of households had less than 0.5 of an acre. With considerable definitional caveats, both the East Pakistan Censuses of 1960 and 1961 offer a figure of 17% for the landless, indicating that we can be sure there has been a dramatic growth in the numbers of landless and near-landless. The increasing supply of labour through the rise in landlessness, partially caused by the subdivision of holdings which reduces the demand for landless labour, contributed to a decline of 50% approx. in the real value of agricultural wages between 1965 and 1975 (Khan, A.R. 1977 p.152; Alamgir 1974). Furthermore, Islam concludes from a review of studies that:

> "...although there is some evidence of increased labour demand consequent on the introduction of new technology, it is clear that it has failed to make any impact on the real wages of agricultural labourers." (GPRAB, Min. of Ag. May 1978. p.130).

Public Works Programmes

It is true that the Rural Works Programme (RWP) and the Food for Work (FFW) programme have contributed to rural employment by, for example, generating 100 million man-days of work in 1976/77 (Institute of Nutrition 1977a), and in the process have created considerable physical infrastructure. Rural Works is an important part of A.H. Khan's proposals, and several findings should be carefully considered in this connection. Due to local corruption the estimates of man-days generated are probably exaggerated. Those most in need do not always benefit: according to the Institute of Nutrition Survey findings (1977b) 25% of RWP workers in Comilla owned more than 2 acres, and 45% of workers in the FFW programme (designed solely for the destitute) had alternative means of support. The seasonal nature of rural works—dry seasons when there is certainly a need for work —cannot easily cope with the needs say of perenially flooded families during the rains. Finally, the operation of these programmes has been dominated by the local rich acting as brokers and contractors with government and agencies seizing the opportunities for an extension of patronage and a new form of economic diversification. This process is reinforced where there has been no prior organisation of the rural workers involved in the programmes, and where funds are channelled through *thana* and district councils, which are dominated by richer farmers and local entrepreneurs (Alamgir 1977). Perhaps this is the inevitable price of achieving higher levels of agricultural production (on the lands of the rich) as a result of the infrastructural provision.

Comilla Formula: Appropriate for other Regions?

In many of the criticisms of the replication of Comilla through IRDP, there seems to be a key element which has been overlooked—the possibility that the institutional structures and programme content which emerged and took shape over a decade in Comilla *Kotwali Thana* were inappropriate when transferred to other regional situations. (Even in Comilla, of course, the cooperatives have been frequently criticised by the founders and members of BARD for their domination by local elites, and the tubewell/ inputs diffusion programme for its continuing dependence on subsidies (Faidley & Esmay 1976).

Regional Statistics on Landholding

It is interesting to note that despite the considerable number of rural surveys in Bangladesh on landholdings and tenancy, there is very little published presentation of regional statistics. In the Land Occupancy Survey of 1977 commissioned by US AID the nearest we get to regional data is a series of tables based on seven Dinajpur Sample Villages and seven Chittagong Region Sample Villages with no further indication of location or characteristics. The summary report of the 1978 Land Occupancy Survey also aggregates the data to the national level. The exception to this pattern is the publication by BARC (1978b). It made no pretence to a rigorous sample, with the limited resources at its disposal, and the report wisely cautions about the reliability of its data (a refreshing admission in itself). The results from the 10 (out of 20) agricultural districts presented in the report are tentatively revealing, because the Comilla region is far from typical. Comilla District shows the smallest proportion of landless labourers (26.1%), the smallest proportion of sharecroppers (13%), the largest number of owner-cultivators by a large margin (54.6%) and the smallest but one proportion of big farmers (more than 5 acres) (6.3%). Dinajpur by contrast has 31.1% landless, 25.6% sharecroppers, 28.5% owner-cultivators and 14.9% big farmers. Khulna and Rajshahi display similar patterns to Dinajpur.

It would be unwise to attach too much significance to those patterns because of the nature of the data-base, but these variations between Comilla and other areas do conform to other pictures of regional variations gleaned from village studies (Adnan et al 1974-78; Arens & van Beurd 1979), location-specific research and some historical data (Wood 1977). On this basis, one may speculate upon the significance of the variations in agrarian social relations, only hinted at in these figures.

The Peasant Economy Error

Until recently, the two major contemporary sources of understanding agrarian relations in Bangladesh have been aggregated data collected during the 60s and 70s, and the research based on Comilla. These sources have produced two generalisations: the absence of a substantial class of landlords and rich landholders, the low level of absolute average size of holding and the predominance of owner-cultivators (74.5% of cultivating population in Comilla *Kotwali Thana*, and 13.2% for the district (BARC 1978, p.20)). From this and the aggregated data it has usually been concluded that "the rural economy of Bangladesh is best described as a peasant economy based on small family farms operated primarily with family labour" (Abdullah & Nations 1974, p.9). However, if we consider the Dinajpur picture, only 41% are owner-cultivators conforming to this 'peasant economy' description, 37.9% are sharecroppers and 21.1% are big farmers (although this latter category in the case of Dinajpur may be a little misleading (BARC 1978b, p.30).

The Comilla programme of co-operatives was based on the assumption that class divisions within the 'peasant economy' were structurally insignificant. Would it be fair to say that both the original plan to expand Comilla in the form of IRDP and the current one for a revival of the Complete IRDP is based upon a similar assumption? The analysis of the Comilla area contained further assumptions: that while surveys indicated that the KSS's were dominated by rural elites, these elites did not constitute a class since the division of holdings between sons ensured a process of cyclical mobility or indeed 'cyclical kulakism' (Bertocci 1972). The more pessimistic view of this process is that there is a steady process of fragmentation for all classes—steadily producing more landless, marginal farmers and weakening the position of the richer ones. This process is referred to by A.H. Khan as the collapse of the entire agricultural system, of which the landless are a symptom. Although by no means conclusive, other village-study work shows that even in Comilla *Kotwali Thana* these propositions do not hold—that larger farmers acquire through land mortgage arrangements and debt relations, and the opportunities provided both by superior landholding and better access to development agencies have encouraged this group to diversify their economic activity into the sphere of exchange (rather than production) and professional employment (Wood 1976, now accepted in private communication with Bertocci, and reinforced by the findings of Martius who worked in four other villages).

Thus even, the traditional view of agrarian relations in Comilla itself with its high number of owner-cultivators, and lower numbers of landless and sharecroppers is being challenged. On the basis of regional data, this picture of agrarian relations with Bangladesh must be even more inaccurate when applied elsewhere—especially in the North and West of the country.

Dimensions of Regionality

I have attempted elsewhere (Wood 1977b, pp 2-11) to trace the existence of significant regional variations in the relations between classes, referring to historical data, and do not intend to repeat that analysis here. Part of the argument is that in general, rural class relations in Bangladesh have been characterised by the use of capital in the sphere of exchange rather than production, but that the precise forms of that capital are structured by particular historical conditions and the related variable effects of especially: soil fertility, structure and elevation, but also flood vulnerability, drainage and waterlogging, rainfall, temperature, population density and structure, cropping systems, pattern of livestock ownership, migrant settlers on marginal land and so on. The critical effect of all this is on variations in the incidence of formal tenancy, the use of non-family labour, productivity and land values (irrigated land in Dinajpur is valued at between Tk. 10,000- 15,000 per acre and is probably four times that now in Comilla), nature of indebtedness, patterns of migration, scale and direction of diversification from agriculture, intensification of agricultural production activity, family size, lineage structures, receptivity to various forms of technology (with different capital-labour ratios).

The picture has emerged that the North and the West have more landlords, sharecroppers and landless labourers than the East; that landlords have better access to inputs than owner-cultivators but cultivate intensively on smaller proportions of their land; that sharecroppers and owner-cultivators are being slowly dispossessed of their land; or forced in the case of sharecroppers to cultivate higher proportions of marginal, less productive land in their holdings. Furthermore, it is not correct to argue that all categories of landholdings are fragmenting steadily, and reducing all classes to the lowest common denominator. The holdings of poorer classes are fragmenting and holdings are mortgaged, eventually transferred, or sold as those classes descend the downward spiral. This land is being transferred to other classes and they are surely not also descending. So, can we observe an opposite process of consolidation and concentration? Many of these statements are still hypotheses, with the relation between tenancy, productivity and marketable surplus requiring further investigation. But a general picture of regional variations of this type cannot be ignored and the principle of absorbing these dimensions in policy and planning activities must be established.

Replication of Comilla—the RD-I Test

If we return to the issue of replicating Comilla through the IRDP, it must be conceded that IRDP as originally conceived has not been given a fair trial in Bangladesh. But the question must now be whether the political, administrative, social and economic preconditions exist for the revival of the earlier strategy without modification. By drawing attention to the dimensions of regionality and especially the variations in landholding patterns (size and tenancy) even the attempt to apply a single project formula loses its validity, let alone the contemplation of a revival. Even if the Comilla model was accepted as valid for Comilla, two sets of difficulties are encountered in its replication: administrative/ extension manpower resources and an exaggerated pattern of rich farmer domination over KSSs.

First, it is often argued both for and against the replication of Comilla into IRDP that the speed and partial character of its reproduction (involving inadequate training, lack of credit discipline, scarcity of specialist personnel at the *thana* level and administrative capacity generally) was responsible for the weakness in the programme, and thus was not a fair test. This pace must have been a contributing factor in the emasculation of the programme, but the question remains whether the Comilla model was in fact too costly in terms of the manpower resources (mobilisers, specialists, administrators, evaluators) required for implementation. The essence of Comilla, as I understand it, was the patient mobilising of farmer groups, educating them as to the economic opportunities which were available to them if they mobilised around some lumpy investment. If so many staff and more than a decade were required for Comilla, why should not the same apply elsewhere in Bangladesh? If the justification for the investment in Comilla was that it was innovative, and required understanding of a detailed interaction with the local social-ecology, then surely the same argument must apply (if to a slightly weaker extent) when attempts are made to transfer the thinking to new situations? In this respect it would be more helpful to talk of the *Comilla Style* rather than the Comilla model—that is a method of approaching rural development problems, identifying and resolving them through constant re-adjustment and monitoring of performance. These are the enormous strengths of the Comilla experience which must be retained in new policies.

Secondly, the distinctive feature of the Comilla method was the relationship between the analysis of agrarian relations in Comilla (erroneous though it may have been) and the devising of a response—viz. the TCCA/KSS strategy. The apparent homogeneity of land-owning classes led to the belief that farmers would join together on more or less equal terms to form a co-operative organisation in their own mutual interests. Leaving aside any problems which might have appeared at Comilla, how realistic is this strategy in districts like Dinajpur where owner-cultivators might constitute nearer 28% of the cultivating population instead of the 55% in Comilla, and the proportion of sharecroppers is much higher too? The entire pattern of dependency and patron-client relations is of course determined by these differences in access to land and landholding status. These relations are reinforced by the chronic indebtedness of the majority of the rural population, and occur between close kin members as well. (richer kinsmen will assist poorer *bari* and lineage members with relief— thereby reinforcing dependency—but not with development assistance).

With these two sets of general problems in mind, it is useful to comment on the experience of the IBRD's RD-I project (started in July 1976) in seven *thanas* of Mymensingh and Bogra. Its relevance stems from the attempt to replicate and strengthen the Comilla approach with the assistance of $16 million additional funding.

> "The project would strengthen a number of production-oriented ongoing rural programmes in project *thanas*. To ensure that benefits reach the rural poor, physical inputs and rural credit provided under the project would be channelled through the Comilla type of co-operative (TCCA-KSS). A major effort would

be made to improve the operation and management of these co-operatives. Technical services to the co-operative members would be strengthened. The project would comprise rural works; minor irrigation facilities; *thana* facilities; rural credit; strengthening of rural institutions and services; technical assistance, and monitoring and evaluation" (IBRD—No 837 - BD 31/10/75, p.9)

Despite attempts to involve the BIDS, by March 1979 no independent evaluation had been conducted. However, in addition to the reports of personnel involved in the project, the reactions of an IBRD review in April 1979 are interesting—they are summarised here from a longer discussion.

District and *thana* committees have concentrated more on rural works to the neglect of institution-building and production-oriented components. There has been a severe lack of continuity in staff due to frequent transfers, undermining morale and contributing to staff ignorance of and lack of commitment to the programme. Salary and status does not match up to workloads so that the appropriate quality of staff is not attracted. Data collection is haphazard, unanalysed and not fed back for guiding policy and implementation. A large accumulation of unspent funds (as of 30/6/78, 70% of total releases by GOB was unspent) attracted interest while other projects in the country were starved of local currency funds. Interest earnings were unaccounted for; monthly statements of expenditure were at best eight months late. Project funds for hand-pump credit were misused and the whereabouts of collections unknown. Sale receipts from BADC facilities distribution (funded out of an IDA allocation of $1.6 million) were unknown, and no decision was taken on use of income. Auditors for the financial years 76/77 and 77/78 have ignored these and other financial irregularities.

On rural works, data on progress is inadequate, often totally unavailable. Overall progress is estimated to be at least one year behind schedule. Cost estimates and scope of works have deviated from provision without the knowledge of IIQ management; and especially for *katcha* works a familiar pattern of inflating quantities of work completed and payments made by supervising staff without actual measurements. (A further commitment to a programme of *pucca* road construction is worrying as the benefits are likely to be highly skewed in favour of richer farmers).

On agricultural extension, the T&V system has been introduced into the four Bogra *thanas*. According to a local FAO informant, this has led to considerable confusion although the review mission refers only to some conflict between the T&V approach and the model farmer approach of IRDP. As a way of reconciling this it is suggested that model farmers be automatically included among the contact farmers of the T&V; it is further believed this will strengthen the co-operatives. The FAO informant is highly sceptical, believing that T&V functions to undermine co-operatives precisely because the extension workers are monopolised by model farmers and others who inevitably belong to the richer groups. This must be further reinforced by the review mission's observation that extension staff do not work outside 'office hours' when most KSS meetings are held. The T&V system is based on contact with individual farmers rather than groups of them. Contact between extension and research is weak on agricultural practices, (and is obviously non-existent on other aspects of cultural practices and social relations)!

There has been little compliance with project procedures for the allocation of LLPs to pump groups. The implication from the review mission's report is that LLPs have been going into richer, private hands whereas the intention has been to maximise pump coverage and increase the participation of small farmers. Overall the minor irrigation activity based on STWs, handpumps and LLPs has been low partly due to procurement problems. But the criteria used for allocation has been questionable.

Finally, the most telling set of criticisms refers to the experience with co-operatives. Despite intensive efforts under the project, less than 30% of the farm families in the project area are covered by TCCA societies, and the participation of the TIP societies in TCCA is very limited. Less than 5% of the farm families in the project *thanas* are receiving production credit under the project. Some of the reasons given for this include cumbersome lending procedures, ineligibility due to previous defaults, and over-emphasis on short-term production loans. Small farmers remain dependent on moneylenders for the crucial medium-term loans for irrigation, livestock, etc. Recovery performance on loans is no better than in any other 'normal' IRDP *thana*. Accounts are poorly maintained at KSS level, and long delays in auditing then occur. The use of annual production plans (instead of crop production plans) as the basis of annual credit plans has not been successful, as they are badly prepared on questionable facts and assumptions. The preparation of land registers for many of the KSSs are intentionally or unintentionally inaccurate and certainly there is no appreciation at *thana* level of their potential use in extension works to estimate input and credit requirements, and allocate pumps. Irrigation ledgers for KSSs to protect members from overcharging/ undersupplying were not being maintained. management of fertiliser dealerships by TCCAs is unsatisfactory—there is no forecasting of demand and insufficient notice to BADC to ensure timely supply. There has been no significant improvement in training programmes for model farmers despite earlier requests from IDA.

Commercialism and 'Relief'—the Dual Alternatives to IRDP

Thus RD-I as an attempt to replicate Comilla on a modest scale with lavish resources, nevertheless is described in these terms by an 'in-house' review mission. The longer-term effects on the pattern of rural class formation have yet to be firmly established but it is not difficult to guess at the final general conclusion. Furthermore, these criticisms draw attention to the problems of running such a sophisticated, elaborate, integrated programme with large funding, at speed with limited staff resources in terms of experience and training. It seems clear that considerable leakage of funds—corruption, patronage, access problems, rich farmers' dominance, entrepreneurial opportunism are among other possible descriptions—is occurring. The rural poor are not only unable to benefit, they are witnessing a realignment of class forces—involving the strengthening of alliances between rich farmers, petty landlords, local contractors and elements of the bureaucracy—which is further removing them from any access to state/foreign aid disbursements. As a tool for reaching the rural poor the IRDP formula in Comilla had its limitations, elsewhere it is a disaster even under the favourable, well-lubricated conditions of RD-I.

With its discovery of 'basic needs', its support for 'target groups', and this recent commitment to the T&V extension approach, the Bank 'experts' seem to be recognising the limitations of IRDP. Perhaps as yet unconsciously, their thinking is drifting towards the alternative trend noted in the introduction—input-oriented agricultural policy involving a specialist agricultural service (T&V) to individual farmers and mobilisation of funds through commercial banks and separate 'basic needs' programmes for poor 'target groups'. It could be that there is some sense to this approach within the non-revolutionary framework of the dominant political forces in Bangladesh and their international allies. But it would involve an acceptance of the rise of a special kind of Bangladeshi *Kulak* (probably happening anyway) with a recognition that the dependence of such a class on state support (subsidies, credit, interest rates, pricing and procurement policy) ought to be phased out as more investment was made in the productive opportunities of other rural classes. For those of us who are unsure about 'urban bias' analyses, this seems an unlikely outcome. More likely is a short-term holding operation in which 'basic needs' proposals for poor 'target groups' attract more interest than hitherto from donors; a watered-down version of IRDP involving an expansion of T&V is tried; a curtailing of BADC activity comes about as the procurement and distribution of inputs is increasingly managed by the expanding private, commercial sector; and, as now, state support for the agricultural sector continues, leaving the distribution of the benefits to the contending classes.

One alarming aspect of this process is that all pretence at rural planning in Bangladesh might disappear—that a weakened IRDP might be mechanically pursued (stale air to fill the vacuum?) along with a steady expansion of the T&V approach regardless of local conditions; that this will be accompanied by an *ad hoc* procedure of generating and allocating 'basic' projects to fulfill budgetary and regional quotas: and that finally in the process the role of rural outsiders will become more powerful, with lip-service paid to the possibilities of participation. It is the fear of this prospect which guides the remarks which follow.

Generation of Services: Methods and Style

In the context of the current foodgrain crisis and the hastily-mobilised response to it, impatience with more abstract discussions about longer-term strategies is understandable. But at least part of current vulnerability can be attributed to a failure to involve groups of landless and poor peasants in viable productive activities as well as past failures to distribute minor irrigation facilities on a widespread basis. For any kind of long-term strategic response to past failure and to the probable repetition of natural crises, more attention must be devoted to the methods of providing or generating services. Under conditions of limited resource and administrative capacity the methods must involve greater participation; more flexibility in interacting with specific sets of socio-economic conditions; and an awareness of the significance of regional variations to the point where the style of extension involves more learning and less imposition, and where institutional experimentation is encouraged.

Building Organisations is Necessary

It is clear that the learning process by which rural people put their faith in each other and their mutual organisation to provide basic needs and security rather than the private ownership of means of production, should be stimulated. Where the rural class structure is highly differentiated, however, caution is advised, since there is a danger that the rise of a class of rich peasants/ capitalist farmers will be further encouraged. With the enrichment of this class and the resulting pauperisation of the rural and urban poor, a number of development objectives become disturbed and undermined—not just the distributional ones, but the so-called 'production-oriented' ones too. By recognising that this hierarchy of agrarian social relations exists with more or less intensity in different regions of Bangladesh it is essential that the TCCA/KSS framework should not be universally applied without first considering its viability in a particular context, secondly, experimenting with modifications to suit local conditions without surrounding the overall objectives; and thirdly, assessing the pace at which institutional innovations might be introduced into an area.

Area Planning not just Implementation

The conclusion from this is that we must recognise a distinction between area *planning* and area *implementation*, and further recognise that apart from the Comilla style, we have had mainly area implementation in Bangladesh. In this respect the decision to revive some version of Thana Plan Book of the 1960s should be welcomed, although again we must appreciate that even the proposals for physical, infrastructural projects are likely to be mediated through the local power structure.

An Exercise in Regional Zoning

The problem with such a notion of area planning is the demands placed upon scarce skills resources: scientific, social scientific, engineering, administrative and extension, could become enormous in terms of person-years, and unrealistic in terms of qualifications in the present context of Bangladesh. In such a situation, compromises are necessary. One useful, practical suggestion at this stage would be to commission an interdisciplinary team (Bangladeshi and/or appropriate foreigners) to attempt an approximate regional and non-regional zoning of Bangladesh in terms of some of the variables noted earlier. A considerable amount of relevant material which could be organised for this purpose by asking 'regional' questions of the data and findings already exists (BARC 1978a & 1978b). The intention would be to develop an awareness of the relationship between physical characteristics (the importance of soil structure cannot be over-emphasised), geographical location, agricultural practices and performances, non-agricultural economic activities, landholding and settlement patterns, kinship structures, labour and dependency relationships, population density, fertility, family sizes, nutritional levels and so on. In one sense this is of course a tall order—an entire social ecology of Bangladesh. However, relevant experience exists within Bangladesh through, for example, the work of the BIDS Poverty Study, and the process of zoning in this way will obviously begin with approximations, and continue as an indefinite process of refinement.

Institutional Experimentation

This zoning exercise would in the first instance provide more explanation of agrarian behaviour—both social and technical. Secondly, the zones so identified might constitute the level at which institutional experimentation should be conducted. It must certainly be recognised that to proliferate the number of 'unique' and 'specific' regions would complicate any approach beyond recognition—ten flowers might bloom but not 100. There is also a genuine problem of the relationship between the central administration of rural and agrarian development and a myriad of location-specific institutional forms.

Another way of looking at this problem is to conclude that regional zones in the sense of coterminous space do not exist, but that types of social-ecological situations exist even if they cannot be grouped together on a map. Such a typology, specifying certain kinds of indicators (for instance cropping patterns, labour relations) might then sensitise an official or 'rural mobiliser' towards particular forms of institutional experimentation, and dictate the pace of progress.

In either case, the intention should not be to prescribe in advance a particular institutional form for a particular situation—the Comilla style does require time for a process of institutional innovation. On the other hand, a prior zoning or typology exercise might assist in the early elimination of potential alternatives. Here are some examples: co-operatives of shared interests cannot contain landlords and their sharecroppers; when this relationship is prevalent, try something else. The manager and model farmer 'extension' strategy has a better chance of working in less differentiated situations where crosscutting linkage ties also effectively exist; but in highly differentiated situations managers and model farmers are drawn from the rich and serve the rich. Several conclusions are possible in the highly differentiated situation: have two model farmers, one from the poorer groups; supervise and audit the management more closely, or place more emphasis upon using official extension agents especially for poorer groups. The same kind of thinking should apply to many of the 'critical input' activities mentioned by A.H. Khan. The arrangements for drainage and water management work might be through co-operatives in some situations but be captured by landlords-cum-contractors elsewhere. In this case, there is a choice between official outside management, or through investing more time in building up groups of workers to run their own commercially viable public works contracting agency. If KSSs are envisaged as commercial groups in the private sector, why cannot the same apply to groups engaged in rural works?

No More Target Groups

At this point, it is appropriate to raise the issue of 'target group' strategies. The 'basic needs' atmosphere combined with conclusions about highly differentiated rural societies has led to programmes directed at particular groups of poor *defined by outsiders* in terms of some criterion such as landholding status or nutritional levels. There are several problems here which prevent such an approach being seen as an alternative or simple addition to farmers' co-operatives First target groups are, as has been said, generally defined by outsiders and do not necessarily reflect a meaningful unit of social cohesion. As a result, the programmes in which they are involved tend to be directed towards them as individuals, with often a high relief content. Secondly, target groups consist of individuals who are members of classes and as such are in relation to other classes. Any programme constitutes a tiny proportion of their pattern of social interactions, meanwhile all their other relationships of dependence, poverty and underdevelopment continue, impacting upon the programme. Subsidies are passed on as indirect subsidies to their landlords, unless the cost-share and produce-share arrangements are altered as a precondition—although perhaps allowing some incentive to the landlord to prevent the dispossession of the tenant. Credit to dependants of any kind for production gives creditors the opportunity to foreclose and receive payment. Is it feasible to offer consumption credit in such situations, in order to break the cycle of indebtedness as a precondition for the successful use of production credit?

The Extension Function

Thirdly, the notion of 'target' implies recipient, and schemes *for* people not *with* them. For productive activities to be self-sustaining for any group (and this was certainly a principle of the Comilla style), the group itself only comes into existence as the result of a shared conception of economic opportunities; the group makes a commitment to a particular activity, and participates in all stages of the subsequent creation. Without this involvement and consciousness, the activity is removed from their control, and it is either bureaucratised or dominated by a local contractor/ entrepreneur/ tout. Both these tendencies are routes to alienation. If this is accepted then extension work becomes more than a technical, advisory role —it involves patient analysis of local situations, assessment of local opportunities either in agriculture, bringing together potential groups based on existing horizontal ties and discussing appropriate proposals for work. Only in this way will groups (especially those without initial control over any means of production) ultimately control the activity—through identifying their own needs and opportunities and participating in design and management, training and support from extension agents are required at those stages.

Phases of Co-operative Activity—Start with the Simple

It is also important to distinguish between phases of co-operative activity, perhaps for any group but certainly for non-cultivators or under-employed cultivators. Some schemes are clearly more complex and difficult to manage than others, and require more capital. Untrained, illiterate co-operatives have to operate within the limits of their capacity—current examples are: rearing of chickens and goats; fish tank clearing; cultivation of land by taking it on mortgage (with supporting loans); rice processing and selling, and other forms of petty trading which would be financed at low interest rates from TCCSs. Other larger schemes such as rice milling, handicrafts, require larger capital outlay (which should involve some of their own equity capital), management skills and marketing knowledge. Furthermore, large schemes might be too lumpy for single co-operatives and require federations so that the activity is located at the appropriate level —e.g. brick kilns, manufacturing operations etc.

There is a further area of activity—servicing agriculture—which raises different issues. So far the government bureaucracy—in particular the BADC—has monopolised this activity. Of course in many societies, agriculture has directly provided employment for a population far in excess of the number of cultivators and agricultural workers through the provision of services. This is a great strength of A.H. Khan's proposals and can be extended—co-operative groups should be involved in the entire management of water resources (generation, delivery, maintenance, sale, and continued generation with surface and ground water systems); the distribution of inputs to cultivators; bullock hire; even the extension of credit (through employment in the TCCAs), certainly marketing and processing. However, it is important to remember that the transfer of this activity to the private sector in effect constitutes a challenge and threat to the process of bureaucratic expansion (which as a form of employment is wasteful without being labour-intensive). Furthermore, servicing agriculture is likely to involve dependent classes in direct and continuous interaction with the richer, cultivating classes who are their patrons, landlords, creditors and so on. Therefore, the tested unity of their own groups through smaller, earlier, more independent economic activity becomes an essential precondition for this later phase.

Rapid Rural Appraisal

This discussion of zoning and typology exercises, institutional experimentation, extension functions of mobilisation, multipurpose management advice, and assessment of local capabilities clearly has important implications for the development of rural cadres of mobilisers and extension agents. There has been a tendency in Bangladesh to operate a strict division of labour between rural researchers and extension workers—only the staff at the Academy in Comilla once combined these roles. Their training activity lacked one vital component—namely that extension agents should have dual roles as mobilisers and `students` of their local environment. However, it is possible to think of quick methods of trying to understand a local situation in its social and physical dimensions, without asking all extension officials to become anthropologists. Careful use of local statistics, sensitive case study work, small purposive samples, and so on—supported by *thana* or subdivisional workshops, probing questions of each other and maybe visiting Academy staff—would produce many insights.

Conclusion: for Whom is the *Thana* Rented—the People or the Planners?

This emphasis on style and methods challenges the usual formulations of frameworks and projects. At the very least it encourages more participation in the creation of any particular framework and moves us away from the universal formula. It seems necessary to challenge constantly the concepts and vocabulary of development planning. Terms like 'integrated', 'area development', 'project' and so on, all carry assumptions of rationality and appropriateness whereas the reality is different. This has surely now been demonstrated for IRDP in Bangladesh, and its various spin-offs. But although RD-I has been found wanting, is the same critical examination of similar projects and proposals taking place? Are the various 'area development' proposals the result of paperwork and hypothesis generation in air-conditioned offices, or are they the outcome of participation and a process of interaction between ideas and behaviour? For whom is the *thana* rented? What definition of 'area' is employed—the administrative one of *thana*? Or is 'area' a variable concept according to the appropriate spatial limitations of particular activity? Does 'integrated' refer to an organic package of mutually dependent factors? Or is it a reference to the integration of rural poor into the mainstream of development? Or is it merely (and more usefully) a reference to a looser combination of activities (not interdependent) where phase timing and consistency could nevertheless increase the marginal rate of return on any particular activity—that is, could constitute a bonus? The idea of a project (especially an integrated one in the package sense) presumes a knowledge of the likely effect of one activity upon another. Indeed it assumes an ability to predict the functional relationships between a large number of complicated variables before either knowledge of their nature exists or before the interventionist event has occurred (that is, the implementation of the first stage of the project) which will set in motion all kinds of responses and alter the course of other behaviour. The extent of knowledge required (for a project area) is rarely found anywhere, and certainly not in Bangladesh. The notion of the project must be made more tentative particularly when it purports to be 'integrated', 'area', and comprehensive chain reactions to external intervention cannot be predicted without more use of pilot experiments—part of the extension/learning activity of field officials working alongside peasants and labour. However, it is unlikely that the T&V approach will operate in this way, since it is based on the outdated assumption that a specific agricultural service is desired, consisting of a package of exotic input and technology whose efficacy has been determined prior to any appreciation of a cultivator's position in the social relations of production and exchange. However, the chain reactions to more modest, unintegrated interventions can only be truly assessed after the event, and such knowledge only then applied elsewhere where conditions at least appear similar (perhaps in the same region, or sub-region).

At the same time, while this invective against the contemporary vocabulary of rural development planning appears to strengthen the case for decentralisation, a note of caution is necessary. It is important to remember that too little or too much administrative control can have a similar result. A high degree of administrative control merely strengthens the interests of dominant classes which are likely to be represented by the State and its apparatuses. On the other hand, too little control simply means that locally dominant classes are able to capture the resources which they should be distributing more widely. This inconvenient observation raises the familiar problem of accountability. If the outcomes are similar whether the provision of service is decentralised or not, because the power of certain classes is articulated in both ways (albeit through different institutions with different effects), then the problems for the poor are ultimately similar as well. In neither strategy of provision are the procedures and outcomes accountable to the poor—they do not constitute a social force of any political significance as they are divided and unorganised by virtue of their structural position in the social relations of production.

Finally and more hopefully, perhaps (although only with the assumption of the benign state) it must be emphasised that this discussion of frameworks, methods and style must not unduly focus our attention on institution building, even of the looser, local kind. The relationship between institutional strategies and substantive policy content must be considered carefully. In particular, much thought should be given to policy areas such as food-grain imports, input subsidies, pricing and incentives, distribution and procurement systems—their interrelations and significance for the kind of co-operative activity which is to be encouraged. These policy items have crucial implications for the process of class formation, and may operate in a contrary direction to other efforts. Certainly it requires very close coordination between at least the Ministries of Agriculture and Rural Development. The objective of expanding agricultural production has to be reconciled with a programme of rural mobilisation, and *vice-versa*. In many of the discussions, for example about reducing food-imports and allowing domestic prices to rise as an incentive to increase marketable surplus—the problem of net rural purchases of food-grain seems to have been overlooked. The relationship between such a policy outcome, regional variations in areas of net exporters of food-grain and forms of co-operative mobilisation is the meeting-place of area and national agrarian planning.

Chapter 4

Women and Gender

Introduction

It is too easy for liberated Westerners to become over-despondent when analysing the position of women in societies like Bangladesh. As the academic literature on women in poor countries grows, so it does in Bangladesh, and professional males concerned with development become daily more aware of the significantly different sets of chains which confine all women, but especially the poor. It has been interesting to observe over the last two decades in Bangladesh how prominent women have been in struggles at all levels of society, despite that society being typified by many outsiders as among the worst examples of female oppression. There is an apparent paradox between the view which dwells upon female subordination and the evidence of female social action. While not wishing to deny the systematic disadvantages which face women, especially rural and poor, this paradox arises from the way the initial questions are asked about women. Much commentary proceeds from the question 'How are women constrained?', instead of asking 'What do women do?' The first treats women, *a priori*, as passive; the second as active. These questions were posed in a section in *Breaking the Chains* (Kramsjo & Wood 1992), arguing that we should be concerned with the second form of question rather than the first.

Female Power and Cultural Norms

Nevertheless, the structural position for most women in Bangladesh determines particular cultural forms for the female exercise of power, and establishes the consequences for women who move out of these cultural norms. When men break chains, they can do so in a way which is consistent with the gender norms about men even though they will be confronting other cultural rules about deference: that is, within kin and lineage hierarchies; perhaps against elders; the undermining of traditional norms of respect for educated, professional and officer classes; and of course religion. Indeed, in a certain sense, successfully struggling men are displaying a virility and may earn on this dimension the grudging respect of their opponents. Such behaviour can be seen as confirming an aspect of masculinity, namely, an aggressive stance in the public sphere. Not so for women in a society where the cultural norms, policed by men and acculturated women in the interests of men, define femininity *inter alia* as quiescent, supplicant, nest-building and caring for family members.

Women do exercise power within this definition, and the significance of this power needs to be acknowledged. Sarah White's (1988) study in Rajshahi, was counter-intuitive in this sense, exploring and identifying arenas of power for women where none had been assumed previously by studies written from a Western feminist perspective. Her basic argument is that while the patriarchal thesis draws heavily upon a distinction between public and private spheres in the observation of gender behaviour, we cannot assume that the physical location of behaviour is indicative of who exercises power within these locations. Since we have always known that males participate in the decisions and the allocation of work and resources in the private sphere where some women only reside, should we assume that men are the only ones to participate in the public sphere simply because women do not reside there? In addition to their labour and management responsibilities within the private, family sphere which is the source of much *de facto* power, women who venture out of the homestead only in *purdah* nevertheless engage in market behaviour. They do so both directly through trading in minor livestock, barter and credit to neighbouring families, but indirectly, and importantly, through decisions over land use, buying and selling of land and major livestock, landleasing strategies, government or private loan options, trading produce, investment in business or agriculture, and so on.

However, even with this argument, White would acknowledge that a woman can only exercise such power by remaining apparently within the cultural norms of femininity, acting out the patriarchal rules, playing the game. The power cannot be seen to be hers. This proposition raises at least three issues of interpretation:

- Is this 'acting out' consciously instrumental behaviour by women who have broken the chains in their own minds but are not yet sufficiently confident of success to do so publicly? Women in many societies are forced into hypocritical behaviour of this kind, deliberately stroking the collective ego of the male community rather than risking open confrontation.

- Are the patriarchal rules so internalised, are women so acculturated, that no other route to power can offer itself within the realms of normative possibility? Some might argue that such adherence to a dependent exercise of power is indicative of false consciousness, but surely this is a proposition which can only be asked of one culture by another outside it.
- Can we accept that women's conformity to these cultural norms of femininity is as much functional to them as constraining? That is to say, their lives are organised, albeit within a framework of apparent dominant male interest, as a reflection of key imperatives in the biological, labour and social reproduction of society in which the management of female sexuality is central, is consequently highly valued, and is therefore acutely insecure, requiring physical and cultural rules of protection to avoid the prospect of greater female violation.

These questions can only be resolved empirically, and are posed here to distinguish between women in different situations, with varying sets of objective conditions and related subjective perceptions. The key principle to understanding the character of gender chains is that the cultural norms surrounding the behaviour of women are contingent both for men and for women; and, in this context, there are distinctions to be made between those who break the norms, those who tolerate the breaking of the norms, those who uphold the norms, and whether such upholders are significant or not. It is here that class and gender interact. Under changing material conditions, affecting, of course, the poorest families first, patriarchal rules about women working outside the homestead, that is, the private sphere, break down quickly. This is especially the case for female-headed households. Thus women work as domestic servants in others' households, in the fields as agricultural labour (perhaps increasingly invading male spheres such as harvesting, though not yet ploughing), and in rural works schemes; or they migrate to towns and cities for domestic service or industrial work such as garment manufacture. All of these options represent varying degrees of breakdown of the cultural norms prescribing preferred behaviour for women since mobility and increasingly unpoliced contact with non-family males will occur.

However, in terms of breaking the chains, it is important to appreciate the reactions to such behaviour for poor women. Such behaviour is the outcome of family stress, it is forced upon the family by material circumstances, and in most cases will involve decisions by men as well as the women; indeed, in many cases, men may be encouraging reluctant women. In such situations, there is a sense in which cultural norms are being suspended rather than broken, since even the actors might prefer not to break them. The wider society can tolerate such a suspension, even though it is likely to be permanent rather than temporary, because of its enforced origins. Such suspensions are also individual, incremental decisions whose longer-term structural implications become apparent only slowly. Ironically, those segments of the society which might be the least tolerant of these survival options, which involve a breakdown or perhaps reformulation of patriarchal codes, may be precisely those classes responsible for reproducing the extreme inequalities and the deteriorating material conditions for poor families in the first place.

Such enforced breaking of the gender codes by poor women is to be contrasted with behaviour by women which is seen by men, and perhaps also by other women, as not being a necessary response to economic hardship. Obviously such a judgement about necessity is subjective. But any behaviour by women which involves their collective organisation is likely to be perceived as a gratuitous challenge to male power, with women deliberately moving out of their socially constructed gender norms. Such behaviour occurs when gender relations are perceived by groups of women as chains, to be broken by establishing a new cultural basis for their status within society. This choice of perpetual deviancy from a still-preferred cultural norm about womanhood, in which respect from the keepers of patriarchal ideology has been lost, is therefore made instead of the degrading individual survival options which face poor women, as noted above.

Double Exploitation, Double Day

Studies of women in the Third World, especially of poor and rural women, emphasise the problem of double exploitation through class and gender relations. Failure to acknowledge this problem is a weakness with the WID (Women in Development) position, which has tended to stress the objective of women moving into public, male-sphere forms of employment. But such a move may, of course, merely expose such women to the relationships of exploitation already experienced by poor men in these positions, in addition to their gender inferiority which applies both at home and at work. Female workers on rural works projects may not be treated equally with men by the site managers. Women in the public workforce also experience the double day, since prevailing gender relations do not release them from the primary responsibility for homestead duties. The double-day problem is likely to be even more acute for female-headed households, since such women are more likely to be seeking paid employment outside the homestead. The irony here is that those women least likely to be seeking outside work are those in extended families, where teams of women in effect reside, with the possibility of sharing homestead duties among themselves.

With these two dimensions of double exploitation and the double day, women face the difficulty of establishing a new cultural basis for the way in which they are socially valued, with all the attendant risks for the pioneers, while remaining oppressed within the domestic culture which does not necessarily change in a corresponding manner. Thus any extension of female participation in the male, public-work sphere has to be accompanied by changes of codes not only in that sphere but in the private one too. In short, women have to fight on two fronts: within the home as well as outside it. Central to these struggles is the whole question of rights for women, *de facto* and *de jure*. The male bias in family law is an accurate reflection of the strength of patriarchal ideology in Bangladeshi society, a further superstructural shackle upon the room for female manoeuvre. A woman has to struggle, for example, over the basic procedures of divorce in this Islamic society; over her rights to property and subsistence in the event of divorce or desertion; over custody rights to children and their subsistence and education; over property rights in the event of a husband's death; over inheritance rights in her own natal family; over her rights to the dowry, which accompanies her at marriage and which affects her status in the marital home; over her rights to choose a husband, or at least veto her parents' choice; over her sexuality in marriage, determining the number of pregnancies (rape in marriage has only recently been pronounced a criminal offence in the UK).

Breaking the chains for women involves challenging both what is considered appropriate female behaviour, and socially constructed gender relations. Women are obliged to establish new cultural codes through which they can gain respect and honour, while seeking to overcome the class dimensions of their poverty. Their gender is particularly hampered by survival options in the male public sphere which place them in culturally precarious situations, as well as by the patriarchal hostility to any action which attempts to assert the validity of new criteria for determining their status in the society. Gender discrimination occurs crucially within the arena of the family, where quiescent female behaviour is sustained by the woman's *de facto* weak position in the operation of family law. She has no meaningful guarantees of security outside of her dependence upon the males of both her natal and marital families. Attempts at collective solidarity with other poor females is perceived as culturally deviant even by her own menfolk, whereas the solidarity of poor males can represent an extension of masculinity, and is only considered class deviant. The situation of double exploitation for women implies double deviancy, from gender as well as class. That we have now in Bangladesh so many examples of successful women's movements, and of women's leadership in struggles which sometimes include men, is testimony to women's unique contribution as well as to the extent of the desperation that forces them to risk so much. Other societies could learn a great deal from the strength of such women in Bangladesh.

It is no accident that a 1983 BBC Horizon programme on the social implications of agrarian change in Bangladesh and the Philippines, should begin and end with the women in Proshika groups weeping over the arrest of their menfolk on charges trumped up by the landlords. Their dependency upon their men was starkly revealed. One of my most harrowing moments in Bangladesh was a visit to some poor village women in Kurigram (north Bangladesh) who had nothing to cook until their men returned from searching for work; some would be lucky on that day, others not. For many women, things have to be this bad before they can contemplate forms of action independently of their men. I recall meeting some women in Faridpur, western Bangladesh, who wanted to attend non-formal literacy classes in their village but had been beaten up by their husbands for wasting time. This is the reality behind the urban, middle-class feminist rhetoric of struggle. The chains are indeed wrapped around twice.

Female Labour Participation

The changes in demographic structure through early processes of urbanisation together with the implications of agrarian change for inducing certain patterns of labour migration and changes in the composition of households have inevitable consequences for female participation in labour markets. There is of course evidence of female participation in evolving urban labour markets: particularly the garments industry, domestic services and public utilities. The circumstances of this employment are often highly exploitative of women: precarious; informal; low waged; youth without career prospects; sexual harassment; loss of cultural honour. Such variables are compounded for migrant women. Often women may not even directly receive their wages, or if they do then their menfolk immediately capture them. Women without men (whether single, married but alone, deserted, widowed) are especially vulnerable.

The circumstances of women in rural labour markets have some similarities, but differ in other respects. There is clear evidence that women are generally participating in a wider range of agricultural operations, with the continued exception of ploughing, in addition to the more traditional post-harvest operations. They are going further afield in search of such work, though incurring cultural costs in doing so. The conditions of employment for women in agriculture are also precarious, insecure, and sometimes reflect unequal gender relations within their own households (that is, their labour is contracted by their menfolk, who may receive the wages as a 'family' unit). They may also share the 'bonded' elements that apply to their husbands (or fathers, brothers and brothers-in-law).

Female-Headed Households

Finally, it is important to draw attention to the special circumstances of female-headed households (FHHs), which are increasing for the various reasons noted in the preceding discussion. Although recognised in the literature and by development workers as an increasing phenomenon, there are many conceptual problems with the term 'female-headed household'. It can carry the implication that 'lack of male' equals 'poor' or 'weak'. It can be employed to portray women either as dynamic actors in new structural situations, or as victims. Are FHHs a new form of social organisation in which women assume new, more assertive roles within patriarchal society? Or are they an indication of social and economic breakdown in which women lose the limited (and restrictive) social protection with which they have traditionally been provided, without being able to find a replacement?

The notion of FHH is to some extent in the context of rural Bangladesh an 'outsider concept' which may simplify an extremely complex and varied reality. The term 'destitute' has for a long time been used in the context of rural works programmes in Bangladesh as a euphemism for 'male-absent households' and is one which brings with it a kind of romanticised vulnerability. The term is also culture specific, and Bangladesh is not culturally homogenous. Muslim households are obviously numerically dominant and women's roles and rights vary between Muslim and other communities.

With these qualifications, it has been estimated that over 15% of rural households are headed by women. These are households where the head is widowed, divorced, abandoned or single. In many other households, the formal male head has migrated in search of employment, leaving a woman with responsibility for the household and for agricultural decision-making. If such *de facto* households are added, then almost 25% of all farm households (and approximately 33% of landless households) are actually female-headed. With women's actual rights to property so precarious, despite the formal provisions of Islamic law, FHHs are over-represented among the assetless. In landless FHHs, 66% of women are employed as agricultural labourers. The neglect of FHHs in development programmes has prompted Kumari, 1989, to argue for: a strengthening of women's inheritance rights; targeted income generation activities for FHHs; improved access to credit; specific training to improve human resource capital.

While not denying the value of these suggestions, there are problems in marginalising and labelling FHHs as targets (Wood 1985). There are conceptual problems with using the FHH category as a target group. FHHs can be differentiated by class, age, demographic factors and by the circumstances through which they have come, temporarily or permanently, to be without a household head. The economic disparity among female heads of household makes them the unlikely target of a specific programme. Despite the general poverty of the category, some female heads are not necessarily poor, having their own land or living under the patronage of a wealthy male-headed household related to them by kinship (Ito 1990). There are also practical problems in that FHHs are not spatially concentrated and are located within a wider context of networks of kinship, patronage and community.

Conclusion

This discussion of women and gender has necessarily involved some fundamental issues of analysis since there are profound implications for modes of intervention to support the interests of women. As noted above, there are even problems of one culture reaching a definition of what these interests actually are for another culture. Providers of aid are frequently obliged to wrestle with the tension between a tendency to moral absolutism on the one hand and the meta-principle of sensitivity to cultural relativism on the other. This tension is particularly acute when women and gender issues are being considered. At another level, the difficulty of 'targeting' women, in the sense of artificially abstracting them from their complex social and cultural environment, is especially problematic for project modes of development thinking.

PART II

LANDLESS PARTICIPATION IN AGRICULTURAL GROWTH

Chapter 5

Rural Employment and Patterns of Agricultural Development

Introduction

This chapter introduces a key set of arguments, maintained over the last decade, concerning the possibilities for expanding income opportunities for the rural landless directly in agriculture. Stated like this, the proposition seems obvious, but it had to be advanced in the context of more limited thinking about expanding labour opportunities consequent on the Green Revolution or rural works activity. These arguments are further explored in the remaining chapters of Part II (6, 7, 8 & 9). Chapter 6 was first drafted at the beginning of 1980, as indicated in the Introduction to this volume. It recognised that labour absorption in agriculture as a result of introducing high-tech methods would fall below the annual net addition to the rural labour force, and that this shortfall could not simply be met by industrial expansion or rural works activities. Thus the consequences of pressure upon the land in the context of restricted alternative employment opportunities should be understood in terms of other developments in the agrarian structure, which might be summed up as "agrarian entrepreneurialism". It also recognised the fragmented nature of Bangladeshi farms and the consequent difficulties for farmers in accessing new, expensive and lumpy technologies. It also recognised the inherent weakness of socially heterogeneous cooperatives, which anyway do not have common objectives, and the need to build a strategy around the principle of class divisions, i.e. the poor transacting with the rich. Such transactions could be based upon extending and redistributing property rights, such as irrigation water or other water and road related infrastructure.

Chapters 7, 8 & 9 pursue this line of argument into practice and theory: initially by describing the landless irrigation initiative (7); then through an examination of the social and institutional factors which affect transactions between rich and poor (8); and finally, through the consolidated idea of 'agrarian entrepreneurialism' (9) which explores the prospect of a rearticulation of agrarian structure away from the central institution of the family farm towards the agricultural service company (in which the poor, suitably supported, can significantly participate).

This chapter (5) establishes the basic context within which these arguments can be pursued. In particular, it attempts to draw a contrast between a Bangladeshi and Punjab model of agrarian change, implying that much agricultural policy (certainly in terms of institutional strategy) is mistakenly based upon 'Punjab' assumptions of consolidated, household level, farming units with land as the principle means of production, instead of acknowledging the consequences of extreme fragmentation of farms interacting with high-tech methods and capital.

This chapter concludes on an alternative note, by speculating on the consequences for agrarian structure of organic, as opposed to high-tech, farming methods. Among many implications, there is the possibility that such organic methods will enable small farms to remain viable under household level management and offset the trend towards disarticulation under high-tech conditions.

Low-level Capital Formation in Agriculture

For a variety of reasons, some attributed to the sub-colonial status of East Bengal within Pakistan, in the period up to the end of the 1970s rural class relations in Bangladesh were dominated by the use of capital in the sphere of exchange, in activities such as renting, moneylending and trading, rather than through the expansion of production. This certainly raised the level of absolute surplus value accruing to certain classes, but did not contribute to the development of productive capital among landowners. The failure during this period to develop productive agricultural capital has resulted in increases in landlessness and rural underemployment, and in the depression of real wage rates. This has created a problem of effective demand among the increasing proportion of rural people obliged to purchase foodgrains, thereby undermining price incentives required to stimulate owners of land into raising productivity.

Up to the end of the 1970s, this situation was in effect reinforced by a strategy of state administered subsidies to agriculture, which were intended to stimulate production. The strategy was implemented through the Bangladesh Agricultural Development Corporation (BADC), which controlled both imports of chemical inputs and equipment, especially mechanised irrigation units, and their distribution through outlets at district and sub-district levels. With a threefold combination of subsidies, state monopoly and restricted access, corruption was widespread, alongside high default rates and under-utilisation of equipment. The state, with its monopoly control over external capital, had become in effect an arena within which funds were diverted into modes of parasitic surplus extraction through an alliance of central and local bureaucratic officials with their favoured clients in the countryside.

Disarticulation of the Rural Economy through High-tech, Commercial Agriculture

It was clear in the early eighties that any expansion in agricultural investment would be dependent upon external aid, and both the World Bank and US AID, as the major donors at that time, were determined from their diagnoses and leverage to liberate the rural economy from bureaucratic management and control by encouraging open market systems of food and input distribution. Their strategy, which became the strategy for Bangladesh during the eighties, consisted of: open-market dealing in food (through procurement at market rates, storage in a network of *godowns*, and release of stocks into local markets triggered by a 20 per cent variance in seasonal prices) (IBRD 1979); the phasing out of fertiliser and other input subsidies (for example on various forms of mechanised irrigation); coupled with a major emphasis on HYV technologies in rice and wheat. These efforts concentrated on the improvement of foodgrain supply, with a noticeable lack of attention to the structure of demand for food. The strategy involved the assumption that a combination of support prices for foodgrains, through a generous procurement price, and open market dealing in rural goods (inputs and food) would encourage the landholders to raise the productivity of their land, thereby increasing their marketable surplus to pay the full market price for inputs.

Looking at the performance of the these policies during the eighties, the annual rate of growth in agriculture was about 3 per cent during the first half (contrasted to the 6.5 per cent planned), and dipped in the second half to nearer 2 per cent. A UNDP 'nationalist' study in the late eighties was keen to attribute this to *policy*, though the floods of 1987 and 1988 will have had some impact. The more input-dependent crops, such as wheat and HYV *boro* (winter rice), which have been responsible for the expansion since 1980, have shown signs of losing momentum due to irrigation failure, farmer preferences for rotations, decline in fertiliser/yield coefficients, and early flooding on late harvested crops. The slow rate of growth in foodgrain production is partially explained by slower rates of expansion than targeted in the irrigated area. But even these rates of growth in foodgrain production have been difficult to achieve. Although fertiliser use has risen by over 30 per cent, its price to the farmer has risen by over 150 per cent while the prices of the two major crops, rice and jute, have only risen by about 50 per cent. Such adverse terms of trade to the farmer may well have depressed the potential use of fertiliser over the period. Furthermore, per capita foodgrains availability has still not reached pre-independence levels (160 kg. compared to 177 kg.). In other words, Bangladesh is still further away from food self-sufficiency than it was 20 years ago.

It must be further noted that a significant proportion of the increase in foodgrain production has been achieved through the expansion of acreage devoted to foodgrains rather than a dramatic increase in yields, which seem to have reached a plateau. Thus about 1.0 million acres of jute have been diverted to rainfed cereals; and about 1.5 million acres of winter pulses and oilseeds diverted to wheat and winter rice. This reveals a society under considerable stress, with anxieties about crop diversification and the implications of this cereal displacement upon diet and the import bill. It should also be noted that jute is a labour-intensive crop, but is under continual pressure in international commodity markets from synthetic substitutes.

These agricultural trends during the eighties (and many commentators have made similar summaries and observations) have structural implications for the agrarian system which establishes the conditions under which the poor in Bangladesh must find options for survival. Despite the many institutional and technical constraints to agricultural growth, we must acknowledge the current path of a significant and continuing expansion of capital in agriculture. This consists of further expansion of mechanised irrigation to intensify land use, a marked recent trend in the use of power tillers (with added extensions for such activities as road and boat transportation, or rice milling), a continuation in the use of chemical inputs, the use of conventional tractors, and so on. In this process, there is a proliferation of actors in the agricultural system, which has the potential of reducing the significance of the family farm, whether large or small, tenanted or owned, in the distribution of agrarian power. Boyce, 1987a (p 37), defines agrarian structure as "...the subset of institutions governing the distribution of rights in agricultural means of production, notably land. These rights include not only ownership, but also such arrangements as tenancy and mortgage, which create a divergence between ownership and actual operation". With this definition and the preceding observations in mind, we are witnessing the addition of non-cultivators and non-landowners to this subset of institutions.

As noted earlier, this proliferation is occurring under general economic conditions which constrain families from leaving agriculture for other sectors of the economy. It is also the case that the rise of landlessness cannot be equated with a simple process of polarisation in landholding. Population growth and the multiple inheritance principle in the division of holdings (maintained by the absence of alternative employment prospects) act as a constraint to the development of large farms, and certainly consolidated ones. This combination of conditions entails the extreme fragmentation of family cultivable land into different plots. These are scattered over the village reflecting divisions at the time of inheritance on the basis of not only area, but also of soil type and elevation.

With the continuing importance of irrigation to family survival, through enabling the intensification of land use, it remains imperative that the division of holdings is equal in all respects. The capitalisation of agriculture thus acts as a stimulant to farm fragmentation, although some of the technologies associated with that process are usually regarded as demanding consolidation to reduce lumpiness. Thus, since this fragmentation is structurally inherent to the conditions in Bangladesh, the contradiction between fragmentation and the technological requirements of agricultural growth can only be resolved by separating the ownership of land from the actual operation of it with the new technologies.

Fragmentation of Holdings and New EIG Opportunities: the Bangladesh Model

In this process, non-land packages of capital, especially led by the control over water, become the more consolidated assets. Fragmentation may have been a problem for conventional thinking on farm and land management, prompting attempts at consolidation of holdings, but seen in this way it actually constitutes an institutional opportunity to those who control these non-land packages of capital and who make up the proliferation of actors noted above. While it is true that richer families can more easily diversify into these non-cultivating, but agricultural, roles of delivering such services as irrigation, ploughing, processing and transportation, the principle is established that surplus value from agriculture can be obtained, without owning land, through the provision of services.

Although the central importance of irrigation has been emphasised, because of its significance to the intensification of land use under demographic pressures, control over water is not alone in this process. There are increasing examples of fertiliser delivery and pesticide spraying, undertaken as a commercial service by command area 'managers', often using contract labour. With the increase in mechanised milling and the likely widespread introduction of power tillers, fragmented plots will be consolidated not around individual household units (the 'Punjab' model), but around the owners (single or collective) of production, processing and transportation technologies (the 'Bangladesh' model). The concept of 'command area' is not then restricted to irrigation sources. Whether coterminous or not, areas of land will be in the 'command' of other packages of mechanical and chemical technologies, and vertically organised marketing arrangements. Even without considering a large-scale processing sector (for example, large rice and sugar mills, potato cold storage or tobacco management), the existing levels of production and food processing capital can combine in the formation of agricultural 'companies', loosely networked or corporate, interlinked with repair workshops, mechanics services, supplies of spare parts and fuel, transportation and marketing. In this Bangladesh model, the family farm declines as a socio-economic entity while the agricultural system becomes re-articulated through these other units of production and exchange.

At this stage, the analysis should not be pushed too far. Three factors offset this logical outcome in which the Bangladesh farm disappears altogether in favour of agricultural companies which develop a complete control of land use:

- Larger farms are able to allocate larger absolute amounts of land to crops other than irrigation dependent ones, and have less need to intensify land use by demanding services from others.
- New strategies for tenancy and mortgaging have emerged, in which poorer farmers may counteract their lack of capital, or 'lumpy' technology problems in technologically dependent seasons by leasing out land to those sufficiently well networked or capitalised to use it (reverse, seasonal tenancy: Glaser 1989). Mortgaging can also achieve a similar result. Poorer farmers can also lease in seasonally to offset their marginalisation.

- Apparently lumpy technologies can be socially broken down and mediated into the needs of small, individual customers. Lewis (1991) has noted this process for tractors in Comilla.

The importance of this grounded speculation about the evolving agrarian structure in Bangladesh centres on the implications of this scenario for empowerment and employment options facing the poor. The issue here is the purpose of action. There can be no doubt that the key problem is the provision for mass employment at wage rates which deliver at least subsistence incomes. It is obvious that there can be no single solution to this. Urban employment in the industrial and services sectors can only make a limited contribution for the foreseeable future; likewise the rural 'non-farm' sector, which has meant in Bangladesh rural works programmes to support the landless during seasonal troughs in the demand for agricultural labour. However, the potentiality of the non-farm agricultural sector has been largely overlooked, yet it contains opportunities not just for expanded employment in more open-market conditions but also for a realignment of power.

When the concentration of the historically central means of production, land, which has determined power and authority for centuries in the Bengal countryside, is becoming diffused in the manner noted above, then the circumstances in which the struggle takes place are also changing profoundly. It becomes important to understand which battles have become anachronistic, and which retain their relevance to the present and foreseeable future. In the scenario outlined above, does it make sense to encourage the poor into struggles for land reform? The rhetoric about landlords and the need for a redistributive land reform is hopelessly out of date, and we do the poor no favours by clinging to redundant slogans. For example, the process of agrarian change outlined here offers the prospect of steering these agrarian entrepreneurial activities towards the landless, where they can gain rents and profits from expanded agricultural production; that is, not just employment but empowerment to enhance their rights and status in ever-widening arenas. The purpose of action for the landless can thus look beyond the prospects of more field-level employment under intensified land-use conditions for a declining proportion of new entrants to the labour force (where increasing proportions of family labour will be used anyway). At the same time, the landless need not be beguiled into false expectations of becoming landowners themselves. This is not to deny the value of certain kinds of land-acquisition struggles, for example over *khas* land (untitled land in the formal disposition of government) or ponds—but such programmes have to be placed in perspective. They represent in most cases a one-off opportunity in any locality (since the status of such land most often arose out of migrations, from 1947 onwards until 1972), and such land constitutes a tiny proportion of cultivable land resources in the country. While rewarding for those involved among groups and their NGOs, such land reform does not impinge upon the main structural problems for the future.

An Organic Model of Agricultural Development?

The model outlined above, with related employment opportunities and strategic options for the poor to capture non-land assets, derives from the continued application of high-tech HYV approaches to land use. The development of organic approaches to agricultural improvement (in terms of sustained yields, employment generation and ecological sustainability) may, if pursued, produce other sets of consequences for the agrarian structure.

Such approaches are in their infancy but represent a fundamental challenge to the HYV high-tech approaches to agriculture which are disarticulating the rural economy, as described above. Proshika, for example, have embarked upon a programme of promoting smallholder, integrated sustainability which is designed both to have environmental benefits and to provide a degree of economic independence as a platform for other struggles. It consists of integrating homestead horticulture with livestock rearing, small-scale fisheries and organic arable farming where such crop land is still available. If successful, it will have the effect of reducing the dependent interaction of such marginal and small peasants with the market for chemical and some mechanical inputs. At the same time, it is hoped to demonstrate that higher, sustainable yields can be achieved through recycling own waste and crop residues. This will enable poor farmers and homestead holders to obtain and retain agricultural surplus for sale at non-harvest prices and thereby purchase consumer goods without encountering the usual adverse terms of trade incurred when they are obliged to sell at low immediate post-harvest prices to repay credit.

The further impact of such a programme would be to delay the process of landlessness which arises from fragmentation of holdings into non-viable units, and to increase the degree of autonomous management over land use. The traditional peasant role of conservers and trustees of the land for future generations will be rediscovered. The programme may also encourage new settlement patterns where, under appropriate topographical circumstances, small peasants may be prepared to move their homesteads to be adjacent to remaining arable land. If this were to happen, there would be a profound effect upon political and authority dynamics in the village as people moved from marginal sites and away from the control of *para* leaders.

There is, of course, a long way to go with such a programme, but elements of it are already being pursued by different households within the groups. It is supported by considerable on-site experimentation at the Proshika Training Centre at Koitta, which has now expanded its area to include arable land. The possibilities of integrating fish and paddy are also known, and the impact of extended livestock and horticulture upon diet is welcomed. With urbanisation and the proliferation of non-primate cities, the market for such produce is most likely to be significantly extended for those in these scattered urban hinterlands. The various arguments in favour of this approach to agrarian reform all contribute to the central objective of empowerment, especially the disengagement from certain input markets.

Chapter 6

The Rural Poor in Bangladesh: A new Framework?

Introduction

In mid-1980 Bangladesh embarked upon its Second Five Year Plan (FYP). Preliminary drafts of the Plan indicate a substantial shift in government investment strategy toward the rural sector with primary emphasis placed on increasing foodgrain production and improving the employment and consequently the welfare position of the rural poor. Few would argue that a radical transformation in rural Bangladesh to attack the growing problems of landlessness, un- and under-employment, and the polarisation of rural wealth and power—are required to meet these objectives. Critical to this transformation will be the involvement of the rural poor.

The basic object of this chapter is to examine how the needs of the rural poor can be met in the Second FYP and what can be done to involve them in this process. The chapter sets out a series of institutional possibilities which arise from contemporary experiments in Bangladesh; and suggests how government should debureaucratise its approach to rural development to allow a range of institutional possibilities to be created, applied, adopted and so on.

The question of rural development and the institutions which support it is in the final analysis a political one. We begin with a brief discussion of the process of planning and the interest groups involved in rural development in Bangladesh. This is followed by exploration of the need for radical transformation in the rural sector, arguing strongly for agrarian reform. We close with a discussion of possible programmes which might be initiated during the Second FYP period to move toward these essential reforms.

The Political Economy of Planning in Bangladesh

Essentially, planning is a political process, in which pressure of time sharpens the conflicts between different interest groups in the government and the specific constituencies which they represent. It is important for the actors in this process to understand what is happening to them, and to recognise that planning is not a techno-rational exercise despite the use of models, targets and projections. It is to a large extent an artificial arena of bargaining and horsetrading.

Nowhere in the process is this more clear than over the question of institutions and proposals to intervene directly in the social relations between individuals, groups and classes; for this question is about the rules themselves. Changing or preserving rules has costs and benefits for different classes and groups, not all of which are able to participate in the decisions. Ultimately in the prevailing political realities of Bangladesh, formal decisions about such rules will be made by individuals whose accountability to the people of Bangladesh must be low, at least in the short term. These decisions will establish the parameters within which the officials have to work. It is important in the process of developing parameters not to restrict the atmosphere of institutional experimentation which persists in Bangladesh and to press for decisions which can be tested and modified over time. If an institutional discussion is to appear in the plan document—let it be a discussion of a range of ideas, rather than any attempt at a universal formula. Neat institutional wizardry at the top such as the mechanical expansion of Comilla into IRDP, is unlikely to have much impact on the lives of rural people.

In the development of an institutional component to the Plan, a complex matrix of individuals, interest groups and ideologies are involved. Discussions in various arenas—political, technical, academic, commercial, national and international—have occurred regarding the agrarian aspects of the Plan. The relation between the arenas as well as the elements need to be clarified.

The Needs of the Government

A major consideration which must override other decisions is the need of the present government to develop a rural constituency. Various formulae have emerged—total village co-operatives, *gram sarkar* (village government), greater use of union and *thana parishads* (district councils), the expanding role of MPs in the government's local resource allocation system in the distribution of drought food, and on the *Thana* and Districts Development Committees. Whatever the outcome of these and other possibilities, there will remain the question of the implications for the control and distribution of goods and services for agrarian/rural development, the relationship to agricultural co-operatives, and provision for and/or mobilisation of the rural poor. We will return to these issues later.

Land Reform

There is the no less political and no less fundamental issue of land reform. Two opposing positions on the efficacy of land reform appear in the present debate. One is that sharecropping relations and absentee landlordism depress the potential productivity of land by reducing the incentives to farmers to apply inputs and make land improvements (Jabbar 1977). The other is that land reform is decisive neither for efficiency nor for equity, (Gisselquist 1979) since the differences in input use, output and employment absorption across farm sizes or between tenure classes would not significantly cope with the expansion of production or employment that is required. Of course, the question of land is much more complicated than these aggregate notions affecting the fortunes of families through transfers, inheritance, indebtedness, and control and use of other means of production. Relations to land also affect access to goods and services, and constitute the principal material basis for the assertion of political influence and authority. On these grounds alone land reforms ought to be considered. But for the same reasons, we cannot expect the Government to contemplate a land reform programme since it depends on the prevailing rural inequality and structures of power and authority in the countryside to maintain the rural 'stability' necessary for attracting more overseas investment to expand the industrial sector. Can the government seriously afford to desert this rural constituency of landlords, landowners and entrepreneurs for an incoherent mass of tenants and labourers? The need for a secure rural political base overwhelms any other arguments in favour of land reform.

Bureaucratic Interests

As represented through different ministries and agencies, the bureaucracy has established crucial areas of patronage through its monopoly over inputs and services to agriculture and rural development. Furthermore, ministries and agencies internally represent separate territories of patronage involving the reintroduction of an informal market in which the difference between the subsidised price at any level of distribution and the price at which the commodity or service is in effect sold does not return as public revenue to the ministry or state as a whole, but supplements individual incomes. In this process, classes of rural clients are affected differently—the poorer have no access at all, the richer pay for their access and have opportunities to pass on these costs to others and extract a profit too. Bureaucratic allocation has certainly strengthened the positions both of officials as well as rural entrepreneurs and landholders, and in so doing has contributed to the processes of polarisation and differentiation in the countryside. The point is that the bureaucracy has a vested interest in preserving the existing arrangements for the distribution and supply of scarce goods and services— it is the basis of both mini-empires and the accumulation of wealth by some. In the face of these interests, national leadership has to compete for influence and support in the countryside with few resources at its disposal. To some extent this is the basis of a struggle between different factions, the bureaucrats/officials and the politicians/entrepreneurs, under the watchful eye presumably of the military whose volatility of recent years has been reduced by large, external sources of funding independent of other budgets. The current anxiety of politicians seems to stem from their realisation that they have so far failed to capture significant resources at the *thana* level to develop a more secure support base in the countryside. It is in this way that a commercialisation/private sector strategy is attractive. It is an ideological resource in this struggle between the different interest groups—and it must be doubtful whether there are any technical advantages which will survive its adoption.

Commercialisation Strategy

In addition to the discussions of land reform, there is the build up of external donor pressure—to liberate the rural economy from bureaucratic management and controls by open-market systems of food and input distribution. Both the World Bank and US AID have recommended a reduction in government involvement in the rationing system and a closer targeting of ration throughout to the poor; the limiting of government involvement in the grain market to instances where open market prices deviate plus or minus 20% from planned price levels and the reduction of fertiliser subsidies. It is important to recognise that these policies can only be undertaken if liberal supplies of food and agricultural inputs are available. Given the magnitude of the programme of ensuring adequate aggregate supply at the appropriate time and place, the present proposals may involve unrealistic assumptions about the development of a physical and market infrastructure and perhaps about the magnitude of funding available. Also, there is a major untested assumption in these plans—that a combination of support prices (through a generous procurement price) and open market dealing in rural goods (inputs and food) will encourage the landholders to raise the productivity of their land, increasing their marketable surplus to pay the full market price for inputs. Most of the evidence on agrarian relations in Bangladesh points to higher rates of return on capital investing in usury, leasing and trading. Returns have been reinvested in similar ventures, including the purchase of land for on-leasing to sharecroppers, which under existing arrangements results in suboptimal production. On the other hand, if the assumption is valid (and especially if a supply of goods and services was not forthcoming) we could expect sharecroppers to be dispossessed at a faster rate and small owner-cultivators to be more dependent upon credit (usurious or institutional) in order to compete successfully in the market for inputs and where they failed to obtain cheap sources of credit *at the right time* for the productivity of their land to decline. A consequence of these tendencies would be a rise in the rural landless population.

These problems along with the uncertainty of achieving favourable supply and delivery conditions do not make this trend in policy thinking very attractive for both production and equity objectives. It runs the risks of leading to greater rural exploitation, a further increase in the number of landless, higher proportions of rural unemployment, a reduction in effective demand for both agricultural products and inputs, and even a decline in production. Of course the commercialisation strategy is very attractive to the urban and rural entrepreneurial classes alike and will encourage richer landholders to diversify their usurious returns into sub-*thana* level trading. In this way the strategy would reinforce the direction of class formation in Bangladesh.

Integrated Rural Development Programme (IRDP)

In developing an institutional strategy in the Second Plan, it cannot be forgotten that a major set of institutions for agrarian development and the expansion of agricultural production exists through IRDP. Several reviews and critiques of IRDP (SIDA 1979, p.134, footnote 1 for references; SIDA 1979 pp 133-162 and Wood 1980a) exist. From this literature the consensus of problems is: IRDP is a diluted version of Comilla, missing in most cases the irrigation and rural works element (Khan 1979); co-operatives were established without adequate preparation by external mobilisers; membership in co-operatives is low and they are dominated (even in Comilla) by rich farmers and landholders, who acquire a disproportionately high share of inputs; regional variations in landholding/tenure relations involving more differentiation than in Comilla compound the problems of replicating Comilla; and, the landless are excluded and sharecroppers discriminated against through dependence on land as collateral for loans. A less familiar problem but one relevant to arguments developed later, is summarised in the following extract from the SIDA (1979) report:

> "... an important implication of the quantitative difference between the Comilla experiment in Comilla (sic) and the IRDP should however be observed... Within the district of Comilla, insufficient demand for grain never became a serious problem: the output of rice could, as we have seen, more than double in ten years, and the surplus that arose could easily be sold outside the district, with only marginal effects on the general price level. If, however, lack of effective demand is the perhaps most important long-term obstacle to agricultural growth in the country as a whole...then a new dimension is added to the problems of replicating the Comilla experiment all over Bangladesh. In practice the whole IRDP approach towards agricultural development is today based on the assumption that all problems originate from the supply side, while a recognition of the key role played by demand makes it imperative to abandon excessive emphasis on production and instead concentrate on the income distribution and employment creation objectives." (p.151)

Position of the Ministry of Local Government and Rural Development (LGRD)

Because the issue of effective demand has not received the political prominence of increased production themes, the LGRD faces the prospect of having IRDP (co-ops rather than rural works) destroyed or substantially modified by the commercialisation strategy. The plight of the Ministry is symptomatic of the way in which the planning process is failing to address the real questions of rural poverty. If the link between effective demand and production (let alone human rights) is established, LGRD would have a central role to play in the overall production strategy, through developing institutional strategies for different classes of landholders, tenants and landless; setting of realistic targets for input supply as well as realistic procedures for their distribution; developing attainable targets for production expansion assuming effective demand and deriving employment targets required to generate the appropriate level of effective demand to ensure favourable producer prices. Such employment targets would constitute the minimum level of hope for an acceptable performance.

Rural Mobilisation—Government and NGOs (excluding RWP and FWP)

A further dimension to the current thinking concerns the role of quasi-governmental programmes and the work of non-governmental organisations (NGOs). Since Liberation, a variety of institutional approaches have been developed and tested in Bangladesh. *Ulashi*-type (see glossary) programmes which have relied on mobilising the rural people, using the theme of self-reliance, to participate in infrastructure/rural works projects. The *Swanirvar* programme is attempting to create an awareness amongst rural people about the development possibilities in their individual area on the basis of their own resources supplemented by proper utilisation of under-utilised resources, physical and human, official and non-official (Chashi, 1976, p.177). The organisational vehicle used in this process is the *gram sarkar* or village government—a village governing body composed of representatives from separate class-based village groups. A third set of activities, carried out by a variety of private agencies, are also active; the ASARRD projects in Comilla; BRAC; Proshika; and the Grameen Bank Project in Tangail.

Each of these approaches, although having problems, provide some suggestions for possible government action. The *Ulashi* approach was directed at the creation of rural infrastructure. There is evidence, however, that such schemes are exploitative of rural landless labour—both in remuneration and long-term benefits and that they do not foster the creation of long-term village institutions. While rural mobilisation around major infrastructure projects is important, other preconditions are required to assure that labour, especially landless labour, shares in the increased productivity generated by their efforts.

Swanirvar, on the other hand, does have long-term organisational implications in the ideas of *gram sarkar*. The system was planned to be a non-bureaucratic approach to rural development with central government and the *gram sarkars/Swanirvar* villages operating on equal terms. The system involves the *gram sarkar* in mobilising village level resources and developing and forwarding proposals to the central government apparatuses (rather than *vice versa*) on what external support is required. Most evaluations of *Swanirvar* conclude that considerable inflows of external resources have occurred, but that official sponsorship has undermined the basic thrust of the programme. Organisational proposals are being devised externally and the benefits from the expansion of the means of production in the village through 'joint' activity (usually the work of landless labourers with little reward) are accruing disproportionately to the holders of the land, the traditional ruling class in the villages. On the positive side, the programme has functioned to some extent in opening up the thinking about institutional possibilities for rural development in Bangladesh—counteracting the totemic influence of Comilla and the bureaucratic weight of IRDP.

As a result, a variety of other non-government programmes have emerged (including ASARRD and TVD projects in Comilla, Proshika, BRAC, Grameen Bank Project.) The main common features of these programmes should be noted since they in effect tackle the relation between expansion of production and the generation of effective demand. Excepting TVD, the programmes work exclusively with poorer farmers and the landless and in some cases only the landless; they depend critically upon the creation of 'natural' internal solidarity within groups which is 'tested' in some way before constituting the basis of collateral for loans. The programmes involve functional and informal education; the involvement of outsiders (programme workers) in defining a group, determining its form of activity and management structure is minimised, thus avoiding preconceived models of organisation.

Although a number of models are available, central government has not decided on a formula for local government. It is still left with the problem of devising a local government system which at least looks inclusive and participatory, but which is either parallel to or incorporates the existing co-operative structure. The final decision will reveal the extent to which the distribution of goods and services, mobilisation of rural labour, and management of production and product markets (including procurement) will become structures of political patronage (overt or covert) or remain in the hands of the bureaucracy. In a sense *gram sarkar* and the other rural development programmes constitute the offer of an institutional theme to the government which has been sponsored but not adopted.

A Scenario of Possible State Policy and Action

From this brief and incomplete review of the various elements and arenas of discussion about the form and content of rural institutions in Bangladesh over the next Plan period and beyond, a further dimension emerges: namely the significance of their combination, and the extent of contradictory or complementary themes.

There is an emerging literature of the role of the state in Bangladesh (for example, Hussain 1979; Westergaard 1979) and perhaps some consensus around the petty bourgeois (urban and rural) basis of the state. This is expressed in the dominance of antediluvian/usurious capital in agrarian relations (Wood 1977b), combined with diversification into trading, through the expansion of a 'comprador' bourgeoisie with the penetration of private and public metropolitan capital into the urban industrial and commercial sector. These classes are incorporated into a metropolitan dominated manufacturing sector and a primary commodity market, maintained by a class of bureaucratic and military leaders who benefit directly or indirectly from these processes. (Directly to the extent that they monopolise the allocation of foreign aid and establish the rules and conditions under which other classes are to accumulate from these resources). Although there are many who would now accept this description, if true then the most likely outcome of this process for the rural sector, despite the objections of some enlightened senior officials, would be a combination of ten features:

- limited commercialism
- a retained but reduced role of IRDP and BRAC
- the possible insertion of a total village level co-operative below the Union Parishad complicating the position of the IRDP co-ops and TCCAs;
- stronger representation of elected (i.e. usually rural rich) representatives at *Thana* and District levels—providing renewed arenas for struggle between politicians and officials at those levels;
- a retention of existing forms of central domination causing frustrated aspirations among national level politicians;
- an associated reduction in the ability of officials in the central apparatuses to protect the erosion of authority and patronage of their field staff.

Alongside this:

- no significant land reform;
- a pattern of input distribution (including irrigation facilities) emphasising HYVs and wheat which will exacerbate the polarisation of rural classes and fail to match the rate of growth in demand for labour;

- inadequate separate employment generation programmes;
- finally with the commercialisation strategy a possible decline in the proportion of the development budget to agriculture (goods and services i.e. extension, training and research).

If this picture of the possible combination of policies for the rural sector emerges during the Second Plan period, the implications for both production and equity are disastrous.

The Argument for Agrarian Reform

In 'Food Policy Issues' (IBRD 1979), the authors list the physical constraints to the development of agriculture as: scarcity of cultivable land; limited yield potential; climatic hazards and the need for dry-season irrigation; improved drainage and flood protection. However they attribute the slow growth of agricultural output to: weaknesses in agricultural support; services; inadequate provision of input supplies; and constraints stemming from the agrarian structure. Their solutions take the form of open-market dealing in food phasing out fertiliser (and other?) subsidies; and a major emphasis on HYV technologies in rice and wheat. All focus on efforts to improve food grain supply with a noticeable lack of attention to food demand.

In assessing this strategy, it is important that the negative effects of a supply-orientated HYV strategy be considered. Given the present distribution of land and access to improved inputs which exist in rural Bangladesh, it is highly likely that larger landowners will benefit significantly from the strategy. While production may increase initially, it is questionable whether prices of labour demand will be sufficient to sustain production. One could hypothesise a situation where production has increased but not enough to absorb new entrants into the rural labour pool. In fact projections of labour demand developed from the output productions in the Second FYP documents indicate this will be the case. They predict that the rural labour force is likely to grow by four million persons during the plan period, with only one quarter of this new labour supply being absorbed in agricultural production.

What does this mean for the supply-orientated approach currently under consideration? Increasing numbers of rural employed would result in a further decline in rural wage rates, a reduction in the ability of the poor to buy foodgrains, lagging foodgrain demand in the face of increasing supply and eventually a decline in real foodgrain prices, the basic incentive for the adoption of HYV technology. SIDA in its recent report provides a more in-depth analysis of this scenario and offers the following policy recommendation:

> "Self-sufficiency, defined as so and so many tons of domestically produced foodgrains is a completely meaningless concept unless it is related to the capacity of the people to purchase the growing levels of production. Demand tends to generate its own supply, but the opposite is not equally true, and the *key to success in a production-oriented strategy in Bangladesh is to concentrate on efforts to increase effective demand.*" (SIDA 1979, p.96 [my emphasis]).

Issue of Transforming Non-Productive Capital

This critical relationship between effective demand and levels of production has been a historical reality in rural Bangladesh since the colonial period. Its presence has led to socially unacceptable forms of rural capital investment such as usury, rack-renting and sharecropping (Wood 1977b; Westergaard 1978). The rates of profit from these activities have traditionally been higher than investing in agricultural production and have led to dispossession of small farmers, depressed production and employment opportunities, and depressed agricultural wages.

These problems are not recent and have shaped the form and determined the results of state intervention in agriculture since 1972. Such interventions, which include belated attempts at land reform and the production/open market strategies recently forwarded by the World Bank, focus on the supply side of agricultural production, assuming that small marginal farmers act as independent producers with complete control over the factors of production they require. These approaches fail to note the adverse terms of trade which exist in rural Bangladesh between small marginal farmers and rural elites, the alternative forms of capital investment available to the latter group and the adverse consequence that an increase in capital controlled by elite rural groups will have on the existing terms of trade. To develop the appropriate mechanisms for state intervention, the complete system of rural relations including the position of the state, access to the services it provides and the present system of resource allocation must be considered.

State Intervention and Access Problems

If we take the question of state intervention, it is generally the case that the bureaucratic management of rural distribution systems results in cultural and class-based problems of access. For example, present programmes favour those who understand the system, generally the more literate, educated, socially and spatially mobile; those experienced in dealing with government; and those who can afford to invest time, transport costs, and fees required to make appropriate connections.

To some extent these are underlying issues behind the general dissatisfaction with the experience of state intervention in the rural economy since liberation. The government has relied on state agencies like BADC, IRDP and the Extension Service to deliver goods and services into a complex agricultural system in which price, availability and timeliness are critical both to the rate of profit and the propensity to invest in new technology. However, because of the informal markets which developed around this structure, it failed to spread the benefits of the new technology to those cultivating classes who would use it most productively and exploit their land most intensively.

Avoid Over-Reaction

These criticisms of the state intervention experience should not result in advocating a commercialisation strategy. If the state were to withdraw suddenly and dramatically from its involvement in the agrarian economy, the potentially most productive classes of smaller cultivators would be left to the mercy of market forces. Problems of transforming rural usurious capital into productive capital, and the issue of effective demand among the marginal peasants, sharecroppers and landless would not be resolved. The institutional vacuum left by such a withdrawal would likely strengthen the present forms of rural capital investment rather than dissolve them—with rich peasants devoting more of their capital to trading in inputs, mortgaging, land acquisition and leasing as victims of the market were steadily marginalised and pauperised.

Even a partial withdrawal, a compromise in which the IRDP co-ops are retained in their present form, and open market systems of food and inputs compete with licensed grain dealers and BADC depots and dealers respectively, is unlikely to resolve these problems. In this latter case it is likely that operators of both systems would benefit from scarcity in aggregate terms and scarcity in the sense of timeliness of supply. This would apply critically to credit, water and fertiliser markets on the input side.

Room for the State to Manoeuvre

If state intervention itself has these effects anyway, is there any sense in which the state can withdraw from the rural economy without exacerbating the present contradictions in the agrarian system? From the earlier discussion, it follows that a central precondition must be to embark seriously upon programmes which tackle the question of effective demand for agricultural products. The strategy behind such programmes is derived from extending the understanding of rural property rights in Bangladesh. To date, property rights are usually conceived of in terms of land, and the literature is preoccupied with land reforms involving either ceilings and distribution; distribution on the basis of land to the tiller; providing greater security of tenure and/or measures to prevent distress sales of land. Accompanying all these 'privatised', individualistic alternatives is a standard list of associated conditions to be met concerning provision for credit, water and inputs.

However, property rights in land are rights to benefit from the productivity of that land. In this sense the state has also had rights in a proprietor's land—through taxation and rents (sometimes indirect) in return for ensuring the rights of the proprietor (protection through coercion and law). In the case of Bangladesh, East Pakistan and even Eastern Bengal before it, the state has never been in a sufficiently autonomous relation to the dominant rural classes to challenge their ability to dispossess other dependent classes.

Extend Notion of Rural Property Rights

But can we envisage other property rights that could entail a right to productivity of land in a predominantly agricultural economy at a rental value equivalent or beyond that derived from leasing-in or from agricultural labour? Rights where the rules and customs surrounding their possession are not so deeply embedded in the social structure, partly because the potential significance of these rights (or means of production) are only just being realised in the context of technological innovations. The major example of such rights in Bangladesh is of course water, especially in the dry season with new forms of irrigation. The entire activity of constructing and maintaining the infrastructure for the various kinds of irrigation, the operation and maintenance of the machinery involved, the allocation of water between cultivators, flood control and drainage works—the performance of all these services constitutes the creation and possession of means of production, required to be used in conjunction with land. The list can be easily extended to self-employed gangs undertaking land improvement as well as crop operations, the manufacture and maintenance of farm equipment, the control and use of processing equipment, post-harvest jute and sugar-cane processing, control over input distribution and so on. Additional property rights somewhat separate from agriculture, and therefore not involving the same interaction with landholders and cultivators can also be envisaged: fish tank cultivation, rural works, rearing of non-draft livestock, construction of buildings, trading, brick kilns, rural light industries and so on. It is important to remember though, that some of these rights are currently held by women in cultivating families.

It is this broader conception of rural property rights where customary rights are weaker which can provide the state with the basis of an orderly retreat over many years. *I am suggesting that state intervention in rural and agricultural development should be guided by the principle of establishing (as a precondition of its withdrawal) property rights other than land in the hands of those from whom effective demand is required in order to stimulate an increase in agricultural production.*

This proposition recognises at least the outer limits of room for manoeuvre for the state; deals with the rural system as a totality by combining the issues of effective demand, production and to a lesser extent equity; reflects the limitations of state intervention but avoids the disastrous consequences of an immediate private sector solution in the prevailing system of rural class relations. This is the key to a programme of agrarian reform which because it does not immediately confront the question of land reform may be within the realms of possibility. It represents in my view the least that should be done both to ensure the viability of the supply-side strategies envisaged by the Ministry of Agriculture (MOA), and to reverse the process of rural poverty.

Does this approach assist the creation of productive relations? More equalisation of the rental share from the productivity of land via the landless possession of property rights (e.g. water) means a shift in the terms of exchange between the factors of production surrounding agriculture toward more equal bargaining between the sub-sectors of rural economic activity. The near-monopoly position of larger landholders in establishing the price (generally low) for goods and services will be challenged. Landholders will be obliged to respond by optimising the relation between their factors of production. This logic is similar to the commercialisation approach (reducing subsidies and allowing food prices to rise) except here there is provision for generating effective demand through a wider distribution of the product.

It is a corollary of this strategy that poorer farmers and tenants must continue to be cushioned from the precipitous effects of this shift in the terms of rural exchange. This is a critical principle guiding possible institutional outcomes. However, it should be the case that sharecropped land too would become more productive under these conditions where the sharecropper was part of a group which was receiving non-land rentals. The likelihood, though, is that sharecropped land would steadily disappear under these conditions leading to increased land productivity (i.e. the leasing capital relation would be transformed). However, increases in the productivity of land will generally have this effect anyway.

These observations raise the issue of whether such a strategy would exacerbate the concentration and polarisation of landholding; and if this is a possible result, does it matter? The effects of concentration would be offset by the proliferation of significant property rights other than land, and may even lead to a greater rationalisation of the use of land. However, the risk of assisting the rise of a *kulak* class is unacceptable ... to me at least. Thus a feature of this strategy must be to invest greater resources in the protection of small farmers and sharecroppers particularly against the emasculating effects of usury. This protection can take several possible forms which are suggested in the next part of this chapter.

Finally, there is a possible outcome of this strategy which is of long-term concern. The initiation of such an agrarian reform strategy may establish the social and political preconditions required to extend the state's room to manoeuvre on key issues such as land reform. Securing non-land rural property rights provides the material basis for more solidarity and mutual trust among poor cultivators and landless, a requirement for autonomous political participation. To the extent that the holding pattern of these non-land property rights is held through group, association, union and co-operative arrangements then meaningful political pressure would exist from a numerical majority to socialise the holding of land. As such the political structures which emerge constitute the preconditions for the eventual socialisation of land.

Implications for Action

Combined Rural Policy

Programmes for the generation of effective demand involving the participation of those from whom it is required, have to be pursued in conjunction with the supply-side objectives to increase production. The methods of supply and the emphasis to be given to the various categories of technology (e.g. low-cost, quick yielding, minor irrigation activity) need to be determined in accordance with this strategy. The need then is for *a combined rural policy* which will reinforce rather than detract from agricultural production.

Subsidise Technology Used by the Poor

A further general issue on the supply side concerns the question: which classes use which technologies? There is persistent evidence that the largest subsidies apply to the technologies from which the richer farmers benefit most (Edwards et al 1978). As a general principle then, and as a matter of aggregate pricing policy on inputs, the pattern of technological use must be established—subsidising those technologies which smaller farmers use as a greater proportion of their total investment. Lifting all subsidies on fertiliser or other inputs is not going to help the poorer farmers. Since the strategy is to retain the small farmer as a beneficiary of rises in land productivity—this issue is central.

Focus on Poor Group Activity which is Exclusive

The relation between technologies and classes is not only relevant to supply and subsidy strategies. It raises an important question about the extent to which an activity is exclusive to particular classes. If poorer groups are associated with particular technologies and activity which the richer classes avoid, then support for such activity is less likely to be undermined by the rich. This criterion should be a guiding principle of the strategy. A clear example of such exclusiveness must be physical labour—so any programme involving a high element of personal physical labour would support the strategy (Greely 1979).

Co-operative Investment, not just Saving

An emphasis on co-operative *investment* must be another principle of this strategy. This affords the greatest protection for the poor membership. Exclusive reliance on individual loans leaves the individual vulnerable to relationships of personal dependency with a patron. While some commentators might apply this point more to women, because of their often weaker social position, it applies equally to men. Co-operative investment on the other hand is more likely to prevent cheap credit being leaked to service immediately pressing, expensive and usurious (rather than productive) obligations.

Pace of Group Formation and Internal Composition

A further guiding principle of this strategy must be the *pace* at which groups develop and receive support in the form of loans, services, training and so on. Evidence of the group's solidarity must be a criterion for such support, if the principle of Co-operative investment is to be meaningful. A group with tenuous solidarity is more vulnerable to takeover and manipulation by landlords and patrons functioning as brokers between the fieldworkers and such groups. In discussions on this strategy, the assumption has been that a group might consist of the poorer kin in a *para*, lineage or extended family. However, the homogeneity of the group members in this sense may be its weakness since they would all be dependent upon the same patron (e.g. from the strong *gosti*), and therefore jointly vulnerable because of this shared relationship of dependency. A heterogeneous group membership (with respect only to this variable) may protect it further, since the separate dependency relations would be a weaker threat to the attempts to develop mutual trust among themselves. The NGOs have been particularly aware of these issues of solidarity and pace, and some of them have tried to avoid the 'target-group' mentality (Wood 1980a). A group which is encouraged to take on too much too soon either collapses or simply transfers its dependency syndrome to the field worker instead of being responsible for its own decisions and investment.

Phases of Co-operative Activity

The strategy also needs to consider phases of co-operative activity—identifying the easier, perhaps more autonomous projects first as a means of developing mutual trust within such groups before moving on to the more delicate task of interacting with larger landholders through the provision of goods and services to them (Wood 1980a). It should always be a guiding principle in rural development to begin with what people already do. With small farmers, for example, there are well-established patterns of mutual aid: sharing of bullocks, joint cultivation of adjacent plots and mutual help on land improvement. From this basis other activities can develop such as collaboration on form-filling, joint meetings with village extension agents (VEAs) on technical issues, sharing of pumps, livestock management, grain processing (involving female members especially in pooling site, equipment as well as labour).

Defence of Vulnerable Land Rights

In addition to earlier comments on alternatives within IRDP, there might be a policy of extending rural banking in a role of protection for victims and potential victims of usury and eventual expropriation of land. The formation of sharecroppers' co-operatives supported by a land mortgage bank system could achieve this. An individual small landholder in trouble would mortgage his land to the co-operative instead of to his creditor. The co-operative borrows the value of the mortgage from the bank which is passed through the mortgagee to pay off his creditor without incurring crippling interest charges. The mortgaged land is cultivated by the co-operative with the mortgagee as a member until such time as it can be released by the mortgagee (at the lower bank rate of course).

The same co-operative would also attempt to become the collective village (or *para*, the unit is immaterial) sharecropper. Part of their role would be to convince landowners (who may be small and absentee/infirm, as well as large and absentee or non-tilling) that if land is held by the co-operative for a minimum of 'x' years (enabling improvements and incentives to invest) a one-third share to the landowner would yield a higher net return since it would be based on the increased productivity of the land. In this way a real possibility exists of extending irrigation to more sharecropped land. Even without an alteration of the crop-share, there is now evidence that irrigated sharecropped land matches owned/irrigated land for productivity. But the crop-share has to be altered to create the incentive.

Short-Term Consumption Credit

Since these arrangements would be a form of co-operative investment, the risk of leakage of benefits to patrons/creditors would be reduced. If the rural banking system was also able to provide short-term consumption credit guaranteed by the co-operative then the cycle of usury and unproductive capital would be counteracted. This issue of consumption credit, tied to the group or co-operative's guarantee, is of general relevance to all these problems of dispersing property rights. Without it, the leakage problem threatens the entire strategy of transforming usurious capital relations into productive ones.

Water Rights—Large-Scale, Surface Water

Water clearly requires a longer and separate discussion to the brief remarks which follow here. The first problem is the expansion of large-scale gravity-flow irrigation systems, referred to earlier in the remarks on canal digging. Experience with these systems elsewhere (e.g. in India) where they exist in the context of private landholding, shows that the public nature of the goods and the associated services cannot be effectively absorbed within the competitive structure of relations between individual farmers, even if they are in the same class. Examples of problems are: stealing and hoarding water; reluctance to host field channels routing water to the less favoured plots of one's neighbours; tail-enders; responsibility for maintenance; and ensuring timeliness of supply. The establishment of large irrigation departments to manage these systems has in effect placed this critical agrarian means of production in the hands of the bureaucracy and the state. This has resulted both in severe issues of corruption/market manipulation down through the hierarchy of engineers and the absence of integration with the agricultural agencies and the agricultural requirements of the locality.

Again, the strategy being proposed here has the potential to address these problems. Although the technology is lumpy and cumbersome, it does have the virtues of harnessing natural resources (surface water), being labour intensive and reducing the dependence on imported technology. The RWP and FWP function to retain this means of production in the control of the state, providing it with a resource of patronage to its petty-bourgeois contractor allies and through the unpredictable dispersal of below-subsistence wages to the landless. But the potential contribution of surface-water exploitation to the expansion of effective demand could be much greater.

The Formation of Irrigation Management Groups

Irrigation management groups composed of the landless (and poor cultivators with additional employment/income needs) may in effect function as 'brokers' between the public attributes of this form of irrigation and the private form of use. The possession of such water rights, involving the creation, control, management, allocation and maintenance responsibilities of the physical infrastructure, not only provides these groups with an income but places them in the position of arbitrating between the competing claims of different farmers. To the extent that poorer farmers (owners or owner sharecroppers) are members of these groups so will the problems of biased distribution in favour of the rich be reduced, as the group's management decisions would reflect the needs of these members.

Water Rights—Small-Scale and Ground Water

The more divisible irrigation technologies do not face these sharp contradictions between the public nature of the good and the private use of the service to the same degree. This is especially so if aspects of the technology are mobile (for example, pump sets). But where mechanisation is involved, then an additional set of relationships would obtain between a group and either government agencies or private enterprises around supply, repair, maintenance, fuel etc., and over the location and sinking of borings. The reform of the BADC and ensuring its accountability is an integral part of the state guaranteeing the property rights on these forms of divisible and mechanised irrigation technologies.

Water Rights—Potential

Finally, the potential for establishing such a pattern of water rights in Bangladesh ought to be enormous since it is generally assumed that there is sufficient water to irrigate a further seven million acres. This can only mean that the pattern of water rights in these areas, especially with respect to non-traditional technologies, has not become institutionalised with highly-developed vested interests around them. This lack of institutionalisation is a considerable resource. Therefore a guiding principle, not only on the irrigation aspects of these proposals but over other activities too, is to focus on those rights and in those areas where the rights are not so established. With water rights, a working hypothesis would be in the flood prone/famine prone areas where the perennial systems are difficult to maintain, involving a corresponding higher use of mobile technology such as low-lift pumps.

Conclusion

These final references to water rights and landless groups selling irrigation highlight a contemporary problem of agrarian policy in Bangladesh. The government is being urged by the World Bank and other international agencies to embark upon an immediate programme of 'flooding' Bangladesh with minor irrigation technology, based on the principle of selling and renting the hardware in the open market at subsidised rates. The quantities of STWs and LLPs through IDA credit (supported by an FAO/UNDP project to retrieve the DTW programme in the North-West of Bangladesh) promise to be enormous. Over the next three years, numbers of 50,000 for LLPs and 150,000 for STWs have been proposed. Without the widespread adoption of a countervailing strategy as outlined above, this prevailing policy on irrigation will quickly establish a new class of waterlords and effectively deny any possibility of significant room for manoeuvre for the government over this critical asset. The same problem exists over a whole range of assets and commodities.

The motivation for this dramatic policy of rapidly expanding the supply of minor irrigation technology is a preoccupation with securing a higher rate of growth in agricultural production and corresponding reduction in the level of food aid. It is the argument of this chapter that far from being a recipe for growth, this policy will establish the preconditions for a long-term steady decline in the economy by concentrating the pattern of ownership and profit. A rise in agricultural production under these conditions could only be secured through an expensive procurement policy to maintain price incentives to the potential producers of a net marketable surplus. In this case all that would have been achieved would be a switch from food aid to capital aid to fund the procurement strategy, accompanied by an elaborate rationing system to distribute food in the countryside to those excluded from income-generating activity—in short the extension of a relief economy. I think the onus of refuting this conclusion now lies with the perpetrators of the privatisation policies.

Chapter 7

Provision of Irrigation Services by the Landless: An Approach to Agrarian Reform in Bangladesh

Introduction: The Problem of Landlessness

Bangladesh is one of the most densely populated rural regions of the world. The population growth rate is also one of the world's highest at around 2.3% (Arthur & McNicholl 1978). The ramifications of population growth throughout the agrarian structure are pervasive: larger families living off the same size or reduced size of holding; fragmentation of holdings; dispossession of infrasubsistence peasants; decline in the availability of land for sharecropping; increase in the number of landless; enhanced value of children as income-earners and providers of old-age security; depressive effect on wages; and undermining of patron-client dependency relations which offered minimal security to labour, with a reserve army of labour contributing to the commercialisation of labour relations.

Elsewhere I have argued (in Part I, Chapter 2) that, in general, rural class relations in Bangladesh have been characterised by the use of capital in the sphere of exchange rather than production (renting, moneylending and trading) which has effectively raised the level of absolute surplus value accruing to certain classes, but which is not currently contributing to the generation of productive agricultural capital. This failure to develop productive agricultural capital resulted in a form of polarisation during the 1960s and 1970s which involved a dramatic rise both in the numbers of landless and the levels of rural unemployment, thereby creating a problem of effective demand for the required expansion in agricultural output, especially foodgrain.

Growing landlessness is less of a problem if alternative rural employment opportunities are available. Unfortunately, this is not the case in Bangladesh. The annual population growth rate is such that 45% of the population is under 15. The rural labour force is increasing at about 4% a year, i.e. approximately one million men and women. This increase far exceeds the number of additional employment opportunities available in agricultural and other sectors. The World Bank estimated that during the first half of the 1980s approximately 2.3 million new entrants to the rural labour force will not be able to find gainful employment. The rapid increase in unemployment has a direct impact upon agricultural production. Limited employment means that a substantial segment of the rural population must limit food purchases. If production increase outpaces demand this could result in low producer prices, the growth of large government-financed food surpluses under conditions of persistent malnutrition and even pressure to export food to reduce internal debt.

Food Production Strategy and Expansion of Minor Irrigation during the Second Five Year Plan

The response to these problems in Bangladesh was contained in the advice given to the government by the World Bank and incorporated by the government in its 1980-85 Second Five Year Plan (2FYP) and the longer Medium-Term Food Production Plan (MTFPP) (IBRD 1979). Essentially the strategy was based upon a significantly expanded investment in agriculture and rural works, but on conditions deriving from the analysis in the 'Food Policy Issues' World Bank report. The report lists the physical constraints to the development of agriculture as: scarcity of cultivable land; limited yield potential; climatic hazards; and the need for dry season irrigation and improved drainage and flood protection. However, it attributed the slow growth of agricultural output to: weaknesses in agricultural support services; inadequate provision of input supplies; and constraints stemming from the agrarian structure. Any expanded investment was dependent on external aid and both the World Bank and US AID were determined from their diagnosis to liberate the rural economy from bureaucratic management and control by open market systems of food and input distribution. This is not the place to investigate further the wider implications of this strategy which consisted of open-market dealing in food, the phasing out of fertiliser and other input subsidies (on various forms of irrigation), coupled with a major emphasis on HYV technologies in rice and wheat (Wood, 1980b). These efforts concentrated on the improvement of foodgrain supply with a noticeable lack of attention to the structure of demand for food. The strategy involved an assumption (untested) that a combination of support prices (through a generous procurement price) and open market dealing in rural goods (inputs and food) would encourage the landholders to raise the productivity of their land, thereby increasing their marketable surplus to pay the full market price for inputs.

However, despite these issues, the 2FYP envisaged an unrealistic annual average increase in foodgrain output of 6.5% to be achieved principally through the development of minor irrigation and doubling the supply of fertiliser. The irrigated area is expected to double from 2.59 million acres to 7.22 million acres, representing 32% of the maximum feasible cultivation area of 22.51 million acres. The greater proportion of this increase (45% according to the MTFPP) is to be through shallow tubewells (STWs), with deep tubewells (DTWs) 32% and low-lift pumps (LLPs) approximately 20%. Deep tubewells are unavoidable in some areas such as the North-West, despite the problems experienced through earlier programmes of organising farmers' irrigation groups; and the increase in LLPs is technically limited by the availability of additional surface water resources. Since the slow rate of irrigation development in the previous plan periods (1973-78, 1978-80) had been attributed to the low level of utilisation of equipment in the field due to a policy of overgenerous subsidisation involving heavy public sector costs and inefficient, corrupt implementation, a policy of privatisation was adopted. Privatisation represented the attempt to deploy lines of credit, based in the first instance on US $40 million from IDA, and to sell STWs and LLPs to any individual farmer, contractor or rural entrepreneur, or groups and cooperatives who could appear in the market with the collateral on which a loan could be extended. No quotas were established even for co-operative ownership based on the Comilla-derived IRDP-KSS system (i.e. the Integrated Rural Development Programme consisting of village level farmers cooperatives—*Krishi Samabaya Samitis*—federated at the *thana* level).

Provision of Irrigation Services by the Landless—an agrarian reform response to the problem of weak effective demand

If, as argued above, the constraints to production were not just technical but social, embodied in the structures of tenancy, moneylending, tied labour and the related problems of effective demand stemming from precapitalist forms of servitude, then this emphasis upon transforming the significance of a major rural means of production—irrigation water—should not be conceived independently of these other aspects of the rural political economy. The dangers of the privatised minor irrigation strategy was the concentration of these new, relatively un-institutionalised water assets in the hands of those who already possessed rural capital either in the form of land or moneylending and other commercial forms of accumulation. Such a pattern of control over water might ensure that large farmers obtained water to expand their production into another process. But can such a strategy work in a rural economy whose structure is so inegalitarian on the basis of highly differentiated and dynamically polarising access to land? As one test of these propositions, the issue of water quickly assumed prominence in the context of the major programme to sell minor irrigation technology as part of the food production strategy (Bhuiyan, 1984).

The Landless Irrigation Experiment

With this point of departure, the action-research consisted of an experimental programme with groups of landless and near-landless workers in rural Bangladesh owning minor irrigation equipment and providing an irrigation service to farmers with plots of land in the command areas, in exchange for payment either in cash or crop-share. This 'landless irrigation' programme occurs within the framework of the Bangladeshi non-government organisation, Proshika, which has assisted in the mobilisation of landless groups (based on self-definition) all over Bangladesh, and which has supported these groups through programmes of consciousness-raising, technical training and small loan support. The involvement of Proshika in this experiment was encouraged by the GOB Ministry of Agriculture, for whom I had elaborated the initial policy option (Wood 1980b). However, two other Bangladeshi NGOs have also participated to a lesser extent—the Bangladesh Rural Advancement Committee (BRAC) and the Grameen Bank (a private poor people's bank). The Bangladesh Rural Development Board (BRDB—a division within the GOB Ministry of Rural Development) is also committed to its own version of the experiment commencing in the 1983-84 season. The material presented here is based upon my analysis (with Proshika's R&D cell) of 83 group projects in the 1981-82 season. Subsequent analysis based upon data up to 1988 appears in the separately published *The Water Sellers* (Wood & Palmer-Jones, 1991).

These 83 groups (51 using STWs; 32 using LLPs) are distributed over 11 regions of Bangladesh, innovating under very different sets of ecological and socio-economic conditions. As a result there are many variations in the practices affecting agreements with cultivators, forms of payment, the value of the water fee as a proportion of the crop, size of command areas and their cropping characteristics outside winter *boro*, irrigation practices, rates of pumping, owner-tenant composition of the farmers in the command area, numbers of group members themselves cultivating small amounts of land in the command area, prices of paddy, sources of finance, and so on. The nature of these variations could not be predicted at the outset. The range of possibilities could only be learnt by practice. This means that the groups in discussion with the Proshika R&D cell and myself have had to identify, by a process of trial and error, which particular combination of variables best suits their local circumstances.

Objectives

The programme has a number of interrelated irrigation, economic and political objectives. Most simply, it was intended to facilitate the acquisition and use of STWs and LLPs by landless groups to enable them to sell water to owners and cultivators of land. Behind this lay other purposes: to ensure that landless groups share in the benefits from the enhanced productivity of land to which they contribute by providing the source of irrigation; and to achieve a more efficient use of water by improving the access to irrigation of small farmers who are usually the tail-enders in other systems of distribution for minor irrigation.

Related to these primary irrigation objectives are the attempts to: prevent landlords and rich peasants becoming at the same time waterlords and monopolising this increasingly important means of production; demonstrate that other rural property rights than land can exist from which landless classes can derive both rent and a corresponding enhancement of political status in the community; raise the level of effective demand for expanded domestic foodgrain production thereby sustaining price incentives for cultivators, through ensuring a secure access to income; demonstrate that loans can be extended to poor people for productive use without depending on land as collateral; and improve through practice the independent access of the rural poor to institutions outside the patron-client ones within which they are oppressed.

Importance of Group Members' Solidarity

The organisational originality of this programme is the high level of interaction required between the group members and other classes in the community upon whom they are traditionally dependent. There are plenty of examples throughout the whole Indian sub-continent of small groups of landless men and women engaged in separate, small-scale economic activity. The difference with irrigation is that the groups are obliged to enter into a range of complex negotiations with landowners in order to establish, in the first place, a command area of adjacent plots, followed by the making of formal agreements in which fee income, rates of water delivery, cultivation practices and length of contract (usually five years) are specified. These exchanges with other classes in the community then continue during the irrigation season and beyond into early *aman* (June) until the payments are made, to be resumed again perhaps as soon as late *aman* (late October) if the rains fade early and additional irrigation is required.

In practice, then, relationships between landless groups and cultivators around the provision of an irrigation service is more or less continuous. In addition, the groups have to develop successful working relationships with other institutions especially the local branches of the Bangladesh Krishi Bank (BKB) and the BADC. The BADC, as part of the Ministry of Agriculture, is currently responsible for the supply of equipment, availability of spare parts, undertaking feasibility studies of proposed command areas, and maintaining a zoned distribution of installations. There are obviously many points of access difficulty between the groups and these institutions with local officials who are more used to dealing with richer rural classes. In these circumstances the groups themselves require qualities of internal solidarity and trust in addition to those needed for more self-contained, cooperative activity. To provide irrigation is to participate directly and crucially in the agricultural system. As a result, the groups which have become involved by Proshika in the programme have been selected on the basis of such qualities from a larger number wishing to participate.

All the irrigation groups have been in existence for a few years; they were not formed purposively by outsiders for inclusion in this scheme. The criterion for a group's involvement is essentially political—the proven capacity of its members to stay together under pressure from the richer peasants/petty landlords in the village, and to sustain their small income-generating projects either against similar political attack or during, for example, the extreme drought conditions of 1978. These criteria were not laid down in a formal sense, but local Proshika workers discussed these issues with the groups in a locality, where the strong groups are known. It must also be appreciated that the membership of these groups is not defined by outsiders (i.e. Proshika staff) according to some strict definition of landlessness. Individuals are approached by Proshika workers often through local volunteers or by other groups if they already exist in a locality. The advantages of joint organised action among the village poor are discussed, and individuals are left to decide whether and with whom to form a group. In doing so, the 'target-group' terminology of 'landless' is modified by broader notions of kin and friends who are poor, who occupy similar positions in village dependency structures, who may have land but only a precarious hold over it, and who are, therefore, in the same *dynamic* of rural poverty even though an individual's exact location in it may vary at any one moment in time.

It is the recognition that this dynamic occurs for some households not others which brings such people together, rather than an exact similarity of position. This is why target-group strategies as such are blunt instruments of rural development. For the purposes of our various sponsors we attempted a definition of a 'landless group' which was:

A group with no control over the means of production or distribution; landless or marginal farmers with no assets; fishermen with no implements; rural artisans who lack working capital or raw materials; families who sell their manual labour; women of the above groups.

In practice, in Proshika, the emphasis for group membership has been on the use of one's own physical labour. Significantly, therefore, holders of small amounts of land have also been involved in the groups and the schemes in all the project areas: 38% of group members in STW schemes have some land; 61.5% in LLP schemes; and 48% in total. 30% of group members have some of their own land in their group's command area. The landholdings are usually tiny and insufficient to sustain the family's consumption requirements, so that wage labour has been a central part of their economic activity.

Action-Research Basis of Analysis

The action-research of this work has included both an examination of the linkages among a network of organisations, as well as an analysis of the financial and technical aspects of a group's performance. The exercise of learning, reporting, analysis or evaluation has been endogenous to the activity itself, where the method of reporting had to be consistent with the actual requirements of carrying out the work. The categories of data to be recorded were designed to provide a self-questioning and learning experience for the different levels of actors in the system at the time of producing such records, so that their (our/my!) own behaviour could be immediately informed and modified where necessary. There are general lessons here for project interventions to prevent the removal of knowledge vital for immediate use from its point of 'production'.

As far as possible then the data on which the analysis of performance depends are not based on recall but on the records that the groups must anyway maintain of their income, expenditure, loan arrangements, fee arrangements, agreements with farmers, and so on. Groups also kept records of fuel consumption and hours of pumping by maintaining a log manual. Independent measurements were made of machine efficiency. In addition a proforma was devised (on the basis of many hours of instructional learning through discussions with groups in the first, highly experimental season) for the group to complete jointly with Proshika workers; and Proshika workers were issued with a set of guidelines on which to base reports on wider issues such as employment, demonstration effects, the implications for poor farmers and tenants, variations in fee arrangements between projects, and so on.

From our first season's experience (1980-81) there were some problems which appear to be genuinely one-off and can now be regularly transferred to new irrigation activity: setting a date for the making of agreements, obtaining loan sanctions, and a deadline for the date of equipment installation especially in the case of a shallow tubewell where farmers have to be convinced of the availability of water approximately six weeks ahead of the season to commit themselves to seedbeds. However, it must also be recognised that in each landless group and with each command area there is generally some scope for improvement either in the size of the command area, the irrigation practices followed together with rates of pumping, and in pricing arrangements for the service.

Review of Performance

A number of variables has affected the performance of this social experiment in providing an irrigation service to farmers. However, since these factors are not necessarily common to both groundwater and surface water technologies (STW and LLP respectively) they are discussed separately at this stage. Table 1 provides a summary breakdown of the STW and LLP groups by region. It indicates the scale of the programme in different areas and the distribution of success and failure as measured narrowly in financial terms. If depreciation on equipment is excluded from the cost side, 75% of STW groups and 78% of LLP groups emerged as financially successful. This meant that 13 of the STW groups and seven of the LLP groups were in some difficulty. Several of these were concentrated in particular areas, which indicated a regional dimension to their problems. The exclusion of depreciation from the calculations for the presentation of these results is justified on the grounds that some of the groups show a net loss which is below the annual figure allowed for depreciation on equipment. Net losses which are marginal in this sense give a slightly misleading impression of the capacity of the groups to meet all their debt servicing commitments and the initial, one-off acquisition of essential capital equipment, additional to the pump set such as tools, drums, the construction of a hut, and so on.

Groundwater—STW Groups

Although there are some dramatic financial successes, the average net return per STW group scheme is much lower than for LLP schemes (Tk. 1256 as against Tk. 6906). However, the extent of innovation required by the groundwater technology from water sellers and farmers alike must also be recognised. At this stage of groundwater development in Bangladesh, there has been little experience in any of the STW localities of irrigated HYV *boro* rice cultivation. New command areas and new concepts of water management, field operations and applications of inputs have to be brought into an optimum combination for the specific conditions obtaining in a locality. Local knowledge has to interact with the technical specifications of the equipment and new varieties. The key variables which account for the pattern of STW performance are: size of command area; diesel cost as a proportion both of income and operating costs; rates of diesel use per acre; and forms of agreement between pump groups and water users. Other variables whose general significance is less clear, though their presence can be detected in particular cases of strength or weakness, include: first season capital costs; salaries for group operatives; oil and grease; spare parts; repair services; payments of water users' fees in advance; and the efficiency of the pump equipment.

There is a strong association between command area size and group performance. The average command area for successful groups is 14.24 acres as against 9.71 acres for unsuccessful groups. Overall the average command area size was 13.39 acres rising to 13.69 acres for 34 groups for whom 1981-82 was their first season. This indicates significant learning from the first season of the experiment when delayed dates of boring and equipment installation resulted in suboptimal command areas. For the 1981-82 season, groups entering the programme for the first time were informed of these problems through Proshika workers, workshops or word of mouth from other groups, and a deadline for installation was established and adhered to. This gave the farmers the necessary confidence to commit their plots to a command area (i.e. to implement their agreement to do so) by investing in HYV seedbeds, plot preparation and digging field channels. The achievement of a 13.69 acre average in their first season shows that with correct timing a group does not have to creep towards its optimal command area over a period of years.

The first variable which limits the significance of command area size is the form of payment between the sellers and the purchasers of water. Our evidence indicates that those groups receiving payment in the form of a 33% share of the standing crop were the most successful as measured by gross income per group, gross income per acre, or net return per group and per acre. In all cases where such a payment was made, the agreement included an insistence that the cultivators perform the full related package of cultivation practices (i.e. weeding, fertiliser and pesticide application). Although reliable yield figures (independent of those derived from crop shares received which suggest a range of per acre yields from 257-316 *maunds*) are not available, the high income per acre figures for the groups with this arrangement suggest that it also provides the greatest incentive for the cultivators to maximise the productivity of their land. Other forms of payment—including fixed cash arrangements, fixed kind arrangements and 25% crop share—have brought successful groups relatively low net returns. In one of the less successful areas (Bhairab, Mymensingh District) levels of income are clearly the main problem since performance with respect to other key variables has been satisfactory.

Diesel expenses represent the largest proportion of operating costs and they were therefore always expected to be significant in determining success or failure. Important subsidiary factors influencing this variable are the rates of machine efficiency and types of soil. Within each region there is a clear correlation between diesel costs as a proportion of income, diesel used per acre and performance. Since both successful and unsuccessful groups had similar diesel/total operating cost ratios (25% and 23% respectively) we conclude that it is the relationship between income and diesel cost which accounts for performance. Groups in the region with the highest net return per acre were found to have the lowest diesel cost/income ratios (between 52% and 60%).

TABLE 1A

LANDLESS GROUPS DEPLOYING STW TECHNOLOGY

Groups	MIRZ	CHAT	BHAI	SATU	GHIO	SHIB	NAGA	MADA	Total Ave.
No Groups	5	1	14	4	3	11	8	5	51
Total Members	73	9	234	83	79	293	232	86	1089
Average Members	15	9	17	21	26	27	29	17	21
Command Area									
Total Acres	59.76	6	183.1	53	39.76	176.5	98.08	68	684.75
Average Acres	11.95	6	13.09	13.25	13.25	16.04	12.26	13.6	13.42
Tot Cultivators	157	24	421	149	80	402	202	167	1602
Ave. Cultivators	31	24	30	37	27	36	25	33	31
Ave. irrigated area per cultivator	.38	.25	.43	.36	.50	.44	.49	.41	.43
After Depreciation									
Net Return No.	3	0	4	3	3	8	5	2	28
Net Return %	60	0	29	75	100	73	63	40	55
Net Loss No.	2	1	10	1	0	3	3	3	23
Net Loss %	40	400	71	25	0	27	37	60	45
Without Depreciation									
Net Return No.	3	0	10	3	3	10	5	4	38
Net Return %	60	0	71	75	100	91	63	80	75
Net Loss No.	2	1	4	1	0	1	3	1	13
Net Loss %	40	100	29	25	0	9	37	20	25
Water Fee	Cash	Cash	7-25%	33%	33%	Cash	33%	33%	
Arrangement			Share 5 Cash 2 Kind	Share	Share		Share	Share	

Key to Names of Groups
MIRZ - Mirzapur/Tangail-D
CHAT - Chatalper/Comilla-D
BHAI - Bhairab/Mym-D
SATU - Saturia/Dhaka-D
GHIO - Ghior/Dhaka-D
SHIB - Shibgonj/Bogra-D
NAGA - Nagarpur/Tangail-D
MADA - Madaripur/Far-D

TABLE 1B

LANDLESS GROUPS DEPLOYING LLP TECHNOLOGY

Groups	ULAN	CHATALPER/COMILLA-D			KHAL	KALK	Total /Ave.
		Rent	Purch	Total			
No Groups	7	8	2	10	12	3	32
Total Members	113	442	26	468	233	44	858
Average Members	16	55	13	47	19	15	27
Command Area							
Total Acres	115.9	427.5	100	468	575.7	53	1151.4
Average Acres	16.6	53.4	50	46.8	48	17.7	36
Tot Cultivators	185	691	236	927	480	83	1675
Ave. Cultivators	31	86	118	52	40	28	52
Ave. irrigated area per cultivator	.63	.62	.42	.50	1.20	.64	.69
After Depreciation							
Net Return No.	1	6	1	7	12	2	22
Net Return %	14	75	50	70	100	67	69
Net Loss No.	6	2	1	3	0	1	10
Net Loss %	86	25	50	30	0	33	31
Without Depreciation							
Net Return No.	3	6	2	8	12	2	25
Net Return %	43	75	100	80	100	67	78
Net Loss No.	4	2	0	2	0	1	7
Net Loss %	57	25	0	20	0	33	22
Water Fee	Cash	7 K	1 C	7 K	11 K	Cash	
Arrangement		1 C		3 C	1X50% Share		

(K = Kind, C = Cash)
Key to Names of Groups
ULAN - Ulania/Barisal-D
KHAL - Khaliajuri/Mym-D
KALK - Kalkini/Far-D

Surface-Water—LLP Groups

The analysis, especially financial, of the LLP groups is complicated by the fact that eight of the groups are renting their equipment rather than purchasing it. In the long term, concern should be focused on the viability of the purchased equipment since the highly-subsidised rented equipment is being phased out under the privatisation strategy. The renting groups are in a single crop area (deeply flooded during the monsoon) with a 15-year history of LLP irrigation. In such areas where there is strong competition for limited water supplies and where water is even more crucial without a rainfed *aman* in reserve, we had hoped that the landless groups might have been able to participate in rental agreements over a longer period. This would have allowed them time to capture water rights from established entrepreneurs by operating under the same favourable subsidy conditions, and would have helped them accumulate capital for a subsequent purchase. Lobbying to this effect both with the GOB and the World Bank has been ignored. In the 1981-82 season, seven of the 32 groups using LLPs had problems. It was expected that LLP groups would appear immediately more financially secure than the STW groups because both rented and purchased equipment was more highly subsidised; farmers in single-crop LLP areas were more dependent on and familiar with irrigated rice production; and the search for a command area could be more flexible as the equipment need not be permanently sited (this may still be an advantage in successive seasons). In Barisal, a tidal area, there were four unsuccessful groups with technical problems: pumping could only occur when the tide was in, thus command areas were too small and pumping rates insufficient to justify the use of 2 cubic metre per second (cusec) machines. These machines have now been switched to 1 cusec, and the 2 cusec machines resold to groups elsewhere. In the case of two other groups, lack of success is attributed to the low income per acre. Our data suggest that the groups were weak in bargaining over the water price with a large number of small non-group cultivators whose own margins on production costs were tight. One of these groups also had major problems with spare parts from an old rented machine. In the final unsuccessful case, a very low command area (8.99 acres!) appears to have resulted partly from the sandy soil and partly from interference by a 'rich man'. The sandy soil requires more water and therefore higher rates of pumping. There was a higher water charge to compensate for this, but this was not fully collected since a dispute existed over command areas with the rich man who prevented the machine from running. This case is interesting, because it reveals that some competition was taking place for rights over command areas in the locality. This is an example of the need to 'fund' a group during its attempts to capture or assert its rights over the provision of irrigation water in particular localities.

The remaining 25 groups using LLPs were successful in producing positive net returns, and the 12 strongest groups are located in Khaliajuri, Mymensingh District (North Bangladesh). All these 12 groups purchased their equipment and produced higher average net returns per group than LLP groups elsewhere (Tk. 16287). Khaliajuri is part of the *Haor* area, a large tract of low-lying land in Bangladesh which is perennially flooded and which is therefore a single-cropped area. In these areas, cultivators have been accustomed to *boro* irrigation performed by the traditional *dhone*. As a result the value of irrigation water has long been accepted by landholders. In some of the areas on the edge of the *Haor* where the ground is higher the crop has been broadcast *aman* (April to November/December) which with irrigation can now be replaced by a *boro*-transplanted *aman* rotation. Such a rotation has a much higher labour demand. This past history of irrigation (including 15 years of entrepreneurs renting LLPs from the BADC) has made farmers willing to commit land to command areas and themselves to long-term agreements.

However, the average size of the command area is 37.91 acres per scheme which might indicate some under-utilised capacity. But there are topographical constraints to expanding these areas. Most of the command areas are in effect extensions of existing *boro* land which has been *dhone* irrigated or low enough to retain sufficient moisture. These areas resemble very shallow, not necessarily circular, bowls (dishes, saucers) and mechanised irrigation is required for the 'sides' away from the low centre. The size of these shallow depressions imposes a natural limit to feasible irrigation. Also the quality of the soil is a factor. Six out of the 12 command areas have a clay constituent enabling irrigation over a wider area of less permeable soil for a given rate of pumping. In these six command areas the average size was 41.61 acres compared to an average of 34 acres for the command areas consisting of the more permeable sandy/loam mix. There is a strong correlation between the clay soil/pumping/diesel use variable and the net return per group. Groups with command areas averaging 41.61 acres, with a clay constituent and 95 hours per acre pumping rate have an average net return of Tk. 21032. Groups in the other category have an average net return of Tk. 10592.

Despite these overall positive net returns to the groups, the workers' fee levels remain low representing approximately 17% of the crop value. But the diesel cost/income ratio to the groups was still 23%. The difficulty of raising the price higher, or even of transforming it from a fixed kind to a crop share payment, is that the current rate reflects the charges under the subsidised conditions of BADC rental arrangements. The strong net return positions of these groups, while bearing the real cost of providing water (through purchasing the equipment), make it obvious that private entrepreneurs were realising huge profits from selling water during the rental era.

Social Implications of Landless Control over Irrigation Technology

This review of the groups' performance has been in terms of their financial viability and some of the technical conditions affecting it. While it is a necessary condition for gaining material independence, financial viability is by no means a sufficient condition of decreasing dependence for the landless and near-landless upon landlords, moneylenders and employers. Indeed financial success may be a misleading indicator of independence if it is merely producing a different pattern or relocation of employment opportunities or if cash surplus is immediately absorbed in the partial honouring of usurious debt obligations. So far this review has demonstrated that the involvement of groups with Proshika support in rural production via the ownership of irrigation assets need not fail. But are there structural implications beyond this with a change in the pattern of social control over irrigation technology? What are the constraints and opportunities for reproducing and expanding the landless control over those non-land, rural property rights?

Enhanced Status and Disengagement from Bondage

In their own evaluation, the groups stressed the impact upon their own status and confidence within the village community. They argued that they are accorded greater respect by the richer and stronger families precisely because they have demonstrated the ability to secure the assets, enter into agreements with the BKB, with the Government (i.e. the BADC) and the farmers, organise themselves efficiently to provide the services, maintain the equipment, honour contracts, and keep accounts and records. All these activities—despite setbacks, attempts at cheating, delays, requests for bribes, and the strangeness of interacting with fuel dealers, spare-parts distributors, mechanics, transport contractors, and so on—have strengthened rather than undermined their confidence. These experiences have been shared collectively among group members, avoiding the isolation which individuals would encounter in a similar process. As a result an important educational and consciousness-raising process is occurring as group members interpret and analyse their experience among themselves, between groups, and with other people in the network such as Proshika workers.

However, as suggested earlier, groups need to have demonstrated some of this potential *before* becoming involved in a programme such as this. The relationships required both within the village and outside it, present opportunities to others to undermine that solidarity. It has to be recognised for example that 48% of all the group members in the programme reported were in debt to local moneylenders, who are usually rich peasants with land in the command areas. Attempts by such land holders to subvert their agreements with the group by putting individuals under pressure have occurred and have been rebuffed. But there is a structural issue beyond this —are profits from irrigation activity being used merely to honour, *partially*, unproductive debt obligations or are obligations completely honoured (i.e. debts paid off, mortgaged land retrieved) in order to sever permanently such relationships? Furthermore is the dividend used collectively for further productive investment or dispersed to individuals for immediate (and understandable) consumption? It is still early days to have convincing evidence on these questions, but an example of these attempts by the groups to use this irrigation activity to disengage from bondage is revealing. In Khaliajuri, Mymensingh District, the change in cropping pattern from broadcast *aman* to a *boro*/transplanted *aman* rotation has substantially increased the local demand for labour. As a result instead of, as before, migrating to neighbouring Sylhet and taking loans from moneylenders to keep their families, from the 1981-82 season no-one from the groups migrated. Some local debts remain but are being completely paid, so that group members' labour was not locally indentured. Furthermore, group savings (from irrigation profits) were being used to help individuals in emergencies so that no group member was obliged to borrow from the moneylenders during periods of sickness and inability to work. This example shows how the unity of their own class within the village has begun to be tested successfully against competing loyalties to kin, patrons, landlords and moneylenders.

Significance of Land Fragmentation

The need for this solidarity extends beyond the delivery of irrigation for a rent to the process of capturing the water rights in the first place. The conditions under which groups secure water rights vary according to: cropping patterns; class relations in prior irrigation arrangements; class alliances with state institutions such as the BADC, the Courts, District Development Committees and local government; attitudes of different classes of farmers in the potential command area; and employment patterns.

However, perhaps the most crucial resource for the landless in their acquisition of water rights is the fragmentation of farms in Bangladesh. With increasing pressure on the land and the system of multiple inheritance continuing through lack of alternative employment opportunities, plots themselves are divided to reflect a balance of soil types between the inheriting sons. As a result, it is rare to find individual farmers with sufficient land in any single command area to exercise a monopsony influence over the price for water or to dominate its distribution. It is true that some farmers connected by kin relations might act as a pressure group, but we have not encountered such organised opposition from larger farmers in the command areas. The centrality of this fragmentation condition, with a farmer's plots being distributed perhaps between several command areas, means that farmers can positively welcome someone else taking on the management of irrigation instead of having to function as water entrepreneurs themselves in order to secure water for their own plots. The controllers of water have a much more consolidated asset than the holders of land in many parts of Bangladesh. In this process farmers will inevitably lose more autonomy over their production decisions, with their cropping cycles reflecting the needs of the plot's command area rather than the needs of the peasant family farm. This clearly represents a shift in the balance of social forces surrounding agriculture, in which we show the landless can benefit. The concept of the 'farm' in such parts of Bangladesh has clearly to be revised.

The Competition for Water Rights

Aside from the advantage of fragmentation, landless control over water is affected by two interrelated dimensions of access: to the command areas in competition with rich farmers/waterlord entrepreneurs; and to the technology in order to realise the value of water. This second dimension of access is a necessary condition for the first. In several of the areas where groups are using STWs, farmers have tried unsuccessfully in the past to obtain STWs from the BADC but they have been defeated by delays, requests for bribes, and other problems. The World Bank's emphasis upon unrestrained privatisation has been a reaction to this. Despite the current generous supply of equipment, in areas of landless group initiative, farmers have not sought equipment from the BADC. Instead they have been watching with scepticism to see if the groups could succeed where they had not. The issue for subsequent seasons is whether contractors or groups of richer farmers will be encouraged to compete for command areas, employing their traditional relationships of access to the BADC? The answer will depend upon the significance of the fragmented farm, and whether absentee contractors or farmers/entrepreneurs with their own land in the command areas can achieve full capacity utilisation of their equipment where the profitability margin on STWs is so tight. Since we predict these problems occurring in the widespread, uncontrolled private/individual accumulation of STWs, the reverse scenario may well appear where the landless group take over such command areas. In LLP areas, the picture is slightly different since the problem has often been to enter an existing market for water in which returns to contractors under rental arrangements have potentially been very high and local BADC officials have had considerable power of patronage. Under future 'purchase only' conditions in an open market situation without formal zoning controls, rights of prospective water sellers will remain insecure. Therefore, the period in which landless groups must acquire water rights is not finite, especially with LLPs. With the larger command areas, groups may fare better than individuals in the water selling business as they can perhaps apply more pressure and sanction in the collection of water fees. Also they can more easily ensure fair and regular allocations of water to the plots of different cultivators (different in the sense of class and status) in the command area.

Relations with Small Farmers and Sharecroppers

It was always recognised that the socio-economic basis for the programme did not consist of a simple antagonistic division between the landless and cultivating farmers, many of whom are small farmers and/or tenant sharecroppers. An objective of the programme is to improve the access of such small farmers to irrigation on the assumption that through their weak political position they suffer 'tail-end' problems. On the STW schemes, 32.5% of the owning cultivators have total holdings of less than 0.99 acres, and 54% less than 2 acres; for the LLP schemes the corresponding proportions are 43.5% and 59%. In total, 18.4% of cultivators in the command areas were sharecroppers. Two kinds of benefits to small owner-cultivators seem to be apparent. First, in the new command areas (STW and

LLP) this additional season, with potentially high yields, has some impact upon the total viability of their holdings making them less dependent upon joining the local supply of under-employed labour themselves. Secondly, in the older more established command areas (of LLPs) where there has been a bidding-down of the price for water in the competition to secure control, the economics of their winter cultivation is being transformed. In both cases, the particular problems of access to water is being managed by the groups, with whom the small farmers can interact on more equal terms.

For sharecroppers, there are two general issues which apply to the whole minor irrigation strategy in Bangladesh—whether the enhanced value of land through irrigation functions to displace them; and the effect of irrigation (its price structure especially) upon share agreements between tenants and landlords, and therefore sharecropper incomes. Obviously, longer time-series data is required to answer the first question. Several permutations exist for the second, involving the distribution of water and input costs on the one hand and the product on the other, and the effect of this distribution pattern on levels of production. While we have plotted these permutations by region, our evidence is not yet conclusive as independent data on yields are required. It is certainly best for the sharecropper when the landlord is sharing in the water cost either directly, or indirectly when a crop share is paid to the water seller before the landlord-tenant division takes place.

With this high incidence of small farmers and a significant population of sharecroppers, will a landless irrigation programme bring their interests together with the landless? The agrarian dynamism noted above reduces the distinction between them, along with the participation of very small farmers from the groups themselves. Certainly if small farmers gain better, regular access to irrigation, then their holdings will become more viable, perhaps arresting their downward fortunes. Thus for expediency alone, the conditions for an alliance between these classes in any locality should exist.

Organisational Innovation and the Socialisation of Irrigation

The groups themselves have evolved organisational responses to cope with the variety of relationships demanded by gaining access to the technology, delivering the irrigation service and receiving payment for it. There have been considerable variations from the initial assumption of one group (15-30 members) per scheme, especially in the case of LLPs. At present, STW groups are mainly operating independently of other groups, but in four schemes, two or more groups have joined together for the purpose of irrigation. In another case, an STW group received the support of other non-participating groups in a dispute with some of the farmers in the command area. However, for the STW groups, the command areas are small and so, therefore, are the numbers of farmers with whom the group has to negotiate over prices, crops, cultivation practices and water distribution.

The situation is different for LLP schemes. The size of the command area is much larger and, consequently, so is the number of non-group cultivators receiving irrigation. Under those conditions different units or levels of operation have emerged: an implementing group (possibly just a reference to the appointment of a driver or lineman from a group); a group signing for a loan from the BKB; and other groups (sometimes in the form of a landless village committee) which in effect collectively commission the activity and to which profit ultimately accrues.

Several reasons have emerged for this network of organisational arrangements. First, there is the issue of collecting water charges from a large number of farmers for which a wider expression of landless solidarity is required than can be displayed by a single group. The highest number of non-group cultivators in an LLP scheme was 105. Recalcitrant farmers using delaying tactics to avoid payment then have a larger village committee with which to contend. Secondly, with the large LLP command areas, possession and management by a single group would effectively exclude other groups in the neighbourhood from ever gaining access to, or benefits from, this activity. There cannot be enough command areas to distribute between all the groups of a locality, especially when several pump units are involved in one scheme. Thirdly, the political implications of different patterns of ownership are important. There is a danger, to our philosophy at least, of small group ownership of irrigation assets reproducing the notion of private property along with petty commodity perceptions, perhaps leading to divisions within those groups if aspirant contractors emerge. With a larger number of groups involved, the ownership of assets is socialised at a wider level, thus strengthening the themes of collective economic activity and united political consciousness. Finally, there are other advantages to such organisational flexibility: economies of scale on drain construction and maintenance, fuel procurement, management of spare parts and repair facilities, and the provision of services such as fertiliser procurement and application; profit can more easily be diverted into other activities; and scheme functionaries (driver, linemen, manager, etc.) can be more adequately supervised by wider, collective pressure.

Conclusion

This chapter has provided an account of the early attempts by some groups of landless and near-landless in Bangladesh, with the support of a Bangladeshi NGO, Proshika, to create and sell irrigation water to cultivators engaged in the production of HYV *boro* rice. The analysis has been based on the single crop but it is recognised that other ways exist of increasing returns from irrigation. These include the enlargement of net command areas, the encouragement of double-cropping, and the supply of supplementary water during the *aus* and *aman* seasons or for wheat. It may also be possible to achieve higher returns on capital investment by using the engine for non-irrigation purposes such as milling or powering lathes in rural workshops. Landless groups could also move into other irrigation-related service activities such as the supply and application of other inputs, and the provision of mechanical and other irrigation backup services (spare-parts, fuel, etc.), through the establishment of rural workshops. These could be set up in any area with a sufficient concentration of minor irrigation, whether provided by the landless or not. In this way, other levels of irrigation and agricultural technology can be captured and deployed as a productive asset by the landless.

This whole process represents a way of extending the agenda for agrarian reform in a society like Bangladesh, which works within the limits of political reality. The struggles over water in rural Bangladesh are second only to those over land, but at this point in its agrarian history, room for manoeuvre exists for the attempt to prevent the concentration of water assets into the hands of a rural elite. The action-research in landless irrigation demonstrates that: the ownership of water can be separated from that of land; the pattern of landholding and plot fragmentation lends itself to that separation; landless groups with support can establish command areas and gain access to the technology; they can use unsubsidised equipment to full capacity and derive an income; they are credit-worthy even without land as collateral; they can maintain agreements with farmers and secure their fee income; they can fend off attempts by some rich peasants and moneylenders to subvert them; they can deploy income to begin to disengage from dependency relationships of indebtedness and indentured employment; their political strength and sense of solidarity is enhanced; and small farmers get better access to water through not being political tail-enders. Finally, this social organisation of surfacewater and groundwater irrigation technologies can reconcile the objectives of growth and equity and thereby contribute to the transformation of rural class structure in Bangladesh, at the same time raising the productivity of land, and creating, through the opportunities for income and employment, a higher level of rural effective demand for the increased output of foodgrain.

Chapter 8

The Social Framework of Rural Exchange in Bangladesh

Introduction

The concept of negotiation refers to the purposive dimension of relationships between transacting parties. It implies a quest for advantage, the analysis of which has been frequently located within the framework of zero-sum games played at the micro-level with rigid parameters. The analysis of such interaction within 'closed' communities was often based upon assumptions of static resources in the context of unchanging, low-level technology, prompting a generation of literature on non-ideological factional struggles, vertical cleavages, competing patron-client dyads, normative and pragmatic rules of conflict. A variation of these assumptions applied to the study of the purposive interaction between 'closed' communities and 'outsiders' involving notions of cultural discontinuities and 'encapsulation', and the methodological equipment of symbolic interactionism. This conceptual history is familiar along with the critiques of it. However, it should be noted that much of this analysis did originate in the context of a more slowly changing technological environment and apparently inviolate structures of rural hierarchy and authority, in which the more abstract notions of class and class struggle could not capture the personalised and socially intimate character of those structures.

We therefore have to ask to what extent vertical solidarities of the patron-client type are the form of class relations under the conditions of agrarian underdevelopment in Bangladesh, and under what conditions of capital penetration may these forms of solidarity thus alter the rural framework of negotiations and bargaining? On its own, such a question may be too economistic or even technologically determinist. In the context, then, of agrarian change characterised by the increasing penetration of new forms of agricultural capital, it is important to recognise the deliberate attempts at restructuring rural solidarities as the basis of securing control over new means of production with the consequent redistribution of income and opportunities both for access and extended political participation. The attempts described here crucially involve new forms of negotiation and bargaining relationships between rural classes hitherto organised on a hierarchical, patron-client basis.

Landless Irrigation: A Vehicle for Change

The particular attempt we are focusing on consists of a programme with groups of landless and near-landless workers in Bangladesh owning minor irrigation equipment and providing an irrigation service to farmers with plots of land in the command areas, in exchange for payment either in cash or cropshare. This 'landless irrigation' programme is described in more detail in Chapter 6 of this volume.

It is obviously a concern of those involved in this attempt to establish an irrigation service by the landless in Bangladesh to assess changes in the patterns of interaction between erstwhile dependent landless workers and landholding employers away from the vertical solidarities noted above towards new forms of horizontal solidarity in which landless workers from different families become effective negotiating units, not only over the water service but extended across other issues as well. Is the possession of irrigation assets and the provision of water service a successful vehicle for this process of restructuring rural class relations or to what extent is this restructuring a precondition for such possession? This relationship between restructuring and possession appears to be dialectical, as this chapter hopes to demonstrate, but since the value of such possession is only realised through the profitable provision of service involving exchange relations with those owning land then an examination of the negotiation/bargaining aspects of the exchange are necessary.

The programme therefore raises many issues concerning: the nature of class awareness among the group members; the formation and composition of landless and near-landless groups; the forms of struggle between such groups and landowning farmers, landholding employers, landlords and/or local moneylenders; the interaction between such groups and state institutions; the possibilities for restructuring the ownership of agricultural means of production such as water; the material basis for political action (and status); and of course the dangers of incorporation with new entrepreneurs being created rather than rural production assets being socialised. This chapter obviously does not seek to address all of these issues directly, but the thread which runs through them does concern the experience of negotiation and bargaining between the groups and the other classes and agencies with whom they have to interact in order to pursue the immediate objective of securing income from the water-selling activity.

The Action-Research Dimension

From early 1984, a six-component exercise was conducted jointly by Proshika and the Departments of Agricultural Economics and Irrigation and Water Management in the Bangladesh Agricultural University (BAU) in Mymensingh. The six components were: technical aspects of water management; employment (with farmers as respondents); employment (with group members as respondents), collected on a weekly basis for one year); access for small farmers; use of income by the groups; and relations with other classes. The two studies which are most relevant to the analysis of negotiation are the use of income, and relations with other classes. The methods for those studies involve (with samples) structured interviews with individual members as well as structured and unstructured collective meetings with the groups to discuss their experience. In addition, of course, there is corroborative evidence from the routine accounting procedures which monitor performance, the continuous observation and reporting by staff at all levels in Proshika.

Negotiations: Arenas and Access

The 170 groups are distributed over 18 different areas of Bangladesh under very different sets of ecological and socio-economic conditions. As a result, there are many variations in the practices affecting agreements with cultivators, forms of payment, the value of the water fee as a proportion of the crop, size of command areas and their cropping characteristics outside winter *boro*, irrigation practices, rates of pumping, the owner-tenant composition of the farmers in the command area, prices of paddy, sources of finance, and so on. Clearly there are several areas of negotiation and several variables which affect the outcomes. The level of solidarity among group members, as implied in the earlier discussion of class relations and consciousness, critically determines the capacity of the group to participate successfully in these areas.

The organisational originality of this programme, compared with other small-scale income generating programmes for the poor, is the high level of interaction required between the group members and other classes in the locality upon whom they are traditionally dependent. The provision of an irrigation service obliges the groups to enter into a range of complex negotiations with landowners in order to establish, in the first place, a command area of adjacent plots, followed by the making of formal agreements in which fee income, rates of water delivery, cultivation practices and length of contract are specified. These exchanges with other classes then continue during the irrigation and beyond into early *aman* (June) until the payments are made, to be resumed again perhaps as soon as late *aman* (late October) if the rains fade early and supplementary irrigation is required. In practice, then, the relationships between landless groups and cultivators around the provision of an irrigation service is more or less continuous.

In addition, the groups have to develop successful working relationships with other institutions—especially the local branches of the Bangladesh Krishi (Agricultural) Bank (BKB) for credit, and the BADC in the Ministry of Agriculture which has been a supplier of equipment, spare parts, and the provider of mechanical services in some areas, as well as providing feasibility certification to release loans. The BADC's role is now weaker with competition from private dealers, but it has been significant. There are obviously many points of access difficulty between the groups and those institutions, with local officials likely to discriminate in favour of the richer classes. In these circumstances, the groups themselves require qualities of internal solidarity and mutual trust in addition to those needed for more self-contained, cooperative activity.

Group Solidarity: The Key Resource

Recalling the opening discussion on the patron-client form of class relations in rural Bangladesh, it is clear that such attributes of internal solidarity can only be established slowly through the experience of depending upon the fellowship of the poor for small victories which accumulate into wider, more reliable trust. The role of Proshika as a mobilising force is critical. Its presence is a source of strength, an alternative force in the wings giving groups greater confidence: to argue for a water price; to insist upon a routine but equitable distribution system for water; to resist individual claims from farmers for special or free service to offset inflated debt obligations from individual group members; to compel farmers to employ local labour; to ensure that farmers implement the full cultivation practices in order to maximise yields when the payment is a proportion—perhaps a third—of the crop, rather than a fixed payment in cash. However, Proshika itself can become a new form of dependency, an institutional patron with group clients. The continuing significance of Proshika as an external agency assisting the negotiating position of the groups will be discussed later, but in order to try and avoid such dependency the groups which have been included in this programme were deliberately selected on the basis of some proven solidarity from a larger number wishing to participate. All the irrigation groups have been in existence for a few years; they were not formed purposively just for inclusion in this programme. The criterion for a group's selection is essentially political—the proven capacity of group members to support each other in adversity, and to sustain small income-generating projects such as goat-rearing or petty trading, against attack. The 'proof' may appear trivial to an observer but is regarded locally as significant (often by other groups not yet included). One example represents a classic incident of negotiation. A group member still occupied some highly-valued homestead land from which his rich neighbour wished to remove him in order to settle families of his own expanding lineage. The rich neighbour's method was harassment, allowing his small herd of cattle to defecate in the poor member's compound and eat any available foliage. The group protested on the individual member's behalf, and when ignored rounded up the herd, took it to the rich neighbour's front door and remained with the herd until it had defecated all round the front door. This tactic was a success, and the group members proved to themselves the value of mutual support in competition with subordinate loyalty to the class of local patrons. This was the basis for entering the programme which required continuous negotiations with the rich local farmers and landlords.

Land Fragmentation: A Divided Opposition

Alongside the issue of group solidarity, a second general resource for the landless both in the initial acquisition of water rights and the maintenance of their negotiation position over a continued irrigation service is the extreme fragmentation of farms in Bangladesh. With increasing pressure on the land and the system of multiple inheritance continuing through lack of alternative employment opportunities, plots themselves are divided to reflect a balance of soil types between inheriting sons. As a result, it is rare to find individual farmers with sufficient land in any single command area to exercise a monopsony influence over the price for water or to dominate its distribution. It is true that some farmers connected by kin relations might act as a pressure group, but there has been no evidence of such jointly organised manipulation by large farmers in the command areas. The centrality of this fragmentation condition, with a farmer's plots being distributed between perhaps several command areas, means that farmers can positively welcome someone else taking on the management of irrigation instead of having to function as water entrepreneurs themselves in order to secure water for their own plots. The controllers of water can have a much more consolidated asset than the holders of land in many parts of Bangladesh. We have observed under these conditions how farmers have to lose their autonomy over production decisions: by conforming to the externally imposed arrangements for serial water distribution (although flexibility usually exists for plot preparation and transplanting); and by accepting as a precondition for receiving water the full set of HYV cultivation practices which includes at least the application of chemical fertilisers and regular weeding. This fragmentation also means that farmers find it difficult to negotiate individual fee arrangements and price levels (even though they may have debt leverage with group member clients), as all farmers in a command area have a collective interest in a standard arrangement. Such arrangements are often anyway established independently of the landless group as well, on the basis of local custom where water selling by rich landholder contractors has been prevalent.

Social Composition of Command Area Landholders

Closely related to the fragmentation of farms is the social composition of farmers holding land in the command areas. Infinite variations are possible here as we learnt from some of our earlier surveys of command areas (a task now beyond us with 170 of them). For a shallow tubewell with a command area between 10-17 acres (with actual size depending on the capacity to make agreements with farmers, the existence of adjacent private contractor competitors, and soil type) the number of separate farmers has ranged from 10 to 38. With LLP irrigating between 20-50 acres (some are certainly underutilised), the number of farmers has ranged from 18 to 97. These farmers own land outside the landless group's command area, and some of them are tenants as well as owners. Clearly there is a variation in the total holding size of those farmers and tenants, which affects their overall position in the local negotiating structures. Landless groups with command areas consisting primarily of small and marginal farmers with a high tenant component to their holdings are less subject to exploitative pressure from patron classes. However, such small farmers have lower levels of capital, tighter margins on HYV production, and are more wary of risk. Often they cannot engage in the full set of practices to maximise yields as they can afford neither inputs nor the labour unless they obtain loan support from private moneylenders (involving in effect a high cost on their production) since their access to appropriate and timely institutional sources of production credit are limited. Sometimes, the irrigation groups have extended loans to small farmers for irrigated cultivation. Small farmers also have the disadvantage (as far as the irrigation groups are concerned) of employing less non-family labour than larger farmers, so that the opportunities for employment of the landless are not maximised. As a corollary then, if a command area consists of a higher proportion of larger farmers, not only does the capital exist to engage in the full cultivation practices, but more non-family landless labour is employed as well. Under certain conditions, this has the effect of pushing up the wage-rates for the irrigated rice season (*boro*) in the locality. This in turn reduces the need for that employed labour to outmigrate and/or take advances from the rich local patrons which previously undermined their bargaining position during the normal peak times of the main rain-fed *aman* rice season. The landless irrigation group cannot precisely exercise a preference over the command area site and therefore the social composition of its landholders, but there is a complex trade-off between greater political equality with small farmer customers and the direct economic prospects which larger farmer customers can offer to the groups.

Presence of Group Members as Cultivators in Command Areas

A further aspect of the social composition of the command area landholders concerns the number of group members themselves cultivating land in the command area. Although we have referred so far to 'landless' groups, this is more a term of convenience than an accurate definition. Target groups cannot be so easily defined (Wood, 1985b) and the determining principle of membership is the use of one's own physical labour. In the dynamics of landlessness in rural Bangladesh, there are many marginal and sub-marginal landholders who rely on the sale of their labour for most of their income. However, they may be clinging to small plots of land at any one snapshot in time. The membership of those groups working with Proshika cannot realistically be restricted to the pure landless, especially when a degree of self-definition is encouraged to strengthen solidarity. The distinction between pure landless and *prantik*—a marginal peasant—is more significant to agrarian statisticians than to the actors themselves. Therefore, it is often the case that some group members have plots in the command area, and they are in a position to set standards for customer behaviour of non-group cultivators if they are likewise marginal or small farmers, or if the command area is not dominated by a few large landholders. Obviously such group cultivators represent greater loyalty to the 'landless' irrigation service, especially as they are also entitled to any dividends which are distributed among group members at the end of the season when the fees have been collected.

Location and Cropping of Command Areas

There are further characteristics of command areas which affect the bargaining relationships between water-sellers and users. The structure of exchange is influenced by the location of the command area. It may be located near the homesteads in the agricultural lands of those richer landholders who are direct employers or landlords of group members, or at some distance outside the immediate spheres of patronage. In one example (*Rajnagar Gorib Unnayan Samity*), the command area of 27.5 acres is two miles away and all the farmers are from outside the group's village. At the same time, however, the management problems are also greater since the pump operator and lineman have to be on duty for the season (approximately four months or more); the equipment has to be maintained and guarded, or delivered from the group's village frequently; the distribution system has to be supervised and disputes with farmers sorted out; the farmers' cultivation practices have to be monitored (since the payment is a crop-share); and the crop-share fees (5/16 of the standing crop) have to be collected and brought to the group members' homesteads for threshing.

The cropping pattern of the command area is also significant. Rotations in Bangladesh are sensitive to minor differences in elevation interacting with flooding and drainage patterns, and with soil types. However, the attraction of much dry season irrigation to farmers is the unpredictability of the other seasons and the precarious yields which accompany them. Lowlying land is the familiar problem and in some areas the flooding is so extensive that the land is a single-crop area relying on residual moisture and traditional *dhone* irrigation. Thus, where other seasons are non-existent or at best precarious, the introduction of mechanised, more widespread irrigation is especially significant. While the competition for those new water rights may be intensified, actual control of water becomes a crucial instrument of leverage—the basis for negotiating the water price, insisting on full cultivation practices and ensuring the employment of group or other landless labour. It should be added that a single crop area also represents an incentive for farmers to maximise their yields through investment and careful practice. If on the other hand, the command area is situated in an area where the main rain-fed rice crop (*aman*) is secure, then irrigated cultivation and double (or triple) cropping is more of a bonus the larger the size of farm of which the plot is a part. Under such conditions, there is less incentive for the farmer to transplant paddy as opposed to wheat (which has a lower value when considered as income to the group), or to invest properly in fertiliser and weeding. In this case, the irrigation group has a harder task in counteracting these tendencies, and fewer employment opportunities are available to them as individuals. At the same time, it must be remembered that wheat demands less water, so there is an offsetting possibility of extending the command area if adjacent competitors do not exist. Of course multiple cropped areas of this kind are only a problem for bargaining if the prevailing water fee arrangements are crop-shares rather than fixed cash payments (fixed kind payments are rare). However, the level of cash payments can be expected to reflect the relative strength of buyers and sellers in the market.

In this context, a particular problem has existed for low-lift pump (LLP) command areas which, although often low-lying and single cropped, may have been served for the last two decades by highly subsidised LLP sets rented from the BADC. The fees (fixed cash), while representing substantial income to the contractors, have therefore been artificially low. With the government and World Bank-induced withdrawal of subsidy and rental arrangements on LLPs, it has been difficult to overcome the customary expectations of farmers and raise the charge to an economic price.

Market Conditions: Exit, Voice and Loyalty

The idea of a landless group providing an irrigation service has of course been as strange for farmers as for irrigation and agricultural planners. The groups, with the help of Proshika, had to persuade potential clients of their capacity to participate in agriculture via the sale of water. This task of persuasion was easier, along with subsequent agreements, in the absence of neighbouring competitors among large landholders and water contractors.

In one STW case (*Bangla Bhumihin Samity* in Ghior Upazila) where the land was virtually single-cropped, the landowners were initially highly sceptical. However several factors assisted the making of an agreement. First, a local private contractor scheme had numerous problems attributed to machine breakdown and interrupted diesel supplies. This not only contributed to the farmers' general scepticism but also reduced their loyalty to the existing scheme. Secondly, the farmers were reassured by the presence of some of the group members as cultivators in the potential command area. But finally, the group offered to compensate the farmers if their production fell below the yields gained on private schemes. This offer of compensation was clearly a risk, but it was based in part on the mechanical skills of their pump operator member who had been trained by the BADC, and had operated pump sets before, elsewhere. The risk paid off. After two successful seasons of providing an uninterrupted irrigation service, landholders from neighbouring private schemes are now eager to switch their land into this group's command area. This example raises the whole issue of the market for irrigation in a locality, and how it affects the interrelationship between Hirschman's categories of 'exit', 'voice' and 'loyalty' (Hirschman 1970). The distinction between the technologies is relevant here both with respect to their mobility and the size of potential command area.

The STW can irrigate on average about 15 acres of paddy, but many variables affect the actual size in a particular location. A number of these installations can therefore operate in close proximity enabling farmers with plots scattered between them to switch their options, and even reallocate their 'boundary' plots to the competing adjacent scheme. Here, there is an exit option. Clearly the water-seller has a strong interest in maintaining a stable clientele by establishing long-term agreements. (The groups with Proshika make legally-recognised contracts for a five-year period to cover the projected period of repayment of the loan on the capital asset.) Farmers are more likely to enter those agreements if alternatives are remote—e.g. through the effective zoning of tube-well installations to prevent overcrowding, which is meant to be a function of the BADC. Furthermore, the efficacy of those agreements will depend not only on the solidarity of the 'landless' groups (where they are the water-sellers), but also on the credibility of the legal sanction if the agreement is broken. This credibility depends upon the access of the groups to legal services, which Proshika's presence as a supporting institution certainly reinforces. If committed to a scheme in this way, then voice rather than exit is likely to characterise the exchange between seller and consumer. But in extreme cases of 'voice', some groups have transferred the STW into a mobile asset by writing off the cost of the boring and irrigation drains, and moving the handset to a new location (or even selling it to another group in another area altogether). In such cases, it is the groups who are exercising an exit option.

With LLPs, which are mobile (within certain obvious constraints like water sources), the exit option for the groups is easier and therefore a more credible threat to recalcitrant consumers. However, consumer loyalty is likely to be quickly replaced by voice if there is evidence of competitors offering a service at more favourable rates. This was an early problem when trying to capture command areas which had previously been served by subsidised, rental machines. At the same time, individual water users in LLP command areas have fewer options to switch their plots to other command areas, since those areas are larger, thus reducing the spatial scatter of a farmer's plots between different command areas. Where LLPs are operating in single crop areas, farmers have fewer inter-seasonal options for exit as well, for example by withdrawing from *boro* and concentrating their production on *aman*.

We have no experience with DTWs yet, but they are distinguished from the preceding two irrigation technologies in the immobility of the equipment once installed. Until now, DTWs have been operated either by individual or family contractors. The high subsidy content has permitted an additional option to Hirschman's categories: exclusion. It is however expected that as the subsidy is scaled down, there will be greater incentives to maximise utilisation. The problem for landless groups and this technology is that their own threat of exit as a response to unwarranted voice and the failure of farmers to honour agreements is not credible with the immobility of the equipment and where loan commitments have to be met. The initial tentative steps to use DTWs in the landless irrigation service do not yet involve purchases, but the takeovers of defunct rental agreements with the BADC. There is however an argument that landless groups should also take advantage of the large subsidies available on the purchase of this equipment. At the same time, as with LLPs, individual water users have fewer options to exit by switching their plots to other command areas, although this can be possible for boundary plots. Whereas farmers collectively can force a LLP operator or group to shift site to make way for a competitor, this option is much more difficult with DTWs since it would involve a change in the ownership of the equipment itself and the rights to water which accompany it.

There is a further dimension to this market issue which applies to many areas of Bangladesh, not just areas where landless groups are providing an irrigation service. This concerns the competition between different rather than similar technologies—especially the DTWs and STWs. The issue becomes, as it were, three-dimensional, since it refers not just to surface area competition but water tables. The effect of a DTW in an area of STWs in some (though not all) geological structures is to drain the high aquifers. Although annual replenishment is likely, the DTW can dry out a STW in late season or make pumping less efficient and more expensive. Under such conditions boundary plots are likely to be switched from STW command areas to DTW ones.

Risk Sharing and Payment Options

Moving from the characteristics of command areas and the general market conditions for irrigation, there are other aspects of the exchange between water-sellers and users to be considered. First, there are issues concerning payment beyond the distinction between crop share, fixed cash or fixed kind. Any water-seller using mechanical pump sets, and the landless groups are no exception, has to risk large amounts of fixed capital and recurrent operating cost expenditure before obtaining any return on investment at the end of the season. There is a cash flow problem, with interest to be paid on the balance loaned from the BKB, or the Proshika revolving loan fund when flood loss prevented some groups from repaying instalments to the BKB. The 1984 floods reminded us that some group water sellers were still bearing all the risk of irrigation until the harvest. Earlier experience had led Proshika to encourage the groups to negotiate risk sharing arrangements with command area farmers whereby they would contribute to ongoing operation costs (mainly the cost of diesel) in exchange for a reduction in the final fee. Such negotiations were often 'successful' except that a drop in fee from one-third to one-quarter crop share was frequently the price. These arrangements are quite common with private contractor service, and often involve farmers actually purchasing diesel and buying dedicated pump time on request. However, this leads to anarchic command area management and reintroduces problems of access and tail-enders for the poorer, politically weaker farmers. Even with an advance payment system less extreme than this, the water seller's control over water distribution remains threatened especially if poorer farmers are still unable to make the same deposit as larger ones, who may then demand preferential treatment as in the *Rajnagar Gorib Unnayan Samity*. In these circumstances, poor farmers may require loans from the group (at prevailing bank interest rates plus a change to cover costs) to contribute to the group's operating costs. But a command area consisting of a few large farmers would pose an alternative threat of loss of autonomy over water distribution. Clearly these arrangements again depend upon the social composition of the landholders in the command areas. Groups which have enjoyed a record of success and which have built up their own savings (such as *Bangla Bhumihin Samity* in Ghior) rather than disperse net income to individual group members as dividends, prefer not to receive any advance payment and instead to maximise profits through the unaltered crop share of paddy at the end of the season. This is evidence of considerable solidarity among the group members to behave collectively in this way and trade off greater security for profit.

Groups' Use of Income and Structural Change

The issues of payment, risk-sharing, and use of savings bring the discussion to other aspects of income use and its impact upon relations between the landless in the groups and local landowning employers, landlords and moneylenders. The analytical framework which guided the study consists of a distinction between two categories in the use of income by the groups: where the structures of dependency are not challenged by patterns of expenditure; and where there are implications for structural change in the relations between these rural classes. It is very easy to understand the pressures from poor group members for an immediate share of the joint income for urgent consumption needs although we have seen resistance to such pressures in *Bangla Bhumihin Samity* (above). It may also be true that certain kinds of immediate consumption will have implications for structural dependency. If by consuming profits in the form of rice, the group member releases himself and his family from an annual reliance upon an advance of rice from an employer (which would normally be paid off in the peak season for labour—thereby undermining the worker's capacity to wage bargain), then clearly part of the relations of structural dependency has been altered. This use of income would constitute a minor individual disengagement from debt bondage, but it will not have direct implications for redistributing control over productive assets in the local societies. If income was used in other ways to open up, for example, hitherto prohibitive forms of expenditure such as dowry payments, ceremonial expenses, house improvements, additional clothing, then such expenditure would be structurally less significant even though we could not judge it to be illegitimate by the norms of the society or human entitlement. It is, after all, difficult to ask the *poor* to defer gratification for the sake of a longer-term, distant improvement in their structural position and bargaining relationship.

As part of the second category (income use with structural significance), group decisions about collective use could be crucial for the future of individual members and their families. Is it possible for groups of landless to deploy their profits from this irrigation service in such a way as to release their members from specific dependency relations such as indebtedness, mortgaged land, bonded labour service, insecure tenancy arrangements, inflated rent or share obligations and low wages (e.g. by developing a strike fund or mutual loan funds for members in absolute distress)? In this way, as water selling continues in successive years so individual members will be less subject to pressure from individual patrons to provide extra, cheaper or more timely water at the expense of other water buyers. Are profits being deployed: to create alternative patterns of ownership or control over land, through the setting up of land mortgage banks, sharecropping cooperatives and joint land purchasing; and to capture other services and technological levels in the agricultural system (e.g. fertiliser procurement and application or repair maintenance workshops)? The *Bangla Bhumihin Samity* are not planning individual dividends after the capital loan to the BKB has been paid off, but have decided to purchase jointly a mechanised boat for freight and passenger transport. The *Rajnagar Gorib Unnayan Samity*, on the other hand, distributed 48 *maunds* of paddy among its 16 members this year. Clearly the demand for immediate individual dividends relates closely to the employment opportunities which exist for group members during the season and the wage levels which they have been able to negotiate.

Water Service: A Leverage for Employment

It is interesting to note an example in the *Char Moheshpur Biplab Samity* where the irrigation service has been used as a leverage to secure employment and increased wage levels for group members. When the farmers ask for water for puddling (plot preparation) and transplanting (where the response is more flexible preceding the main serial arrangements for the main, regular distribution) the group members set their own employment, including the later operations of weeding and harvesting, as a precondition for starting the water service. This exchange of water for employment is not strictly contained in the formal agreement, but constitutes a regular, informal aspect of the bargaining relationships. It is often the case, as reported by the *Rajnagar Gorib Unnayan Samity*, that group members providing an irrigation service have greater skills in HYV labour operations (especially plot preparation and transplanting) and are therefore preferred by farmers cultivating HYV irrigated *boro* rice. This group has been able to exercise an influence over non-organised labour in the morning market-place where labour is hired and where sometimes daily bargaining over wage rates takes place. All groups met so far report that with the irrigated *boro* season wage levels have increased, in some cases by a factor of 2.5 for this season.

Impact of Agency Access on Group's Negotiating Position

Finally, in this review of variables affecting negotiations between water sellers and users, the relations between the landless groups and agencies outside the water exchange should be considered. While this discussion should not be diverted into one of access, it has to be acknowledged that the more successful these access relations, the stronger the position for the group in the village. This applies not only to securing loans from the BKB branch, which is currently Proshika's major problem with the whole programme despite agreements at national level with the BKB headquarters management, fulfilling the BADC's feasibility criteria while avoiding the payment of bribes, and obtaining the supply of equipment, either from BADC itself or from a private dealer, but also to the continuous transactions with mechanics, spare parts dealers and diesel suppliers. A major cause of irrigation failure for the mechanised minor irrigation technologies (DTW, STW and LLP) is machine breakdown, lack of spare parts and repair facilities, and interrupted fuel supplies under conditions of general scarcity. In most of the cases examined so far, the groups have been successful in these continuous and immediately relevant transactions despite also having continuing problems with BKB. In most cases, long term contracts have been made with local small town/bazaar mechanics and fuel dealers. At first the dealers demanded cash before releasing stocks, but they are now sufficiently confident in the groups to offer fuel on short term credit which assists the groups' cash flow problems and enhances their position with the farmers who are more convinced that the service will not be interrupted and threaten their input investments.

The groups also report that they are now being treated more respectfully by these agencies, which raises *en passant* an interesting question about data. What indicators of respect can we accept? Two of the groups, without leading questions and independently of each other, were impressed when their members were offered a chair and even tea by the local BKB branch manager in one case, and the local diesel supplier in the other. However, it would be wrong to dissociate these access relations from the background of Proshika support. Many of the longer-term contracts with dealers and mechanics have been mediated by Proshika, and the local Proshika workers quickly intervene if relations break down. The threat of their intervention is often sufficient to keep the relationship running smoothly, except with the BKB where Proshika's presence appears to be continuously necessary. The reliance on Proshika and the consequent reliability in the operation of the pump sets has a positive impact upon the landless water sellers' negotiations with landed water users, but we all know that the crucial test comes when Proshika can withdraw without weakening the bargaining position of the providers of a commercially viable irrigation service to their erstwhile landed patrons.

Conclusion: Where Patrons become Clients of their Clients

The participants in this programme (the group members and the staff of Proshika as well as some observers) are concerned to assess its material and political progress. Its objectives can be interpreted at different levels, and will be valued differently by participants and observers. However, a basic concern of Proshika's total mobilisation efforts with approximately 7000 groups through 35 area development centres is the development of a consciousness, organisational capacity and confidence in joint action based explicitly on a sense of class position and the common experience of subordination in the social relations of production and exchange.

Despite the strength of kin relations cross-cutting highly differentiated landholding categories, and the way in which these are reinforced through Islamic ideology of attachment to the lineage, the patron-client form of the class relation, which has frequently occurred with extended lineages, is being replaced inexorably by a steady commoditisation of rural exchange relations. There may be several explanations of this process but significant among them is the penetration of new forms of agricultural capital and modern varieties of crops, especially rice and wheat. As may be expected, the ownership of these new capital assets in agriculture is partially concentrated in the dominant landholding classes. But it is also the case that with break-up of farms through multiple inheritance in the context of rapidly rising rural population, a significant proportion of individual farmers are unable to utilise, alone, less divisible assets—whether irrigation equipment or tractors, threshers and multi-purpose power units (for irrigation, ploughing, threshing and milling). It is even difficult for many farmers to own their own draught animals. Under these conditions, rental practices abound. Exchange relations as a necessary condition of production are intensified.

At the same time these relations appear to be increasingly disentangled from the more familiar, highly personalised, long term and localised exchanges between land and labour. It must also be recognised that with an increasing supply of underemployed rural labour, relations themselves are becoming more single-stranded and transactional. With labour relations though, the commoditisation process is not complete in two senses: first there is a continuing use of a small core of *rakhal* (bonded labour); and secondly larger farmers are still interested under appropriate cropping conditions in giving advances to labourers in lean seasons to secure their labour in peak seasons, when labour might be scarce and expensive and yield gains through timely operations cannot be sacrificed. But in areas where the introduction of irrigated *boro* rice (and wheat to a lesser extent) has been significant, then the lean seasons for employment are not so lean. Most of the Proshika groups in such areas (whether owning irrigation equipment or not) report higher wage rates not only for that season but for others as well. The reduced dependence on advances improves the workers' bargaining position, and strengthens their confidence in joint action, including those forms of action—such as an irrigation service—which requires continuous interaction with the employing classes.

Thus changes in the structure of exchange aids the possession of assets such as irrigation equipment, which in turn of course contributes to a further favourable transformation of the terms of exchange. It is in this way that structure and possession must be seen as dialectical: the social relations of production interacting with the forces of production. Furthermore, as noted above, the new forces of production in agriculture in Bangladesh are prompting a density of rental transactions and therefore room for manoeuvre for new categories of actors to enter the arenas of competition. For poor landless labourers in this situation of expanding agricultural capital there is, therefore, room for manoeuvre in two senses: at the level of their own consciousness induced partly by the commoditisation of exchange relations; and in the contradictions between the lumpiness of mechanised technologies, stimulated by the requirements of more divisible variable inputs such as new seed varieties dependent on chemical fertilisers, and the fragmented landholding patterns of rural Bangladesh which prevent the single ownership of capital assets by a majority of farmers.

In these circumstances, it is possible to redraw the map of patron-client relations, unhindered by the propositions of 'structural fragmentation', to approach a more reciprocal pattern of exchange relation in which the class consciousness and organisational capacity of the poor favour their capture of certain lumpy categories of new agricultural capital. Although the analysis of negotiations between water-sellers and water-users identified constraints as well as resources for water-sellers, their experience does indicate a route whereby patrons can become clients of their clients.

Chapter 9

Agrarian Entrepreneurialism in Bangladesh

Introduction

The intention in this chapter is to speculate—on the basis of recent performance and policy trends together with some preliminary findings—about the entrepreneurial character of agrarian change in Bangladesh, and thereby to gain reactions not only from the context of Bangladesh but also on the application of similar arguments to West Bengal. This exercise involves an element of stocktaking and a selected summary of recent developments in the agricultural sector. As for significance, the central argument is not yet well-researched so no particular claims are made for the presentation. Many of the propositions still have the status of hypothesis, representing the beginning of a sequence of work rather than the end of it. However, if the propositions have a ring of truth, then their implications are highly significant.

The basic question is: under the fragmented landholding conditions of Bangladeshi peasant agriculture, will the increasing capitalisation of agriculture transform the units of production away from the family based farm into a system of vertically and horizontally integrated agricultural 'companies'? This suggests a particular interaction in which the organisation of production is disarticulated and re-articulated in a manner which reduces the significance of landholding in the realisation of agricultural surplus. Related to this process is the contention made by Glaser 1989 and others that expanded opportunities in agriculture and agricultural services express an increasing preoccupation with productivity alongside continuing relationships of surplus extraction through rents, interest rates, land acquisition and low wages. Indeed we could push the argument further and suggest that the mode of surplus realisation is shifting towards investment in productivity and agricultural services away from rent and usury extraction, agreeing with Adnan that 'technical limits' exist in such forms of exploitation 'compatible with the reproduction of exploiter and producer' (1984 p.24). Such a shift would reflect not only the donor sponsored intrusion of capital and an assault upon state monopolies, but also would be a response to demographic trends and imperatives to use land more intensively.

These propositions should be connected to a decade or so of attempts to conceptualise a 'development of capitalism' in South Asian agriculture. This is a substantial debate, much of which is now familiar. The important issue for my purposes is the distinction but also linkage between investing in agricultural productivity and investing in agricultural services. Banerji's original distinction between 'forms of exploitation' and 'relations of production' (Banerji 1972), and Alavi's use of the distinction between 'formal' and 'real' subsumption of labour under capital (Alavi 1982) have *inter alia* conceptually liberated us from a simple-minded and teleological 'development of capitalism' thesis. Banerji's contribution has been to remind us that the function and meaning of relations of exploitation such as sharecropping, tenancy, usury and wage labour can vary according to the overall reproduction context in which they occur, so that 'modes of production' cannot be simply 'read-off' from the presence or absence of indicative relations of exploitation. Rudra (1978) expanded this point. Alavi (building on Laclau's critique of Frank [1971] and Brenner [1977]) is distinguishing between the parasitic functions of capital in appropriating from producers through exchange activity without changing the forms of exploitation (formal subsumption) and capital which transforms directly the relations of production, where relations of exploitation may keep their form but change their meaning (real subsumption). Adnan (1984) takes us a step further by identifying the 'phenomenal forms' through which capital 'takes hold of production itself' through an examination of the conditions under which interlocked market activity changes the productivity of land or labour rather than remaining parasitic. For some, the innovative interlocking of markets is a definition of entrepreneurialism.

My particular concern, then, is to see how far the expansion of agricultural services activity takes the form of interlocked markets in which the control over the productivity/parasitic options for land use shifts from poorly networked owners towards a new entrepreneurial class of well connected dealers and contractors some of whom may be richer landowners themselves, while others may be more urban (commercial or official) in their origins. The notion of interlocked markets refers to factor markets being linked by non-price mechanisms with restrictions on who sells to whom and with different prices for different participants (see also Griffin 1979). Furthermore, it is important to recognise that where richer landowners have widely scattered plots, their interlocked market strength especially to secure productivity objectives may only be restricted to a small proportion of them. Thus, even for richer landowners seeking productivity increases on many of their plots, a surrender of control over cropping rotations, cost variables and timing of operations (involving capital and labour inputs) may become necessary.

Population

In mid-1986, the population of Bangladesh was approximately 104 million, with a rural population close to 90 million. This population has grown from 42 million in 1951, therefore by 139% which compares with India's 117%. The annual rate of population growth is considered to be about 2.5%. The overall area is 144,000 sq. km, of which 63% is cultivable. The population density of cultivated area is therefore 1146 per sq. km.. For a population, which is 90% rural and mainly dependent upon agriculture, this is very high. However, the significance of this figure can be modified slightly by recognising that about 40% of the cultivated area is double cropped and 8% triple cropped. This population density, together with rising landlessness has not resulted so far in a dramatic increase in urbanisation. Definitions have changed over time, particularly with the new *Upazila* headquarters being classified as urban. Discounting for changing definitions, the rate of urban growth since 1974 has been about 6% per year. There is a marked gender imbalance in this process with a sex ratio of 151 men per 100 women in Chittagong, 139 in Dhaka and 133 in Khulna, with an overall sex ratio in the country of 106. This gender imbalance reveals a process of rural males in search of urban work (and seasonal imbalances are of course much higher), leaving women (as mothers, sisters, wives or daughters) to survive from rural production or labour. Perhaps the most general consequence of population pressure is not urban expansion but rural settlement on marginal lands liable to flooding especially in the disaster prone coastal areas.

Poverty and Landlessness

Average per capita income is $150 (compared to India $260, Sri Lanka $330, and Pakistan $390). However this is far from equally distributed.

With a predominantly rural population, highly dependent upon agriculture, poverty is a function of access firstly, to cultivable land and secondly, to employment opportunities in the countryside. With the narrow industrial and manufacturing base, there is no immediate prospect of a rapid expansion of urban employment. The ramifications of the population growth noted above throughout the agrarian structure are pervasive. Larger families are living off the same size or reduced size of holding, so that there are about 3 million farms with holdings above 2.5 acres and in excess of 3 million farms with a smaller acreage. The high fragmentation of holdings reflects the kin-ecology outcomes of multiple inheritance (in effect between sons), so that over 60% of farms consist of more than 6 plots, with 10% of farms having more than 20 plots.

With this increase in fragmentation and the decline in family farm size, infrasubsistence peasants are increasingly dispossessed through mortgage, debt and distress sale. The viability of such smallholdings increasingly depends upon the availability of land to sharecrop, but with double cropping such opportunities have increasingly become seasonal rather than annual, and involve higher risk investments in modern methods to which the poor have precarious and expensive access. The increase in landlessness which has resulted from this process of dispossession and insecure tenancy has been variously estimated, with some consensus that over 60% of rural households do not possess enough land to ensure their subsistence and have to rely upon wage labour in agriculture, rural works or outmigration. The loss of land by these classes does involve a modest accumulation of land by other families who are in a position to acquire land either through an initial strong position to accumulate from modern agriculture, or through moneylending, or by diverting urban incomes into rural capital investment. Although this is not a simple process of polarisation, the Land Occupancy Survey of 1978 (the last major land survey) reports the top 10% of landowners holding 49% of cultivable land, and the bottom 10% of landowners owning 2%. In this context, it has been estimated that 62% of rural people receive only sufficient income to satisfy 90% or less of the minimum recommended daily calorific intake.

The rise in landlessness along with the decrease in average size of holding to precarious levels for the majority of rural landholders has increased the private value of (male) children as income-earners and providers of old-age security. This has had a depressive effect on rural wages except perhaps in the peak seasons of new production. At the same time, with the general cultural constraints on all but the poorest women to seek waged employment outside the household, daughters-in-law have now been transformed from labour assets in the hitherto self-sufficient family farm to consumption burdens with a consequent dramatic increase in the practice of dowry and the value of the payments in order to 'compensate' the receiving family. Furthermore, the steady increase in those seeking work in the rural labour markets (including the poorest women) has created a reserve army of underemployed labour which has functioned to break down patron-client dependency relations between poor peasant/labourer and landlord/employer through which minimal security to labour had been offered. This has intensified wage exploitation, while 'freeing' labour at the same time.

Employment

Employment statistics are very inaccurate since they cannot adequately capture the complexity of labour relations in agriculture, in particular its seasonality and the use of family labour. Most importantly, women's labour in the household goes virtually unrecognised despite the major productive contributions made by women in food processing and preparation. The 1981 Census abandoned any measure of 'unemployment', but reported that among active males for Bangladesh as a whole: 58% were in cultivation; 4% in manufacture; 11% in business. 14% of urban males were in transportation; 12% in manufacture; and 24% in business. By contrasting the 1974 Census with the 1983-84 Labour Force Survey, there appears to be a 30% increase in the employed male population, but none of this increase went directly into agriculture: shops absorbed about 40%; and manufacturing, transportation and services took about 15% each of the increase. With such a high rural proportion of the population, open continuous unemployment is rare. The problem is underemployment either in the sense of intermittent access to employment opportunities among those required to sell labour power (especially due to seasonal variations in the pattern of labour demand in agriculture) or in the sense of low productivity of family labour on smallholdings with limited access to inputs. Various micro-studies at village level indicate that about 33% of the agricultural labour force is underemployed in these ways. Under present technological/tenure conditions however, this proportion of the agricultural labour force cannot be completely pulled out of agriculture (even if employment opportunities existed in other sectors) without damaging output since the demand for labour at peak harvest/transplanting times remains strong.

These issues of seasonality and under-employment, together with the recognition that a high proportion of rural workers are either landless or combine family farm and outside work, have drawn the attention of government, donors and NGOs towards a sharper focus on employment generation programmes either in rural works (such as FWP/IRWP) or in rural manufacture and services through group or family based activities. The World Bank projects the growth of the labour force at 3.2% per year for 1985-90, rising marginally thereafter to 3.5% by the end of the century which would add about one million new entrants to the labour force each year. Much reliance is being placed on the non-farm sector (basically rural works) as a panacea for the problems of labour absorption within agriculture (Hossein 1987). However, there are contradictions with such a strategy. Under the conditions of rural landlessness noted above, and the consequent high proportion of the rural total population dependent upon purchasing food, unless purchasing power exists in the form of rents and incomes then it is difficult to see how an incentive price for agricultural output can be sustained. The existence of food for works and vulnerable group feeding programmes may provide distribution and crucial periods of employment for some, but they are also reliant upon distributors of food aid which undermines the objective of incentive prices for farmers. In short, reliance on the food aided non-farm sector undermines the level of effective demand for domestic food production rather than contributing to it.

Agriculture—Progress and Policy Issues

For a variety of reasons, some attributed to the sub-colonial status of Bangladesh within Pakistan in the period up to the end of the 1970s, rural class relations have been characterised by the use of capital in the sphere of exchange (such as renting, moneylending and trading) rather than production. This has certainly raised the level of absolute surplus value accruing to certain classes but has not contributed to the generation of productive agricultural capital among landowners. The failure during this period to develop productive agricultural capital has resulted in the form of polarisation noted above involving increases in landlessness and rural underemployment and the depression of real wage rates (Wood 1981). This had created a problem of effective demand among the increasing proportion of rural people obliged to purchase foodgrain, thereby undermining price incentives required to stimulate owners of land into raising its productivity (Wood 1980b). The consequent shortfalls in foodgrain to meet basic needs consumption has been met by food imports and food aid (together totalling 2.6 million tons in FY85 or 18% of domestic production, although projections for subsequent years indicate a lower import content which might now have to be revised upwards to respond to the floods in 1987).

Until 1980 and the second FYP, this situation was in effect reinforced by a strategy of state administered subsidies to agriculture formally intended to stimulate production. This strategy was implemented principally through a virtual state monopoly (the BADC) which controlled both imports of chemical inputs and equipment (especially mechanised irrigation units) and their distribution through outlets at District and *Upazila* levels. With subsidies, state monopoly and restricted access, corruption alongside high default rates and under-utilisation of equipment was widespread. The state, with its monopoly control over external agricultural capital, became in effect an arena within which funds were diverted into modes of parasitic surplus extraction through an alliance of central and local bureaucratic officials with their favoured clients in the countryside.

The response to these problems was contained in the 'advice' given to GOB by the World Bank in Food Policy Issues (December 1979), and incorporated by GOB into the 2FYP(80-85), and the longer term Medium Term Food Production Plan (15 years). Essentially the strategy was based upon a significantly expanded investment in agriculture and rural works, but only under certain institutional conditions. The Food Policy Issues report listed the physical constraints to the development of agriculture as: scarcity of cultivable land; limited yield potential; climatic hazards; and the need for dry season irrigation and improved drainage and flood protection. It attributed the slow growth of agricultural output (7.3% for the entire 70s) to: weaknesses in agricultural support services; inadequate provision of input supplies; and constraints stemming from the agrarian structure. The latter was a euphemistic reference to the points noted above: capital in exchange/circulation rather than production—i.e. exorbitant rents, insecure tenancy, moneylending at usurious interest rates, and the control of input supply and rural works employment through monopolistic patronage. It was clear that any expansion in agricultural investment would have to be dependent upon external aid and both the World Bank and US AID (as the major donors) were determined from their diagnosis and leverage to liberate the rural economy from bureaucratic management and control by encouraging open market systems of food and input distribution. The strategy therefore consisted of: open-market dealing in food (through procurement, storage in a network of godowns, and release of stocks into local markets triggered by a 20% variance in seasonal prices); the phasing out of fertiliser and other input subsidies (e.g. on various forms of mechanised irrigation); coupled with a major emphasis on HYV technologies in rice and wheat. These efforts concentrated on the improvement of foodgrain supply with a noticeable lack of attention to the structure of demand for food. The strategy involved the assumption that a combination of support prices for foodgrains (through a generous procurement price) and open market dealing in rural goods (inputs and food) would encourage the landholders to raise the productivity of their land, thereby increasing their marketable surplus to pay the full market price for inputs.

The 2FYP therefore envisaged an annual increase in foodgrain output of 6.5%, to be achieved principally through the development of minor irrigation and doubling fertiliser supply. The irrigated area was expected to expand from 2.6 million acres to 7.2 million acres, representing 32% of the maximum feasible cultivation area of 22.5 million acres. The greater proportion of this increase was to be through STWs (45%), with DTWs at 32% and LLPs at 20%. Despite the problems experienced through earlier programmes of organising farmers' irrigation cooperatives around DTWs (problems which still persist in contemporary UK supported programmes), the use of the technology is regarded as unavoidable in areas with low water tables. The increase in LLPs was technically limited by the availability of additional surface water resources. Since the slow rate of irrigation development in previous plan periods had been attributed to the low level of utilisation of equipment in the field due to a policy of over-generous subsidisation involving heavy public sector costs and inefficient, corrupt implementation (Boyce 1987a pp. 228-248) (to which should be added the lack of demand incentives to use capital productively), the privatisation strategy was extended to the sale of irrigation equipment through liberal lines of credit (initially from IDA) to any individual farmer, contractor or rural entrepreneur, or groups or collectives who could appear in the market with the collateral on which a loan could be extended by the National Commercialised Banks (NCBs).

Looking at performance over the second plan period, agriculture production as a whole grew at about 3% per year, with foodgrain output slightly higher at 3.2% (by contrast with the 6.5% planned). However, since 1985, performance has declined to nearer 2.5% (with 1987 probably lower as a result of the flood affected *aman* on which hopes had been pinned). The more input dependent crops such as wheat and HYV *boro* which have been responsible for the expansion since 1980 have shown signs of losing momentum (due to irrigation failure, farmer's preferences for rotations, decline in fertiliser/yield coefficients, and early flooding on late harvested crops). The growth rate might therefore dip to 2%, contrasted with the 3FYP target of 4%. The slow rate of growth in foodgrain production is partially explained by the rate of expansion of irrigated area which has risen to 22% of cultivated area (from 11%) instead of the 32% targeted. Drainage and flood control is now supposed to cover about 15% of inland cultivated area (compared to 5% in 1977), although the 1987 floods have exposed some of the weaknesses of these constructions in retaining water once embankments have been breached.

Water development (drainage/flood control plus minor irrigation and surface water schemes) has attracted an investment of $1.2 billion since 1977, representing about 55% of development expenditure directly in agriculture. Further, irrigation related expenditure occurs through the support of irrigated crops. Therefore water development does represent the cornerstone of the policy for the expansion of agriculture. The drainage and flood control investment is designed to support the rainfed crops and allow an expansion of HYVs to *aman*. However, there are problems in achieving even this rate of growth in foodgrain output. While fertiliser consumption has risen by 30% over the 2FYP to about 1.2 million tons, it has done so during a period in which subsidies have steadily been withdrawn to the point where Urea (nearly 70% of the total fertiliser package) has risen in price by nearly 150% while the prices of the two major crops (rice and jute) have only risen by about 50%. Such adverse terms of trade to the farmer may well have depressed the potential use of fertiliser over the period. Furthermore, per capita foodgrain availability has still not reach pre-independence levels (162 kg compared with 177 kg). In other words, Bangladesh is further away from food self-sufficiency than it was 17 years ago. Secondly, as indicated by the irrigation discussion, much of the increase in foodgrain production (40%) has been achieved through the expansion of acreage devoted to foodgrain (rather than a dramatic increase in yields, which seem to have reached a plateau). This has involved a diversion of jute acreage (.85 million acres) to rainfed cereals and a diversion of winter pulses and oilseeds (1.45 million acres) to wheat and *boro*. There are currently critical policy anxieties about crop diversification, and the implications of this cereal displacement upon diet and the import bill. However, there remain ambitious targets for cereal expansion in the 3FYP, to be achieved mainly through further costly irrigation at the expense of these other crops (although there has been a significant expansion of potato). The irony is that if Bangladesh were to achieve self-sufficiency in foodgrains, not only would it displace these other important crops (jute remains a vital earner of foreign exchange and constitutes the main source of cash income for many farmers and indirectly their labourers) but it would also signal the end of food aid (about half of total food imports—1.3 million tons in 1985) which would end the FFWP and

Vulnerable Group Feeding (VGF) programmes which, despite criticism of them, do to some extent reach the poor, seasonally unemployed.

There are many distributional issues, implied in the earlier discussion on poverty and landlessness, which are hidden beneath this condensed review of agriculture. Smaller farmers have difficult access problems to inputs whether they are subsidised and therefore controlled by an unsympathetic and corrupt bureaucracy, or available at a price on the open market. Furthermore, their access to institutional credit is hampered by lack of collateral, where some of their holdings may be rented or mortgaged out against private debt. Local moneylenders charge interest at about 120%. Despite these problems, there is evidence both from the Agricultural Census in 1977 and more recent surveys (the Water Master Plan Organisation (MPO) in 1985) that smaller farmers irrigate higher proportions of their holdings and use HYV seeds and fertiliser more intensively upon a higher proportion of their land. Very often this performance is achieved through intra-household transfers of income from off-farm employment. The effects of these changes upon tenancy is also complicated, with rich farmers sometimes renting out land during HYV/*boro* to avoid investment costs.

Non-Agricultural Economy

While the expansion of agriculture, and rural infrastructure to support it, is obviously central to development in Bangladesh, it is also clear that the labour force is expanding beyond the realistic capacity of agriculture to absorb, even with crop diversification and the development of entrepreneurial capital in agricultural services, food processing, transportation and marketing. The potential for the expansion of non-food agricultural products such as jute or tobacco is constrained by the need to grow food, and this also sets limits to the utilisation of such raw materials in manufacturing. Although there has been an exportable surplus of agricultural goods such as jute, tea and leather, it seems clear that room for manoeuvre has to be sought in manufacturing and services. Such developments have been so far been concentrated upon exports such as: seafood (shrimps and prawns) and garments; the 'export' of manpower to the Middle East; and some services (consultancies, banking etc.). The problem of course is to enter sectors where output can be traded at internationally competitive prices and/or supplies an expanding domestic market, where the constraints are severe, with an average of 70% of rural household income spent on food.

Since independence, agriculture has only fallen from 55% to 47% (at constant prices) as a proportion of GDP. However, there has not been an equivalent rise in manufacturing which has only expanded its share by 0.4%. The shift has been towards services such as urban housing and public administration, which conforms to superficial expectations. During the 70s, there was an annual rate of growth in GDP of 6% (but starting from a devastated low base). During the 80s, this has slowed down to 3%. The rate of growth in manufacturing might have been expected to be faster, but in fact has only been at 4%, with services not much higher though there is likely to be a significant unrecorded element in transportation (a rapid expansion of rickshaws, buses and trucks). The deficit on the balance of payments has been largely covered by foreign aid and more recently IMF credits. The terms of trade have worked against Bangladesh to the extent that in the 10 year period to 1983, the movement in import and export prices has virtually halved the buying power of exports. At independence, raw jute and jute products accounted for about 90% of export earnings. This has now declined to 70% at constant prices, and 50% at current prices. Leather exports initially rose quickly but have now tailed off; tea exports have grown steadily. The expansion of garments and frozen seafood as an export component has, however, been dramatic: from 7% to 31%. The export of garments could probably have been much higher had it not been for import restrictions in the USA and Western Europe. With its natural gas deposits and the domestic manufacturing capacity for Urea, Bangladesh is likely to be exporting Urea competitively over the next few years. However, in two other sectors, handicrafts and textiles (especially silk), which should be internationally competitive, exports are at present negligible.

This brief review of the non-agricultural sector, together with the earlier discussion of employment, indicates the severe constraints to labour absorption and capital formation in the recorded sectors of manufacturing and urban services, with the growth in public administration stimulated by policies of decentralisation offset by privatisation strategies in the public distribution systems. At the same time, the data are consistent with an unrecordable expansion of activity in agricultural services, transportation and marketing.

Institutional Strategies for Agrarian Management

During the 1960s, perhaps the most famous rural development experiment was the Comilla cooperative model among an apparently homogenous class of small farmers in the densely populated Comilla area. These institutions formed the basis of a nationwide institutional strategy for rural development in the form of the IRDP, which in 1982 was reconstituted as the BRDB, and the RPP was added to it in recognition of the problems of increasing landlessness. There have been many criticisms of the limitations of the cooperative strategy under the class differentiated and land scarce conditions of rural Bangladesh, especially where these problems are intensified in the areas outside the Dhaka-Comilla belt. To the extent that cooperation of farmers was required around lumpy technologies like DTWs, the high fragmentation of holdings and extreme competition between farmers for land, water and inputs resulted in both a skewed distribution of benefits and/or a severe under-utilisation of equipment. Many of these cooperatives existed on paper only, to enable their leaders to gain access to cooperative assets for private use. This has been well documented, though these institutions persist as the only national network of savings/input/equipment use/information organisations recognised by the GOB. It has been clear that many of the problems among KSS farmers have been reproduced in the BRDB landless and women's cooperatives. It is also clear that at the *Upazila* level, federations of these cooperatives have become incorporated into the political patronage arrangements. Poor farmers and landless labourers involved in these organisations regard both the KSS managers and officials with deep suspicion, and stories of corruption and misappropriation of funds abound. The other major GOB institutional network in agricultural development has been the BADC, whose limitations were discussed above.

The recognition of these problems, which includes for the national leadership the problem of establishing secure political support in the countryside, has produced a number of institutional and policy responses since the mid-70s. This has consisted of a search by successive governments (reflecting of course particular class and sectional interests) for an ideological/institutional formula of rural management, which has now resulted in the *Upazila* decentralisation. This process is intimately connected with the attempts at a controlled return to democracy, beginning in the rural periphery. However, this programme of decentralisation has profound implications for the formulation and implementation of development programmes. The long range intention is that planning and technical capacity should be located at this level to assist local representatives in allocative decisions using centrally dispersed block grants funded through the development budget (84% supported by external aid). Although expenditure headings and proportions are specified, substantial discretion is allowed for the spatial distribution of this expenditure around the *Upazila*, and within village locations. The political competition for these resources has in effect been restricted to the class of patrons in the locality which does not lead to optimism about distributional and equity outcomes. However, most rural development programmes, whether drainage and flood control, road building, or income generating/employment activities, are now to be implemented through this framework; with a parallel programme of extension and credit support to agriculture through the BRDB and specialist departments organised at the same levels and subject to the overall supervision of the UNO.

Some Indices of Expanded Capital in Agriculture

Crop Trends

As indicated above, during the 2FYP 1980-85, the rate of growth in foodgrain production was 3.2% p.a.. About 85% of the increase in foodgrain output has originated from wheat and IRRI *boro*, and the latter especially has relied almost entirely upon HYV seeds, modern mechanised irrigation and the use of chemical fertilisers. The annual rate of growth for IRRI *boro* was 14%, compared to *aman* (1.3%) and *aus* (-0.7%) (World Bank 4/85 Table 3.1) The corresponding annual rate of growth for wheat was 29%, but of course from a low base. Thus the proportionate contribution of IRRI *boro* to overall foodgrain output rose during the 2FYP from 14.7% to 21.2%, while *aman* declined from 56% to 50.3%, and *aus* from 25% to 20.4%. Wheat rose from 3.7% to 7.7%. From the early 70s to 1983, cropping intensity rose from 134% to 155% (net cropped 21.3 million acres, gross cropped 32.9 million acres). Although the direction of this growth in output has been achieved through a significant expansion of acreage at the expense of other crops (rather than through dramatic increases in yields), the important issue here is that it has been dependent upon new technologies—especially mechanised irrigation and chemical fertilisers.

Mechanised Irrigation

As indicated above, the expansion of irrigated area, although below target, has been the central variable in the crop trends noted above. The following table shows the growth in mechanised units, and in the case of LLPs, the switch from rental (BADC) to private ownership:

Year	LLP		DTW	STW	
	Rental	Private	Operating	Sold	Cum
79-80	37381	-	9795	17551	24458
80-81	31688	2206	10131	17551	42009
81-82	28117	9594	11491	26465	68474
82-83	17619	21973	13794	39145	107619
83-84	9308	34307	15519	30537	138156
84-85	8337	42324	16901	22630	160786

Source: BADC

There are many sources of inaccuracy in these figures, with sales (especially for STWs) invariably exceeding those in actual operation. But they do give some impression of the scale of expansion, and corroborate also the decline in the rate of sales for STWs after the peak of 1982-83 which probably reflects a phase of rationalisation with the most favourable sites being occupied. (For further discussion on these processes of rationalisation see Wood & Palmer-Jones, 1991). By 1985, mechanised minor irrigation covered 69.5% of all irrigated land, and 15.5% of net cropped area. DTWs' share of modern irrigated area had risen from 26% to 31%, but for our purposes it is important to remember that these remain heavily subsidised and allocated to 'cooperatives' through the BADC monopoly. By 1985, LLPs irrigated 48% of the modern irrigated area (compared to 73% in 1979), a decline which reflected the rise of STWs. The share of modern irrigated total area served by STWs rose from 1.3% in 1979 to 21% in 1985. The problems of calculating acreage for STWs should be noted since there are wide technical variations in command area capacity along with institutional problems of under-utilisation in private hands. Our overall argument is not that all such equipment is yet in optimal use, but that as the imperatives to invest in land productivity develop so their management will be increasingly divorced from the monopoly of individual, irrigating farmers. The figures for STWs not only indicate significant expansion in this sector of mechanised minor irrigation, but also reflect that sales are now dominated by the private sector. BADC sales are falling rapidly, as intended by the 2FYP policy. Subsidies for purchase have now been totally withdrawn. Furthermore, initial subsidies on drilling and installation services by the BADC have been withdrawn, and these services are now mainly provided by the private sector, along with mechanical services, equipment accessories, spare parts and fuel supplies. Finally, STWs (and LLPS) are predominantly sold to individuals either as farmers (intending to under-utilise, sell off surplus capacity, or [rarely] with consolidated holdings of a size compatible with capacity), or to contractors/entrepreneurs where the irrigation of their own land is of marginal or no interest. It is clear that much scope yet exists for the expansion of STW groundwater irrigation both on hydrological and comparative organisational grounds. It is also clear that a buoyant re-sale market for STWs and LLPs exists, with command areas opening up and closing down.

Fertiliser Distribution

As indicated earlier, fertiliser consumption has expanded over the 2FYP period by about 30%, during which time subsidies have been steadily withdrawn as part of the privatisation policy. Costs of fertiliser have been rising at three times the price of major agricultural output such as rice and jute, which might be a contributory factor to recent downturns in use, though there are other agronomic reasons as well. However, the restructuring of distribution is more important to our argument. Alongside this withdrawal of subsidies, there has been a strategy of transferring the distribution of fertiliser from the public (BADC) to the private sector. This strategy was initiated in a joint project with US AID in 1979, and was designed to break the BADC's monopoly over the entire distributional system from international procurement to the *Upazila* level, where private dealers were still registered and under the BADC's control, with a sales territory restricted to each dealer's own Union. Under conditions of increasing demand and continuing scarcity, corruption was widespread and an important source of bureaucratic incomes.

The privatisation experiment was started in selected parts of the country with the introduction of large-scale regional government warehouses known as Primary Distribution Points (PDP) where wholesalers and dealers can buy stocks at fixed prices. There are now about 60 PDPs across the country. Under the old system only licensed dealers could buy fertiliser from BADC *godowns*, but since 1979 a trading licence is issued to anyone who guarantees to buy at least 10 tons of fertiliser every four months. In this way, many dealers can now in principle enter the fertiliser market—requiring less capital and fewer bureaucratic connections. However, we should not assume a total absence of interlocking especially since BADC retains control over the flow of fertiliser (through its monopoly of purchase from ports and the three domestic factories), a flow which is manipulated as part of its rearguard defence against privatisation. There have been interruptions of supply to PDPs, which effectively restricts the number of dealers who can enter the local markets. This in turn restricts the number of retail outlets, and has contributed to a rise in retail prices over official wholesale prices. In this process, localised dealer monopolies are occurring, ensuring superior farmer access for those in interlocked markets (e.g. exchanging fertiliser access for rights to market the harvested marketable surplus).

Draught Power

In overall terms for Bangladesh, mechanised draught power remains a significant likely option for the future—an object, then, more of speculation from existing pockets of activity and from the emerging constraints on animal power. Gill's study of Farm Power in Bangladesh (1981) argued that conventional tractor technology was too lumpy for Bangladeshi farmers with small and scattered plots. At the same time, he recorded that about 50% of farmers had their own, adequate draught power with the rest dependent upon hire. We can assume that more are now dependent on hire. T. Rahman's study of a village in Faridpur revealed the extent and complexity of such arrangements (1985). The 1977 Agricultural Census recorded ownership of 3,422 tractors and 1,709 power tillers, but use was higher with ratios of 10.3:1 and 7:1 respectively. Such access is gained through hire. The pressure to extend the use of mechanised draught power will be very strong: as cropping intensities increase leaving less land fallow for fodder; as chemical inputs including herbicides reduce the amount of nutrient or safe fodder; as new varieties reduce the output of straw; and as cropping intensity increases the demand for draught power and land use, animals will compete for scarce arable land—a variant of the food/feed issue. The apparent technical limits on tractor use has prompted interest in the hand power tiller, which has the potential of a multi-purpose power unit for ploughing, pumping water, small-scale milling, electricity generation, transportation (with small trailers) and powering lathes in small workshops. Power tillers appropriate for Bangladesh-type conditions were first manufactured in Japan, and now South Korea and China. They are heavily promoted by importers, but a moratorium on imports was imposed in August 1986 to provide space for the emergence of a domestically produced tiller. Whether domestic or imported, I predict a rapidly expanding market for such small-scale mechanisation which, through its multiple power capacity, will be linked to irrigation and other seasonal uses. The ratio between ownership and use will be crucial, though perhaps less interlocked than tractors. However, amid the excitement about the structural implications of power tillers, 'conventional' tractors should not be dismissed—hence the earlier references to 'pockets of activity' and 'apparent technical limits'. Lewis (1991) found significant tractor use in Chandina, Comilla, where tractors were introduced in the late 1960s. Although casual observation and expatriate rumour links tractors solely to transportation and haulage, it is evident that their capacity is hired out to farmers for ploughing in both winter for *rabi* crops, especially potatoes where there is expanding HYV production, and during May/June for transplanted *aman*. Lewis estimates that such ploughing services realise Tk. 3500 profit per week. It is clear that tractors are in intensive ploughing use for the potato crop, when land preparation in the first mild, dry weather conditions pays enormous dividends in size and quality of potato yields. Lewis examines the ways in which these contracts are organised and arranged, with village brokers emerging to bring enough farmers' land together to reduce the lumpiness of a tractor's visit. Other linkages are appearing between different services: ploughing, delivery of fertiliser or irrigation equipment and fuel, and transportation of produce sometimes linked to advance purchase as with potatoes. Lewis suggests the following categories of actor in these networks: town-based brokers who are linking tractor rental to their other services, deploying their existing range of business contacts; medium or large farmers who need to function as village brokers to bring the land of other farmers into contract in order to have their own land ploughed, but taking commission and building up access to other services as a result; and farmers already deploying other technologies such as irrigation which offers ready-made units of clients to receive tractor ploughing services.

Structural Implications of Expanded Capital in Agriculture

Despite the many institutional constraints to agricultural growth in Bangladesh (noted by Boyce and many others) the actual and likely developments in agricultural production, together with a steady expansion of small-scale mechanised rice-milling and the increased density of foodgrain market transactions (partly stimulated by the PFDS), we are observing a proliferation of actors in the agricultural system. This proliferation has the potential, it is argued, of reducing the significance of the family farm, whether large or small, tenanted or owned, in the agrarian structure. Boyce (1987a, p.37) defines agrarian structure as:

"the subset of institutions governing the distribution of rights in agricultural means of production, notably land. These rights include not only ownership, but also such arrangements as tenancy and mortgage, which create a divergence between ownership and actual operation."

With this definition in mind, it is therefore argued that the present character of such growth as there is, involves the significant addition of non-cultivators and possibly even non-landowners to the subset of institutions noted above.

This proliferation is occurring under general economic conditions in Bangladesh which prevent a significant exodus of families from agricultural activity whether as farmers or labourers. It has been observed that the rise of landlessness could not be equated with a simple process of polarisation in landholding, since population growth and multiple inheritance division of holdings (maintained by the lack of alternative employment/career prospects) act as a constraint to the development of large farms, and certainly consolidated ones. It is, of course, true that richer families can more easily diversify into the non-cultivating agricultural roles noted above. However, such diversification does not leave cultivating brothers as sole owners of large farms, since inherited land is never released in this way. As is well known, this combination of conditions in Bangladesh entails the extreme fragmentation of household land into different plots. These are scattered over the village area as a result of multiple inheritance practices in which not only area, but soil type and elevation is equally divided. With irrigation and the prospect of more intensive rotations, it is even more imperative that the division is equal in all respects.

The increasing importance of irrigation to farm viability (especially small farms) stimulates the contradiction of intensifying fragmentation in response to technological opportunities which demand consolidated land units to reduce lumpiness. Societies with different socio-historical conditions (including the availability of alternative sources of labour absorption) have been able to achieve such market or technologically determinist consolidation (the two phases of enclosures in the UK, or the Punjab in India). The problems for Bangladesh are as much an outcome of colonial history and more recent inferior status in the international economy, as noted above, as of the interaction between human and physical ecology. But since, therefore, this fragmentation is structurally inherent to the conditions described here, its contradiction with the technological requirements of agricultural growth in Bangladesh can only be resolved by separating the ownership of land from the actual operation of it with new technology. Under these conditions, non-land packages of capital, especially in the form of control over water, become the more consolidated assets. In this sense, fragmentation may have been a problem to conventional thinking on farm and land management, but as has been remarked in previous chapters, it actually constitutes an institutional resource to those who control other categories of capital. They can realise surplus from their control over the increases in the productivity of land which is denied to (or shared under adverse terms with) individual farmers who have fragmented holdings and no direct access to the technology. This is the resolution of the public goods issue for ground water and for surface water in the minor irrigation sector. While this has applied more strongly to LLP and STW technologies, the generally acknowledged failure of attempts to solve public goods management in DTW systems through cooperatives of user farmers (see also Boyce 1987b) has brought DTWs into the discussion of organisational alternatives involving proposals for public sector management (very unlikely) and for leasing to private water management companies (much more likely).

Water is clearly the lead issue in this analysis because it is so central to any strategy of agricultural growth. While not losing sight of the broader arguments concerning agrarian entrepreneurialism and interlocked markets, the significance of this process of separation of ownership from operation should be described further with reference to mechanised irrigation. From my own research on command areas (see Wood 1982, 1984a, and with Palmer-Jones, 1991), the number of farmers in STW command areas between 10-17 acres ranged from 10 to 38. With LLPs irrigating between 20-50 acres, the number of farmers has ranged from 18-97. DTWs are similar to LLPS in this respect. Glaser 1989 has reported one farmer in 'her' village in Singra (Natore, Rajshahi) as having plots in 18 of the 31 different STW command areas. While 18 was exceptional, lower figures of 10 were not. The logic of production decisions on these different plots becomes more a function of the efficiency criteria for managing the irrigation service than of family farm subsistence and exchange needs. Increasingly, the farmer responds to the rotation imperatives, timetable of operations (involving labour hire), use of hybrid seeds, fertiliser and pesticides (a particular public 'goods' problem) as set by the ownership and management practices of the irrigation source. If payment for water is in the form of crop-share, then water-sellers have a strong incentive to use their leverage to ensure that yields are maximised through adequate labour inputs and investment in divisible inputs. If payment is a fixed cash sum, then the farmer anyway has a strong incentive to maximise yields.

Control over water is not alone in this process, with increasing examples of fertiliser delivery and pesticide spraying (see Lewis 1991 for case studies), undertaken as a commercial service by command area 'managers', often using contract labour. With the increase in mechanised milling and the likely widespread introduction of power-tillers, fragmented plots will be consolidated not around individual household units (the 'Punjab' model) but around the owners (single or networked) of production, processing and transportation technologies. The concept of 'command area' is not then restricted to irrigation sources. Whether coterminous or not, areas of land will be 'in the command' of other packages of mechanical and chemical technologies, and vertically organised marketing arrangements. Large-scale food processing, such as automatic rice mills and sugar mills, are not currently significant, as they are in other countries, but their presence would extend this principle still further where a stable supply of paddy and cane would link such large-scale processing to local production. This process can already been seen to some extent in potato cold storage, and has long been the case with tobacco. However, even without the large-scale processing sector, the existing dimensions of production and food-processing capital can combine in the formation of agricultural 'companies' (loosely networked or corporate), interlinked with repair workshops, mechanical services, supplies of spares and fuel, transportation and marketing. In this way, the concept of the Bangladeshi family farm disappears, as the agricultural system becomes re-articulated through these other units of production and exchange. At its most extreme, this interpretation of the process would mean farmers engaging in reverse tenancy on their own land, without the bargaining power to insist upon adequate subsistence rents from the agricultural service companies or networks of entrepreneurs. They become indebted workers or supplicant tenants, finally losing possession to these entrepreneurial interests. The re-articulation of the agricultural system would then be complete.

Of course, the implications of such an extreme structural scenario would be enormous for agricultural policy in Bangladesh. Presenting the argument in this manner is deliberately provocative, since it develops our antennae for understanding the direction of agrarian change. However, there are various ways in which this speculation should be qualified.

First, the emphasis of this analysis has been on irrigated foodgrain production: wheat and IRRI *boro*. This reflects the importance of multiple foodgrain cropping to the viability of smallholdings in Bangladesh and to the overall prospects for agricultural growth. However, the significance of irrigated production varies by class of farmer, with larger farmers having more options in other seasons both for *aman* and other crops such as jute and sugar, while smaller farmers are more dependent for their own subsistence upon the irrigated season. This helps to explain why the intensity of irrigation (modern and traditional) is higher on smaller holdings than larger ones, despite access and tail-ender problems. But as larger farms become smaller through multiple inheritance, the reliance of these divided holdings upon irrigation will also become stronger. In the meantime, larger family farm members can diversify into agricultural services (that is, join the entrepreneurial networks) and thus accumulate indirectly from rises in productivity rather than through maximising the intensity of their own cultivation.

Second, but related, is the emergence of new strategies for tenancy and mortgaging. This is where forms of exploitation may appear to remain the same, but their meaning for production relations has altered. With irrigation, there is more seasonality to tenancy arrangements. Precise rationales are contingent upon very specific conditions reflecting: perceptions of seasonal risk which are intimately related to predictions of flooding, therefore elevation of land; the owner's plot portfolio—number, fragmentation, spatial distribution, soil and elevation types; the owner's production capacity—draught power, cash flow, timely access to labour and variable inputs; the owner's management capacity (a function of plot portfolio, production capacity, family composition and position in interlocked markets). Thus larger farmers may lease out land during IRRI *boro* if high risk is perceived, or they are unwilling to commit investment but see the potential productivity of tenant cultivation, while retaining it for *aman* if low risk is perceived. But the strategy can equally be inverted under other sets of conditions, retaining land during IRRI *boro* and leasing out during *aman*. Likewise, poorer farmers now engage in reverse tenancy (confined to *boro* season) where the conditions above prevent them from HYV cultivation themselves. Clearly they are not 'landlords' in such arrangements, and because of other extant patron client ties are likely to be in a supplicant position for rent bargaining. Larger farmers thus lease-in too for IRRI *boro* cultivation as a form of 'consolidation'. Mortgaging is performing a similar function and responds to similar conditions, though especially connected to cash flow. Larger farmers in urgent need of cash for some investment may mortgage land out even to poorer farmers who thereby gain temporary access to cultivable land. The reverse relation is more common. However, these arrangements have become extraordinarily complex in Bangladesh: use of mortgaged land for cultivation may be seasonal rather than annual, and can be leased back. Within both these tenancy and mortgaging options, the structure of accumulation for larger landowners can be surplus extraction via rents and interest (leasing out or mortgaging in) or gaining value-added through investing in productivity, or a combination (e.g. when leasing in during IRRI *boro* but using interlocked market advantage to pay below market rents in the reverse tenancy). Glaser has found these variations between 'her' village and seven neighbouring villages in the same ecological conditions (there is no determining single/multiple cropping area variable). In 'her' village, Samitigram, 26 of the 31 STW command areas are organised by groups of cooperating farmers and 'managers' with many counter-checks on price and quality of delivery since each farmer has plots in several command areas. Thus Glaser found that linkages between factor markets had loosened under irrigatedgated conditions, which had made higher output and returns possible for all water users though at differential rates. However in the *aman* season, factor markets remained interlocked as before (rents, interest and wage levels). By contrast, in the other seven villages, irrigatedgation has brought about "an enforcement of factor market linkages. This further facilitated extractive practices by the powerful, while allowing for some overall productivity increases." (Glaser 1989, Ch. 1). This analysis gives some insight into the other options which farmers have, to manipulate their use of land for rent and productivity purposes, thus avoiding some of the extreme implications of the 'loss of control' argument outlined above—at least in the short-term.

Thirdly, and in an empirical sense more tentatively, it is important to 'explode' the concept of technology (the approach pursued by Lewis). Rather than confine to a narrow class of emerging agrarian entrepreneurs our notions of participation in the opportunities created by the package of irrigatedgation technologies and agricultural services, these processes should be desegregated into chains of transactions and activities. Opportunities to participate are therefore wider, albeit dominated by interlocked factor markets so that returns are not equally shared. Equipment is not just manufactured and operated by end-users. It is marketed, transported, maintained, repaired, and re-sold (in highly fluid [!] STW markets). Brokers are involved, linking supply to demand via networks and interlinked exchanges. Fuel and spare parts have to be supplied, involving other chains. Likewise with divisible chemical inputs, many levels of transactions occur. This suggests that we should understand agrarian entrepreneurialism as a more widespread social phenomenon, and apply some of the informal sector analysis to it (i.e. entry conditions and capital requirements alongside a rejection of dual sector models because of linkage). It is also the case that, with a high import content and a donor supported economy, at certain levels these transactions include government officials despite the policies of privatisation. The involvement extends to *Upazila* officials in alliance with local politicians (entrepreneurs and patrons) who are deploying public funds for roads, embankments, canals, and market construction. These all impinge upon the way in which entrepreneurial opportunities are distributed in the locality—both spatially and socially.

This third qualification to the central argument, along with the central argument itself, opens up other possibilities for those concerned with the advancement of the poor. The problems of rural underemployment and effective demand for any expanded food production have focused attention (somewhat belatedly after Wood 1980b) upon employment programmes for the landless and near-landless, with an almost panacea reliance upon the non-farm sector. However, the process of agrarian change outlined here offers the prospect of steering these agrarian entrepreneurial activities towards the landless, where they can gain rents and profits from expanded agricultural production, as an alternative to the acknowledged but inadequate increase in agricultural labouring opportunities which give access to higher incomes for some of the landless. The provision of irrigation services by landless groups associated with Proshika, BRAC and Grameen Bank has been one example of this process (see Wood & Palmer-Jones 1991).

PART III

RURAL WORKS: DEVELOPMENT OR WELFARE

Preamble

The four chapters in this part of the volume have all been written in the context of those rural works programmes in Bangladesh, funded by the Swedish International Development Authority (SIDA). These programmes (the Intensive Rural Works Programme followed by the Rural Employment Sector Programme) were essentially distinguished from other rural works programmes by the objective of greater participation and more secure employment rights for labour. Therefore they had social as well as infrastructural or physical objectives. Chapter 10 was the first in the sequence, written on the basis of field studies for the programme to investigate the reality of landless labour participation. It concludes with a number of recommendations which were pursued by the programme, sometimes in conjunction with NGOs (such as the direct contracting with rural workers' federations, which were called 'labour-contracting societies').

It was during this assignment that I became concerned at over-reliance on the concept of 'target-group'. This concern was extended theoretically into the notion of development policy labelling, on which I collaborated with Bernard Schaffer (before his death in 1984) and others to produce a volume of papers (Wood ed. 1985a). My case study is included here as Chapter 11, and sets the issue of poverty-targeted rural works programmes in the broader policy context of rural development labelling in Bangladesh. In particular it draws attention to the process whereby the programme relies upon targeted poor people to adopt the labels given to them, and to use them as the basis of mobilising themselves to secure their entitlements from the programme.

The third chapter (12) in this Part was initially prepared as an elaborate set of lecture notes, to highlight the parallel relationships between physical structures and social institutions on the one hand, and engineers and social scientists in the decision-making process on the other. These notes originate in mid/late eighties and precede the subsequent concerns about the flood action plan in Bangladesh. Some of the issues raised in this chapter were therefore prophetic when applied to an analysis of the flood action plan in these terms, especially the relative priority given to: 'participation' of affected populations in planning infrastructural projects (see Adnan et al 1992, which reports a study on this issue originated by CDS at Bath which also fielded two members of the study team); and social science as opposed to engineering findings. These notes have been written up especially for this volume, using the opportunity to make deliberate references to the flood action plan. Finally, these SIDA funded programmes encountered the familiar problems of official rent-seeking and weak structures of public accountability, despite attempts to empower workers directly in order to discount the deficiencies of the formal institutional map (i.e. central government departments and local government). This paper was prepared for an edition of the Institute of Development Studies (Sussex) Bulletin (1988), and appears here as Chapter 13.

Chapter 10

Landless Labour Participation and Mobilisation in Rural Works Programmes

Introduction

I was retained as a Consultant by the Swedish International Development Agency (SIDA) during April 1983 to provide an input into the Donor Annual Review Mission looking at ways of improving the participation of landless people in the Intensive Rural Works Programme (IRWP), and of ensuring longer-term benefits as a result of the IRWP, through the development of other forms of income generation and asset ownership. I was invited to perform this function on the basis of previous association with those objectives in Bangladesh rather than prior familiarity with the IRWP. The discussion which follows is therefore the product of considering the IRWP intervention in Bangladesh in relation to this wider experience. What follows is a combination of judgments which reflect primarily a desire to be constructive, even when the negative effects on the programme's objectives of the political reality in Bangladesh are identified.

Room for Manoeuvre

The exercise is essentially one of determining the room for manoeuvre, and if the judgements are not always encouraging, it is worth remembering that false optimism can often do more harm than good. But the search for political, social and technical options to pursue the objectives concerning the rural poor in Bangladesh must continue. However to do that meaningfully, the constraints cannot be ignored even when they are outside the immediate control of both the rural poor and the programme staff. It is precisely the distinctiveness of IRWP's objectives which makes the discussion of these constraints (i.e. the context of the programme) necessary.

'Target-Group'—the relation between 'them' and 'us'

The literature which the programme has already generated by way of reports and plans employs the term 'target-group' with reference to objectives and interests. This is unfortunate, not just for refined semantic reasons but because of what it reveals about the relationship between 'us' and 'them'. In this respect the IRWP has adopted the language of other development agencies. It conveys the notion of what 'we' can do for 'them', and contains the assumption that the relationship essentially consists of stimulus-response, with outsiders initiating change through the creation of opportunities for 'them' to utilise. That is of course a premise of all rural development interventions, and is certainly part of the picture. But there is another side to the relationship, where 'they' create through action a structure of opportunities to which we respond with policy confirmation and material support. This involves 'us' listening and participating with 'them', not just 'their' participation with 'us'. How much of such listening has happened from the inception of this programme is a matter of concern. The issue is one of appropriate style or continuous method for a programme with these objectives.

Furthermore, the use of 'target-group' to refer to 'them' contains the radical assumption that the most significant aspect of the perceptions of poor rural people is that they constitute a social class with common interests. But this is not true; family survival, and the relationships through which it is precariously pursued, are uppermost in people's minds. This gives rise to other forms of solidarity and identification which run vertically through the hierarchies within the *baris* and the *paras*. Poor rural people may be a target but generally they are not a 'group' and this is part of the problem concerning our participation with each other.

Finally, and here we are all guilty, it is not clear who the 'target' is. We use general categories like landless (and women), assuming we have identified a clearly-recognised rural social entity to which special support and resources can be directed. But the society is dynamic with rapidly changing and continually enduring social relations—today's small peasant is the landless of tomorrow, so that in terms of connection and co-operative effort people will define themselves into meaningful social units according to dimensions beyond those contained in a term like landless. If this reads like another unhelpful academic comment, remember that if the programme is to rely on group organisation this has to take a form within which people feel comfortable and where there is mutual trust. A 'group' of labourers brought together by labour *sardars* from different parts of a village is more likely to reflect complex patronage arrangements and dependencies than a sense of group solidarity which could continue once the season's earthworks have finished. To assume otherwise could be dysfunctional and dangerous to the programme's objectives.

In view of these remarks, the term 'target-group' will be deliberately placed in quotation marks, but it is also recognised that other descriptive terms have their limitations too. The important issue is to ensure that the descriptive terms which are used are consistent with real, social possibilities of collective mobilisation among the rural poor.

Modes of Intervention—two-way participation

As suggested above, those issues of terminology are important because of their implications for modes of intervention, the methods and processes by which the programme's objectives can be achieved. This report cannot just contain a concrete list of substantive proposals through which the landless of either sex can improve their material and social conditions. Indeed, such lists can be found in the 'North Faridpur—General Planning Report' (July 1982), the 'Kurigram Thana—Upgraded Planning Report' (March 1983) and elsewhere. The problem is not lack of ideas 'for' them, or even 'by' them, but through listening and assessing the combination of constraints—to devise ways to pursue such ideas. Thus the methods of achieving participation and asset-generation have to include participation itself. In their other work outside their involvement in IRWP, the two NGOs Saptagram and Proshika are standard-bearers of such approaches. But it is important to know whether the IRWP can organisationally adjust to the requirements of such approaches in order to pursue its 'target-group' objectives beyond the present stage of hesitant resolve. The point is that methods and style are as important as content in achieving the programme's objectives, and this has profound implications for developing the capacity of the programme to address its objectives in a sincere way.

Context—Objectives and Capacity

Objectives of IRWP

These were defined in the IRWP Plan of Operation 1981/82—1983/84 as:
  a. To increase direct and indirect employment and other income opportunities in the short and long run for the landless labourers,

marginal farmers and women from such households (i.e. the 'target-group'). Such households comprise 65-70% of all households in the country.
b. To improve the infrastructure in 100 flood-prone *thanas* by implementation of rural works thereby promoting agricultural production. Marginal, small and medium farmers with less than five acres, who own about 57% of agricultural land in Bangladesh, would be the main beneficiaries, though the incomes of large farmers and traders would also increase.
c. To review and assess measures needed to strengthen and improve the organisation and administration of the Rural Works Programme within local bodies and to work out solutions and make recommendations for a nationally replicable framework within which to achieve a) and b).

Furthermore, the Plan of Operation would commence with an Initial Implementation Stage (IIS) of about one year. Included as the main purpose of the IIS was a commitment:

"to undertake a comprehensive review of the RWP set-up within the concerned local bodies and to initiate measures in order to: ....

ii. increase participation by the local community and, in particular, by the target-group, in the planning, implementation and utilisation of IRWP schemes through the assessment of appropriate institutional arrangements and improved methods of scheme selection.

....the selection of schemes for the IIS would be guided by the following criteria:

i. the programme of works planned and implemented each year will aim at a wage component of at least 70% of the total cost.
ii. a substantial share of the schemes included in *thana* plans will be of a kind that directly benefits the target-group in addition simply to providing direct employment."

Distinctiveness of Programme

From this summary of the objectives, it can be seen that IRWP is a distinctive programme of intervention through rural works by virtue of its orientation to the 'target-group' of landless labourers and marginal farmers (male and female). It is worth emphasising that the programme invites a judgment of its progress against those objectives rather than simply the provision of temporary employment during the lean months of the agricultural cycle.

Contradiction between 'Target-Group' Objectives and Dependence on Local Bodies

Those objectives also instantly contain problems. The major one which will have to be confronted by the Review Mission is the apparent contradiction between diverting resources to the 'target-group' and implementing such a redistribution through a continued dependence upon the local bodies. Clearly such a commitment to the local bodies is part of current government policy, but it will be argued at several points that this represents the single most significant constraint in the room for manoeuvre to achieve the 'target-group' objectives. The evidence is so strong that the Mission would be undermining its sincerity of purpose if it fails to address the implications of this contradiction.

Patronage versus Representation

From this problem several other issues follow concerning the distribution of power between institutions (Government and rural). It is likely that by depending on the local bodies for certain categories of scheme implementation (and such categories will be expanded after the local elections in the upgraded *thanas*), the power of those local bodies is being reinforced through their effective control over scheme project committees. By placing such significant resources at this level the patronage rather than representational character of the local bodies and their leaderships is likely to be stimulated. Such a process intensifies problems of local accountability and indeed access to the institutions of planning and implementation. At the same time it strengthens the capacity of the local elite to manage labour (its recruitment, work on site, and payment) on its own terms, not only during the rural works season but outside it in the agricultural season too.

We will return to these arguments, but they serve to illustrate both the dynamism of rural social relations, and the problems of partial interventions like the provision of temporary, seasonal employment opportunities into total, highly interconnected situations. Expectations about behaviour have to be considered not just in relation to the Programme but to the other rural social mechanisms through which families attempt to arrange their survival.

Capacity of Programme to Meet Objectives

A second set of constraints concerns the capacity of the programme to develop its 'target-group' objectives and deal with these institutional problems. The programme demands an integration between engineering and social science approaches to the development of rural infrastructure. However, these approaches require different time scales both in terms of information needs and the process of intervention itself. The more rapidly the programme proceeds, the greater the weight which is effectively attached to the engineering criteria for scheme selection and implementation. In this respect, therefore, the issue of alignment between the rate of projected expansion of the programme and the conditions required to address the 'target-group' objectives of the programme will have to be faced by the Mission. At the same time there is the problem of the working relationship between engineers, other technical inputs (for example feasibility studies on proposals such as bank sericulture), social scientists-cum-rural animateurs and, most importantly, different groups of rural workers, artisans and farmers, of both sexes. Each part of this network has to be exposed to the strengths and limitations of the other. One precise response to this is to develop a capacity for rapid rural appraisal (RRA), where field workers—perhaps recruited from research projects undertaken—develop an intimate local knowledge and in conjunction with local people engage in a continuous process of local analysis and problem solving, in which engineering and other contributions are subjected to critical review by the consumers and possible beneficiaries of the infrastructure. Such an approach to planning is much more location-specific and 'target-group' based than the present union and *thana* plan strategies, and the long and sterile route of feedback between local demands (usually of the local elites only) and engineering responses is short-circuited. Some case examples of this approach will be discussed later in the chapter.

Distribution of Risk

A third set of problems arising both from the objectives and the constraints noted above concerns the distribution of risk between participants and organisations engaged in pursuit of the programme's objectives. At best the programme is a limited intervention in people's lives. At the same time it is becoming increasingly clear that even the limited objectives of ensuring workers' rights during the period of temporary seasonal employment requires forms of struggle and militancy on the part of workers which potentially jeopardises their other relationships of survival. Programme staff, let alone donors, have a limited role in providing support for such struggles (even though they represent a means of achieving the objectives). But at the least the programme, supported by WPW and MLGRDC, must firmly resist the victimisation of workers, their exclusion from schemes, their physical intimidation even if this means confronting and sanctioning the local bodies at union and *thana* level through which these attacks on workers frequently appear. Where other local organisations exist and include within their programme the development of workers' rights through education, meetings and the organisation of claiming, the Programme must recognise that such organisations are consistent with pursuing the 'target-group' objectives and should therefore receive public statements of support. The immediate, urgent and highly significant case of Saptagram in Kotwali Thana, Faridpur constitutes a test of the programme's resolve and sincerity, and its capacity and willingness to support Saptagram will be important evidence for the Mission. Certainly both Saptagram and the Programme's Field Officers in Faridpur are not yet convinced of such support. This case of Saptagram will be discussed in more detail later, but a failure to confront this victimisation of site workers on the schemes of the Badarpur-Saltha road will not only leave the women workers weaker but may also undermine the wider range of Saptagram's involvement with the women from landless families, and some of their menfolk. It has to be remembered that Saptagram was encouraged to participate in this way by the Programme. Neither the site workers nor the Saptagram workers can walk away from this problem. The risks are very high. There are ethical questions here about responsible enticement of others into action, and the donors should be convinced that the programme is serious in its intention, and is prepared to follow through the implications of its objectives, before considering a further round of expansion beyond the 1983/84 commitments. Saptagram represents a test of resolve for the donors and the government more than anyone else.

Participation—Problems of Claiming

Workers' Rights

There are different levels of rights for rural workers outlined in the programme. The first and most obvious is the opportunity for the landless and destitute of both sexes to work in the IRWP schemes. Secondly, to ensure that such work leads to the receipt of fair wages. Thirdly, that priority is given to labour-intensive schemes to provide those opportunities. Fourthly, that workers share in the longer-term benefits accruing from their labour through related production activities. Fifthly, to ensure these possibilities, workers should be involved in the planning and implementation of work.

Workers as Claimants

The first issue then is to review the problems of workers in achieving these rights. A key theme running through this discussion is that the Programme cannot ensure the provision of those rights without the help of the workers themselves, organised as claimants. This is partly the result of dependence upon the local bodies for implementation, and partly the related systems of labourer recruitment and management.

The field officers in the programme and the NGO workers with whom they are associated (Proshika in Madaripur and Saptagram in Northern Faridpur) were well aware of these issues. From discussions with them, organised groups of workers, and non-organised workers in those areas, the following picture emerged.

Project Committees—instruments of patronage

The Project Committees (scheme-based for the *thana* schemes and ward-based for the Union schemes) are effectively an extension of the patronage system of the Union Chairman. They either become the PC chairman themselves, or nominate both the chairman and the secretary (when they should be elected by the committee once constituted), and other members are either unaware of their membership, or of their duties, or have been selected for strategic political reasons by the U.P. Chairman. The cynicism which is applied by those local leaders and landed classes to the Programme and its objectives is revealed especially over the central issues of payment rates and measurement of earth work—where cheating is common to all the schemes visited.

Recruitment and Management of Labour

The problems of claiming rights on payment and measurement are related to the ways in which labour is recruited and managed. The Ministry of LGRDC Circular No 5, para 16, p 9 states:

> "Function of the Project Committee is to mobilise labourers for the execution of the scheme as per list of destitute maintained by the Union Parishad in their assessment list of the payers and gang them up."

In no case visited has reference been made to such a list as a basis of recruitment. If it did exist as the key to significant employment opportunities, then workers would no doubt have difficulty in ensuring that their names were included in the list. That would be the first of many problems of access.

Labour *Sardar* System

However in practice, when NGOs are not involved, labourers are organised through a labour-*sardar* system. The main source for this information is from the Dapada Bheel in Boalmari Thana, Faridpur, but corroboration was found elsewhere. These *sardars* are in effect sub-contractors of labour. They are not necessarily landless themselves. Often they are educated sons in households where they will inherit land after division. They either have or develop connections with local leaders, finding out when work will be available and where. They create small gangs of perhaps ten labourers who may be dependent upon them in various ways; who may be scattered across the village without previous or intimate connections among themselves; and whose association with each other is likely to be temporary for the season. These *sardars* function as an intermediary between the labourers and the union chairman/project committee. At the site itself, they do not usually work but act as supervisors. Payment to the labour gang is divided to include the *sardar* and in addition the *sardar* receives a payment from the Project Committee chairman. Payment to the gang is made via the *sardar*, which may often involve him travelling off-site to the PC chairman's house. He thus controls the flow of income (and most importantly information about income) to the labourers, and has an interest in maintaining this system of recruitment and management.

Credit and Control over Labour

There is also a credit dimension in these relationships which further circumscribes the freedom of labour to mobilise. Labour *sardars* frequently function as money lenders to 'their' labourers, giving them or their families advances to be worked off during the earthworks season. These arrangements can be institutionalised further, with the Project Committee paying advances to the labour *sardars* for this purpose. In the case of the Gorai Embankment at Dumain in Baliakandi Thana in North Faridpur, families of labourers living about four miles from the scheme received advances from the Project Committee to bring the male members to work. The daily payment to the workers on site is Tk. 10 per day, and the family members of site workers come on measurement days to receive and take home the balance arising out of measurement.

It is difficult to assess how widespread these arrangements are, but they certainly highlight the structural problems of rural indebtedness in relation to the Programme. By definition the programme is operating in the lean season when income flows are low. Daily advances are a reflection of the need to provide immediate income, but they also function to postpone and thereby facilitate cheating on measurement. How dependent are the labourers and their families on loans from village moneylenders to 'fund' low daily rates, low overall rates, and delays in measurement? With interest rates high, are remaining pieces of land being lost through mortgage? Thus politically, behaviour is constrained; and materially, family conditions actually deteriorate when delays and cheating occur.

Fragmented and Vulnerable Labour

Thus these apparently potential units of a claimants' organisation are actually highly fragmented groups of vulnerable labourers without much contact with other labour groups on site, recruited and brought to the site by other *sardars*. At the same time, this system is convenient for the Project Committee in executing the scheme, especially as it is also instructed in Circular No 5, para 16: "To ensure implementation of project according to time schedule." Through this system (and other mechanisms to be discussed) a quiescent labour force is maintained in a manner which really undermines the capacity of labourers to claim their rights as set out above.

Pace of Investment Reinforces Patronage

This situation is reinforced by the pace of investment envisaged in the programme, since the Project Committees (leaving aside their own political inclinations) would face impossible problems of mobilising a large and continuous labour supply without recourse to such a network. As a result, the system ironically functions to provide workers with limited access to the employment opportunities while at the same time effectively denying them any participation in the programme even as claimants for the rights embodied within it.

Family Survival Strategies Undermine Militancy

Alongside this system, there are other structural factors which affect the capacity of labourers to claim. Perhaps the most important concerns the way in which poor families attempt to survive. Under conditions of increasing landlessness, pressure on the availability of land for sharecropping, and the general evidence on rural under-employment, families rely on landlords and richer farmers to provide employment for both men and women, land for leasing, and consumption credit in the form of advances. It is true that such patrons may sometimes be from within the same *bari*, but they are nevertheless connected at other levels in the structure of rural power to secure access to inputs, the registration of recently acquired land, procure licences and leases for business activities, leverage in court disputes (e.g. over land), and so on. In this pattern of exchanges, Union Council members have always occupied a strategic position. Labouring families are likely to increase the precariousness of their survival if during the earthcutting months of the year they challenge the rural hierarchy upon which they depend not just for the rest of the year, but even for access to the earthcutting work itself. Thinking of rural people as individuals independent of their families and capable of easy militancy is a mistake. There is a range of calculations to make which determines the reality of their room to manoeuvre.

Systems of Payment—village and site compared

There are further problems for claiming which are more specifically related to the payment and measurement aspects of the programme. It is interesting that the practices of payment reflect patterns of payments for agricultural work with a combination of daily payments and an end of 'season' pay-off, through a share of harvest in agriculture and through measurement in rural works. Leaving aside judgments about the implications for social justice, such delayed payments in agricultural work occur between parties with highly personal and long term—over generations—exchanges between them. In that sense there is a trust that expectations will be honoured. Unfortunately, given the system of labour recruitment and management on site, no such trust exists in those earthworks schemes. The level of morality in the exchange is lower, and it is easier for employers to contemplate defrauding workers on site—especially over payment of the final balance after measurement.

Patterns of Work—a resource for cheating

This process is further reinforced by the patterns of work on site. There is much evidence from my visits that workers frequently travel some distance for site work (more than three miles), and sometimes they are fully migrant in that they stay on site for the duration of their 'contract'. If they are walking to the site daily, and are receiving daily wages—technically 'advances'—then they are less effective in maintaining the pressure on the Project Committees to ensure timely and accurate measurement. In effect they 'disappear', either into other work (other sites or agriculture), or returning to claim is not considered worthwhile because of expectations about cheating. Acceptance of a daily advance becomes a satisfying strategy, weakening the position of others who are trying to insist on proper measurement. The capacity for Project Committees to defraud under these conditions is enhanced by the irregular and temporary participation of workers in the scheme, which is encouraged by daily payments. Sometimes it is alleged that Project Committees will deliberately organise a high turnover of labour for the purpose of defrauding on measurement. This was claimed in connection with the now-suspended work on the Kumar river excavation in Pearpur union, Madaripur. The Project Committee was insisting on lengthening the working day (which, with a proper management system, should not matter but it does when the safe or guaranteed payments appear in the daily form) to reduce the attractiveness of working on the scheme for long periods. Thus irregular, temporary, semi-migrant and high turnover labourer involvement, in combination with daily payments, prevents the labourers from establishing a sustained and united connection with the scheme, and therefore undermines the possibility of being effective claimants of the piece-rate set for earthcutting which relies on measurement.

Fictitious Muster Rolls

This weakness of 'connection' to the scheme is reflected in keeping of muster rolls of workers. The legitimate maintenance of this register is central to the objective of ensuring accountability on payments and measurements, although it is only a necessary rather than sufficient condition in the case of measurement. There is plenty of evidence throughout the programme that those muster rolls are substantially fictitious in the names entered, thumb prints given (since no mechanism can really exist for checking these), dates of attendance, labour gang association (workers do move in and out of gangs), and borrow pits worked (since with delays in measurement, they are inevitably working in new ones).

Earthwork Schemes as Private Business with Intimidation of Workers—Kumar river

The case of the Kumar River evacuation reveals further problems in claiming, even under conditions where Proshika have organised groups in the locality over the last two years (before these schemes) and where 50 out of approximately 140 workers on the site at different times come from 20

Proshika groups. The temporary, semi-migrant turnover of non-organised labour has made it difficult for the organised workers to claim. The daily rates have been variable, ranging from Tk. 8-13 depending on demand in the locality. The Project Committee chairman is the Pearpur Union Chairman—he employs four people to act as site supervisors; his son and nephew run the scheme like a private business. Measurements on completed work have been delayed (he is accused of deliberately waiting for the rain to confuse the dug levels in the borrow pit). The workers fear that he will remove the dam across the head of the scheme to flood the excavation before measurement; so they are having themselves to guard the dam at night as to prevent this happening. Workers from the Chairman's own village (not immediately local) are hired on a simple daily basis at lower rates even than those working under the more normal daily advance measurement arrangement where the rate offered is Tk 165 per 1000 cft. The Chairman has threatened to file cases in the *thana* against troublemakers, and some of the more outspoken members of the organised workers are in fear of personal attack. The Chairman's threats are considered seriously because of a reputation established on earlier programmes such as FWP. The Secretary of the Project Chairman personally handled the allocation from the TDC Chairman (the Chairman of Khoajpur union and 'boss' of the President's Canal scheme on the eastern side of the Arial Khan river). The secretary did not know, nor had he met, the other members of the committee.

Organised Workers can Affect Bargaining

However, the presence of organised groups of workers can affect the bargaining relation between committee's employers and site workers even when they are only a proportion of the total workforce. In the President's Canal scheme of Khoajpur Union in Madaripur Thana, Proshika organised groups for the first time in response to the scheme. At first, in total there were 300 workers, but half of these left quickly because payments were too low and irregular when compared to harvesting of *boro* and sowing for early *aus/aman* and jute. Out of the remaining 150, 90 come from groups which were being organised with Proshika's encouragement. The Chairman is under pressure in his Union from a radical political party and along with other leaders has received threatening letters. Although his *bari* is in the Union, he regards it as unsafe for him to be there, so he remains in Madaripur town, where he is also TDC Chairman. The secretary of the Project Committee (a kinsman) recruits and supervises labour and does the measurement, but the Chairman negotiates payment with the *sardars* and pays out in Madaripur town. The labour *sardars* therefore have to travel about four miles and cross the Arial Khan river. The Chairman initially offered Tk. 150 per 1000 cft. When the organised groups protested, he offered the full rate to them but asked them not to communicate this offer to the other organised workers. Eventually, under continuing pressure, he conceded the full rate to all the workers (although it has not actually been paid yet); finally, because he is seeking electoral support in the *thana* election he went (as the current TDC Chairman) to the Kumar river scheme and told the workers that he was paying Tk. 200 per 1000 cft and that they should struggle for it in their own scheme. However, the evidence from the President's canal scheme is that the cheating is now occurring not over the rate but over the measurement (i.e. the fall-back position).

The Experience of Claiming—the Case of Saptagram and the Workers on the Badarpur-Saltha Road, Kotwali Thana, Faridpur

Significance for ARM and Proposed Timetable of IRWP Investment

This review of the structural conditions which determine the capacity of workers to claim rights in the programme should consider the ongoing case of Saptagram and the female site-workers in two women's schemes on the Badarpur-Saltha road. It is likely to dominate the ARM's discussions on 'target-group' objectives. Groups of women and men from landless families have been organised over several years by Saptagram Swanirvar Parishad around independent economic activities, especially sericulture, as a basis for carrying on a programme of informal education and consciousness-raising. The significance of their experience of claiming their rights as workers in the IRWP scheme is the depth of resilience, solidarity and extent of struggle required when even established organisations of workers clash with locally vested interests. When considered together with the issues discussed above on problems of participation, it should be recognised that some of the proposals for 'target-group' participation brought together from various sources (e.g. Proshika reports) into the 1982/83 Assessment Report pp 79-89 are the outcomes of sincere frustration rather than realistic predictions about behaviour. In particular, it will be argued after this discussion of Saptagram that the timescale required to establish meaningful conditions for participation has to be extended, with profound implications for the timetable of IRWP investment commitments.

Dramatis Personae

On the Badarpur-Saltha road there are six schemes under the *thana* section of IRWP of earthcutting and earthwork. Two schemes have been specially designed for women's employment. 75% of the workers have been organised by Saptagram (425 women out of 550). These are organised by Saptagram into approximately 30 groups from two of Saptagram's project areas—Bonogram in Kaijuri Union (where the road is located), and Komorpur in Ambicapur Union. In the men's schemes (which occupy an important but subsidiary role in this story), there are approximately five groups organised by Saptagram in each of the four schemes, which means that 50% or 450 workers have a Saptagram connection from Kaijuri and Kanaipur Unions.

Before the IRWP or other earthworks schemes, the women were crop trading, doing small homestead earthwork, and rice husking. The men were largely engaged in brick fields. Prior to Saptagram, labour gangs were organised by *sardars* from *paras* as described above.

The Chairman of Kaijuri UP is Niru Mia. His village home is in Kaijuri Gram where he has 200 *bighas*. He 'inherited' his position from his father. He is a retired Service holder (suspended as a Food Controller). He lives in Faridpur town, from where he conducts the union business. He appointed all the Project Committees. However, it was understood when the special women's schemes were designated that the women's committees would be formed on advice from Saptagram. Saptagram submitted a list for each Project Committee, but half were ignored so that four Saptagram group members are on each committee of members but without any executive position. The chairperson of one of those Project Committees is Union Member, Mamataz Begum, and the Secretary Rashid Mollah is regarded locally as the right-hand man of the U.P. Chairman and a landlord. The secretary of the other women's Project Committee is also a landlord and 'close to the Chairman'. The field co-ordinator of Saptagram in Faridpur is Sultana Begum.

The Story

The women's scheme started on 26/2/83 and the men's on 28/2/83. At first the rate offered to the women was Tk. 5 a day. This was refused and subsequent bargaining raised it to Tk. 10. On 2/3/83, the second day of the men's schemes, the men in all four schemes went to the Project Committees to get the work at the full rate. They were twice refused, the second time on site. There was abuse and some jostling. Some Saptagram workers were present. At that moment, a Swedish team of visitors with Hassan, the Executive Engineer (XEN), arrived in Bonogram, the Saptagram Project Centre near to the site. Sultana Begum called the Saptagram workers from the site, and the earthworkers also came and made their case to the people gathered in Bonogram Centre. It was tape-recorded. The XEN promised to inform the Circle Officer (CO) to insist that the Project Committees employed them according to the provisions in the circular. On the next day they were employed on the scheme, but after a few days they were told they would be paid at the rate of Tk. 140 per 1000 cft. At the same time some Project Committee members broke the 'tell-tale' pillars in the borrow pits to confuse the measurement.

After the Swedish Mission had left Bangladesh, the Project Committees sacked workers from four women's groups and some men's groups. In total, about 400 men and women were expelled. Around 15/3/83, 150 male workers marched to the DMLA to complain. The CO received news of this march and intercepted them outside the Luxury Hotel in Faridpur and said he would settle the matter and make sure they were reinstated at the proper rate. On 20/3/83 after nothing had been done, Sultana Begum went to the Sub-Divisional Officer (SDO) to describe the situation and complain about it. He called the CO. The CO in the discussion referred to Sultana Begum as a 'labour *sardar*' and said she ought to be complaining to directly the Project Committees. Sultana Begum and the SDO reported to the CO that labourers had been complaining directly to them. The CO replied: "Saptagram are stirring up labourers to make complaints, but actually the labourers are being paid correctly".

Just before these events, on 11/3/83, a Project Committee member beat one of the group-based earthworkers, and the group resolved to attack the PC member. When Sultana Begum received this news, she informed the DC of Faridpur of the assault and the impending response. The DC refused to discuss the situation with her without an appointment, despite the urgency and possibility of injury or death. Sultana Begum returned quickly to the group to try and defuse these particular emotions among the group.

Thus during late March, Sultana Begum met and explained the situation to the XEN, Faridpur and by telephone to the Women's Programme Officer, IRWP,.in Dhaka (Mrs Shamsun Nessa), and to the Director of Saptagram in Dhaka, Mrs Rokhaya Kabir. Sultana Begum then contacted the army vigilance team in Faridpur to explain about the irregularities in the programme. The team maintained it had no role in cash programmes, but it was urged to raise the whole issue at the next meeting with the DMLA. Sultana Begum felt she was being denied access to institutions which should be offering support to the workers in their attempts to claim their rights. She went to a reporter of a local Faridpur newspaper who agreed to write a news item, and contacted the National Press.

During this time the PC leaders continued to abuse the workers, paying low daily rates and cheating on measurement. Every day labourers were reporting to Saptagram workers, who were trying to keep records of workers and work done in different schemes. Finally, the Saptagram workers suggested the traditional mechanism of peaceful protest in Bangladesh—a *gheraoh*.

On 3/4/83, 400 workers (men and women) marched to the CO's office (since these are *thana* schemes and formally his responsibility). The CO was not present, but 150 out of this 400, including some of the Saptagram PC members went on to the DMLA and the DC. A written petition was submitted to the DC and the DMLA said he would not receive a large group but they should send their representatives on the following day. On 4/4/83, Sultana Begum with some of the workers went to the DMLA and explained the continuing situation. From 5-6/4/83, work partially stopped across all the schemes. The DC sent a two man team to do a site investigation: the District Office Engineer and the ADC.

On 7/4/83 the team was expected first at the Bonogram Centre at 4pm. At 3pm the UP Chairman arrived with some supporters. He started issuing threats to Saptagram staff and the labourers gathered there, but also tried to take some labourers aside to persuade them to report in favour of him and his appointees on the Project Committees. Items of food were offered in this cause. In the gathering there was one worker from Saptagram, Islam, who was from Kaijuri village and had been recruited on the Chairman's suggestion. When the Chairman saw him here, he started to abuse him. A woman Saptagram worker intervened to protest at this treatment and was herself threatened by the Chairman who advanced on her with his stick. When the ADC arrived on the road in his jeep, he was met by Sultana Begum who complained about the Chairman's behaviour. They proceeded together to Tambulkhana Haat which is next to the road and one of the schemes. The chairman was now present at the *haat* and had been joined by the CO. The chairman made an attempt to monopolise the ADC's attention but the ADC avoided this and talked directly with the labourers.

The ADC then made an order to the CO that after technical staff had measured the borrowpits, a duplicate copy of the measurement should be handed over to the labour *sardars* or group (wherever appropriate) as a basis for them claiming their wage rights from the PCs. (This order had not been carried out before the Payment Enquiry on 27/4/83). The ADC said nothing about the behaviour of the Chairman, nor did he address himself to the seething discontent among the workers. Sultana Begum, afraid of retaliation and counter retaliation between the workers and the Chairman's party, went then to the police station to request a police presence in the locality, and a unit was duly sent that same evening to prevent any fighting.

On 8/4/83, all the PCs, obviously over-ruling the by now outcast Saptagram members, sacked all the labourers who had demonstrated. Their names would not be entered on the muster rolls, so that 50% of names would either be false or somehow missing. (During a visit to the road on 18/4, the IRWP Field Officer intercepted Mamataz Begum, the Chair of one of the Women Scheme PCs on her way back from receiving the measurement from the *thana* technical staff. At that time—after the completion of a section of work—she was carrying mainly blank muster rolls which contained only one or two names with thumb prints. This means that the names of many who had completed the work had not been entered, and that other fictitious or favoured client names would subsequently be entered.)

Since this time, the schemes are operating at a much slower rate since they are only employing the unorganised workforce, although about 50 out of the 400 sacked have been re-employed. Only about 200 in total are working, without proper daily wages or measurement. Copies of the measurements as ordered by the ADC have been requested but not given. A further petition has been submitted to the DMLA and two enquiries have been called.

The Payment Enquiry was held on 27/4/83, on the site at Tambulkhana Haat, with workers from all schemes present. The enquiry was convened by a member of the Vigilance Committee, Raquibuddin Ahmed, an ex-Municipal Chairman. The women and men from the different schemes put their complaints to the Convenor and office committee members: the XEN (Hassan), and a reporter from the *Daily Ittefaq*, Delwar Hussain. The results of the enquiry are awaited, but the evidence was accepted. The Union Chairman and CO were not present. A law and order enquiry is due to be held on 28/4/83. (This was subsequently held, but with inconclusive results which could not lead to any action.)

Resilience Needed even under Favourable Conditions of Preparation

By looking at this case study in some detail, the real implications are revealed of asking landless workers of both sexes to come forward, organise themselves in conjunction perhaps with supporting agencies and claim their rights not only to a narrow set of benefits arising from the immediate earthwork but also to longer-term benefits which might be stimulated by the programme. The events largely speak for themselves, but it is worth emphasising that this scale of struggle and personal resilience for staff and labourers alike is necessary even under the favourable conditions of preparation with Saptagram's presence in this community for several years, the existence of groups with cohesion and strong mutual identity, an earthwork scheme running through the community and connections through the Director (as well as Field Director) with the local establishment in Faridpur as well as in Dhaka. Most other conditions of preparation will be/are less favourable than this.

Piecemeal Recommendations Lead to Incorporation

This is why the ARM must be cautious and critical when considering piecemeal recommendations for improvements in landless participation. Not only would such recommendations be extraordinarily difficult to achieve, even if they were approved by MLGRDC and circulars changed, but there is no guarantee that their implementation would lead to the desired goal. Anybody with an insight into the rural power structure in Bangladesh cannot be convinced that new 'target-group' criteria can be applied: by whom? with whose knowledge? Neither is it credible that local ward members of UPs can organise a group of landless to elect representatives to Project Committees who will not be their own indentured labourers; that the Project Committee will 'form small groups of labourers who intend to work in the scheme' from which representatives can be selected onto Project Committees; that the Project Committees will train labourers on RWP rules and operational procedures. The analysis here does not lead to those conclusions. The proposal to experiment with Project Committees formed with 75% representation of the target-group and the remaining 25% nominated by the UP is likely to encounter similar problems when we remember that Saptagram had nominated the members in the women's schemes. It is clear that the crucial actors in the committee are its Chairman and Secretary. Where no NGO exists in the scheme's locality, and the even more unlikely situation of there being an effective labour organisation without an NGO, selection of landless representatives through public meetings cannot be considered a serious proposition. The labour *sardar* system alone will undermine that, let alone the other rural networks which will operate in a public meeting. In the 'normal' Project Committees outside the limited experiment, the obligatory co-option of one member of the labour force means either a labour *sardar*, or a frightened, isolated, vulnerable individual who will hardly need to be paid off....

Strategies for Participation

What are the implications of this analysis for room to manoeuvre in relation both to existing sets of rights offered in principle to workers through the IRWP (albeit through the established delivery/recipient system of programmes which did not emphasise 'target-group' objectives) and to the broader longer-term intentions of the programme to generate incomes and asset accumulation among the landless rural poor of both sexes?

Can Local Democracy be Learnt?

If there is an absolute parameter that the programme has to stay, for planning and implementation purposes, within the existing local government institutions supported by the local level bureaucracy, then little can be proposed beyond the straw-clutching efforts in the various reports available to the mission. There is an argument which was deployed 20 years ago with force and passion by Akter Hameed Khan (founder-director of BARD at Comilla), that the practice of local government and local democracy is a learning process. Provided that elections took place on a regular basis, the electorate would eventually learn the error of electing landlords, entrepreneurs, contractors, touts and so on. My worry about this proposition is that judging by the failure to do otherwise in such established democracies as the US and the UK where networks of power, accumulation and employment are less closely and personally defined than in rural Bangladesh, such learning may never express itself for the structural reasons discussed above. In my experience in rural Bangladesh, learning is not the problem: action without jeopardising the survival of your entire family is.

Alternative Receiving Mechanisms and Adjusted Time Horizons

Therefore, proposals for action have to be responsible and incremental, the rhetoric and impatience of outsiders tempered by the recognition of who is actually bearing the risk. At the same time, proposals which amount to incorporation and co-option will function to postpone the objective since they result either in total identity for those co-opted, indirect forms of labour management and further divisions (vertical and horizontal) among the landless similar to the current labour *sardar* structure. So there is a dilemma—advocating caution on the one hand but consistency with the objectives on the other. There can be no magic formula at this point to unravel the knot. However, there are ways forward as long as ambition and realism are allowed to remain in touch. In the specific case of this Programme that means insisting on alternative 'receiving mechanisms' to local bodies and adjusting the time horizons of the projected investment and expansion to allow such receiving mechanisms to develop and take shape. These are two significant parameters, and changing them is a precondition for pursuing the programme's 'target-group' objectives. The two parameters must be taken together. A formula whereby action-research testing of ideas and proposals in particular areas for subsequent replication in areas where the programme has already expanded will again mean that alternative institutions more conducive to reaching the objectives will be in the position of catching up and competing with local bodies, local bureaucracy and rural vested interests which have already developed an established presence in the programme. This problem is relevant not just at the implementation stage, but in the process of scheme identification, appraisal, land selection itself (at the planning stage). If the programme is serious about its objectives, it cannot logically run ahead of its ability and capacity to achieve them. But in the case of programmes with target-group objectives (as opposed to other objectives) that means waiting and planning in a different way. The pace is not set by the 'programme', but by the capacities of the people who have been assigned a priority in it. It is the task of the staff of the Programme both to assist in the development of that capacity and to identify the points at which it is realistic to match it with appropriate levels of investment.

We must Act—a landless group member

We can also recognise that there is a willingness and indeed eagerness amongst landless men and women to develop this capacity. In a meeting with some group members organised through Proshika who were working on the Kumar River excavation, one man said: "You can only help us in a very limited way. We have to do it ourselves. We have to expect victimisation, but we have to move. The chairman will intervene to break up our unity, and those isolated as militant will be subject to continual harassment, court cases etc.. We cannot achieve much in a day, but we have to go step by step." Outside the context of this programme, this is being said all over Bangladesh. The determination exists, but what chance do they have if the capital is pumped into the countryside before they are organised to compete for it?

Two Approaches by NGOs

So the argument has to be that there is no short cut, no quick route to the preparation and mobilisation required. The problem with rural works programmes is that their planning and implementation requires a high intensity of interaction between local interests (not all necessarily distributed among unequal classes—e.g. fishermen and farmers), workers who might develop longer interests in relation to the physical structures, and technical inputs. It is difficult to imagine landless groups (of either or mixed sexes) participating instantly in such a process. It has been a working principle of many NGOs engaged in strategies of developing material autonomy for landless groups through economic activities to start with schemes that are relatively self-contained (livestock rearing, handicrafts, rice trading) before moving into forms of activity which require more interaction with other classes with the possibilities of competition, conflict and confrontation. More solidarity is required at this stage. This is incremental.

However this is not the only model, since there are other NGOs which are not concerned with the short-term acquisition of material autonomy through capturing levels of economic activity, but with mobilising workers directly around the contracts and exchanges which are already part of their income sources—employment and tenancy. This involves the mobilisation of groups into immediate struggle with other classes and interests.

Compare Approaches

Most NGOs work in effect with a combination of both models but in sequence—direct confrontation and struggle (over for example, *khas* land) comes later on the agenda. But the two strategies exist and have different implications for the speed with which this programme can develop. It would be interesting to compare the longer-term preparation with the instant struggle strategy. Proshika might be encouraged as part of its current involvement to examine its comparative experience between Khoajpur Union (groups organised at the inception of the scheme) and Pearpur Union (groups organised over three years prior to the scheme) in Madaripur Thana. On similar lines, a comparison between Nijera Kori and Saptagram and/or Proshika using the Delta Development Project experience in Khulna could be made.

Avoid Incorporation

The common theme with these strategies is the attempt to avoid the exploitation of the rural poor through their incorporation into the prevailing institutional network. This is not easy and the Proshika recommendations in the Assessment Report seem to be a lapse, arising perhaps out of sheer frustration. All groups engaged in economic activities usually have to interact with some institution (the landless irrigation groups have to deal with BADC, the BKB, as well as landowners and dealers in fuel, equipment and spare parts). This is not the same as incorporation, where interests become captured by the power of others because the worker has little of significance to exchange beyond, in this case, individual labour power. So if the Project Committees are to be 'invaded' then either the local government involvement has to be removed *de jure* (difficult, if not impossible), or it has to be circumvented and neutralised *de facto* (difficult, but not impossible since it would rely on people's action rather than a shift in entrenched government policy).

Parallel Rural Works Lobby—rural workers federations

Since mobilisation around specific issues is more likely than around abstract principles, then a strategy of developing a parallel rural works lobby might be seen as one objective of group and inter-group organisation. (One objective among many others, since landless organisations have many other things to think about in addition to developing their share of the implementation and subsequent benefits of rural works programmes).

Shadow Project Committee—a counter-information system

At the initial stage, organised groups of workers would seek to perform a Shadow Project Committee function in the way that Saptagram has partially attempted by monitoring work on the site. This would involve maintaining alternative muster rolls of all workers, whether in organised groups or not, in order to defend unorganised workers against the labour *sardar*-patronage network and demonstrate the value of worker solidarity. It would require developing alternative measurement skills among the workers themselves, and recording the volume of earth cut from borrow pits on a daily basis. Alternative payment *khatas* would be maintained, recording payments made against dates. Workers would need to read their own name, dates of entry and numerical *taka* entry. These elements would represent an initial counter-information system to be used in disputes such as those encountered by Saptagram and Proshika in Madaripur, and by the groups which are respectively associated with them.

Continuous not just Seasonal Unity

However, groups and federations of groups in a locality would have to maintain their cohesion and solidarity permanently, year on, not just for the earthwork season. Continuous, not just seasonal, unity is necessary to sustain their organisation for earthwork seasons, to engage in other struggles over agricultural wage rates, land and tenancy rights and to constitute a viable institution within which assets can be produced, reproduced and rents received for their use. Such unity is also necessary in order to establish over time, a legitimacy or at least credibility both for the counter-information system, and for their rights to use it and therefore represent workers on site, in enquiries, in court. Such federations would early on require subscriptions from members to generate a fund to pay for paper, training, legal fees and so on. Such a fund should not be financed externally as that would weaken (as it always does) internal discipline and structures of accountability and democratic decision-making.

Direct contracts to 'Target-Group' (earthworks)

Through this continuous and routine action to claim the rights as they currently and formally exist within the programme, this rural workers' federation would seek as a further objective to become itself a viable organisation which could undertake direct contracts for earthworks schemes. Allocating work through contracts rather than Project Committees is open to unions and *thanas*. From within its membership the groups most likely to be involved in the work would constitute in effect a collective PC for the duration of the scheme. But crucially when the scheme had finished and such PCs disbanded, the groups would still be part of their rural workers' federation, supporting other workers, working on other schemes themselves, maintaining the counter-information system. There should be many examples where direct contracting to the 'target-group' is possible on technical grounds, under conditions of appropriate landless organisation. One specific proposal to test these possibilities would be the first HBB part of the Badarpur-Saltha road. This may be outside IRWP's category of activity (unless it can be 'adopted' for the purpose) under the *Zilla Parishad*. It is reported that it has to be widened but no contractor will now undertake it because of expectations of labour trouble as a result of troubles further down the road. Since Saptagram groups are well-established, can they take the contract direct to do the work? The award of such a contract to these groups would be a concrete sign of government support for the principles of workers in rural works programmes claiming their rights, and developing further access to income-generating activities.

Direct contracts (*pucca* constructions) Supported by Artisan Training

In addition to self-contracting on earthworks, these rural workers' federations can undertake, in the longer term, direct *pucca* works contracts for bridges, culverts, even metal roads and building construction. Although some of these categories would remain outside IRWP, this would be an example of IRWP support for rural works groups, initially associated with it, to engage in wider productive work. Artisan training support would be required for masons, carpenters, road laying, and perhaps even design for small scale, simple schemes. Such training would be organised by the rural workers' federations eventually in a permanent location, and it might be oriented specifically to the sons and daughters of families in the federation. Such training would require considerable funding since some kind of apprenticeship/periodic release system would have to be developed. This is an example of contributing to the creation of social infrastructure of skills and capacity as the basis for the longer term ability of the 'target-group' to claim and extend its rights in all programmes of infrastructure development. The access to *pucca* works depends on skills not just mobilisation—each are necessary conditions for the other.

Rural Workers' Federation—entry into local planning

But the possibilities do not stop here. The issue of local planning is not separate from this. Planning the development of any resources is always more of a political than technical process. One of the major problems of participation is: who sets the agenda? Who is in on the ground floor? At present with all forms of planning in Bangladesh (and almost everywhere else), the majority and the poorest are provided with some work at the end of a long process of decision-making. The problem with a 'target-group' oriented programme of rural infrastructure development is that the rural infrastructure itself is designed in response to the criteria of those with the political strength to enter the process early on. The reports on the planning process for IRWP in Madaripur, the Northern Faridpur *thanas* and Rangpur certainly confirm this.

Rural workers' federations would have to develop their strength to enter this process so that they do not only capture more of the work, but determine the value of that work to the community which they increasingly represent. This would be a process not only of reacting to schemes proposed by others, but of initiating ideas themselves within the community. There are many earth roads in Khoajpur Union, Madaripur Thana which would not have been constructed if local people had a chance to voice an opinion, since the paths were perfectly adequate, whereas the roads have removed valuable homestead land especially from the poorer *baris*. The embankment around the Dapada Bheel in Boalmari Thana,

Faridpur might have taken a different route—but the poorer *baris* were unable to defend parts of their homestead and paddy land, or have been left on the 'wrong' side of the embankment. These examples have become cases of reaction, but rural workers' federations can themselves identify infrastructural needs within the community which will be more conducive to wider interests, especially where at the outset schemes are proposed which have the potential to deliver wider secondary and tertiary opportunities for the 'target-group'. In this way, the criterion of 'value' is changed. Planning will always be politics. The IRWP should encourage this process, legitimise it, and provide meaningful support by authorising schemes which have arisen through this form of social planning.

Scheme-based Intensive Area Development

Avoid Isolated Interventions

The line of thinking outlined above can be given a more concrete form by considering some examples of how the process might have occurred (or might still occur) from the field-visits made in Faridpur. How, therefore, such a process can lead to the generation of other sources of income through the development of control over different kinds of rural productive assets. However, it is necessary to combine this discussion with some preliminary observations about the way in which interventions by different government agencies have to be co-ordinated in a locality. If an earthworks scheme (with or without a *pucca* element) is to lead directly to other income and asset-owning opportunities for the 'target-group' then the schemes should not be implemented in isolation from the other requirements to create the opportunities. No engineer would build a vehicular bridge without vehicle-bearing roads leading to it. The same principle has to be applied to schemes with intended 'target-group' benefits. An irrigation canal should not be constructed without beforehand preparing for a 'target-group' participation via landless ownership of irrigation assets through the negotiation of command areas, rates of payment for water, lengths of agreements with farmers and an assured mechanism for receiving credit and acquisition of equipment.

From this illustration, there should be, therefore, an integrated process of local mobilisation, planning and implementation around a precise geographical area of infrastructural and related opportunities. This is not area development at the *thana* or union level, but at the scheme level. Thus it is better to focus IRWP resources and complete all the elements that are required to realise the full 'target-group' potential of an infrastructural intervention, instead of spreading the schemes thinly on the ground where the multiplier effects of co-ordinated investment do not occur. Rural workers' federations would, in the early stages at least, operate at this location-specific level.

Local Analysis and Problem-Solving

Secondly, this concept of small-scale, scheme-derived area development brings back into the approach the question of capacity referred to earlier. It is in this context that local analysis, problem-solving, negotiation and delivery skills are required of programme field staff in locations where NGOs are not performing these functions. Social engineers who are prepared to be the servants not masters of the area's 'target-group': working in close association with rural workers' federations; assisting with technical advice on the feasibility of different options; arranging the technical skills required, access to funds, suppliers of materials and legal services. Again the NGOs might be able to experiment with this approach, with IRWP supporting the appointments of such 'social engineers' and using them later to train other regular IRWP field staff who would be engaged in similar work in all the scheme-specific areas of the Programme. Such training should occur through attachment to a particular area where this process has begun. Regular cross-fertilisation of ideas and experiences between such field staff (and NGOs where involved) will always be necessary and should be encouraged by IRWP.

Rental Income from Structures

Thirdly, the circumstances have to be recognised where the income source to the 'target-group' (beyond the wages from earthworking) derives from the structure itself and not from secondary or tertiary potential. Although it is not feasible in practice, the most obvious example of this would be the collection of road tolls. But flood control embankments, drainage and irrigation canals, for example, are all providing an immediate and direct service to a specific collection of consumers—permanently raising the value of their land in return for, at present, a very temporary source of income to the earthworker. The redressing of this imbalance has to be an objective of the Programme and the rural workers' federations. There has to be some kind of exchange between the consumers and providers of the service. This extremely sensitive issue is nevertheless quite central to the Programme's 'target-group' orientation. Such exchanges have to be either in the form of taxation or rent, and they ought to be a feature of many of the schemes on the IRWP agenda. Arranging these exchanges through formal taxation is not likely to perform the function of recycling and redistributing value to rural workers, since they will have no control over the collection and dispersal of revenue. There is only one possibility in this connection where the Government leases the right to collect tax on enhanced land values in the same way as it leases out the rights to collect tax from fishermen on inland water. At least in this way the rural workers' federation (assuming it took over this function too, in time) would have its rights confirmed by the government over the structure and a recognition that it provides a continuous service for which payment is legitimately due.

Maintenance is an Addition to Service

However, such a device may not be realistically within the room for manoeuvre, and the essential requirement is that the exchange be immediate and local. It is important to accept that maintenance is a service over and above the service which the structure in position provides. Whether it requires maintenance or not, rents should still accrue to it. In practice most earthworks structures do require maintenance which reinforces the perceived need for the structure and should facilitate the charging of rent for it. Thus rents should be made up of two components, although the justification for one is more obvious than the other and will be the key to the experience of rural workers' federations in attempting to charge rents. These options might be tested either in the schemes discussed below, or in other areas where preparation and group mobilisation have occurred so that a rural workers' federation in effect exists.

Mutual Participation

The case studies used in the research include river and canal excavation schemes, embankment projects, road constructions and a small tank project (full descriptions not given here) and show how an integrated process of mobilisation, location-specific multi-functional planning with the support of social science and technical inputs from Programme staff, development of productive exchanges between different classes of rural people in which property rights to significant rural means of production were disaggregated between the 'target-group' and others, might have occurred. They emphasise how this recycling and redistribution of the value-added from investment in physical structures has to be achieved on a slower time scale than that envisaged in the projected expansion of the Programme. These aims must be achieved through a combination of effective organised pressure from groups and federations of rural workers, and supportive policy responses by the government and the Programme. The cases researched did not of course include all the possible sets of opportunities, but the same kind of thinking (or approach) can be applied to *haat* ownership management; rural electrification earthwork and pole contracts; brick production (although the established competition is strong); bus co-operatives with preferential quotas on new roads; freight co-operatives on similar lines; rural workshops to service irrigation and equipment (Proshika are proposing an action research project arising out of its landless irrigation programme); and perhaps an entry into the boat sector.

However, the construction of physical structures can always occur in a faster time frame than the development of social institutions. Unlike other programmes, the IRWP should be governed more by the pace of the latter than the former. That is the challenge, and some patience is required.

Conclusion

Relation Between Experiments and Replication

In trying to develop these approaches as a basis for pursuing the 'target-group' objectives of the programme, there remains the question of the relationship between experimental activity and the replication of that experience and method. The programme has already relied critically on two NGOs—Proshika and Saptagram—in the IIS in Faridpur; it is possible that other NGOs such as Nijera Kori will be invited to participate in Rangpur and perhaps elsewhere. Does this mean that the IRWP is dependent upon NGOs to pursue its 'target-group' objectives? Is the involvement enabling the programme to avoid confronting the major institutional constraints noted above? Are NGOs, like Proshika, testing implementation strategies with a view to changing policy and reorganising the government rural works system? Or are they actually involved, like Saptagram, in confrontation and struggle which might influence policy but via a different process? Are the implications of the general approach outlined above that the programme is dependent upon NGOs for the mobilisation and continued support of landless organisations? If that is the case, then the pace of the programme would have to be determined by them, but is such a proposition realistic within the room for manoeuvre or will the MLGRDC resist this threat to its monopoly over rural works interventions? These questions have to be resolved and the answers are likely to determine whether the programme ceases, continues by abandoning its objectives, or whether concessions are allowed by which approaches outlined here can be pursued.

Consistency with Present 'Target-Group' Work

It should also be clear that nothing is being suggested here which is not already happening in embryonic form. It would be nonsense to propose something outside the working practices and philosophy of the organisations in Bangladesh which are directly addressing these 'target-group' objectives. The strategy outlined above for an organised counter lobby leading to permanent and workers' federations which also develop secondary and tertiary opportunities is only trying to point out the longer term potential of landless group formation. This strategy is based upon a firm rejection of proposals which will amount to the incorporation of workers into the prevailing institutions. Their presence (for example in Project Committees) will provide some legitimacy for those institutions but their power will be weak without the support of an organised counter-lobby to defend them throughout the year.

Should BRDB Capacity be Tested?

Can the IRWP really expect to find the capacity for this strategy of avoiding incorporation from within the government system? Are the BRDB landless co-operatives now supported by the World Bank (who want NGOs to train BRDB field staff for the aspect of the RD-2 programme) a vehicle for the IRWP objectives? Should there be experimental work testing their capacity to pursue a genuine mobilisation of landless interests around the rural works sector, while avoiding incorporation? Although it was a mistake to commit the programme beyond the IIS to include eventually 100 *thanas*, even the currently planned expansion (somewhere between 19-30) would require an understanding of the capacity of larger agencies than the NGOs to develop the 'target-group' objectives. In the landless irrigation programme, the BRDB has undertaken to support 80 group-based schemes from the next irrigation season (1983/84). It is important to test whether the BRDB system can match the NGOs, if opportunities for the landless are to become widespread in the prevailing but therefore challenging political reality of Bangladesh.

Criteria for *Thana* Selection

The programme can only proceed in a manner consistent with its objectives if *thanas* are included on the criteria of NGOs and/or BRDB willingness to continue or commence the experimental combination of activities in the sequence outlined above. At this stage these conditions have not even been fulfilled in the 'experimental' *thanas* of the initial implementation stage in Faridpur. Therefore, more consolidation of existing commitments is definitely required before new ones are entered into. Saptagram may now be unwilling to continue a further relationship with IRWP—such feelings have certainly been expressed by its workers. Proshika is in a different position, and a concrete proposal for extending their contract with IRWP for a further year can be made. This proposal may then constitute a model for similar relationships with both other NGOs and a BRDB experimental involvement to test the wider capacity of government. But no further comments beyond a small number of *thanas* selected for these reasons should be made.

Conditions for NGO Involvement: the Proshika Example

A proposal concerning Proshika is based on a continuation of the President's Canal and Kumar River Schemes in Khoajpur and Pearpur Unions of Madaripur Thana. However, if Proshika is to consider attaching its normal work of mobilising landless groups around economic activities and the struggles with other classes over land, village decision-making and so on, then it needs to be convinced that other schemes which will be appropriate to the interests of the landless will be encouraged in the area. In the year of this research false expectations were raised. It needs also to be certain that every effort will be made by the programme to engage in joint planning with the landless groups themselves, and that any scheme is at the outset conceived in terms of the wider opportunities it can provide for the 'target-group'. No NGO with a reputation for working among the landless can be expected to settle for less than this.

Chapter 11

Targets Strike Back—Rural Works Claimants in Bangladesh

Introduction: Target—Groups and the IRWP

Target-group vocabulary is strongly represented in development programme language in Bangladesh. The political and moral imperatives of reaching the 'poorest of the poor' have focused attention on two designations: 'landless' and 'women'. Clearly these designations are too broad to be manageable and further restrictive labelling has occurred. In *Bangla* this has been reflected in a shift from *bhumihin* (lit. landless) to *bittaheen* (lit. assetless), partly because the landless are not always what they seem (they may be absentees, sons of landlords/landholders, service-holders, businessmen with other assets). Similarly, the policy preoccupation with women (mainly rural) from the mid-seventies seemed to be adequate as a poverty-focused strategy until the evidence emerged of these programmes being dominated by richer rural women (Feldman, Akter and Banu 1981). Women (*mahila*) as a target-group are now qualified by the term *bittaheen mahila*, which is creeping into official language.

This chapter is concerned with the deployment of these designations in the 'target-group' focused Intensive Rural Works Programme (IRWP) in Bangladesh. It should be stated at the outset that this programme has attempted to address more directly some problems of rural poverty than many other examples of rural development interventions. If particular difficulties are known to exist, it is partly due to the intensive monitoring and special investigations which have occurred within the programme and partly because the criteria of judgement have been stricter against the 'target-group' objectives of the programme. Moreover, it should also be stressed that this paper does not represent an evaluation of the programme. The purpose here is to examine the programme-client relationship from a labelling perspective.

Many of the problems encountered by the clients of the programme can be attributed to the behavioural assumptions which form the basis of the programme and which are embodied in its target-group designations. Perhaps most significantly, rural labour (male and to some extent female) is treated as a freed commodity in search of non-agricultural work at prevailing market rates at certain times of the year. Secondly, the designations 'landless' and 'women' are assumed to be sufficiently accurate indicators of the organisational solidarity required to insist on market rates for payment even though that payment is administered through the local government institutions which have traditionally prevented the participation of rural labour in order to exploit it. (Wood 1976; Jahangir 1979; Hartmann and Boyce 1982; Rashiduzzaman 1968; 1982; Sobhan 1968; Alam 1976; Solaiman and Alam 1977; Rahman 1981). Thirdly, as a result, the notion of client participation has been confined to a process of claiming with the expectation that successful claiming is possible. The difficulties faced by the rural workers in claiming has become the single most important constraint to the programme's advancement, and may indeed lead to the virtual abandonment of the 'target-group' objectives in this programme form. In this case it is becoming apparent to the programme initiators that the target-group labels are a misleading guide to the behavioural opportunities for rural workers, with the result that the physical infrastructural features administered through the bureaucratic, political and commercial rural hierarchy have assumed organisational prominence in the programme.

The Problem of Landlessness in Bangladesh

The context of this target-group terminology has to be appreciated in order to understand both the commonsense aspects of terms like landless and the importance of capturing the target-group theme via apparently meaningful policy interventions such as 'rural employment generation' and 'income-generating projects'. Bangladesh is one of the most densely populated rural regions of the world. The 1981 census claimed a total population of approx. 100 million, although this is probably exaggerated. However, allowing for this, the average population per square mile of cultivable land is nearly 1700 persons. The population growth rate, for whatever reasons of private gain over social cost, is also one of the highest at around 2.5%. An estimated 91% of the population live in rural areas, and approximately 80% are considered to be directly dependent on agriculture for their income. The ownership of land has become increasingly concentrated with 11% of rural households owning more than 52% of all land, while over 30% own no land at all. This 30% together with those owning less than half an acre constitute about 48% of the rural population.

This situation is accompanied by high rural under-employment and a decline in real agricultural wages. With the annual population growth rate, 45% of the population is under 15. The rural labour force is increasing at about 3.2% a year, approximately one million men and women. This increase far exceeds the number of additional employment opportunities available in agriculture and other sectors. The World Bank estimated that during the first half of the 1980s, approximately 2.3 million new entrants to the rural labour force will not be able to find gainful employment. The rapid increase in unemployment has a direct impact upon the prospects for expanding agricultural production. Limited employment means that a substantial segment of the rural population must restrict their food purchases. The per capita income for Bangladesh was reported in 1981 as equivalent to US $140 (World Development Report 1983) but if the concentration of incomes in urban areas is taken into account then the average *rural* income would be considerably lower. It has been estimated that 62% of rural people receive only enough income to satisfy 90% or less of the necessary daily intake of calories. (ILO 1977). Yet by the end of the 1970s the production of major agricultural crops was only 7.3% higher than in 1970 (Clay 1978). This suggests a vicious circle whereby the lack of effective demand acts as a constraint to expansion of production which in turn further restricts the development of employment opportunities in agriculture.

The high rate of population growth combined with this failure to develop productive agricultural capital (Arthur and McNicholl 1978; Wood 1981) have resulted in: larger families living off the same size or reduced sized of holding; an intensified fragmentation of holdings; dispossession of infra-subsistence peasants; decline in the availability of land for sharecropping; a consequent increase in the numbers of landless; the enhanced value of children as income-earners and providers of old-age security; a steady depressive effect on agriculture and other rural wages; and finally, under these conditions of a labour surplus, a transformation of the patron-client dependency ties which offered minimum security to labour.

There can be no illusions, therefore, about the polarisation and dislocation which is occurring. But whose definition of the situation and the responses appropriate to it will prevail? The struggle to interpret this process and offer the most authoritative praxis has been complex and intense. The survival of the military-bureaucratic oligarchy and the class interests represented by it clearly depends upon a target-group presentation of the problem to provide the apparent basis for meaningful state intervention. Such attempts are crucially supported by foreign aid. In this way, a donative discourse might head off the revolutionary alternative in which the depth of rural contradictions will lead inevitably to the final showdown, with the dispossessed rising against their oppressors. However, this too is a discourse which fits uncomfortably upon the social relations of poverty in rural Bangladesh at this time.

Emergence of Target-Group Policy

A discussion of this search for an authoritative donative discourse in Bangladesh will establish the significance of the current landless focus. Naturally there is a history to this process, inseparable from the structural conditions which led to the emergence of Bangladesh. The dependency of East Pakistan and the related underdevelopment of national capitalist classes were not ideal conditions for the effective appearance of a liberal bourgeois ideology to define and legitimate the relationship between state and society. This is the general condition of peripherality and Bangladesh represents an extreme example (Wood 1984b). In this context attempts to construct a cohering ideology inevitably became variants of rural populism. The Comilla cooperative model was the prime example of this process involving as it did a clear set of assumptions about the predominance of small farmers and their behaviour, derived from the Ford Foundation-funded pseudo-scientific inputs from Michigan State University. From 1959 for almost two decades these assumptions monopolised the agenda of donative discourse in the country with a corresponding set of farmers' cooperatives (*Krishi Samabaya Samitis*—KSS) organised nationwide through the Integrated Rural Development Programme (IRDP), which has now been re-titled the Bangladesh Rural Development Board (BRDB). However, the increasing evidence of landlessness and the awareness that the rest of Bangladesh was not like the Comilla region with its relatively low incidence of landlessness, sharecropping and large landlords (GOB/US AID 1977; BARC 1978b; Wood 1981) eventually reduced the credibility of the Comilla labels.

The demise of the Comilla formula has been slow and lingering, although far too totemic for a rapid desertion (Wood 1980b). However, the approach ceased to be politically useful in developing and maintaining a loyal or at least quiescent mass base especially in the post-famine crisis of 1975. Thus, immediately before the August 1975 coup and since then the search for an institutional formula of rural mobilisation and incorporation has continued. The evidence of landlessness together with the crisis of mobilising political support produced therefore the possibility of revising the 'farmers-only' cooperative strategy. After November 1975 until 1981 when Zia was assassinated, this formula had various incarnations: *Swanivar* villages, the *Ulashi* programme and *Gram Sarkar* (see Glossary). These all contain themes of self-reliance, 'grassroots' institutions generating resources supposedly through mobilising the 'voluntary' labour of the village ablebodied (men and women) in re-seeding flood damaged plots, constructing embankments and roads, and digging canals. In practice, landless labourers were recruited through the traditional patronage networks for this work, with the difference that their labour was unpaid, while most of the longer term capital benefits went elsewhere. Significantly, however, most of these institutional experiments did involve the notion of constituent target-groups such as landless, women and youth (usually male)—who were supposed to have a quota allocation of places on the committees, although their representation could never amount to a majority.

In response both to the evidence of rising landlessness and in this atmosphere of institutional experimentation, the IRDP-BRDB started to organise landless groups for men and women from 1977. The greatest initiative was undertaken in the areas under the World Bank financed *Thana* (now *Upazila*—lit. a sub-district) Rural Development I programme (RD-I). A World Bank report was sceptical of these efforts, commenting for example:

"Furthermore, because of the unfamiliarity with the techniques of promotion and development of non-crop programmes for the landless, there was a general lack of commitment from within the IRDP.." (World Bank 1983a para. 5.12). (There is an implicit comparison here with the activities of NGOs in rural development which will be discussed below.) Despite this criticism, the IRDP was instructed in 1982 to mobilise the landless all over the country following the joint GOB/IDA review of IRDP in 1981 and the subsequent creation of a landless cell within IRDP. This is the institutional basis for the Rural Poor Programme which now exists within BRDB, and which is committed to reproducing the Comilla-originated farmer cooperative formula for the landless with the formation of *Bittaheen Samabaya Samitis* (BSS—lit. Assetless Cooperative Organisations). This is a component of the BRDB's programme which is receiving intensive commitment in the areas covered by the World Bank-financed sequel to RD-I, appropriately called RD-II. The RD-II Appraisal Report (World Bank 1983b) observed, however, that:

> "Major problems faced in the implementation of those programs of BRDB have been delays in the registration of new societies, delays in the flow of credit, training and extension support and a lack of management focus in BRDB to implement a more extensive program." (p.20).

Outside the RD-II areas, the BRDB—Rural Poor Programme will therefore be less active in the short-term. (An extended discussion of Government approaches to the rural poor in Bangladesh appears in Wood 1984c, and an edited version appears as chapter 16, part V of this volume).

Significance of NGOs

Alongside the development of this ideological and institutional preoccupation with the landless within government, there has been a proliferation of NGOs in Bangladesh since the early 1970s with programmes focused on landless men and women. Some of these NGOs such as the Bangladesh Rural Advancement Committee (BRAC) and *Proshika Manobik Unnayan Kendra* have expanded significantly throughout Bangladesh with programmes based upon the mobilisation of landless groups around income generating activities and the conscientisation strategies associated with the work of Paulo Freire (Freire 1970). Many other NGOs have followed this lead on a small scale, for example, *Saptagram Nari Swanivar Parishad* which has a particular (though not exclusive) focus on women and has so far been confined to one district, Faridpur. Proshika and Saptagram (lit. seven villages) have both been connected to the Intensive Rural Works Programme, so are mentioned here. A different category of NGO has been *Nijera Kori* (lit. 'We do it our way') which has been less concerned with income-generating projects but more exclusively with generating a consciousness among rural workers to develop their capacity as claimants over a range of issues such as wages, access to rural works employment, occupation of *khas* (government) land, and so on. To place these organisations (among the largest Bangladeshi NGOs) in perspective it is worth noting one estimate of 169 NGOs working among the poor in Bangladesh (ADAB 1984). This prominence of NGO activities has clearly raised the value of the 'target-group' labels and has intensified the politics of capturing the ideological themes which accompany them. The recognition of a problem by the regime and its international allies is not so much evidence of a shift in the relationships of power between rich and poor, as a sense of fear that such a shift might occur unless some response is forthcoming

Two Concepts of Target-Group: Delivery Mechanism v. 'Empowering the Poor'

However, there are contradictions here especially in the case of Bangladesh where the capacity for bureaucratic containment is limited, prompting an official concern for mobilisation rather than just participation or claiming. The preoccupation with 'self-employment' and 'income-generating' activity, which is certainly the direction of policy in the Third Five Year Plan and a significant component of the proposed sequel to the Intensive Rural Works Programme (to be renamed Rural Employment Sector Programme), seems inevitably to involve the development of possession over assets by the rural poor. This becomes a challenge therefore to the structure of social relations of production and exchange through which poverty is presently reproduced. Furthermore, such mobilisation has to stimulate collective forms of activity for several reasons: to organise politically in defence of assets or to claim service; to obtain economies of scale for some small-scale enterprises; to constitute collateral for credit either as a group or for individuals within the group if economic activity is individual (the *Grameen* Bank strategy); to function as a bureaucratically convenient end point in a credit extension system which cannot reach individuals. These are the assumptions on which 'target-group' strategies have been adopted in Bangladesh by various government and non-government agencies. But the question remains whether they are institutional devices for incorporation and containment involving both de-linking and organisational dependency; or whether such a form of mobilisation will stimulate greater autonomy for such classes from the state and its local representatives and transform the meaning of participation as a result?

In Bangladesh at present, different understandings of the term 'group' exist among government and NGO programmes. The main distinction between these different conceptions is the recognition not just of collective income-generating activity but of the *value* of collective solidarity across a broad range of issues. The first category is a 'delivery' concept, whereas the second (associated with the more radical NGOs such as Proshika and BRAC) is committed to 'empowering the poor' through the autonomies of both knowledge and material assets. (See Wood 1986b for further discussion of this distinction). The contradiction for state rural poor policy is that the first category of group mobilisation is required to offset the more dangerous, de-stabilising consequences of deepening mass rural poverty, but there can be no guarantee that such mobilisation will not stray into the second category where NGOs have already shown the way. In this way, there are clearly problems for the regime in trying to establish a safe version of the landless theme as part of an authoritative donative discourse. The attitudes expressed by the World Bank and the government towards the involvement of NGOs in their programmes seem to reveal this dilemma. The Intensive Rural Works Programme nevertheless deliberately and therefore distinctively sets out to use the conscientisation dimension of NGO activity on an experimental basis precisely in order to move rural works interventions from the 'delivery' notion of 'target-groups' with successful claimants as a potential corollary, towards a more complete participatory notion in which 'target-groups' would extend their control over the planning and implementation of the rural works programmes and over the physical assets created by such activity. The problem was that the programme made a series of unrealistic assumptions about the capacity of its 'target' population to act as a 'group'. As a partial and bureaucratic intervention in the structure of agrarian and rural social relations, it had to define the capacity of poor rural people in terms of 'claimants' and, it is hoped, 'participants', on the assumption that the common (but not necessarily 'shared') experience of being landless would separate them from their individual 'stories' (the personal and historical packages of relationships through which individuals manage their survival) to induce collective behaviour. But the label 'target-group' has been dangerously misapplied.

The Issue of Target-Group Membership

As suggested at the beginning of this chapter there is the further issue about the membership of such 'groups'—the nature of the 'target'! It is of course convenient for bureaucratically administered programmes on a national scale to establish a set of criteria. As the overall labelling arguments remind us, it is not *whether* we label, but *which* labels are imposed and by *whom*. In Bangladesh, there is a definitional convergence on the principal of reliance upon own physical labour for the major proportion of family income. But it is also the case that this would include small farmers who do not employ the wage labour of others to any significant extent. Certainly this criterion of using own physical labour would make it difficult to distinguish between wage labour (by oneself) and self-employment, and between farm and non-farm activity. The term 'landless' or *bhumihin* does not express the physical labour variable, and does not therefore have uniform implications for the people thus labelled. *Bittaheen* is marginally better, except that it also excludes those marginal farmers with land but who are nevertheless dependent upon their own physical labour all the time and who share most of the other dependency relations of *bhumihin* and *bittaheen*. Are they to be excluded because they are not destitute enough... yet? If real group formation (alliances, solidarity) is the most likely way in which the poor will counteract the classes responsible for their poverty (the second category of 'group' concept, noted above), then as suggested in Wood 1985a, it is the shared, familiar, local and immediate experience of the same dynamic of rural poverty which is the more relevant basis of social action. In this way people join self-defined and therefore more meaningful groups instead of being grouped by a bureaucratically abstracted feature of their total situation derived from someone else's view of the family's problem and assumptions about their natural alliances. Some of the NGOs have understood this and the criteria of group membership have been relaxed, though without constant vigilance this relaxation can create similar problems to the *bhumihin* label. Finally, it is well worth remembering that it is part of the agrarian class logic that landless labourers can be in bitter competition with each other for scarce employment. The density of cross-cutting kin and neighbourhood ties is crucial here but cannot easily be contained within a universal label for a 'target-group'.

The World Bank's Entrepreneurial Model

The particular role of the NGOs in the development of an authoritative donative discourse is highlighted by the increasing, though guarded, interest in their target-group activities by the World Bank, an influential actor in the policy structure in Bangladesh. In a discussion of rural employment issues (World Bank 1983a), the report noted that:

"The NGOs generally have a good understanding of local institutions and environment and many of them have gained the confidence of the target population by reason of their familiarity with, and presence at, the local level." (World Bank 1983a p.98). This accolade is part of a criticism of government performance in Rural Works Programmes and the Food for Works Programme (FWP). As a result, the government is urged to "formally bring in NGOs for the rural employment creation strategy" (*Ibid* p.99). The report continues: "However, not all NGOs will be equally suitable for participation in such a programme; the Government should establish criteria for selecting appropriate NGOs... The selection criteria may include... the ability to prepare acceptable programmes" (*Ibid.* p.99). It is clearly intended that the NGOs are to be incorporated with their landless discourse on the government's (and the World Bank's?) terms. "...The NGOs can contribute to the tasks of organising the poor for productive purposes and making government services available to them." (*Ibid* p.100 para 2.136).

Both this report and the RD-II Appraisal Report, which launched the Rural Poor Programme emphasise "...equipping the target population with necessary skills and technologies for identified vocations, credit and guidance to solve problems as they arise, and the provision of appropriate services such as training.". These conceptions, this vocabulary, reflect a narrow understanding of the social relations of poverty. They assume essentially that what people lack is capital and skill and that a combination of credit and skill training is necessary to turn all *bittaheen* into individual entrepreneurs. Attempts to expand employment opportunities, capture the rents from technological innovation by acquiring appropriate skills, and to provide credit where necessary are clearly important. However, in isolation from an understanding of power, such attempts are partial and insufficient. The politics of training around these issues are illustrative of the problem. Negotiations between the NGOs and BRDB over NGO involvement in training at various levels for the Rural Poor Programme in RD-II have been protracted. The government and the World Bank wish to restrict the NGOs to a skills input, but their own expertise lies elsewhere—working to form and sustain groups around income activities and struggles for land and wages. This is work which occurs explicitly in the context of an understanding about power, inequality, and dynamic agrarian change. To assume that NGOs are committed to employment creation and income generation within the existing pattern of social arrangements appears as an attempt to re-direct their work into a more 'acceptable' form; to translate the original radical implications of their analyses, their labels and their action into an extension or delivery function with groups existing purely as 'receiving mechanisms'. In this way, the ideological theme of 'target-group' rural development is captured, and converted into a less dangerous, more populist alternative. The 'target' remains certainly, but the 'group' is lost. Poor rural people become objects of policy, not—as the NGOs wish— subjects of it; and they have been de-linked. The explanations of poverty which have been successfully imposed in this process are restricted to the *possession* or otherwise of assets rather than the social processes which may have led to the unequal distribution of assets. This weakness is equivalent to the distinction between analyses presented in terms of social stratification (cases) rather than class relations (stories), whereas the NGOs' understanding of rural poverty in Bangladesh is broadly based upon unequal exchanges between labour and capital (including land) *between* rural classes at village level.

IRWP: A Distinctive Programme of Intervention

These issues (the political conceptions of 'target-groups', and the assumptions about the behavioural capacities of groups and their members) are now considered more specifically in the context of the Intensive Rural Works Programme (IRWP) in Bangladesh. The IRWP is a special programme situated in the Local Government Engineering Bureau within the Local Government Division of the Ministry of Local Government, Rural Development and Cooperatives (MLGRDC). It has to be distinguished from the more familiar and larger Food for Works Programme (FWP), implemented by the Ministry of Food, paying labour directly in wheat from food aid sources. The FWP has increasingly dominated the labour-intensive earthworks projects but is conceived in narrow terms as providing short-term employment opportunities during lean agricultural seasons, and relies explicitly upon the traditional institutions of rural labour mobilisation dominated by the *Upazila* and Union Council patronage networks. The FWP is widely regarded in Bangladesh as extensively corrupt. The IRWP was initiated with broader objectives in 1981 but has shared many of the problems of FWP (while explicitly trying to overcome them). Furthermore, because of the near monopoly of earthworks schemes by FWP (as a result of a government instruction) the IRWP has been increasingly confined to *pucca* constructions (bridges, culverts and so on). However, in the period for this analysis, earthworks were still prevalent.

The IRWP was intended as a major programme for employment and the creation of rural infrastructure, eventually covering 100 flood-prone *upazilas* in eight districts. The objectives as stated in the Plan of Operation included principally:

i. To increase direct and indirect employment and other income opportunities in the short and long run for the landless labourers, marginal farmers and women from such households (i.e. the "target-group") [sic].

ii. To improve the infrastructure in 100 flood-prone *upazilas* by implementation of rural works thereby promoting agricultural production. Marginal, small and medium farmers with less than five acres who own about 57% of agricultural land in Bangladesh would be the main beneficiaries.

The programme was committed to: "increase participation by the local community and, in particular, by the target-group in the planning, implementation and utilisation of IRWP schemes through the assessment of appropriate institutional arrangements and improved methods of scheme selection." The selection of schemes was to be guided by the following criteria:

> "i. The programme of works planned and implemented each year will aim at a wage component of at least 70% of the total cost.
>
> ii. A substantial share of the schemes included in *upazila* plans will be of a kind that directly benefits the target-group in addition simply to providing direct employment."

From this summary, it can be seen that IRWP is a distinctive programme of intervention through rural works by virtue of its orientation to the 'target-group' of landless labourers and marginal farmers (male and female). The programme attempts more than the simple provision of temporary employment during the lean months of the agriculture cycle.

The Struggle for Employment Rights

The programme offers in principle a set of rights to rural workers. The first and most obvious is the opportunity for the landless and destitute of both sexes to work in the IRWP schemes. Secondly, such work should lead to the workers' receipt of fair wages. Thirdly, priority is to be given to labour intensive schemes to provide those opportunities. (The expansion of FWP has already functioned to undermine this right). Fourthly, rural workers should share in the longer term benefits accruing from their labour through related production activities. Fifthly, as a way of ensuring these possibilities, workers are to be involved in the planning and implementation of the rural works schemes.

However the workers, the landless men and women, the 'target-group', have problems in achieving these rights. By managing the schemes through the local government institutions—the 'host' Ministry's own structures—and by depending upon the connected unreformed systems of labour recruitment, the Programme cannot ensure the provision of those rights without the help of the workers themselves, organised as claimants. They are being asked to organise not so much for the purpose of participation in the planning and implementation of schemes, but to struggle in an arena of seeking fair payments from local scheme managers and labour bosses. This is an arena in which they are structurally weak, and a reason of course why the IRWP has involved, on an experimental basis, the two NGOs, Proshika and Saptagram.

This arena consists precisely of the hierarchical relationships through which poor families have to arrange their survival. This programme, in the form of a local rural works scheme such as an embankment or a road, is of course a limited intervention in such people's lives. Yet even to ensure workers' rights during the period of temporary seasonal employment requires forms of struggle which potentially jeopardise those other relationships of survival. Who bears the risks involved in such struggles; struggles which represent a means of achieving the programme's objectives? Programme staff have a limited role in providing meaningful support, yet workers who do struggle, as encouraged, have been victimised, excluded from schemes and physically intimidated by union council members involved in the schemes, *upazila* officials and labour bosses. Where they have a presence, the NGOs Proshika and Saptagram are assisting in organising the solidarity for this claiming activity, but they can be at risk too. For example, Saptagram has worked from some years with men and women in a locality which was therefore chosen by the IRWP for an experimental scheme especially for female landless participants. These women have been substantially victimised by the Union council leadership in response to their struggle for the first and second rights noted above. (See Case Study in Chapter 10 for a description of this process). The failure to confront this victimisation (exclusion, cheating on payment, fictitious lists of workers and beatings) of site workers on a road scheme has not only left the workers in a weaker position but may undermine the wider range of Saptagram's involvement in the locality with the women from landless families, and some of their menfolk.

The Union *Parishad* (Council) is the lowest elected body in the local government structure (and the elections for the *upazila* and *zilla* levels have been postponed indefinitely). The Programme's schemes are managed either by Project Committees established by the Union Council, or (for different categories of scheme) by the *Upazila Nirbahi* officer (the *Upazila* Executive officer—UNO) in the absence of elected *upazila* members (councillors). Most of these Project Committees are effectively an extension of the patronage network of the Union Chairman, and influential Union Members. The Union Chairman himself either becomes the Project Committee Chairman, or nominates both the Chairman and the Secretary when those posts should be elected by the Committee once constituted.

Usually other members of the Project Committee are either unaware of their membership or of their duties, or have been selected for strategic political reasons by the U.P. chairman. The cynicism which is applied by these local leaders to the Programme is revealed especially over the central issues of payment rates and measurement of earthwork (volume of earth shifted for embankment/road construction, canal excavation and so on). Cheating was common to all eight schemes visited in the research for this chapter. By placing such significant resources as the budgets for those schemes at this level, the patronage rather than the representational character of the local government institutions is stimulated. Such a process intensifies the problems of local accountability, and of course prevents wider participation in the planning and management of such schemes—the fifth right noted above.

The Labour Recruitment System

The problems of claiming rights on payment and measurement are related to the internal composition and political solidarity of the labour gangs, and this in turn relates closely to the way individuals are recruited and the gangs are managed. The relevant government circular states: "The function of the Project Committee is to mobilise labourers for the execution of the scheme as per list of destitute maintained by the Union *Parishad*, and gang them up." (MLGRDC No. 5, para. 16 p.9). Yet no reference was made to such a list in any of the schemes visited as a basis of recruitment. If it did exist as a sign of qualification for significant employment opportunities, then merely to appear on the list and be labelled as destitute would be the first of many problems of access.

In practice, however, labourers are organised through a labour-*sardar* system. Even where NGOs are active (in three of the eight schemes visited) 'their' groups of landless have not replaced this system. These *sardars* are in effect sub-contractors of labour. They are not necessarily landless themselves. Often they are educated sons in households where they will inherit land after division, thus illustrating the weakness of the *bhumihin* label. They either have connections or develop them with local leaders (notably the Union members and Chairman) to find out where and when employment on a scheme is available. They create small gangs of about ten labourers who may be dependent upon them in different ways—perhaps scattered across different neighbourhoods (*paras*) of the 'village' (another bureaucratically convenient label) without previous or intimate connections between themselves; and whose association with each other is likely to be temporary for the season. These *sardars* function as the intermediary between the labourers and the Union Chairman or Project Committee. At the site itself, they do not usually work but act as supervisors. Payment to the labour gang is divided to include the *sardar* and in addition the *sardar* receives a payment (unauthorised) from the Project Committee Chairman. Payment to the gang is made via the *sardar* which may often involve him travelling off-site to the Project Committee Chairman's house. He thus controls the flow of income (and, most importantly, information about income) to the labourers.

These labour-*sardars* frequently function as moneylenders to 'their' labourers, which further undermines the capacity of the 'target-group' labourers to organise as successful claimants. Labourers (or their families when the site involves temporary migration from the *bari*, the residential extended kin unit) receive advances to be worked off during the earthworks season. These arrangements can be institutionalised further with the Project Committee paying advances to the labour *sardars* for this purpose. Another arrangement is where the individual labourers receive a daily advance (in effect a daily wage) to be deducted from the piece-rate payment due when the weekly measurement of earth shifted has been made. By definition, the programme is operating in the lean season when income flows are low. Daily advances are a reflection of the need to provide immediate income, but they also enable the measurement to be postponed, to become irregular and thereby the scope for cheating is increased. The labourers and their families become dependent upon loans from village moneylenders to 'fund' low daily rates, low overall rates and delays in measurement. Materially family conditions can even deteriorate when delays and cheating occur.

The persistent structural weakness of the labourers in the rural political economy further undermines their capacity to be effective claimants. Under conditions of increasing landlessness, pressure on the availability of land for sharecropping, and the general evidence on rural under-employment, families rely upon landlords and richer farmers to provide: employment for both men and women; land for leasing; and consumption credit usually in the form of food advances in lean seasons to be worked off in the busy agricultural seasons. It is true that such patrons may sometimes be from within the same *bari*, but they are nevertheless connected at other levels in the structure of rural power. This enables them to: obtain access to inputs; secure the registration of recently acquired land; procure licences and leases for business activities; and exercise leverage in court disputes, perhaps over land. In this pattern of exchanges, Union Council members have always occupied a strategic position. Labouring families are therefore likely to increase the precariousness of their survival if during the earthcutting months of the year they challenge the rural hierarchy upon which they depend not just for access to the earthcutting work itself, but also for support, even relief, during the rest of the year. Thinking of people as individuals independent of their families and capable of easy and bold militancy is a mistake. There is a range of calculations to make which determines the reality for their room to manoeuvre.

Exchange Morality between Schemes and Rural Labour

There are further problems for claiming which are more specifically related to the payment and measurement aspects of the programme. The practices of payment on the schemes reflect patterns of payments for agricultural work with a combination of daily payments and an end of season pay-off. In agricultural work this is through a share of the harvest such as one-eighth of the individual quantity harvested. In the rural works schemes the pay-off is the balance accruing from the final measurement of earth shifted. Leaving aside judgements about the implications for social justice, such delayed payments in agriculture occur between landholders and labourer with highly personal and long-term, often generational, exchanges between them. The complete 'story' is operative. In this limited sense there is a trust; a trust that expectations will be honoured. However, given the system of labour recruitment and management on site, no such trust exists in the earthwork schemes. The level of morality in the exchange is lower. Labour can more easily be treated instrumentally (not of course as a freed commodity) as a 'case', detached from other obligations to it. It is easier for the Project Committee as employers to contemplate defrauding workers on site—especially over payment of the final balance after measurement.

This process is further reinforced by the patterns of work on site. The labourers frequently walk more than three miles for site work, and sometimes they are fully migrant, in that they stay on site for the duration of their 'contract'. If they are walking daily to the site, and are receiving daily wages (technically: daily advances), then they are less effective in maintaining pressure on the Project Committees to ensure timely and accurate measurement. In effect labourers 'disappear' either into other work (other earthwork sites such as FWP and/or agriculture) or because returning to claim their share of the balance owing after measurement is not considered worthwhile because of expectations about cheating. Acceptance of the daily advance as a form of wage becomes a satisfying strategy, which weakens the position of others who are trying to insist on proper measurement. The capacity for Project Committees to defraud under these conditions is enhanced by the irregular and temporary participation of labourers in the schemes, which is encouraged by daily payments. It was alleged by some labourers on a river re-excavation scheme in Madaripur District that the Project Committee (chaired by the Union Council Chairman and run more like a personal contracting business with considerable intimidation of the site-labourers) was deliberately organising a high turnover of labour for the purpose of defrauding on measurement. The Project Committee was insisting on lengthening the working day. With proper measurement arrangements this should not matter within reason, however, if the only safe or guaranteed payments appear in the daily form of advances then the longer day is clearly unattractive. In this way, labourers were reluctant to work on the scheme for long periods. Thus irregular, temporary, semi-migrant and high-turnover forms of labourers' involvement in the schemes, combined with daily payments, prevents the labourers from establishing a sustained (and united) connection with the scheme.

This weakness of 'connection' to the scheme is reflected in the keeping of the muster rolls of the labourers. The accurate maintenance of this register is essential to ensure accountability on payments and measurements, although it is only a necessary rather than sufficient condition for measurement. However these muster rolls were found to be substantially fictitious in: the names entered; thumb prints given (since no accurate and authoritative mechanism can really exist for checking these); dates of attendance; labour gang membership (labourers do move in and out of gangs); and the borrow pits in which labourers have worked, since with delays in measurement, labourers are inevitably working in new ones.

Conclusion: Rural Labour—Objects or Subjects of Policy?

From this description of the claiming process in the IRWP several issues emerge. First, it should be noted that even for those concerned with 'target-group' objectives within the programme, the focus of attention is dominated now by the problems of claiming rather than the wider, more significant, objective of encouraging participation in scheme identification, planning and implementation. The pursuit of such participatory rights has been virtually abandoned. By attrition the 'target-group' preoccupations of the programme have been brought back to a delivery concept of organised groups of landless men and women. The agenda has shifted towards an admission that at best the landless are recipients of rights bestowed or withheld by other institutions and classes outside the 'target-group'. Nevertheless the success of the programme still relies upon groups of landless labourers behaving *as if* they were not constrained by the rural classes who have become the bestowers of these rights in the programme. In other words, the risks of claiming are similar to those involved in the pursuit of the more radical participatory rights.

Secondly, therefore, the success of the programme is premised on the assumption that the landless, once brought together around an earthworks scheme, can act in unison in defence of perceived common interest. The programme is trying to train groups of labourers in the objectives of the programme—the rights to be bestowed or struggled for—and in the techniques of measurement. However, these apparently potential units of a claimants' organisation are actually highly fragmented groups of vulnerable labourers, who do not have much contact with other labour gangs on a scheme site since the different gangs have been recruited and brought to the site by different *sardars*. A 'group' of labourers brought together by labour *sardars* from different parts of the village is more likely to reflect complex patronage arrangements and dependencies than a sense of group solidarity which could continue once the season's earthworks have finished. Indeed family survival and the relationships through which it is precariously pursued are uppermost in people's minds. These relationships consist of other forms of solidarity and identification which run vertically through the hierarchies within the *baris* and the *paras*. These are the social relations of dependency and poverty which the programme, with its limited form of intervention, cannot reach or transform.

Thirdly, it has been convenient for the programme, and indeed the government as a whole, to conceive of a 'target' delinked from such relations. In most programmes in Bangladesh however, this delinking enables the poor to be reorganised as fragmented objects of a policy of partial interventions—the recipients of skill training, credit and services, ghettoed into small-scale income-generating activities, an entrepreneurial model in which significant success could only possibly be enjoyed by a few and thereby absorbed without overall structural change. The paradox for IRWP however was that its delinked designation of a landless target-group was also the basis of a range of expectations about the possibilities for concerted landless labourer action which would convert the rural poor into subjects of policy (through participation) and significantly alter the pattern of control over rural infrastructural assets as the basis for further structural change.

Fourthly, it has been shown here that this designation cannot contain these expectations. It is therefore argued that these expectations can only be realised: when the existence of real groups based on single criterion membership (whether *bhumihin* or *bittaheen*) is not assumed *a priori*; when it is realised that in the context of rapid, dynamic agrarian change, with today's small peasant becoming the landless of tomorrow, people will define themselves into meaningful social units for co-operative action according to dimensions beyond those contained in a term like 'landless '; when the linked package of a person's survival arrangements (the story) is recognised as setting limits to the utility of partial programme interventions which rely upon interacting arbitrarily with only some of these arrangements (the case); when the risk to a rural labourer is appreciated as that story is undermined for the sake of becoming a good participant (or in this programme, a claimant) as defined by the case; when the approach to such rural mobilisation consists of an explicit recognition of the social relations of poverty through programmes of conscientisation and continuous rather than partial support in a manner similar to the experimental involvement of NGOs like Proshika and Saptagram.

Fifthly, there are panacea implications in the last condition which must be avoided. The experimental or action-research involvement of those NGOs in IRWP can be interpreted as a response to the paradox referred to above. But the NGO involvement has revealed that even to perform as good participants in the delivery conception of the programme, that is as successful claimants for employment, payment and accurate measurement, rural labourers cannot be regarded in bureaucratically convenient terms as a de-linked category of quiescent recipients. Struggle even for the 'normal' employment rights in the programme is necessary, leaving aside the more ambitious, participatory rights such as the longer term control by rural labour over rural works assets, direct contracts to rural labour for work on the schemes, participation in local planning and scheme selection. Unfortunately, this involves struggle against the very institutions through which the programme is administered. The NGO involvement has exposed these contradictions embodied in state intervention: between the qualifications required to be designated as a member of the landless target-group and a rural works claimant on the one hand, and the qualifications required to act the role of a successful rural works claimant on the other. The first requires loyalty and the second requires voice; and much of it (Hirschman 1970). Authoritative labelling has failed to reconcile the two; the hegemony is not complete.

Chapter 12

Rural Infrastructure and Social Relations: the Intensive Rural Works Programme in Bangladesh

Introduction

As indicated in the preamble to this Part of the volume, this chapter was originally an elaborate set of lecture notes designed to comment upon the parallel relationships between physical structures and social institutions, and between engineers and social scientists. Although it focusses mainly upon the experience of the SIDA funded rural works programmes in Bangladesh (especially IRWP), it also reflects experience from working with irrigation systems both in Bangladesh and North India. Although these notes were prepared before the Flood Action Plan (FAP) appeared on the scene as a major, donor-led policy intervention, the arguments criticising the primacy given to engineering solutions have immediate relevance to the controversial FAP. In the writing up of these notes for this chapter, therefore, explicit references to the FAP will be made. The IRWP has offered a relevant experience because although engineers were being asked quite explicitly to work within a framework of social objectives, many of them (expatriate as well as Bangladeshi) had great difficulty in coming to terms with both social objectives and the social science based monitoring of performance. Indeed, this experience has prompted us at Bath to mount a short course for engineers on 'Social Science Issues in Development Practice'.

A substantial agenda of issues arises from this experience beyond the general relations between physical structures and social institutions. Engineers often see their work as the 'hard' element in a project in contrast to its 'soft' social aspects. Yet perhaps this attitude could be inverted to reflect the situation whereby physical targets are more easily established and achieved than the social use of them. Engineers (especially expatriate technical assistance ones) are most likely to be working in ignorance of the social context, and are unable, independently, to assess the distribution of costs and benefits, both public and private, arising from technical decisions. The question then arises whether engineers assume that these choices have been resolved for them, or whether they accept a necessary iteration between objectives, information, interests and design. Apparently technical choices such as location, design and materials are nearly always directly political, since they rarely have equal effects on who gains and who loses both during a construction phase as well as afterwards. If this leads us to accept that engineering is political, then the training and preparation of engineers for specific assignments becomes significant, enabling them both to point out the social implications of choices as well as joining in the advocacy for social as well as engineering objectives.

It is clear then that infrastructural investment has differential effects. It is, of course, obvious that societies are divided up in a variety of ways: regional, ethnic/ethnic, natural resource endowment, gender and class. With such divisions, infrastructural choices can rarely have a neutral impact. New means of production and exchange are being created or altered. Flood control and irrigation both produce new land and enhance the value of existing land. Roads, bridges, canals and electrification facilitate markets and trade. Clearly the ways in which people accumulate and survive are affected by such infrastructural investments. Opportunities appear in different ways. New monopolies over bus services and freight transport occur when metalled roads allow for the substitution of a myriad of head-loaders, rickshaw-pullers, small boats along canals, and carts by buses and lorries which require a large capital input by the operator. The value of land changes when it is protected from early flooding, or drained from water-logging, by embankments and canals--but only those landholders with land in the 'command' of such constructions will benefit (and that distribution is not random but the result of negotiation). Cropping patterns on land so enhanced will change. Also the land can be more securely operated, prompting higher investment and yields. Under some conditions, owned small farms will retain viability due to increased land usage and productivity. On other land, owned as part of larger holdings, tenants may be displaced thus undermining the viability of small farmers who had relied upon a proportion of tenanted land in their total operation. Labour patterns can change when a net additional season can be added, or an old unpredictable rotation can be substituted by a new, more secure one. On the scale of land use changes envisaged as an outcome of the FAP, we can expect changes in labour migration patterns. Such changes in labour patterns may alter cash flow patterns for families. If local labour opportunities are more spread out through the year, with the family being less dependent upon the migratory absences of males, then indebtedness and especially female vulnerability and dependency is reduced. At the same time, as raised in the preceding chapters (10 and 11), the labour intensive construction of embankments, roads, canals and water-courses may provide immediate employment, but who has long-term control over these assets as well the indirect benefits from them? As we now know, even with short-term employment opportunities, there are many issues of how workers are recruited, measurement of work, modes of payment, patronage arrangements and cheating.

Infrastructure & the Rural Poor: the Objectives of IRWP

We can note from the following summary of objectives for IRWP that it was not only poverty-focussed in intent, but also constituted a forerunner of FAP in certain respects.

- a distinctive programme of rural works by virtue of an orientation to target group of landless labourers and marginal farmers (men & women)
- attempting more than just provision of temporary employment
- infrastructure to aid agricultural development in flood-prone areas certainly intended via: roads, embankments, canals, bridges, drains, sluices, culverts
- programme originally conceived for earthworks with wage component at 70% of investment cost
- rights offered in principle to rural workers extended beyond short term employment to include: share in long-term benefits accruing from their labour through related production activities; and participation in planning and implementation of such schemes.

Achievement of these objectives was undermined in various ways. The combined effects of corruption and slow disbursement due to bureaucratic delays reduced the wage component as a proportion of the investment budget in 84/85 to 16-27% (assuming to wage rates at Tk. 15-25 per day). The programme was also a victim of other policies: the shift of most earthworks to FFWP; the loss of road maintenance to CARE; major local government reform (i.e. the creation of the *upazila* system) placing more control in hands of local political actors; and GOB political pressure to expand IRWP into new areas, thus diluting the prospects of close monitoring.

The practicality of the programme could also be criticised in various ways, despite its laudable aims. The programme had to be implemented through local government institutions where the poor had no voice, and where technical engineering capacity was weak, especially in relation to water which was significant both for production and any prospect for the landless acquisition of productive irrigation assets. Furthermore, although the programme provided about one million man-days of work in 1984/85, the project cost of providing one man-day was Tk 160, or Tk 50,000 for one man-year (about 10 times the wage rate accruing per labourer).

This prompted a series of questions about the IRWP strategy concerning: comparative employment opportunities of normal rural works programmes; alternative more direct ways of using same aid finance to generate long-term employment—e.g. occupation of *khas* land for cultivation, support for landless participation in whole range of agricultural services and marketing; development of long-term capacity to achieve both short-term employment and/or control over productive assets; loss of opportunities where dominant local classes have secured monopolies over investment resources (land, water, transport) and enhanced their position in relation to rural workers.

The problem with a strategy of poverty-focussed, rural development intervention via infrastructure investment is its diffuse character. Rural infrastructure is especially prone to capture by local dominant classes because its effects cannot be directed so purposively, even when compared to a programme of new technological inputs in agriculture where distribution has been skewed negatively in terms of class and region. Most studies of the social implications of infrastructural development conclude that local landowning and merchant classes are the main beneficiaries, seizing new marketing and transportation opportunities as well as production ones (see for example Blaikie et al 1979, writing on Nepal and M. Hossain's road study in Bangladesh). The IRWP attempted to be an exception by exploiting the potential linkages between infrastructural investment and more permanent opportunities for the rural poor. To achieve this involved challenging the local structures of power embedded in control over land and commerce. The authors of the programme, especially among the donors, were naive about the necessity for this challenge, having in effect a view of the state as benign. But GOB was not primarily concerned with social objectives, rather with the physical aspects of the programme. The LGEB officials have also revealed a clear interest in institutional expansion and the development of patronage networks in the recruitment of staff, as a way of sustaining the system of misappropriation. Suffice to say that the state is not benign. Furthermore, the combination of external aid flows, bureaucratic monopolies over a significant proportion of them, decentralisation of bureaucratic patronage interacting at *upazila* level with local class relations, and the history of low levels of rural (or even urban) capital formation which encourages misappropriation and unproductive forms of accumulation, represent together an extremely hostile environment within which to work alongside the rural poor.

Labour Relations and the Rights to Construction

The availability of construction work has a number of potential structural effects. Post-monsoon & post-*aman* site work can in principle crucially affect labour and therefore dependency relations. It changes the seasonal distribution of work, reduces fluctuations in labour demand, which in turn reduces reliance on advances or outmigration (which is male, leaving women vulnerable to harassment from landowner/patrons). Thus behind any figures for short-term employment generated, lies the possibility of structural effects, where new patterns of work interacting with net additional income can improve wage bargaining and political rights by reducing reliance on advances. One implication of this is that provision of employment at the appropriate time is as important as the number of man-days available. However, the evidence for the 84/85 post-monsoon construction period was: that work was offered too late i.e. not in the period between transplanting and harvest; that only 20% of the post-monsoon allocation was spent, and only 15% of allocation received by labourers. Most of the schemes started in January 1985 instead of October 1984. Post-*aman* work should have been available at beginning of January 1985, but 75% of such schemes started in late February-early March. So despite one million man-days created, the timing was really inappropriate to maximise the structural effects noted above. What is happening to labourers during these delays? Typically advances are being taken thereby reinforcing dependency relations within local patron-client structures.

A further index of inequality and power in these schemes comes from bad record keeping and (or leading to) misappropriation. Thus in addition to correct timing of work, a man-day should mean full payment as per the programme guidelines, supported by matching records (i.e. daily attendance registers and muster rolls). The existence of such records is an index of labour rights for site-workers, and itself constitutes data about the relative power in labour transactions between workers, members of Project Committees (usually local patrons) and officials. The programme's own monitoring exercise revealed in 3 different localities of the programme that in 83%, 100% and 94% of cases, records were either falsified to some degree or not properly maintained. In one case, in Kurigram on a road scheme, 72% of the funds allocated were incorrectly paid out by *upazila* officials. As has been established, workers are recruited through patronage networks which link local patrons to officials, allocating schemes. Despite the existence in the programme of an official wage rate, workers are rarely able to insist on it. Daily or half-weekly advances become the mode of payment, since measurements are delayed both during work and at its completion. In the latter circumstance, workers have to leave the site area for other work before the final measurement to determine that the balance of wages owing is completed. Thus balances owed are rarely paid when the measurement of earth shifted in finally done. In the 1984/85 programme monitoring exercise, no payment was made after measurement in 78% of schemes visited. The labour *sardar* failed to sign a receipt for wage allocation to his team in 79% of cases. A separate bank account for scheme funds (to distinguish such funds from the project contractor's personal finances) was only opened in 31% of cases. With this kind of evidence, site-workers are unable to be more than precarious supplicants in the labour relation. The expectation of opportunities for cheating by local officials and patrons is revealed by the example from one *upazila*, where the sub-assistant engineer tried to maintain correct technical standards (thereby reducing the scope for cheating on payments to workers and costs of materials) and his senior officials arranged for him to be transferred. Site-workers have been too fragmented into discrete patronage systems to compete with this alliance between local patrons and officials, and have not been able to exercise voice and effectively monitor their entitlements. This is why the intervention of NGOs in rural works schemes is so important to contribute towards a sense of solidarity and effective exercise of voice.

The IRWP explicitly wished to exercise affirmative action in favour of women, and attempted to involve women via two levels of intensity: schemes under normal guidelines (as for men); and special schemes more closely monitored and with NGOs involved in mobilisation. But using the same monitoring data for 1984/85, out of 22 normal schemes, only 5 were implemented entirely by women, 2 were almost entirely male, and in the remaining 15 normal schemes 40-50% of workers were men who dominated the work and payment of women working alongside them. There were 9 special schemes, but only 2 operated according to the guidelines (i.e. special amenities like toilets, sunsheds, creches, fresh water). Muster rolls were falsified, men were working, cheating occurred on payments. All this happened despite an average of 15 visits per scheme made by IRWP field officers to check on implementation. A major problem of affirmative action towards women was the lack of scheme continuity. Women were mobilised for one scheme only (which would last for 8-12 weeks), undermining the development of any cumulative solidarity, even when NGOs were involved. Women faced enough problems of stigma and competing household pressures to offset any potential militancy in seeking their rights. This is why the case example cited in Chapter 10 was so important to record.

We can conclude from this evidence about real labour rights that, despite the efforts of some staff, the programme could not ensure the rights which were in principle on offer to site-workers. The underlying reasons for such weak guarantees of employment can be summarised as: the recruitment and management of labour through *sardar* and patronage networks; the credit needs and related dependency of labour upon the patrons who gain access to the management of these schemes and control employment; the problem of partial and seasonal interventions into complex structures; sporadic participation in sitework undermining workers' potential unity; fragmented, competitive labour, leaving workers divided and vulnerable; family survival strategies (necessarily cautious, risk avoiding) within such networks undermining any prospect of militancy by workers in claiming rights; the contrast between agricultural and rural works payment systems involving different exchange moralities (as discussed in chapter 11); and of course the programme's institutional reliance on local elites for implementation.

These problems highlight a central irony of programmes like this which seek to offer workers' a better deal within the context of an otherwise unreformed political economy. The programme cannot guarantee the rights which it is in principle offering. However, for the same reasons, the only prospect for programme success (measured in terms of extending rights to workers) is that responsibility for securing rights to wages and direct contracting opportunities has to lie with the siteworkers themselves. In other words, the programme's success relies upon successful claimants, who to be so have to bear all the risks of such militancy in other structures through which their households have to arrange their daily survival (long after the programme and its schemes have moved on to other localities). This realisation led me to propose the 2 deliberate strategies to overcome this irony, relying upon the continued presence of NGOs (and the solidarity they induce) in a locality where schemes have or may be implemented: direct labour contracting societies; and shadow project committees (to ensure accountability through a counter-information lobby). The labour contracting societies have now become a regular feature of the rural works landscape, and their use should certainly be a central part of the construction strategy in the Flood Action Plan, if its physical plans become a reality. It is important to recognise that both of these initiatives represent a shift of power away from expatriate and local official engineers and their local allies in the political economy.

Acquisition of Direct and Indirect Rights over Infrastructure

Too often, rural works programmes are conceived more in terms of immediate short-term relief for workers rather than their long term development. The IRWP intended also to be distinctive in this respect through, where possible, the acquisition of direct and indirect rights over infrastructural income by the workers involved in the construction, or at least the local community of workers. The activities through which this acquisition could occur were: routine maintenance, pipe casting, culvert construction, fish cultivation, rickshaw cooperatives, and tree plantation management. During the 1984/85 season, there were 201 years of female employment of which routine maintenance was 185; 136 years of male employment, of which routine maintenance was 122. In programme objective terms, it therefore mattered when GOB policy shifted the responsibility for earthwork road maintenance throughout the country to the management of CARE, which has a stronger short-term relief orientation and a more passive view about class structure and local-level political economy. The removal of routine maintenance, managed under this programme, was therefore very significant, especially as it provided employment and income security for destitute women supported by programme intervention (often via NGOs) to overcome cultural resistance to women working in this way. It was also crucial to maintain the link between construction and routine maintenance, since the latter was a signal that roads, canals and embankments require constant upkeep for which the direct beneficiaries of such infrastructure should pay rent, and that the poorest site-workers should thereby continue to derive income from the asset created by their own labour.

This issue leads us back into the more general policy question of the longer term distribution of benefits from infrastructural investment, and how an exchange can be established between the providers of the service (construction and maintenance) and the immediate consumers of it (farmers, commercial fish farmers, transport operators). Physical structures can be regarded analytically as a means of production, from which rents can be derived. How is the ownership of such rural works assets to be determined? Stretches of road and embankments, bridges, canals, and sluice gates could all be owned by the site-workers themselves, or they could be owned by the state in central or local form. At present, neither potential set of owners are effectively gaining rents or incomes as providers of the infrastructural service. These issues will need to be resolved in the context of the physical structures created from the FAP.

With each of these examples of physical structures, the current beneficiaries are those who already own commercially viable land, or who are merchants and transport operators. These classes are gaining a free (aid-sourced) good, yet it is their use of these structures which causes damage and entails the provision of the routine maintenance service. In a fiscally ineffective state with little prospect of taxing such beneficiaries, this process skews the pattern of opportunities arising from infrastructure strongly in favour of such landholding and commercial classes. Indeed we might argue that the presence of aid releases the government from an obligation to fund the provision of this service via taxation, and that in this sense aid is not only contributing to the maldistribution of opportunities, but making it difficult as a result for the state to establish effective ownership over the assets through an unsuccessful assertion of rights to claim tax. However, we could imagine an aid-scarce situation for the provision of infrastructure in which merchant and landowning classes, intent on deploying agricultural capital and variable inputs (fertilisers, pesticides etc.) had a low elasticity of demand for roads, embankments and *pucca* constructions and were therefore more amenable to meeting its costs through local or national taxation (which, short of actual ownership being vested in the siteworkers themselves, could be used to offer siteworkers franchises for maintenance and even tax/toll collection). It will be essential to adopt a strategy on these lines for FAP and other major rural works initiatives to achieve any kind of sustainability. As a parallel, it is worth reflecting on the agricultural sector, where a low elasticity of demand for agricultural capital and inputs is being tested to a considerable extent by the withdrawal of formal subsidies. However, there is a limit to this comparison, with the true picture of demand for inputs being disguised by a generous credit policy and undisciplined agricultural credit practices.

Water Schemes: Social or Engineering Criteria?

The water management issues arising from the experience of the IRWP and its sequel programmes are of central relevance to a contemporary discussion of FAP and the balance of analytic power in the policy process between social and engineering criteria. (I leave others to speculate on the issue of their respective vested interests!) Several years before the invention of FAP, flood control, drainage and irrigation canal components were prominent in the objectives of the SIDA funded rural works programmes, although actual activity was slow to materialise. The issue of participation is central, which is why CDS at Bath in conjunction with 'Research and Advisory Services' in Bangladesh (directed by Shapan Adnan), embarked upon the recent study (Adnan et al 1992) of the FAP with this agenda. Such schemes are in effect reorganising the value of the means of production in the countryside, especially the value of agricultural land. In the process, as noted above, other 'service' assets are being created (e.g. flood protection embankments, water delivering canals).

The small-scale water resources component of the programme retained the landless as the primary target group and small farmer classes as the secondary 'target'. However by ignoring landless irrigation possibilities (still too extraordinary for most engineers), opportunities for involving the landless from the design stage of canals were lost (despite their knowledge of local potential command areas from a social distribution perspective), and other possibilities for direct contracting, maintenance and embankment cultivation (i.e. social forestry) were also ignored.

From an agricultural production perspective, the secondary target of small farmers was supposed to be favoured in the small-scale water resources initiative by ensuring through surveys that command and/or enhanced catchment areas (i.e. areas to be offered flood protection or irrigation as appropriate) have a landownership structure dominated by smaller farmers. The problem with this initiative was that it was conceived in the following sequence: social surveys to screen for secondary target group beneficiaries (i.e. small farmers) followed by an engineering intervention through construction. No further social intervention was envisaged. This strategy could not secure the objectives. The engineering intervention should not precede a social intervention if direct opportunities are to be captured by the small farmers target group. An alternative sequence would therefore be: social analysis (using RRA or PRA); mobilisation of local landless/marginal farmer groups, perhaps over as long a time as 2 years; participation of such groups in negotiations with farmers over the social organisation and final physical character of the proposed engineering intervention; and then the engineering intervention coming last in the sequence, managed according to the prior outcome of negotiations.

Even with the original sequence of weaker targeting and participation with the more limited objectives of identifying areas of small farmer concentration, 'political' problems emerged at the initial social survey stage. The programme socio-economists were placed under pressure to 'reconsider' their feasibility analyses which had rejected areas on the grounds of rich farmer composition but which were favoured by engineering and/or local political actors. From various interviews as I conducted my analysis of this programme, it was clear that senior engineers in the programme were insisting on priority for technical rather than socio-economic feasibility studies, making it hard to retain the equity criteria in the *upazila* planning meetings. These engineers (both expatriate and national ones) were clearly of the view that physical infrastructure has to be created first, and only then should the target group interests be addressed. Their analysis of the role of infrastructure is to generate general, unspecified multiplier effects in the local and national economy, and the content of these effects should be left to the interplay of market forces. They did not accept that their technical decisions should be responsive to social objectives.

Part of the explanation for this position was their frustration at the slow progress on these small-scale water schemes. It is extraordinarily difficult under the socio-ecological conditions of land fragmentation in Bangladesh (discussed in Part II of this volume) to establish more equitable command or catchment areas. In 1984/85, the socio-economists rejected 75% of the schemes proposed by engineers because there were too many large farmers. This highlights the difficulty of reaching the secondary target group via these location-targeted water schemes. Certainly with irrigation (rather than flood protection), the landless strategy under these socio-ecological conditions reduces the significance of a skewed distribution of landholdings as a variable in matching social objectives to engineering ones, since small farmers would be served more equitably and the landless would share in the value-added (as outlined in Chapter 7, Part II and in *The Water-Sellers*).

Of course, there are other, more familiar problems with surface-water management for irrigation. They tend to result in either oppressive bureaucratic management with a dose of corruption, or self-managing groups of farmers which often dissolve into mutual suspicion, disputes and acrimony. Tail-enders are common to both. Private, entrepreneurial farming systems and surface water management do not mix very well, which is why the rich buy themselves out if they can into relatively autonomous tube-well or LLP based production. The problem in Bangladesh is that even large farmers are unlikely to have sufficient adjacent plots to avoid selling water to neighbours if they are to achieve an optimum utilisation of their equipment investment, and with the withdrawal of subsidies that is becoming more necessary. It is difficult for engineers to determine command area units under these conditions.

Irrigation is only part of this small-scale water resource issue. Flood control and drainage raise other issues since the productive value of private land is enhanced. As noted above, the owners of private land enjoy a rent-free enhancement of the long-term productive value of their land, offset marginally for the richer landowner and less marginally for poorer landowners by the loss of land to borrow-pits and embankment sides. There are knock-on effects here for loss of production and employment, especially from road construction rather than dedicated flood control and drainage structures, where the purpose is to enhance the productive value of the remaining land. However, small farmers may be doubly disadvantaged in this process: through higher proportions of any land which is randomly adjacent to scheme sites; and through lack of influence over scheme design and alignment which may result in more of their land and less of richer competitors' being deliberately adjacent to the final alignment. I have seen many such 'negotiated' alignments. There are possibilities for skewing flood control, drainage and irrigation (FCDI) projects in favour of *khas* land, which represents in principle a stronger opportunity of establishing direct control over reclaimed cultivable land by the landless. This is a clear case where an engineering intervention would have to follow social mobilisation, since it is no use enhancing *khas* land first, then looking around for groups of landless to occupy it. The potential occupants have to be ready (through prior mobilisation and conscientisation) to plough, irrigate, transplant or broadcast, and so on. Without that social preparation, an FCDI scheme could remove forever in a locality the hope of land occupancy for the landless. It would be a strongly negative intervention for the target group, with only a trickle-down prospect of some employment as richer landowners established their claim through the plough.

Multiplier Effects

Both in the context of the small-scale water schemes as well as the other rural works schemes, it is worth reflecting on the potential multiplier effects of such interventions, and the influence of the engineering decisions. If the landless were able to enter agricultural production via either *khas* land occupation or irrigation rights, their availability for wage employment is reduced, thus raising either wages, or opportunities for employment, or both, for other workers. Their dependence on landowners and patrons is reduced. They increase their participation and bargaining position in a range of factor markets. They enjoy more economic and political security in the local political arena.

However these potential, positive linkages depend upon changing *inter alia* the present relationship between social and engineering criteria. Instead of separating the engineering cadres from responsibility for the social consequences of their choices, they either need to become amateur socioeconomists themselves, or be prepared to work in close collaboration with the social development agenda and not expect to have the last word. Their work has to be sensitised to the socially differential implications of their technical choices, for example choices for paved and HBB roads, which produce economies of scale in transportation as well as potential efficiency in terms of speed. Such roads (as part of the feeder road programme) drive *tonga* drivers and rickshaw-pullers out of business to be replaced by scooter taxis, mini-buses, buses and trucks. Of course some employment is created in this process at various levels of manufacture, assembly, and use of such technologies, but more labour is displaced. The expansion of market activity will also reorganise employment and perhaps create some net additional jobs, but the benefits are usually skewed. The effect of canals and long borrow-pits for transportation and fishing is less socially differential, at least in class if not in gender terms, since the transportation technology entailed remains less lumpy than high-tech road vehicles. Nevertheless, there are always questions to be posed in any project context: for example, which classes are acquiring small boats in these locations and thereby seizing freight and passenger opportunities; who is erecting the fishing nets or leasing the pits for mass catches?

Environmental and Social Linkages

This chapter has analysed various issues concerning the relationship between social relations and rural infrastructural interventions through an examination of IRWP and other programmes in the road building, construction and minor irrigation sectors. As noted at several points, the discussion here predates more recent discussion upon aspects of the FAP, but reveals the extent to which some of these issues had been highlighted prior to FAP, yet overlooked. Recent discussions with some of the social scientists engaged in the various FAP studies reinforces my original suspicions that much of the commissioned social analysis in these studies has been window-dressing with quite systematic filtering out of inconvenient analysis as the higher levels of decision taking are reached. Related closely to the social development agenda is the environmental one, and this was also clear from the earlier rural works interventions forming the basis of the analysis in this chapter.

As implied several times now, engineering decisions are interventions which are disturbances, with backward and forward linkages, stimulating or retarding productive behaviour, but always with social distribution and environmental consequences. Summary examples of such linkages are:

- Roads, bricks, kilns, mud/clay, land lowering and change of elevation, top-soil composition change, crop bearing character changed for better or worse depending on prior elevation in relation to neighbouring land, prior cropping pattern, status of top-soil sale—distress or commercial?
- Roads, bricks, kilns, fuel/wood, wood markets, transport (river or truck), deforestation (perhaps in another region inhabited by an ethnic minority), therefore ethnic/regional as well as ecological linkages.
- Embankments may enhance the value of some land but waterlog other land as a result; drainage similarly may dry out some land too soon in the season and affect crops; it may also increase downflow and cause flooding elsewhere (do engineers point out the implications of completing a project in one locality upon another?)
- Roads and embankments produce long borrow-pits (in effect canals) which create fishing and irrigation possibilities, but some land surface has been lost (who has lost it; at what price?), at least during the flood season, but perhaps there is moisture retention for other dry seasons, which spreads a farmer's options.
- Groundwater irrigation, choices of technology, effects on aquifers, differential access to different technologies, therefore affected differently by draining of acquifers. Probable annual replenishment in most parts of Bangladesh which may offset long term deterioration, but little help if the cheaper STW goes dry each year. Choices of technology have implications for choices of crop. A policy option could be for more dry

land crop investment instead of heavy re-arrangement of ecology through groundwater irrigation. But sometimes there is no choice but to use groundwater, which leads to choices of groundwater technology by environment and by class.

- Decisions about irrigation and water use have very widespread linkages on cropping, food processing, other processing, centralising processing technologies, moving production away from the family into larger more commercial units which can increase small farm vulnerability. Of course, opportunities also exist, as the agricultural system proliferates, for dispossessed classes to capture other dimensions.
- Mechanised irrigation (and other forms of mechanisation encouraged by roads) entail energy dependency and foreign exchange costs, repair, maintenance, spare parts representing problems but also opportunities.
- Roads encourage use of other inputs into agriculture (fertiliser, chemicals) therefore dependencies, market integration, ecological transformation, pollutants.
- Roads increase intensity of rural-urban exchange, usually unequal, penetration of manufactured goods with impact upon local crafts, but also market opportunities for local produce, but on whose terms? This is not just an urban bias thesis, since certainly in Bangladesh alliances exist between rural/urban classes exchanging land for education and commercial access.
- Choices in location of roads, grade of materials, width, height of embankment, use of bridges all affect the kind of transport, carrier and freight, therefore affects the distribution of opportunities between bus/truck, rickshaws, carts, walking, and boats for obtaining value-added benefits from transportation. Local population therefore being reorganised in terms of opportunities.
- On the other hand, as noted before, roads have to be maintained, therefore long-term maintenance contacts, agro-forestry and borrow-pits, but how to tax class of beneficiaries? If tolls are impossible, therefore tax on freight but regressive as this increases prices especially to net rural purchasers as well as urban consumers, so costs are passed on to consumers and the poor either share that cost, or their demand is elastic, leading to degraded exit, and production incentives lost. Perhaps the solution would be to tax luxury goods only.
- Roads sometimes compete with rivers as means of transport in Bangladesh. Even with river transport, choices exist between power sources (country boats or modern ferries).

- Interference with river flows (embankments, Farraka, various FAP proposals) can affect navigation especially in the dry season, create waterlogging with flood waters trapped behind embankments, denial of necessary top-soil nutrients to agricultural land.

These 13 examples have not been elaborated, but left here virtually in note form. For most readers, these environmental and social linkages are familiar. They are listed here as they do reveal the enormous complexity of apparently innocent engineering choices. It is almost a cliche to observe that the engineer has to live in the society, but it is worth repeating here because so many of the influential engineering decisions being made about Bangladesh emanate from foreign initiated interests and personnel. Such interests may reflect the needs of their governments to show some 'concrete' results of their aid to the country. They certainly reflect the desire by international consultancy firms to prove their relevance to aid-sponsors by devising grand plans which also create further overseas consulting and contracting opportunities. The myriad of studies and designs carried out for FAP also entail a growing number of Bangladeshi counterparts who are enticed to share the expatriate agenda of technical fixes and the rewards which accompany their implementation. This process contributes to a fundamental lack of responsibility to the social and environmental concerns, noted above. Quick, obvious and short-term solutions (bounded by lengths of tour duty or project cycles) are the order of the day. Alas, the FAP has also increasingly poisoned the social science establishment (both expatriate and local) as demands for its tokenist services have been tacked onto the engineering investment interests. This has undermined the credibility of development social science in the country to speak for the social development agenda, and within that to speak for the poor in the arenas where the poor cannot easily be heard. This explains why some NGOs are playing a crucial role in the mobilisation of poor people's interests against certain FAP decisions, but more broadly why some NGOs feel increasingly obliged to substitute for the formal social science establishment by setting up their own policy analysis capacity, collaborating with those remaining academics not yet seduced and coopted.

Chapter 13

Plunder without Danger: Avoiding Responsibility in Rural Works Administration in Bangladesh

Introduction

As indicated in the preamble to Part III of this volume, the discussion in this chapter is based on sporadic fieldwork as a consultant to the Intensive Rural Works Programme (IRWP) and its successor the Rural Employment Sector Programme (RESP) and as a participant in the evaluation of them. The fieldwork has consisted of: visits to project sites; interviews and discussion with siteworkers, field officers, government and aid officials, staff of NGOs which were participating in the programme; and extensive reading of the documentation provided by the programme's own monitoring activity.

The Problem

In a formal sense, an administrative system must meet two requirements which constitute a measure of its competence. It must deliver the goods actually ordered by the political system, and it must do so in a cost effective way. The problem of incompetence then takes two forms. Monopolistic positions can allow administrators to expand costs to the outer limits of any potential budgetary allocation without improving services. Secondly, such positions enable them to alter the way in which the programme is implemented, and thus fail to achieve the goals actually intended. However, this formulation of the problem relies heavily upon a sequential image of the policy process, consisting of discrete stages of goal setting, appraisal, implementation, evaluation, and so on, in which different sets of actors are involved. This is rarely the case anywhere, though the image sustains an important illusion of rationality (Wood 1986a). In countries like Bangladesh, a particular combination of circumstances complicates the picture still further. The nature of the regime, with weak structures of public accountability, strengthens internally the monopolistic position of the bureaucracy as a whole, and senior officials within it. At the same time, the maintenance of such monopoly depends crucially upon its dominant ability to manipulate external aid flows which represent 80 per cent of the country's development budget. In this process, the regime's own interests can be challenged by the sets of ideologies which underpin such aid: whether the privatisation initiatives endorsed by the World Bank and US AID (the major sources of aid to Bangladesh) or the 'target-group' approaches of the Scandinavian and other small bilateral donors in which the 'landless' and 'women' feature prominently. Bangladesh, then, represents an extreme case of a situation found quite commonly in Third World countries, when external aid agencies sponsor projects designed to meet the political criteria and standards of bureaucratic provision determined in their own countries, rather than those of the host countries.

This chapter will explore a situation in which 'target-group' approaches are attempted through extant local government institutions which continue to function within national and local patron-client networks, and which ensure that it is the needs of the local political and economic elite which come to be maximised rather than those of the 'target-group' originally intended by those who initiated the programme (see also Wood 1983 and 1985b). In this process, administrative structures reveal their real rather than their formal content, that is not simply as instruments of the domestic political system or external agency, but as reflections of local political and economic power structures. The interests which emerge in these structures derive from the opportunities offered by administrative office in a society where other avenues of accumulation are limited. Where these interests are unsympathetic to the goals of the external agencies (whether privatisation or 'target-group'), a large gap will emerge between objectives and achievement.

This gap can be explained, therefore, by understanding the administration of the programme as a set of opportunities for state officials to service political and economic objectives in a manner which reorients activities to serve local rather than externally set goals. In this way, we can begin to develop a more complex view of the sources of administrative competence than one which confines itself to the inner structure of the bureaucratic system alone, and which therefore approaches the problem of reform in the same way. By looking at administration in this way, we can then attempt to identify the real constraints which affect the ability of well-intentioned liberal donors to deliver benefits to politically and economically disadvantaged groups, and thus enable them to set more realistic objectives and to pay more attention to the institutional framework through which they make their interventions.

In this chapter I will look in particular at a public works programme in rural Bangladesh. Among the many ways of describing Bangladesh, the salient features for the purpose of this analysis are: a deltaic region with regular flooding, wandering rivers, rich alluvium siltation; particular areas which are prone to extreme flooding; minor variations in the elevation of plots within the same locality which are significant in determining farming practices. The major social transformation of the last 30 years has consisted of a rapid rise in the proportion of rural landless to the point where they now constitute approximately 50 percent of the rural population; the formation of indigenous NGOs working with the rural poor in consciousness-raising and income-generation programmes has occurred over the last decade; approximately 80 per cent of the development budget is funded out of foreign aid; and there is evidence of considerable bureaucratic corruption and misappropriation stimulated by a virtual government monopoly over the use of external funds.

In 1981 Swedish, Norwegian and Danish aid was used to set up an Intensive Rural Works Programme (IRWP) in Bangladesh which had explicit long-term target-group objectives for the rural poor. This involved the building of rural infrastructure (roads, embankments, canals, bridges, drains, sluices and culverts) to improve rural productivity and economic security, using labour intensive methods in which wages would represent 70 percent of total investment cost and in which women from landless and marginal peasant households would be explicitly included. The programme also wished to extend rights to these workers beyond short-term wage employment through access to related production activities and, to ensure this, to involve site workers in the planning and implementation of the schemes.

The Institutional Framework

The programme was administered through the Ministry of Local Government and its Works Programme wing, which was later renamed the Local Government Engineering Bureau (LGEB). At the local level this coincided with a major local government reform in which more authority and resources were decentralised to *upazilas* or local government sub-districts. These structures involved *Upazila Nirbahi* Officers (UNOs: governmental officers deployed by the Ministry) and elected chairmen. These have responsibility for the *Upazila* Planning Committees and for nominating the members of the Project Committees set up to manage the IRWP schemes.

Within these structures, engineers exerted considerable influence at all levels, and tended to justify decisions involving 'hard' expertise on physical aspects of the programme on purely technical grounds. Yet the exercise of these technical judgements invariably involved choices which had significant social and distributional consequences by making available new land, enhancing the quality of old land, altering the pattern of labour demand and increasing access to internal and external markets. These choices represented sets of opportunities, within authoritative local government structures, where political alliances could be struck or sustained between local state officials and rural patrons, in which they were able to reproduce and extend their respective interests, even within the context of a technical engineering programme designed by well-intentioned outsiders rather than with the rural poor.

But with the employment generation objectives embedded in the original programme, we cannot explain the outcome solely with reference to the public structures involved, since these had to interact with private sector networks in order to produce the labour required for the schemes. Labour was recruited through *sardar* and patronage networks which were connected to the local administration; workers were dependent for credit and other needs on the local patrons who were also managing the rural works schemes; in the rural underemployment conditions of Bangladesh, workers are also divided among themselves, competing for scarce employment opportunities and thus vulnerable to wage exploitation. Under such conditions, they are unlikely to pursue their official rights in the programme with any degree of militancy. Different labour gangs and individuals only participated sporadically in sitework, thereby undermining any prospect of unity. Thus the members of the 'target-group', originally chosen as the prime beneficiary of the project, were economically and socially dependent on the very classes of rural patrons who exerted direct control over the recruitment of labour. Yet mobilisation of labour was a central feature of the programme, designed to serve the interests of rural landless and near-landless workers, rather than those of the rich and powerful elements in the community.

The Results

It is therefore not surprising that in the event, the redistributive objectives were not actually achieved. The eventual wage component as a proportion of the investment budget was between 16% and 27%, assuming one million man-days of work and wage-rates of Tk. 15-25 per day. The investment budget represented only 58.8% of the total budget, so that no more than 10-15% of total project funds actually went to site workers in the form of wages. Further, given the seasonal pattern of labour demand in the region, it was essential that the labour demand be created at the right time in order to reduce seasonal fluctuation and the reliance of workers on credit from local moneylenders and employers. Yet in the event, most of the schemes were started too late, so that the additional employment did not maximise the wider effects upon the structural position of the workers in the local political economy.

The remainder was absorbed by contractors, staff costs, equipment, training, expatriate technical assistance and misappropriation. In 1984/85, for example, the cost of each workday created was Tk. 160, approximately ten times the average daily wage for sitework. It is also clear that these high overheads did not have the effect of guaranteeing that the longer term objective of distributing control over assets to the poor would be met. In addition, the programme's own monitoring exercise established that labour records (an important index of power) were not being properly kept, and as a result workers were being systematically underpaid. Further, those schemes intended to provide women with special opportunities involved very substantial use of male labour, inadequate provision of special amenities included in scheme guidelines, falsification of Muster Rolls and cheating on payments. And these 'irregularities' occurred despite an average of 15 visits per season by field officers, supported by expatriate counterparts, to monitor 'socio-economic' performance.

Implementation of the original flood-control, drainage and irrigation canal objectives of the programme has been slow, along with the realisation of linkages to 'target-group' interests (including those of small farmers). In principle, opportunities exist from the design stage of these schemes for the landless to be involved directly in irrigation, maintenance, embankment cultivation (or social forestry) and in self contracting for the earthworks. But here a struggle has clearly been taking place even over the more limited objective of identifying areas of small farmer concentration for flood protection through embankments. On the surface, this takes the form of a struggle over the interpretation of data, but this masks underlying political conflicts. Here a group of socio-economists is responsible for conducting screening surveys to identify the areas predominantly occupied by small farmers where embankment/canal construction should take place. But they have been placed under pressure to 'reconsider' their findings in areas where senior engineer/officials have committed themselves technically to another scheme, often in conjunction with the interests of local political actors.

This initiative to skew the benefits of rural works construction towards the 'target-group' of small farmers threatens both the interests of rich farmers in the area as well as the monopoly of control over earthworks projects by the cadre of engineering officials at local and HQ level. These officials insist that priority should be given to technical rather than socio-economic feasibility studies, and that their technical decisions should not be constrained by any previously established parameters of social objectives. But of course, any particular alignment of embankments gives the adjacent owners a rent-free enhancement of the long term productivity of their land, while the land lost to embankments is likely to represent a higher proportion of the fixed productive capital of small farmers than of large ones. In the last analysis, it is the *Upazila* Planning Committees which have the final say over scheme design and alignment, and small farmers have far less influence here than rich ones.

The result has been a process of 'negotiation' on these committees to reduce the random nature of plot distribution relative to the embankments, so that more of the small farmers' land and less of the richer farmers' land is deliberately adjacent to the final alignment. This process stimulates a flow of bribes to engineering officials and their 'brokers' among the community. Thus here again the intrusion of a 'market' into a supposedly bureaucratic (in the sense of the 'objective' use of unambiguous data) and technical allocation based on redistributive criteria has marginalised the interests of the poorer farming households.

Further, the excavation and re-excavation of fishponds associated with the construction programme represents an opportunity for the acquisition of long-term assets by the landless which the IRWP programme did wish to ensure. Yet in this and other programmes in Bangladesh, these opportunities have not been realised because of manipulation by local patrons with the connivance of local officials. Funds were used for excavation without honouring the accompanying provision that the earthworkers have long term leasing rights to the ponds. In the IRWP, of 27 re-excavated fishponds in one of the programme's districts, at best three could be superficially described as managed by the 'target-group'. Yet it was clear that even these had been infiltrated and were dominated by the non-landless. This experience is similar to the attempts by the Bangladesh Rural Development Board (BRDB) to mobilise landless groups around fishponds.

Problems of Administrative Responsibility and Accountability

The IRWP was formulated on the assumption that a given administrative structure inevitably tied into national and local political networks could be turned into a redistributive instrument simply through the external establishment of a set of allocative criteria favourable to the poor and landless, reinforced by a small cadre of expatriate field staff with monitoring functions. This belief stems from what has been a very general set of assumptions about the utility of state as opposed to private institutions as benign vehicles for economic intervention, reinforced by a formalistic understanding of bureaucratic practices and outcomes. The limitations of these assumptions have been explored elsewhere. Yet the experience outlined above demonstrates very clearly that these assumptions cannot be taken for granted, and that the nature of both the administrative structure and its political context will have a decisive impact upon the extent to which the original intentions come to be translated into reality.

The objectives of the IRWP involved using infrastructural investment to create short and long term economic opportunities for the rural poor. Yet such objectives could not be achieved without challenging the local structures of power embedded in the control over land and commerce, while the conditions for such a challenge were very unfavourable. Substantial external aid flows were provided through monopolistic bureaucracies with state patronage decentralised to the *upazila* level. Here it interacted with local class relations where the low level of capital formation encourages misappropriation, thus creating an extremely hostile environment within which to work alongside the rural poor.

In this context, it is hardly surprising that the senior officials and engineers in the Ministry were mainly concerned with achieving the physical objectives of the programme. They had a clear interest in institutional expansion, the accompanying development of patronage networks through the recruitment of staff, and the possibility of misappropriation at the 'going rate' from aid-sourced programmes which, as we have seen, meant that the intended beneficiaries received only a negligible share of the assets and incomes, while an excessively large proportion went into 'overheads' rather than actual construction. These transfers and distortions have been made possible through the retreat into a bounded sphere of professional competence in which engineers have exercised a virtual monopoly over certain decisions on technical grounds, and, in the process of asserting the primacy of technique, have attempted to disguise the political implications of the resulting decisions. Indeed, the whole notion of a Planning Committee at the *upazila* level contributes towards an image of rational authoritativeness to throw a shadow across reality (Wood 1985c).

But apparently 'technical' choices concerning location, design and materials are nearly always directly political because they rarely have equal or neutral effects upon those who gain and those who lose both during and after construction. Thus accountability based upon a narrow definition of technical and engineering criteria and sectoral responsibility ceases to be credible. As a defensive ideology, that definition breaks down when the linkages to other sets of relationships can be so strongly observed. In the real world, the failure of the social scientist and the engineer, both as aid and government officials, to communicate their interests (well-intentioned or not) and their institutional constraints (assuming good intentions) to one another has resulted in the exclusion of the landless and marginal farmer households from many potential benefits, alongside proliferation of opportunities for private accumulation by an alliance of rich farmers, contractors, dealers and engineering officials in government service. Friends in senior positions in the Indian Administration Service have always referred to their own Public Works Department (PWD) almost affectionately as 'Plunder Without Danger'—clearly the failure of responsibility in Bangladesh has produced a similar result.

Consequences and Implications

In soon became evident to the donors that the programme was not achieving its intended objectives. However, this has not led to cancellation but instead to the redefinition of objectives. Certain externalities assisted this process. First, it became government policy to shift the greater part of the most labour intensive of the rural earthworks to the Food for Work Programme (FFWP) which is the main vehicle for administering food aid in the countryside and, despite large-scale misappropriation, provides some employment relief in the lean months of agricultural employment. But it makes no attempts to mobilise siteworkers around their wage rights or to offer longer-term control over the assets created by their own labour. In no sense is it a programme seeking structural change. Secondly, the government has also shifted the administration of road maintenance, which employs siteworkers in localised small-scale patching, to CARE, the US relief organisation which often undertakes contracts for US AID supported programmes. US AID is supporting District road construction in Bangladesh, and CARE has experience of rural works through its management of the FFWP.

At the same time, there had been constant pressure from the government throughout the programme to expand operations into other areas defined in the original proposal as flood-prone. This may be interpreted either as a genuine desire to spread the effects of infrastructural investment, or as a further set of opportunities to service its patronage networks at the local level. Whatever the interpretation, some expansion did occur, thereby weakening any capacity within the programme to supervise both the technical engineering aspects of the work, and any attempts to achieve 'target-group' objectives through the vehicle of rural works.

Both these sets of conditions assisted the process of redefining objectives by default, as it were. However, this was more formally achieved during the period of formal evaluation, which took place at the end of 1985. Prior to the evaluation, the sense of failure among some of the committed expatriates had already prompted an exercise to search on the one hand for institutional alternatives involving NGO support for siteworkers, and on the other for broader strategies of employment generation not confined to the construction of earthworks infrastructure, and not, therefore, confined institutionally to the LGEB and the local government/*upazila* structures. A draft Phase 2 project proposal therefore already existed before the evaluation. The severe criticisms of the evaluation mission unwittingly lent support to two outcomes not intended by this search, amounting simultaneously to redefinition and legitimation. The IRWP was renamed the Rural Employment Sector Programme (RESP) with two components: the Infrastructure Development Programme (IDP) and the Production and Employment Programme (PEP). This is the current programme. The IDP represents about 90% of programme funds, supporting infrastructural schemes (feeder roads, small-scale water schemes), training and institution-building. Although, of course, it employs labour, the primary justification for this component is now the general value of infrastructural development to increasing productivity, security of land and transportation. The idea of rural works as a vehicle for wider and more long term benefits to the 'target-group' has been lost. The level of investment has been retained through a process of redefining the objectives. But the donors can satisfy erstwhile liberal consciences and radical lobbies with PEP, which concentrates on the 'progressive' parts of the earlier search: more broadly conceived employment generation ideas (not just earthworks), innovation with NGOs and groups of people 'mobilised' by the government's Rural Poor Programme (RPP) in the BRDB; credit support and technical advice networks. There are strengths and weaknesses with these strategies, but this is not the occasion to discuss them. The point, here, is that PEP represents approximately 10% of the new programme funding.

If we now ask whether any other outcome was possible, we first have to recognise the fact that it is utopian to expect either enlightened donors or individual Bangladeshi officials and engineers to bear the responsibility for delivering the intended rights to employment (not just the work itself, but also conditions and wage rates) and new assets to the poor. We have seen that where a programme is to be implemented through existing governmental structures which uses locally selected Project Committees, this effectively precludes any enhancement of workers' rights. In these circumstances, compliance with the externally set goals can only be enforced through the efforts of the workers themselves. This means that success actually requires the mobilisation of claimants who bear all the risks of militancy in seeking fair wages and conditions since they also have to maintain their relationship with employers on whom they depend for survival at other times of the year. Thus the burden of establishing these rights (and of ensuring that the programme's objectives are met) has been shifted onto those with the least power and the most to lose.

This problem can only be modified within the context of a centrally administered scheme where there are actual innovations in the mode of construction itself. In the existing schemes, for example, the emphasis on one-off, single season projects produced a lack of continuity in employment opportunities across successive seasons which was particularly disadvantageous for women, who were not so free culturally to travel to schemes in other localities to work in an area of strange men. It is difficult to mobilise and risky to be mobilised on such a short term basis. Employment practices therefore tend to change only when a new technological activity has been introduced, such as pipe-casting for culverts, or where the traditional structure such as an embankment or an embanked road has been completed and longterm maintenance is required.

This, therefore, suggests that any serious attempt to alter an existing system of distribution through an economic programme must look very closely not only at the administrative and political structures likely to be involved, but at the implications of the choice of technologies available to achieve the given physical objective. In the case examined here this was not done because a whole series of invalid assumptions were made about the political, economic and social context within which they were operating.

Conclusions: The Alternatives

Using a public sector strategy (LGEB, *Upazila* Planning Committees, engineering officials and nominated Project Committees), controlled by social forces opposed to the genuine advancement of the 'target-groups' identified by the donors, was therefore an inappropriate means of achieving their objectives. This option clearly presupposes the existence of a regime sympathetic to these goals, and a state apparatus organised in such a way as to be able to deliver the services involved, something which manifestly did not exist in Bangladesh or in many other similar countries. Simply attempting to retrain and reorient the perspectives of officials, to give more authority to those responsible for ensuring that the social as opposed to the technical objectives will be met, will not resolve the problem of goal displacement in such a hostile environment. To increase the number of outsiders with supervisory/monitoring roles would only create more political and institutional tensions, and reduce the proportion of income spent on the poor, given the immense expense of maintaining expatriate technicians in the field. Hence it is clear that the problem of administrative reform in such contexts requires the ability to use more appropriate agencies to perform the functions required.

1. Privatisation

The most obvious alternative to the public sector would have been to use local private contractors more extensively in earthworks schemes, in addition to their use in *pucca* schemes such as bridge and culvert construction. In principle, the earthwork schemes could be put out to tender and adequate supervision established to ensure that technical standards were maintained. In this way many of the overheads and leakages, which have occurred in the actual project, might have been avoided. Would it be possible under such arrangements for more embankments, roads, and canals to be constructed with a much larger share of total expenditure being paid out as wages? This option, of course, presupposes that local contractors with the requisite skills and capital could be found, and that the donors could organise an effective tendering system to enforce minimum cost standards without a dangerous loss of quality. It is by no means certain that this could have been done in this part of Bangladesh, since few firms existed with the requisite technical capacity and all of them were directly tied into the political patronage networks. The end result might have been expensive and incompetent, with the work given to the local contractor willing to pay the largest bribe to the local authority with the task of administering the local tendering system. The experience with local contractors on *pucca* schemes confirms this, certainly with respect to quality control where constructions have collapsed after cheating on estimates for materials. The opportunities to 'fund' local political alliances and patronage networks do not disappear with this option.

Thus the privatisation option presupposes the possibility of establishing an effective local agency capable of ensuring that the private agencies meet minimum standards, and this cannot simply be taken for granted. It is only the problems with the public sector strategy which compel the exploration of such private sector options. But transforming the state's involvement from a management to a regulatory role does not overcome all the problems of administrative interest and competence noted above. This also presupposes, of course, that the state itself would have been willing to give up the patronage involved in public sector provision and to accept the alternative method. This, of course, raises questions about the degree of leverage available to the donors and the legitimate use of it. Given the heavy dependence on aid spending to sustain political support, one may assume that the regime would have been loth to reject a project out of hand because it was operating through the private sector, especially with the pressure from major donors like US AID and the World Bank in conjunction with commercial lobbies inside Bangladesh, to commercialise other sectors, such as agriculture. Under such conditions, small agencies such as the Nordic donors only have leverage if they make common cause with the larger donors. The dilemma for Nordic donors is that they have been operating in an environment which, partly because of these pressures, has become increasingly hostile to the pursuit of responsible public sector management or regulation, while being concerned with the policy areas of equity and redistribution where private sector solutions will only institutionalise still further the power of local patronage networks over the poor. Privatisation would, in effect, involve surrendering the attempt to 'target' a specific stratum of workers (and thus to provide especially for women) or to enforce regulations as to trade union rights, conditions of work, and the payment of fair wages.

In the event, of course, the attempt to impose such conditions in an environment characterised by a large labour surplus and extreme social dependence was bound to fail, even when administered by a state agency supposedly directly responsible to the donor through the government. Further, the cost of attempting to monitor these conditions only reduced the level of employment—the cost of keeping one expatriate in the field would have been sufficient to add many dozen local workers to the payroll at the going wage rate. A substantial increase in the demand for labour for public works in itself could have been expected to push up the local wage rate and thus benefit not just the workers actually incorporated in the scheme, so that the negative consequences of the loss of direct control could be offset, to some extent, by the benefits incurred through the increase in overall employment.

2. Local Non-Governmental Organisations

Given the concern with redistribution and 'empowering' the poor, a more promising alternative has involved the attempt to provide resources to local NGOs supporting the groups which the programme wishes to help. In the last decade, a large number of indigenous NGOs have been set up, working with the rural poor in consciousness-raising and income-generating programmes. Organised through institutions like BRAC and Proshika, local groups have been involved in self-contracting with irrigation schemes and have been able to meet necessary technical standards, to meet their credit obligations on time, pay salaries to group members working on the schemes, and make net profits (see Wood 1984a, and Wood & Palmer-Jones 1991). Since 1984, the IRWP and now RESP have worked through NGOs not just on the issue of claiming employment rights, but also on the creation of self-contracting labour gangs, as an alternative to Project Committees and private contractors.

Thus, while operating through the existing private contracting network can only serve to reinforce the existing power structure, using such groups can serve to develop higher degrees of competence among the poor and dispossessed. Here, resources go direct to the 'target-group' and leakages are minimised. The problems involved in using this method then arise out of the scale and spread of the groups actually in existence. It is easier to envisage poor groups, supported by NGOs, being more capable of sinking and managing local tube wells than building large and technically complex earthworks in a difficult physical environment. This probably means that the NGO option (the credentials and competence of such NGOs also have to be carefully monitored, since they, too, can be used as fronts by the rich and powerful) must be confined initially to relatively simple and small scale projects. It also suggests that considerable efforts should be put into finding effective means of getting resources to them and providing them with the organisational and technical competence required to ensure that they can do the job without damaging their autonomy.

In this process, these organisations contribute not just to the technical but also to the political competence of the poor. With resources behind them, they will begin to be able to challenge the political monopoly of the existing patrons, and be in a position to exercise more effective surveillance over the performance of the public sector agencies as well. At present, as we have seen, the inefficiency and corruption of the political and administrative elite is a function of their monopoly of information and control over all of the developed networks in the society. Only by enabling alternative sources of social organisation to develop and create an effectively articulated 'civil society' can this control be challenged and a genuinely responsible administration created. Paradoxically, therefore, we can see that it is only by creating effective autonomous structures outside the state that the autonomy and honesty of the state itself can be guaranteed. However, a strong note of caution is required. NGOs should in no sense be regarded as a panacea for this process. There are contradictions, in which NGOs can become incorporated as mere delivery arms of the 'soft' parts of government and donor programmes. As their own support services to the poor expand and become more complex, so their organisations may become more managerial and routine in style, and less inspirational. The political competence of the poor as a source of accountability for state practice will only arise through the long haul of poor groups creating wider federations and unions to represent their interests as an organised class.

PART IV

FISH AND POVERTY

Preamble

As indicated at some length in the introduction to the volume, from 1988 I started to work with a number of colleagues in the fish sector in Bangladesh, initially with fishpond culture but also later making an small input into open water capture fisheries thinking. Both of these exercises were in association with ODA-funded projects, and the Dinajpur (NW) fish culture project was in collaboration with GOB Directorate of Fisheries (DOF). The lengthy Chapter 14 which follows constitutes an overview of the many production, technological and extension issues which had to be confronted when the original technical orientation of the project had to be modified to address a poverty-focused social development agenda. Since the work in this chapter includes the contribution of a small team of colleagues, my role should be explained in order to justify its inclusion in this volume. In the introduction, I have described the invitation from the UK-ODA to participate in the project as a consultant social development adviser. This invitation derived partly from the difficulty of having such a Technical Cooperation Officer (TCO) assigned to the project alongside the 3 person ODA team dealing with site construction of the Fish Seed Multiplication Farm (FSMF) and the extension strategy to find a market for its production of fingerlings.

My involvement led quickly to a shared critique of the project idea with the ODA extension TCO and some of his Bangladeshi colleagues in the project. Together, we identified both our ignorance of various elements in the social and institutional context of the project and its locality, and we designed a programme of studies which could lead to a revision of the project design to meet a poverty focus more closely. The

ODA and the DOF agreed to this programme of studies, and the ODA funded appropriately. I brought a team of researchers together from CDS at Bath and from the Bangladesh Centre for Advanced Studies (BCAS) in Dhaka, together with a freelance person. These studies were carried out as per timetable during 1989-early 1990, supervised by regular visits by myself, especially during drafting of reports. In April 1990, I summarised all these studies and together with the ODA Extension TCO (R.Gregory) addressed a number of findings and issues in a composite report ('Off the Page and into the Pond') which formed the basis of a project policy workshop between DOF, its parent ministry, project staff and ODA personnel. Since then, there have been many demands for copies of this report as it was not on general release. An adapted version of it therefore appears here to represent my poverty-focused inputs into the fish sector. The work of colleagues has been attributed in the text where their contributions have been summarised, though I am responsible for any points of emphasis and interpretation. Two of the studies on trading (carried out by D.J. Lewis, supported by Gregory and myself) have been re-written to appear as a separate book, jointly authored by the three of us. We have also published a small article on some of the extension ideas.

Chapter 15 is simpler to explain, and perhaps to read! On one of my visits to the fish culture project, I was asked by the ODA to examine the appraisal documents for the Third Fisheries Project (to be funded by the World Bank with UK-ODA and other bilateral support). This project concentrated on the stocking of large freshwater water bodies in order to stimulate the employment of fishermen and add to production in the context of declining wild fish stocks. Again, the original conception of the project was a technical fix without much consideration of the socio-economic issues which would affect the distribution of benefits from a considerable investment. This chapter constitutes my commentary on these issues, and is written as a critique of the appraisal document, thus referring to it quite closely. However, the line of argument stands on its own without requiring the reader to have a copy of the appraisal document at their side.

Chapter 14

Off the Page and into the Pond: Fish Extension Strategies

1. Introduction

The Dinajpur fish culture project at Parbatipur, (for this chapter 'the project') has production and distribution objectives in promoting the development of fish pond culture in N.W. Bangladesh. The project as designed followed a model for stimulating fish culture in the private sector which has been pursued elsewhere in Bangladesh with variable results; namely, the construction of a fish seed multiplication farm, (FSMF) to be operated in the public sector, producing hatchlings and fingerlings for stocking food fish ponds. The project in its early conception was designed almost entirely within this production objective. Indeed, there are probably many involved in the project who see it solely as the construction and running of a public sector FSMF, operating under the authority and management of the DOF. The initial partners in the project were then the DOF and the U.K. ODA as the donor. The ODA has been concerned for several years to improve the poverty focus of its rural development aid in Bangladesh. It has responded to criticism of its involvement in the Neemgachi fish hatchery project by seeking to promote the distribution objectives in this project at Parbatipur. In particular, it is concerned to achieve a poverty focus to the extension, training and technological development activities of the project.

Design Weaknesses: production

However, the project in its original design had several weaknesses. It underestimated the dynamism of the private sector in nursery pond, fingerling rearing activity, thus reducing the justification for concentrating FSMF production primarily on fingerling output but increasing the case for producing and selling hatchlings to service the expanding nursery pond sector. The private sector beat the public sector to the draw!

It was also unaware of the complex trading and transportation system for bringing 'imports' of fingerlings and *dhani* (some wild, some cultured) into the area via Parbatipur station (ironically, because it was situated adjacent to the designated site for the FSMF) and other sub regional stations. Although proportions are difficult to project, on certain assumptions it is estimated that if the Parbatipur FSMF was at full targeted production this coming season it would contribute about 5% of total demand of fingerlings in the project area. This figure would be reduced to around 2% by the time the FSMF is in a position to meet the specified production targets. However, the prognosis that such targets can be met in the public sector must anyway be qualified by the inability of public sector FSMFs to operate efficiently.

The conclusion from this is that the trading networks in fingerlings will continue, that private sector mini hatcheries will be developed (if ecological constraints like ground water iron contamination can be overcome), that private nursery pond operators will be interested purchasers of FSMF hatchlings if they can get access to that production before it is dominated by the bulk buying of wholesalers operating from Parbatipur station. The successes in the private sector (an objective of the original design) will continue to undermine the viability of the FSMF.

Design Weaknesses: distribution

On the distribution side, the weaknesses are the inadequate provision for extension and training, and the absence of a credit line to promote directly private sector activity among the rural poor—men and women. At the same time, the familiar assumption that the rural poor can be involved in fish culture by acting in groups to gain access to *khas* ponds under new GOB policy initiatives has to be qualified by the intense competition for these potentially valuable assets. These problems are more acute in the N.W. than elsewhere, with a more highly differentiated class structure and a greater domination of larger land holders/lords over their tenants and semi bonded labour.

This realisation has prompted personnel advising the project to explore the fish culture system in its entirety from hatching to marketing of food fish. In this way, the project can identify the classes of people involved in the different stages of fish growth, and the entrepreneurial opportunities for income and employment generation among the poor. The proliferation and actors, roles and entry points for profit taking and rent seeking require analysis to understand the distributional implications, to see how the social range of opportunities could be extended and to see how the project might support such opportunities.

Key Role of Fingerling Traders

As this thinking developed in late 1988—early 1989, with my interaction with the extension personnel of the project (Gregory and Golder), the idea emerged from them that the key role of fingerling traders in the fish culture system might provide the opportunity to link a poverty focus to the objectives of stimulating private sector food fish production, and overcome at the same time some of the extension limitations of the project.

Therefore, the Wood and Gregory 'Proposals for a poverty focused extension strategy and supporting socio-economic research inputs' (April 1989) included the proposal to investigate the role of the fingerling trader as an extension agent of the project (and therefore of improved fish culture practices), supplying his customers (the food fish pond holders) with information about stocking and pond management. At the same time it was realised that other dimensions of an 'extension menu' could be pursued; *bhittaheen* nursery pond businesses; women and homestead ponds; the familiar leasing by the poor of *khas* ponds; as well as direct support to private sector fish culture.

Need for Project Partners

It was further realised that the project would not be able to pursue these distributional, poverty focused objectives in fish culture with its own limited extension capacity. Partner organisations working with the rural poor in the region would have to be identified and cooperative arrangements with them established. Such organisations are NGOs and the BRDB-RPP (with the commercial banks a more recent entrant to our thinking).

Natural Resources Information

It was also recognised that the project had much to learn about the natural resource base in the area in terms of spatial distribution of different types of ponds and the characteristics of their ownership, topographical condition and utilisation. Clearly it would not be possible to launch into a comprehensive survey of such an enormous resource base, but some methodology had to be established for building a database in the project on these issues and interacting with compatible data sets amongst potential partners over time.

People's Portfolios of Competing Activities and Credit Needs

We were also aware that any project runs the risk of tunnel vision—seeing all people as potential fish culturists and understanding rates of adoption solely in terms of technical knowledge, resource and credit constraints. However any intervention enters a complex system of opportunity costs and trade offs for individual households. People have portfolios of activity with different weights of risk and return attached to them according to their overall entitlement position. Obviously the agricultural cycle is a major factor in this and some assessment of season labour and credit constraints had to be made. It was also increasingly clear that credit played a central role in many transactions within the extended fish culture system, as well as affecting the portfolio decisions of individual households.

Institutional Capacity as Key Constraint

Accordingly the Wood and Gregory April 1989 report proposed a series of studies to inform the project on these issues, and the Bangladesh Centre for Advanced Studies (BCAS) joined this exercise. The Centre for Development Studies at the University of Bath provided some of the personnel for these studies. TOR were drafted for all these studies and work started in mid 1989. The strategy all along for the studies was that the institutional capacity of potential partner organisations to work with the rural poor was the key limiting factor. An assessment of their quality and spatial distribution could then guide the location of some of the other studies, and provide essential institutional information.

These studies were summarised by me for the document "Off the Page and Into the Pond" which was presented to a workshop in April 1990 consisting of donor and DOF officials. For this chapter, I have selected the following five summaries (in section 2) as reflecting more of my input, or as they relate to the overall arguments about understanding fish culture as a socio-technical system.

> 'Trading the Silver Seed' A study of fishseed
> trading practices in N.W. Bangladesh.

The food fish network; production, harvesting and trading in N.W. Bangladesh

Women participation in fish culture in Greater Dinajpur District; constraints and possibilities.

Q&D data mapping of natural resources relating to fish culture in 4 villages of N.W. Bangladesh.

Credit and the development of fish culture in N.W. Bangladesh.

At the same time the interaction between Wood and the project extension personnel produced a further list of issues which were arising from the studies and the general acquisition of knowledge about fish culture and social dynamics in the region. Three of these have been grouped together as a strategy discussion under the third section of this chapter. They are:

Production Issues

Technology Development Issues

Extension Issues

The fourth section sets out agenda for action research, proposed to the workshop, which stands here as a useful summarising agenda of future work.

2. Summaries of Studies

2.1 Title: *TRADING THE SILVER SEED: A STUDY OF FISH SEED TRADING PRACTICES IN N.W. BANGLADESH.*

Author: Dr D.J. Lewis and Mr R. Gregory.

Date: Dec. 1989

Summary

In many parts of Asia declining natural fish stocks in rural areas, (through overfishing and environmental disturbance) are compelling more farmers to manage resources in ways which increase fish production. Fish culture in ponds normally follows a simple pattern; the fish seed is released into a pond in which conditions are controlled to suit the fish seeds' needs; the fish are fed; and then finally 'harvested' at an optimum size and time. This study is concerned with the first steps in this process—the production, acquisition and transportation of seed. It looks in detail at the elaborate network of fishseed producers, collectors, and transporters/traders serving the N.W of

Bangladesh It represents an interdisciplinary effort involving sociology and aquaculture.

The idea of a fish 'trading network' forms the conceptual basis of the study which allows a spatially diverse set of participants or actors to be described and linked together through their relationships. It can be used to express the movements of fish, people and information around the trading system. The methodology therefore consisted of:

- unstructured interviews with key initial informants and broad category of actor identified
- structured questionnaire interviews of actors, revealing trading practices, socio-economic issues, and fish culture knowledge and its use
- participant observation, travelling with actors to gain knowledge of their choices, risks and skills
- case studies of example actors to reveal a more total context of their behaviour
- regular market sampling of fish seed at Parbatipur station to identify points of origin, species, quantities and prices. (These data were collected in conjunction with BCAS and some of it is reported in the seasonality study)
- scientific analysis of transportation technologies and conditions
- a video recording of 'dominant' model tracing the progress of fish from their source in a Jessore hatchery to their deposit into a food fish pond in Thakurgaon

The study involves the introduction of some new terminology in order to convey the complexity of relationships and transactions. Some are actors' terms, some derive from more specialised technical vocabularies.

Chapter 1 presents the biological and technical information required for an understanding of the practices of fish culture in Bangladesh, covering the collection of wild seed, artificial spawning in hatcheries, conditioning and transportation.

Chapter 2 deals with the geography of the fishseed trading network supplying the N.W. of Bangladesh and presents the 'dominant' model in which fish seed is first either produced in Jessore or collected from the river in Bogra and transported in large quantities (drums or *patils*) by train to Parbatipur station for onward dispersal in the N.W.

Chapter 3 introduces the social relationships which structure the fishseed trading network. Different categories of actors in the network are identified. The network is broken down into a series of eight transaction points, and each of these 'points' is described and analysed with a diagram.

Chapter 4 analyses social and economic themes arising from the network: profit, risk, decision-making choices, network entry, credit and indigenous knowledge.

Chapter 5 focuses on the implications of the trading network for the project. It concludes that the fingerling production capacity of the FSMF is of doubtful value, given the evidence of rising private sector nursery capacity and the efficiency but also credit content of these trading networks which will not easily connect to the fingerling output of the FSMF. Any fingerling production from the FSMF is likely to be dominated by the *arotdars* (wholesalers) of Parbatipur station and reinforce their dominant position. There is no poverty focus by this route. However, the FSMF's hatchling production role will be important for the developing nursery ponds (if credit constraints can be overcome e.g. by selling to NGOs which supply group nursery pond businesses). The FSMF should also have a role in stimulating the development of more efficient, competitive private mini-hatcheries. The extent of success in this way will, over time, force the FSMF to change its functions towards training, extension and broodstock services, and away from production. (See other papers in this workshop collection for further analysis of these issues.) The idea that small traders can be deployed as extension agents to help overcome the project's extension limitations is fully supported.

To expand on some of the issues raised in Chapter 5: it is very clear that hatchery operators in Jessore and elsewhere sell hatchlings or *dhani* to nursery pond operators on credit, and the absence of this provision has been a constraint to public sector performance. The extent of 'tied' relationships between wholesalers of fingerlings and village traders intensifies the difficulties for the FSMF to target its fingerling output away from or independently of such relationships. But in the absence of credit, wholesalers of fingerlings would also buy up the FSMF stock of hatchlings and dominate their trade to nursery ponds (and perhaps dominate other aspects of nursery pond businesses). Steps to target the sale of hatchlings to NGOs and RPP working with groups in the nursery pond sector will therefore be necessary. Once the demand falls even for the FSMF hatchling production due to successful local, private competition, the FSMF may turn its attention to the breeding of other species—especially those species that might be regarded as poor people's fish (such as Chinese carp), or for open water stocking.

The labour intensive network involves an unequal distribution of risk for effort. Village traders and drum splashers have higher risks and more effort for comparatively low returns. Other actors such as *arotdars*, hatchery owners and nursery pond operators have consolidated their positions well.

The expanding network offers a range of direct and indirect entrepreneurial opportunities for enterprising people wishing to enter. However, the emphasis on skills and information as the precondition for network entry indicates that the project's extension and training role is of central importance to raising the level of fish culture practice and assisting poverty-focused 'target groups' to gain entry to the network and generate income.

The FSMF project has been designed without an adequate understanding of the social, economic and technological environment in the area, and in production terms has misplaced emphasis upon fingerling rather than hatchling output, and has encountered severe problems of iron in the groundwater. The project now finds itself in the position of trying to fit its existing specification across a complex set of relationships which have only been identified subsequent to design and which reflect impressive indigenous ingenuity. There are important lessons in project creation for ODA as well as GOB in this. A change of emphasis (signalled in earlier reports by Wood, and Wood and Gregory) is advocated towards training and extension and away from local production, especially fingerlings.

Chapter 5 (5.3) concludes with a set of recommendations for this shift of emphasis:

1. Creating Extension Traders

To service food fish pond operators with information, thereby using trainers' time more efficiently. Traders consider their sales would improve in this way. Traders may be encouraged to enter 'sharecropping' arrangements with some of their more commercial client pondholders, thereby gaining access to some of the incremental production. If supported by credit, traders could overcome pondholders' seasonality constraints by more readily offering credit to them as part of a transaction. Although there are some problematic issues, a pilot extension trader strategy (ETS) is recommended by training a small number of traders and monitoring their success with a sample of pond operators.

2. Stimulating Homestead Pond Production

Maintaining a poverty focus within ETS by preparing traders to target smaller food fish producers who farm homestead plots. There would be fewer status differences in such transactions, although the traders will likely be more reliant on extending credit and will need therefore to be supported.

3. Encouraging Short-term Seed Storage for Small Traders

Small traders to be encouraged to experiment with homestead pond (*happas*) for short term storage of seed, thus increasing the value of hatchlings to *dhani*.

4. Support for Local Nursery Pond Development

Among private, NGO and RPP group operated ponds to cope with increased hatchling availability from FSMF. This would generate more employment and distribute incomes better (as well as facilitate wider distribution of fingerlings to food fish pondholders) than present design plans for the FSMF.

5. Collecting Pituitary Glands

Through NGO groups as an income-generating activity for target groups. At present hormones for breeding are scarce and clearly a constraint to private sector hatchery development.

6. Providing Credit at Key Points in the Network

This will be necessary for targeted 'bulk' purchase of hatchlings (impossible to provide for individual 'unorganised' fingerling traders from the FSMF), for the ETS with NGO and RPP organised traders, and for targeted poor, homestead foodfish pondholders.

7. Experimenting with Widening the Small Trader Role

Although there have been status constraints in maintaining a division of labour between the functions in the network up to fishing and marketing food fish, these constraints may be slowly disappearing, enabling traders to become fishermen and vice-versa, to extend their participation in and access to more of the value-added in expanded, intensified fish culture.

8. Emphasis on Field Research and Demonstration

This is an obvious reference to the style of applied research and training within the project.

2.2 Title: THE FOOD FISH NETWORK: PRODUCTION, HARVESTING AND TRADING IN N.W. BANGLADESH
Author: Dr. D.J. Lewis with Md. Shahid Ali
Date: March 1990

Summary

This study examines the relationships between fishermen, pond operators and market traders. Following from the earlier fry trading study, it examines the processes through which food fish are cultured in ponds, harvested and marketed. The project area is a fish deficit area and locally produced food fish are supplemented by stocks brought by commission agents from other parts of the country (e.g. Barisal, Khulna and Chittagong) and marketed via Parbatipur *bazar*.

A qualitative, anthropological methodology has been used for the study, based on the collection of representative case studies, to capture the complexity of diverse actors and relationships embodied in the foodfish network. These date are supplemented by the Seasonality study in which quantitative market data have been collected on species, prices and volume by month. The network concept has been retained as a means of describing and analysing these production and trading relationships. The data were collected during 6 weeks fieldwork (Feb./March 1990). After identifying the main categories of network actor through consultation with project staff and existing literature, a further classification of fishermen, traders and pond operators was formulated through the use of participant observation and informal and semi-structured interviewing. The research locations included food fish ponds, *bazar*, *haat* and fishing communities in five *Upazilas* around the project area (Parbatipur, Chirirbanda, Khansama, Badarganj and Kaharole). Following from this, case studies were collected from each category and sub-category and a selection of case material forms the basis of this report, which must at this stage be regarded as a 'workshop draft'.

The study argues that there are two types of pond fishing activity: the primary harvest in which large food fish are caught, and the secondary harvest in which small wild species are collected. These two types of fish are destined for different consumers, with the food fish consumed primarily by urban, well-off families and the smaller wild species forming an important component of the diet of the rural poor.

There are three main categories of food fish producer in the project area, reflecting low input village production on the one hand and a growing interest in more commercial food fish production on the other.

These categories are:

1. *Large farmers:* owning a village *pukur*, receiving minimal inputs, fish consumed by immediate and extended family members, important for wider set of economic and social functions (bathing, washing clothes, irrigation, farm animals etc.—see Natural Resource Study). Fish production is more of a 'hobby', and its control contributes to the maintenance of patronage and other social and political relationships.

2. *Small and medium farmers:* with small *digi* ponds, practising a low intensity form of fish culture, selling fish in the market from time to time when they have cash needs.

3: *Commercial Pond Operators:* with entrepreneurial ventures, owned or leased, individual or group, and becoming more common especially in well-connected areas around the main *bazar*. These operators are keen to acquire better fish culture knowledge.

The terms of trade between fishermen and pond-operators are changing in favour of the latter. The previously predominant share system for primary fishermen is rapidly disappearing in favour of purchase by fishermen at 20% less than the market price, and is moving for some commercial pond operators towards a wage-contract arrangement. This removes the fishermen from a share of the incremental output, and forces them to compete with new entrants over the price of their services, driving down their income per pond fished. Therefore, as food fish production becomes more commercial, fishermen may suffer a deterioration in their economic condition, especially with newcomers.

Fishermen sell food fish (cultured fish) to *arotdars* (wholesalers) in the town *bazars*, who sell the fish on to market retail traders (*pikars*) who buy at a uniform price, retail initially in the *bazar* and then sell a deteriorating unsold surplus in village *haats* during the afternoon and evening. Ice is available at town *bazars*. The *arotdars* also purchase the 'imported' species of food fish on a commission basis and therefore dominate the local markets.

Five reasons are suggested for a decline in pond stocks for small wild fish, two of which are linked to expanded, cultured food fish production, raising therefore the possibility that 'poor people's fish' might actually be declining as food fish production intensifies.

The following implications are noted for the project:

1. *Geography of food fish farming* (with the importance of proximity to market outlets in town *bazars*) sets potential spatial limits to the expansion of the commercial pond operator category. Similar spatial constraints (in this case proximity to communication routes) are observed for the expansion of nursery ponds.

2. *Marketed food fish are produced and consumed mainly by and for wealthier groups.* This means that by simply improving the quantity and quality of food fish, the project is unlikely to benefit the bulk of the rural population. The expansion of intensive food fish production may entail a decline in the availability of poor people's fish. However many food fish are marketed before they reach optimum size for market sale (around 1 kg) and are therefore bought by the less well-off. The sale of smaller food fish may result from smaller pond operators' cash requirements or from ponds which

have dried out. There is therefore a case for investing in and targeting attention on smaller pond operators and ponds which dry out. Since *Rui* and *Mrigal* do less well in small ponds, the project should encourage stocking of Chinese carp, *Catla* and *Tilapia* to target consumption opportunities more on the poor. (However, see Women's Study for the objections to *Tilapia*.) It must also be remembered that expansion of the supply of food fish may bring prices down and extend the social range of consumption, but then falling prices could reduce production and re-establish a price equilibrium. It is suggested that the project adds a study of the patterns of consumption and availability of wild fish species to its future research agenda.

3. *Disincentives for ponds to be farmed more intensively remain:* production for use rather than exchange ('hobby'); co-ownership is frequently cited as a constraint (though see Women's Study for a comment on potential for women as co-operators between co-owning families), though it also functions as a risk sharing strategy; and underutilised ponds in fish culture terms may be performing a portfolio of other key functions of much greater significance to the owner (see Credit Study for further examination of this point). Since group leasing is the only access route for the poor to food fish production, the viability of such arrangements need to be tested by the project through a simple piece of action-research on a group lease of a co-owned pond.

4. *The Extension Trader Strategy (ETS)* outlined in previous project reports is supported here from the perspective of the pond operators. Since there are a few fishermen who are also fry traders, it is suggested that a pilot action-research project should explore the potential for linking the two roles further in a 'bundle' of pond culture services, such as selling fry, harvesting, advising, marketing, provided by fishermen traders to pond operators. This would strengthen the economic position of fry traders and fishermen and assist the development of improved pond culture practice in the N.W. Bangladesh.

2.3 Title: WOMEN PARTICIPATION IN FISH CULTURE IN GREATER DINAJPUR DISTRICT: CONSTRAINTS AND POSSIBILITIES

Author: *Ms. Bodil Maal with Shahid Ali*
Date: March 1990

Summary

Since a key objective of the Dinajpur Fish Culture Project is to involve the poorest section of the population in fish culture, the TOR for this study requires a description of the position of women from landless and poor households in fish culture in the region. Special attention was to be paid to women's involvement in the small ponds close to the homestead, and the significance of distinctions between classes of household and communities (Muslim, Hindu and Adhivasi). As the project possesses a limited extension component, the institutional capacity of the government and private organisations are considered, especially the quality of NGOs' and BRDB-RPP support for women in landless and marginal farmer households.

The study was hampered by the unavailability of the NGO/BRDB and Natural Resource Studies (the latter now 'taken over' by the project for data analysis and report writing) which were intended to guide and focus the Women's Study with essential background information. The structure of the work had to be altered to identify geographical areas with clusters of ponds and to explore the institutional capacity of GOB and private organisations as well as to examine women's involvement in fish culture. The study is based on interviews with poor women, and members of 15 NGO and BRDB staff. Six different *Upazilas* were visited: Parbatipur, Chirirbanda, Khansama, Birol, Thakurgaon and Badarganj. Six Hindu and ten Muslim households were interviewed in detail about their small pond management, and nine women groups involved in fish culture through NGOs.

The report essentially consists of five parts (across seven chapters) which are summarised below:

1. The Context for Women's Involvement in Fish Culture

Aspects of the ecological, cultural, religious and socio-economic situation in the project area are presented, with special attention paid to the caste and *purdah* system which constrain women's productive activities.

The family's status and social position is important, and the source of income has often been considered more important than its size. Although trading is a prestigious occupation in Islam, Bangladeshi Muslims with their strong cultural inheritance from the Hindus view trading in a less honourable way, and fishing also has low status. Likewise, a family will resist their women working outside the homestead unless economic necessity really forces them. Poverty increases the number of female-headed households (approx. 25% among landless households) whose women cannot afford to live up to caste and *purdah* standards and thus have greater room for manoeuvre to enter fish culture. However, the study has found that the poorest section of the population often lived in areas with low numbers of ponds and little access to them. Although the *Upazilas* nearest to Parbatipur appear to have the highest concentration of ponds, it is difficult to find small ponds and poor people in the same place. Also in Greater Dinajpur District it seems that in areas where Hindu and Adhivasi women are more concentrated, and more inclined to relax *purdah* restrictions, there are fewer ponds and therefore fish culture opportunities for them. The areas nearer to the FSMF are more culturally traditional, and it will be more difficult to get female extension workers to move around between villages and supervise/advise other women.

2. Individual Women's Involvement in Fish Culture

Three case studies of how women in individual households are engaged in the cultivation of fish in small homestead ponds is presented. These case studies represent geographical spread and cultural variations.

In fish feeding, women use the household products (rice bran and cowdung) whereas men purchase fertiliser and oilcake in the market. It is possible, from some case study evidence, that women of co-owning pond families may cooperate more easily over pond culture than their menfolk. While it is not degrading for men to engage in earthwork on their own ponds, it is so for women. If the males are at home, they buy the fingerlings from the traders, but if alone, women will also buy directly from the traders, except in more traditional areas. Trading occurs during the *aus/braus* rice harvest so that men are frequently absent. (See Seasonality Study for corroboration of this). However, there is concern that traders will cheat the women buyers. In the early trading season, poorer families have no savings to buy fingerlings and need credit or have to buy later in the season when prices are higher. Poor people, remember, do not receive credit from traders. If their ponds dry out, this late purchase of fingerlings means that the fish have less time for growth. Women often save the money to buy fingerlings. All these issues link problems of seasonality, employment, savings, credit, stocking, quality of seed and poor people together in a way that increases risk and problems of access for small-scale stocking. The villagers also prefer to produce the more expensive carp species and are not attracted to *Tilapia* for several reasons, of which the main one is that fish culture is seen as a cash crop so value in the market is paramount. Where fishing occurs for consumption and a surplus is caught, women prefer to distribute among other family members without insisting on immediate exchange. These informal markets involve the exchange of fish for other goods such as vegetables. If the expansion of fish culture results in an increase in informal market transactions for fish, then the nutritional status of village people would directly improve and give women greater control over the product of ponds. This is reinforced by the status objections to selling in the market (which is why many pond operators sell their product to fishermen or traders). Women seem to be quite well-informed about the economies of their household ponds.

3. Women's Group Involvement in Fish Culture

Six NGO women groups involved in fish culture are described, and the special constraints and opportunities which occur when groups of women take up the cultivation of fish.

Some of the issues applying to individual women are repeated in a group context. Some groups (e.g. with Caritas) are multi-class, and are dominated by the richer women members. Women groups have access problems to larger ponds (usually located further from the house) both through *purdah* constraints, but also controlling other uses of the pond (pumping water for irrigation, washing animals and so on). The consistent protection of NGOs is required for any women group's involvement in *khas* ponds. Such ponds also require re-excavation, which poses additional status problems for women over the need for credit or grant. The guarding of such ponds cannot be done for similar reasons, so that cooperation of husbands is necessary. It is more difficult to combine fish culture in large ponds with household duties, so that women of all communities have greater difficulty in fishing such ponds. If mixed groups are formed to overcome such problems, the women are likely to lose control. All these constraints reinforce the strategy of women using smaller ponds nearer their homesteads. In groups of landless women, there are usually widows and destitute, female-headed households. These women are more prepared to cross cultural, class and caste barriers to survive, and therefore will have fewer constraints in earthwork, work inside the pond, buying fertiliser and oilcake from the market and fishing. They have greater room for manoeuvre. Small ponds can represent viable fish culture opportunities for poor women but they need training, technical support and substantial, short-term credit.

4. Institutional Capacity of NGOs and BRDB-RPP to work with women in the Project Area

Some smaller, but potential partners are not reviewed in this summary.

Caritas have 85 groups in fish culture, six of which are women's groups. Two were visited and both turned out to be multi-class groups, leasing a pond from one of their richer members. Although Caritas has a strong fish culture programme, the women are dominated by men and richer households. Out of 75 **RDRS** groups in fish culture, 18 are women's groups. Of these, 12 are only involved in feeding, while six are engaged in the full sequence of activities, but in smaller ponds nearer to homesteads. Later in 1990, RDRS will start to re-excavate 82 ponds in size range of 0.5 to 3.0 acres. The smaller ponds will favour women's involvement. Mini-hatcheries at village level are planned. RDRS at present assist in fingerling transportation and the provision of nets. **Dipshika** have about 150 women's groups in three Unions in Birol and Kaharole Upazilas. It targets the poorest classes. While there is little involvement at present in fish culture, some homestead ponds are used by women for fish production. More important is the potential. In its health programme, Dipshika has trained 18 female grassroots workers, supported by four female fieldworkers. Such workers, or a similar model, can work with women in fish culture. The **Grameen Bank** support for women's groups management of larger, *khas* in Neemgachi is an important experience to monitor for the Parbatipur project. **CARE**'s work with women is more in the health sector, and is less targeted on the poor. The **Gender in Development Forum** brings together 10 smaller NGOs in Greater Dinajpur and Rangpur Districts—this could be an important institutional route for the project's work with NGOs on training and extension. The **RPP**'s work with poor women remains hampered by weak organisational capacity, and over-bureaucratic style, preoccupied mainly with loan requirement issues.

5. Recommendations and Proposals for Action-Research

- extension through the use of fingerling traders
- use of Gender in Development Forum to encourage small NGOs
- extension through female grassroots workers of Dipshika in Birol and Kaharole Upazilas.
- extension through a female entrepreneur in Dangapara village, Khansama Upazila
- cooperation with RPP in Chirirbanda to explore organisational capacity to work with women
- cooperation with the Lamb Hospital, Parbatipur to help landless groups to excavate ponds and cultivate fish in Parbatipur. (About 20 such groups exist)
- technical and monitoring services to RDRS, with initial examination of stocking policy and profit levels
- appointment of a woman co-ordinator by the project to oversee these proposals

2.4 Title: QUICK AND DIRTY DATA MAPPING OF POND RESOURCES IN 4 VILLAGES IN N.W. BANGLADESH

Authors: Mr R. Gregory and Mr S.S. Alam.
Date: April 1990.

Summary

The N.W. of Bangladesh is rich in water resources and fish culture is now starting to gain momentum in some areas as natural fish supplies dwindle. The natural resource (NR) study which is described in this summary was carried out to examine the different types of pond use in areas close to the projects facilities.

A project such as Parbatipur, if its staff are to be involved in meaningful extension and training, must become familiar with the different characteristics of ponds, management situations and integrated use which currently exist in the area. Only then would project staff be in a position to offer sound advice in a variety of different field circumstances encountered. From the large amount of data collected, attempts were made to categorise ponds into those which are suitable for fish culture and those which are not.

The NR study, as well as providing an interesting insight into integrated pond use also sets out guide-lines for the project to consider as it plans its own NR inventory and impact assessment programmes.

The study was *not* designed to accumulate huge amounts of data on the whereabouts and characteristics of ponds throughout the region, (which represents an enormous task). Instead it tested a methodology of collecting 'quick and dirty' (Q&D) qualitative information in the field, with a view to refining a data collection framework through which natural resource data collected over long periods of time, could be accumulated and retrieved by the project or its clients, in a systematic way.

Methodology for natural resource study

The NR study was designed to collect data in two forms:
1. A quick and dirty (Q&D) survey of all ponds within a selected area.
2. Case studies of selected ponds from the Q&D survey.

Criteria for selection

The following criteria were used for the selection of the four villages studied in the *upazilas* of Parbatipur and Chirirbander.
  Village 1. **Nowdapara**, Parbatipur
    Situated close to the project site.

Village 2. **Indrapur**, Parbatipur
    NGO active in fish culture in the locality.
Village 3. **Mohadhani**, Chirirbander
    BRDB active in fish culture in the locality.
Village 4. **Mondalpara**, Chirirbander
    Developed nursery pond resources.

Results

Through analysis of all (221 cases) pooled data from the Q&D data mapping exercise, various trends in pond characteristics and use patterns have been identified. The more significant ones are presented below.

Single variables

1. Virtually all ponds are harvested for fish, be they cultured or wild. Wild fish resources remain more important than cultured fish for the time being. Fingerling culture to date is very localised.
2. There is a predominance of individually owned ponds in the villages studied.
3. The majority of ponds in the area studied are small; less than 20 decimals in size.
4. More than half of the ponds maintain enough water to grow fish all year round.
5. Less than half of the ponds studied flood regularly and over half have complete embankments.

Multi-variables

1. Small ponds are predominantly used for wild fish capture as their primary activity. There appears to be an increasing tendency to stock ponds as pond size increases.
2. Wild fish ponds generally flood and are not completely embanked, whilst food fish and fingerling ponds are normally flood free and have complete embankments. Human bathing and domestic uses are more commonly carried out in ponds which do not flood and have complete embankments.
3. Food fish are raised more frequently in ponds which hold water all year round.

4. There seems to be an increasing frequency of pond co-ownership as size increases. The majority of small ponds of five decimals or smaller are individually owned.
5. On average 70% of co-owned ponds were still used for wild fish harvesting, whilst only 28% were stocked deliberately.

Compatible activities

The following combinations of activities could be said to be compatible with fish production:
  Wild fish harvesting and duck rearing.
  Wild fish harvesting and irrigation.
  Wild fish harvesting and animal washing.
  Wild fish harvesting and jute retting.
  Wild fish harvesting and irrigation.
  Food fish culture and duck rearing.
  Food fish culture and fingerling production.
  Food fish culture and human bathing.
  Food fish culture and domestic use.
  Fingerling culture and duck rearing.
  Fingerling culture and human bathing.
  The following combinations do *not* appear to be compatible:
  Food fish culture and wild fish harvesting.
  Food fish culture and animal bathing.
  Food fish culture and irrigation.
  Fingerling production and wild fish harvesting.
  Fingerling production and animal bathing.
  Fingerling production and irrigation.
  Wild fish harvesting and human bathing.
  Wild fish harvesting and domestic use.

There appear to be strong links between the practices of wild fish harvesting and animal and agricultural activities. However, food fish culture more often takes place in ponds which are used for human bathing and domestic purposes.

Village profiles and case studies

Following analysis of pooled data, village profiles were prepared detailing all the NR data collected for each case. Analysis of the Q&D data enabled the study to select 24 ponds suitable for case study work. The following characteristics of ponds were treated in this manner.

24	ponds varying in size from 4 - 400 decimals
3	ponds which dried up each year.
4	ponds which flooded in 1989.
4	ponds lacking complete embankments.
4	*khas* ponds.
4	co-owned ponds.
3	leased ponds.
1	unused pond
12	ponds which were used for food fish culture.
6	ponds which were used for wild fish harvesting
4	ponds which were used for fingerling culture.
2	ponds which were not used for fish.

The case studies provide an insight into the decision making and perceptions of pond operators which were not reflected in the Q&D data mapping exercise. (These are presented in volume II of the NR study).

Implications for the project and recommendations

The NR study we believe, has implications for the project beyond pointing out some useful correlations in pond use patterns. The research team are of the opinion that the Q&D method of data mapping a village could be adopted by the project as a way of targeting extension and training efforts.

Scoring villages

A system for 'scoring' villages was tried out during analysis of Q&D data from the four NR villages. The 'scores' seemed to endorse the teams' feelings towards the potential for fish culture development in each of the villages studied. The method of weighted scoring is, tentatively, recommended to the project as a tool for extension and training planning.

Baseline data storage and monitoring

The computerised database prepared for the project can, with slight modification, be adopted by the project as a way of storing baseline data in any areas it wishes to work in. The database could also be used to assess the impact that project programmes have on the development of resources in target villages.

Larger scale studies

The team working on the study was surprised at how quickly, unambiguous data could be collected by adopting the Q&D data mapping approach. The project could conceivably conduct its own NR inventory of much wider areas if it so desired.

On average, Q&Ds in each village took three people two days to complete. By slimming down the study to what variables we now know to be significant this time could conceivably be reduced.

The project should cooperate with its NGO and BRDB partners and standardise data storage management techniques. In this way, the project and its partners would be able to store and retrieve data collected by each others officers. Short training in data collection methods would enable NGO and BRDB staff to collect acceptable data for databasing.

2.5 Title: **CREDIT AND THE DEVELOPMENT OF FISH CULTURE IN N.W. BANGLADESH**
Author: J. Allister McGregor with S.S. Alam
Date: March 1990

Summary

The principle which underlies this report is that credit is not a detachable component of development projects. It cannot be simply dealt with in a technical manner in isolation from other aspects of project organisation. Considerations about the viability of a credit intervention are inextricably linked to questions about the organisation of the target population, its competing uses for scarce labour time and capital, and the organisational capacity of the project of which credit is to be a part.

The investigations on which this report is based were carried out in December 1989, involving fieldwork in the project area, visits to fish culture projects elsewhere in the country, and discussions with NGOs and Banks both in the field and at HQ level in Dhaka.

The report consists of five chapters. After the introduction: it situates the importance of considering credit in the rural economy of Bangladesh; reviews the credit policy environment in the country; examines the relationships of fish culture, credit and the rural poor and reviews this experience; and analyses the organisations lending in the fish culture sector. There is a final section with conclusions and recommendations.

It is argued that unless the project can arrange or negotiate relatively untied access to credit for rural poor fish culturalists, both its production and distribution objectives will be jeopardised. Without a coherent credit strategy the project will fail. Since the project has no resources to contribute directly to a credit intervention to promote fish culture, it does mean that the project will have to depend heavily upon and work in close collaboration with some of the organisations which are able to provide credit in the region. There are three categories. **First, it is likely that the project will have to work most closely with NGOs.** In the project area, Caritas already have a well established programme, and there is little need for more than consultation with the project. This directs the project to working most closely with RDRS and some of the more credible smaller NGOs. RDRS are currently reshaping their credit system which provides opportunities to place fish culture within that agenda. **Secondly,** the brief review of BRDB in the report confirms that the RPP remains poorly organised, with inadequate credit provision for fish culture and ineffective training.

For the time being, it is recommended that the project adopts a passive stance towards BRDB with advice and training on demand. Thirdly, it is essential that the project attempts to work with the Banks. Only the largest NGOs have enough funds to be able to run their own self-financed credit schemes, and even these funds are small compared to those in the banking system. **The project must therefore endeavour to establish a line of credit with one or a number of banks, for the promotion of fish culture among the rural poor.** However, with the increasing gap between the banking system and the rural poor, it may well be that the banks will not be prepared to consider a proposal unless the ODA is able for example to provide guarantee funds.

Pursuing the task of improving the banking system's attitude towards fish culture, the project along with the DOF should enter into negotiations with the banking system (from the Bangladesh Bank down to the branch level) to change the organisation of credit provision for fish culture. More flexibility is required to extend lending beyond food fish production to fingerling and hatchling production, and even for trading.

The report argues that since the project cannot itself act as a credit agency, it is forced to adopt a **facilitating role** whereby it will attempt to promote and support a fairly direct and constructive relationship between rural poor fish culturalists and the organisations which might be able to provide them credit. **The key feature of the facilitating role is that it will involve the project in providing support to both sides of the** relationship: i.e. to the lending organisations as well as to poor fish culturalists. This facilitating role can be broken down into four constituent parts: lobbying, training, advisory and monitoring.

Lobbying

  i. The project should lobby credit disbursing agencies on behalf of the rural poor, to change credit provisions and practices so that they are more appropriate to the requirements of the activities.
  ii. Equally, the project must lobby fish culturalists to ensure that they are aware of and respond to the needs of the lending organisations.

Training

  i. As a direct part of its facilitating role, the project should provide training for the staff of credit disbursing organisations, assisting them to become more familiar with fish culture practices. For example, the training of bank workers in the assessment of fish culture loan applications.
  ii. The project should provide training for would-be fish culturalists to ensure that they have an adequate level of skill and knowledge before they take loans to embark on fish culture activities. As part of this, the project should provide 'training of trainers' courses for NGOs and even the RPP.

Advisory

  i. The project should be prepared to advise organisations as to what provisions and practices are appropriate and how to deal with specific problems in relation to fish culture.
  ii. The project should actively provide advice on fish culture practices to the rural poor to ensure that they are able to make profitable use of and repay loans extended to them, paying particular attention to their overall portfolio of competing demands for credit and labour time.

Monitoring

  i. The project should establish systems for monitoring fish culture practices in the area to ensure that it is aware of the changes which might affect the relationship between fish culturalists and credit organisations. This might include changes in technology, markets, environment. Some of this information might arise from the cumulating database on natural resources. The project should disseminate information to be used by fish culturalists and credit organisations alike to adapt their practices and procedures.

Conclusion

Finally, the above discussion implies that it is incumbent upon the project only to encourage rural poor involvement in those parts of the fish culture cycle in which they can be fairly sure of success. If they are encouraged into activities in which even small numbers fail, then loans will not be repaid and the relationship with credit disbursing agencies will be placed under a strain. In dealing with the banks especially, the repayment performance of project sponsored fish culturalists must near to exemplary. As noted in the report, greatest emphasis to date has been placed upon involving the rural poor in food fish production. Yet the discussion in chapter 4 raises doubts about whether this is always an appropriate area of investment for rural poor groups. Other studies in this series confirm this. It is recommended that the project adopt a highly cautious strategy in its promotion of rural poor involvement in fish culture.

i. If the project is able to confirm that fingerling production is the safest and most appropriate area of involvement for the rural poor, then initial project effort should be concentrated on that.

ii. As a parallel, an important supporting role to this, the project should work with its partner NGOs to organise traders in the fish culture cycle. The possibility of establishing revolving loan funds among traders should be experimented with.

iii. In the first instance, only a few well established rural poor groups who meet the conditions set out in chapter 4 should be encouraged to take up food fish production. The performance of these groups should be carefully monitored and the expansion of food fish production for the rural poor should only take place when the lessons of these experiments are clearly understood.

iv. In the meantime, the project should provide passive support to non-poor food fish cultivators.

v. From a long term perspective, the project should also monitor mini-hatchery development in the area with a view to possible rural poor involvement, if and when conditions become appropriate.

Undoubtedly the facilitating role is a less resource intensive one for the project than would be the case if a direct line of credit were to be established. The amount of resources that will be needed to perform the role adequately should not, however, be underestimated. In the introduction to the report, the question was raised as to whether the project currently has enough resources to fulfil this role. In the conclusion, the answer to this question is no. **It is recommended that the ODA and DOF seriously consider making an appointment for an officer (at a reasonably senior level) who will be primarily responsible for the management of the project's credit affairs.**

3. Strategy Discussion

At the end of each section, the key questions posed for the project are highlighted.

3.1 Production Issues

The Department of Fisheries paved the way for the successful artificial breeding of carp in the fish farm environment within Bangladesh.

Their success stemmed from the Fish Seed Multiplication Farms (FSMFs) of which they had over 100. Each major sector of Bangladesh was served by a large FSMF and several smaller ones. However, this was not the case in the NW of the country as there was no regional aquaculture centre in this sector.

The New Parbatipur Farm had operational hopes of stimulating fish culture development in the NW through two avenues:

> Production of hatchlings and fingerlings to encourage local farmers and organisations to enter or adopt a more commercial approach to fish culture.

> Through training and extension to farmers and organisations who may or may not have purchased seed from Parbatipur.

3.1.1 Lessons from the past

It is beyond dispute that government farms are very successful in stimulating private sector involvement in small hatchery and fingerling businesses. Increased availability of seed is then thought to stimulate the intensification of food fish culture in an area.

Through examination of developments around Jessore and Raipur following the establishment of large government farms, we can to some extent predict what is likely to occur at Parbatipur once production has started.

However, the long term role of these large farms is ambiguous. Having helped create thriving private sector hatchery and fingerling businesses through their own production and training and extension activities, the government farms continue to stay in competition with these more commercially orientated businesses. Unable to offer credit or other services which might attract new or loyal customers, the government farms find themselves in a position where they struggle to sell their produce and are eventually criticised as being ineffective.

These principal FSMFs have all gone through an "evolutionary" process which can be summarised as follows:

i. Development of the facility as a FSMF with a training role.
ii. A period of successful sales of both hatchlings and fingerlings to the private sector.
iii. Successful stimulation of the private sector to produce seed fish. Increased competition from the private sector for the sale of fish seed. Technology development hampered by strict target production levels.
iv. An increasingly less active phase as the private sector requires less in the way of fish seed and training requirements. At the same time the public sector finds it is unable to sell fish as its operational costs are higher than those of the competition.

After phase (iv) there is a danger that the farm becomes "dormant" if not "extinct".

To avoid the onset of (iv) the centre must be given the mechanism for providing a continued support to the industry. This would, ideally, make use of the infrastructure installed to make (i)-(ii) a success.

For this to be possible it must incorporate design features which allow for changing usage. Thus as the start of phase (iv) the centre might switch its production from largely hatchlings to broodstock. This would leave much of the hatchery infrastructure available for technology development work. At the same time many nursery ponds could be put to other uses like the rearing of non carp species or the process of selecting genetically better carp specifically for the NW region.

If this is to be achieved then there must be an allowance in the centre's budget to permit these activities which will in turn enable the DOF to continue to play a development/support role to the aquaculture industry in the area.

> Parbatipur, if it is not to suffer the same fate as other major farms must be allowed freedom from mere numerical targets and given the flexibility to attempt to produce what the local area needs are, as those needs change.

3.1.2 Centre design and an integrated approach to production

The new facility has been designed with a degree of flexibility so that the old FSMF (now upgraded) can be run next to a modern fish production centre which in turn will have the infrastructure for production, training and technology development.

It is well known that the integration of livestock and/or crop production with fish culture can in many situations increase the productivity of an area. The Parbatipur farm may be a more profitable operation if it were to adopt this approach and utilise its embankments and fallow areas around its ponds. Such a farm would have a desirable demonstration effect on local farmers visiting the site.

The old FSMF could be used as an integrated aqua/agriculture unit including the production of ducks and a variety of fruits and vegetables.

This raises two immediate issues:

> The farm would have to have outlets for its other produce such as duck eggs or fruits.
>
> The farm would have to employ staff capable of running multi disciplinary, integrated systems or be supported by specialists in the respective fields of livestock and crop production.

> Should the Parbatipur farms (old and new) attempt to become fully integrated farm systems, run upon commercial lines or should they remain exclusively fish production units?

3.1.3 The NW of Bangladesh as a special case for aquaculture development.

It is evident that the NW of Bangladesh is rather different from the other sectors of the country due to three major factors:

> The proximity to the Himalayan mountain ranges in India and Nepal results in a great amount of sand deposition by the rivers flowing through NW Bangladesh. This results in soil conditions which are not conducive to year-round aquaculture as the ponds tend to dry up seasonally.
>
> There is a more widespread occurrence of iron and aluminium contaminated ground water which can make fish egg incubation difficult.
>
> The climate is more extreme than other areas of the country. The winter can be very cold (water temperatures as low as 10 C) and the months of March and April extremely hot and dry. Both extremes effect the growth performance of

fish. Flooding in the form of flash floods can also occur in the wet season causing severe erosion to earth structures.

> Should the farm be looking at new or improved varieties of fish specifically for the NW of Bangladesh?

3.1.4 Targeting Hatchling Production from Parbatipur.

Hatchling production from Parbatipur is targeted at 200 kg. Through close association with fingerling producers, the market for hatchlings can be predicted on a season to season basis. Production could then be adjusted for example: to try and produce more *Catla* or Grass Carp hatchlings once private hatcheries are capable of producing sufficient Silver Carp and *Rui*, in the area.

In this way the Parbatipur hatchery could continue to contribute to aquaculture development in the area, even in the presence of a strong private hatchery sector.

The production of fingerlings from hatchlings is an attractive proposition for farmers with even limited resources as it can be far more profitable than food fish production, and far less risky, (if hatchling quality and pond management are good). The key steps necessary to produce fingerlings at a profit can be taught to even an undereducated farmer in a short time.

The likely people to adopt this activity will be small entrepreneurs with a small capital base (5,000 taka) and who have access to nursery ponds, either owned or leased. The project, theoretically has the opportunity to target its hatchling production to poorer individuals who, if they could become established early enough could benefit greatly from this activity. What is unlikely to work is trying to get marginal farmers established once fingerling production has already become competitive in the area.

It is suggested here that the project make a conscious effort in this regard, perhaps through NGOs rather than merely selling to whoever turns up at the gate. The better off entrepreneur should not be ignored but the project should try to ensure that hatchlings do reach the poorer farmers as well. That may mean setting aside hatchlings for NGOs or poorer individuals who have ordered fish and in the situation where competition for hatchlings exist, favouring the marginal producer before the richer entrepreneur. Economically, this may be a poorer option for the production staff as it may involve selling smaller quantities of seed per unit sale, which is time consuming and more expensive in terms of transportation materials.

The Farm and Hatchery managers of course have their targets to reach and can be forgiven for not caring where their produce is sold, just as long as it is sold. In addition, this target-orientated approach to their work often means that these officers are not interested in training or demonstration activities during their peak working periods as this may affect their production for that year.

> The farm should propose specific targets in terms of its hatchling production i.e. targets for each species produced, and the hatchery staff be freed from the responsibility of finding markets for their production. This work could be carried out by an extension orientated officer, whose main job is to target and distribute hatchlings from the Parbatipur farm. The production staff should be encouraged by their superiors to train and demonstrate their work even if this may result in a lower production from the farm.

3.1.5 Targeting Fingerling production from Parbatipur.

It is less easy for the Parbatipur project to imagine its future role as a fingerling producer. The transportation of fingerlings into the area is exceptionally well developed and will probably continue to run independently of Parbatipur's production. (In 1989 it is estimated that 100 million *dhani* and fingerlings entered the NW from areas further South such as Jessore and Bogra).

Local fingerling production seeded from wild hatchlings from the Jamuna, is established in some areas and hatchling production from Parbatipur should encourage more fingerling producers. This could further reduce markets for project fingerling production.

The transportation of large numbers of fingerlings over long distances is risky and problematic and best left to the hundreds of individuals who carry small quantities of seed, regularly. Anyway, the project does not have the capability to transport fingerlings to the remote areas which may not receive such efficient service from the private seed transportation network.

Neither has the project, at present, an extension capability to enable it to find or create markets for some of its fingerling production. This would probably only be a temporary solution in any case.

Fingerling production for the release in the open water stocking programme may offer an outlet for the farm's production but Parbatipur is neither central to the programme nor realistically able to transport fingerlings of the required size over long distances. It might be possible to stock some of the smaller local *beels* but these are not included in the proposed flood plain fisheries policies currently under review and the transport constraint would still exist.

> The workshop should debate the above ideas and suggest ways to avoid the contradictions in project design which might cause the training and extension component to work to the detriment of the farm's hatchling and fingerling production. The farm's hatchling production could eventually work against its fingerling production viability as the sale of the former would encourage the private sector to produce fingerlings thus making the sale of Parbatipur fingerlings difficult.

3.1.6 Brood Stock facilities.

Given the extensive facilities planned for Parbatipur, the farm could be in a position to support local hatchery development through offering brood stock services. Private hatcheries may well find shortages of brood stock a constraint to development, more so than in other areas, due to the high percentage of ponds which dry completely each year in this area. (Data collected during the natural resource study suggest that at least 40% of ponds dry up in the Parbatipur area in a normal year. Further North this percentage is likely to be higher).

The project could theoretically assist hatcheries in this by holding a stock of maturing fish through periods of drought and selling them at the onset of each spawning season.

Alternatively, immature brood stock of over a year could be sold to hatchery producers for breeding the following year. However, problems encountered during the transportation of large numbers of brood fish are likely and the operation may be rather risky for the project to undertake.

The benefits of such a strategy to the private sector may go beyond mere convenience as the project would have some control over the quality of brood stock used in the area. At present, many hatchery owners will breed any mature fish that they can regardless of size or condition.

> This is asking for a very flexible approach to utilisation of the farm's facilities although some DOF farms do produce market fish and this is only one step on from that.

3.2 Technology Development Issues

3.2.1 Problem solving

The Parbatipur Farm's facilities if ever relieved of their production role in the area could further benefit aquaculture development in NW Bangladesh by being utilised in an experimental way to help solve local aquaculture problems. Such on-station work would need to be constantly reorientated through information brought to the farm either directly by farmers or more indirectly through extension staff or NGOs.

For the farm to be able to conduct realistic trials or experiments with direct relevance to farmers, a part of the site, (perhaps the old FSMF site), could be adapted, to mimic rural conditions locally.

Some key areas for attention have already been identified and these are briefly described below as examples of the type of problem solving, of direct relevance to aquaculture development in the region, that could be carried out at Parbatipur.

3.2.2 Iron treatment of ground water

Iron contamination of ground water supplies in the area is common and this may be a reason why hatcheries have been slow to become established. Attempts to transfer hatchery technology from Jessore and Bogra, neither of which have serious problems of this type, have usually failed in the NW.

The Parbatipur hatchery facilities could be utilised to find cost effective ways of treating ground water for high levels of dissolved iron and lay down guidelines for private hatchery operators to follow on the most effective way of mixing contaminated ground water with pond water to reduce the problem in any given situation.

3.2.3 'Micro' hatcheries

If we dispense with ground water entirely and look to using surface water as a medium in which to produce hatchlings, opportunities for marginal farmers to set up very small hatcheries begin to exist.

It should perhaps not be forgotten that hatchlings can be produced in 'low tech' situations such as *happas*, hatching nets and portable jars. Although inferior to mini hatcheries situated over clean ground water supplies, such appropriate systems may be affordable to the marginal farmer. Such farmers may have access to a small number of brood fish but be unaware that hatchlings can be produced outside of the expensive seed multiplication farms he sees around him. Hatching success rates of, say, 30% may not be acceptable to commercial or government farms but could still represent a huge income to the marginal farmer.

Whereas it is unlikely that such small producers would ever contribute very much to an area's total hatchling production, they should not be ignored by the project and a programme should be started on ways of maximising output from 'micro' hatchery situations.

3.2.4 The RDRS Treadle Pump

The treadle pump, pioneered and successfully promoted by RDRS may have direct application in some small scale aquaculture situations. Costing less than 500 Tk to buy and install, and producing up to 50l/min, the role that this pump could play in aquaculture development is as yet undetermined.

Possible uses for treadle pumps are listed below:

'Micro' hatcheries could use them to supply single or pairs of hatching jars with groundwater or pond water.

Fingerling producers could use them to fill nursery ponds prior to the first rains, and help towards producing early seed.

Mini ponds, which normally dry up each year, could be kept full by regular topping up by a treadle pump, which could mean better results in homestead ponds in the area.

Clearly, the treadle pump should have several applications to marginal farmers in aquaculture. and the Parbatipur project could test these opportunities on station before extending appropriate advice.

3.2.5 Integrated systems

As mentioned in the first session, the project has the potential to adopt a 'farming systems' approach to its production. As yet, little deliberate integrated aquaculture is carried out in the area but the Parbatipur project could accelerate the adoption of such farming practices by demonstrating profitable integrated fish systems.

Such an area would have a degree of experimentation about it. Integrated systems which appear viable after on station trials, could then be extended to local producers to attempt. Close liaison with the Dept. of Agriculture Extension and agriculture orientated NGOs would be necessary in this case, in order that the project keep up with local agricultural practices as they change.

3.2.6 Extensive and intensive aquaculture options

Fish culture is possible under a wide range of different conditions. Ponds may be cultured in a purely extensive manner, where little manipulation of stock or environmental conditions is necessary. Intensive culture of fish relying totally on inputs and stocking represents the other extreme. Most fish culture in Bangladesh lies somewhere between the two. The more intensive a culture system becomes the more investment in technology (e.g. cages, aerators) it requires to remain stable. The project could carry out trials to compare rates of return for pond holders in different risk categories to ensure worthwhile returns on technological investment.

3.2.7 Stock improvement and new species

The facilities at Parbatipur could also be utilised to carry out work related to the genetic quality and growth performance of fish stocks as well as exploring relevant technology developments necessary for the production of new species as they become popular for culture in the N.W. such as catfish or sex-reversed *tilapias*. The Parbatipur farm could play a major role in solving production teething problems before disseminating technology to private operators. As with its production role, the type of development work done at Parbatipur would change as local needs demanded and would require constant reorientation from field information.

3.2.8 Fry/fingerling transportation

Fingerling transportation in the project area is extremely well developed, partly because of the lack of local seed production. However, the project could help this sector to develop further by investigating ways of improving quality and survival rates of seed during and after transportation.

3.2.9 Marketing of food fish

The project could experiment with ways of maintaining the quality of food fish following harvesting in its transportation to and storage at the market. This may be a present constraint to the spread of intensive fish production in areas distant from potential markets, e.g. town *bazars*.

3.3 Extension Issues

As a development project, the Parbatipur project faces a dilemma in its training and extension responsibilities. Generally speaking, most of the better fishpond resources are in the hands of richer people and by purely using pond ownership as a prerequisite for extension or training attention, the project would exclude the poor and its development endeavours would largely benefit better off individuals. The emerging poverty-focused orientation of the project becomes the principal criterion for designing extension and training priorities. The project argues that the fish culture system is not confined to food fish production, where the principal private resources are likely to remain in the hands of richer classes, but consists crucially of a chain of other activities from hatching to the final marketing of food fish. It is argued that such 'service' activities are more easily targeted towards the poor (either as groups or individuals) than food fish production itself. In this way, it is hoped that an expansion of fish culture in the region will benefit both those who provide the services to food fish producers (before stocking and after fattening) as well as the food fish pondholders themselves. Of course, there will be some limited opportunities, even within the hostile political economy of the region, for poor groups to take *khas* ponds or lease private ponds for nursery and food fish production. In such cases, the poor operate in the system as groups rather than as individuals. Extension is therefore conceived both as organisational innovation and the transfer of knowledge.

> The project proposes to collaborate with partners in the project area to involve the poorer members of the society in the expansion of employment and income generation opportunities arising from developments in fish culture.

3.3.1 Limitations in present design.

The project in its present form has little scope to carry out an effective extension programme in the project area. Limited numbers of extension staff with restricted mobility, together with the absence of a credit line, means that regular support to various actors in the fish culture systems is impossible in all but the immediate area around Parbatipur. The project must therefore face up to the fact that it must: either significantly bolster the numbers of its extension personnel, or develop working relations with partners (among the DOF District and *Upazila* staff, NGOs, BRDB and the local Banks, or resign itself to having a more passive field role. A further option to work with 'informal' extension agents is discussed in (3) below.

3.3.2 Liaison with DOF staff, NGOs, BRDB and others.

One way to alleviate the project's extension constraints will be to work closely with organisations or departments which have their own extension programmes. In this way, the project would be providing technical support to extension services already in existence. This technical support could take several forms: advice on hatchery techniques and pond management; delivery of seed; advice on credit requirements; establishment of a uniform database on pond types and patterns of use for comparative analysis and monitoring; specific action-research services on particular technical and organisational innovations; transfer of knowledge gained through experimental research under 'technological development'; convening of workshops for partner extension personnel to facilitate collective problem-solving.

> Representatives from DOF, BRDB, Banks and NGOs should discuss their own constraints to more effective extension support to fish producers and propose ways in which the project could support them at field level. The merits and feasibility of compiling a natural resource database of ponds throughout the project area should be discussed.

3.3.3 Extension Menu

UTILISATION OF INFORMAL EXTENSION AGENTS

An alternative approach to the problem of extending information to food fish producers spread over a wide area is to use the networks which already exist in terms of seed supplies and food fish harvesting and experiment with ways of harnessing the mobility of these operators. The rationale for this approach is that since many fish seed traders and fishermen are already travelling between points of supply and fish ponds, they can become the vehicle for carrying basic extension messages. It is possible that traders could therefore be trained in basic aquaculture practices and be motivated to pass on this knowledge to their clients. This concept is more fully explored in annex 1. ('The Utilisation of Fish Seed Traders in Aquaculture Development in Bangladesh'), the earlier research design report, and the Fish Seed Trading study.

HATCHERY AND NURSERY POND BUSINESSES

The areas where a small number of extension personnel can make a significant contribution is by encouraging and supporting local hatchling and fingerling businesses to become established. While the project would expect to pursue this strategy in conjunction with the potential partners, it may also be possible for the three satellite seed farms to organise extension teams to operate in a similar way in each locality. However, without purposive targeting through poverty-focused partners, these opportunities will be dominated by the better off entrepreneurs. The project would therefore encourage the satellite farms to share these poverty-focused objectives in identifying potential private sector hatchery and nursery pond businesses.

The viability of small-scale private hatcheries, and the expansion of nursery pond businesses (especially outside the existing areas of concentration) should be discussed in terms of markets, technology and investment issues. The extent to which such businesses represent opportunities for organised landless groups of men or women must also be considered.

SUPPORT FOR OTHER 'SERVICE' ACTIVITIES IN FISH CULTURE

As noted in the introduction, many employment and income-generating roles exist in the fish culture system from hatching to the purchase of food fish by consumers. Using the findings from the studies, the project needs to assist partners in identifying opportunities to support producers and traders (technically and with credit) in: transportation, storage, equipment, stocking choices, assessment of quality, pond management, feeding, preparation for transport and markets, and so on.

> For a coherent policy toward fish culture, there is a need to establish a consensus on the key roles in the system, how they are differentiated socially, and the organisational prospects for supporting such 'service' roles.

OPPORTUNITIES FOR WOMEN IN THE FISH CULTURE SYSTEM

The Women's study reveals some diversity of experience in different parts of the region. The ethnic/religious variable seems to be as important as class in this diversity. It also appears that in the more culturally favourable areas there is a lower density of ponds. While there is, as yet, no extensive involvement of women in nursery pond businesses and food fish production, some NGOs are working with women's groups in this way. Women also participate in the fish culture system through their use of homestead ponds, preparation of nets, and as key decision-makers in household dietary preferences. They are not involved directly in trading and retailing. Despite the important gender dimension of poverty-focused objectives, the project and potential partners need to be realistic in assessing the room for manoeuvre to expand the opportunities for women.

> These questions have to be posed: Which are the most realistic activities for women in an expanded and intensified fish culture system? Are their present roles restricted at best to the production of use-values? What commercial opportunities are possible for groups or individuals beyond current household roles?

LEASING BY THE BITTAHEEN OF PRIVATE, UNDER-UTILISED PONDS AND *KHAS* PONDS FOR FOOD FISH PRODUCTION

While this aspect of the extension menu appears to be the most common among poverty-focused NGOs in Bangladesh, it is not a panacea option even in an area of *khas* pond density. The local political economy and power structures hinder access to such ponds by the poor, as well as the retention of them after they show profitability. The Grameen Bank have acknowledged this problem with its 'fish factory' approach of managing ponds through its own line staff. Other NGOs (such as CARITAS and RDRS) provide intensive organisational support to 'their' groups. However, groups and combinations of groups have demonstrated some success, and the Project should assist these efforts whenever possible. Its extension personnel should advise its partners on all aspects of risk attached to such ventures, including, of course, the availability of market outlets for commercial production. Credit issues will be important, as well as an understanding of comparative rates of return on other uses for scarce capital. Combining technical knowledge with social knowledge and the opportunity cost of capital may result under some circumstances in advice **not** to proceed with a lease and investment. The key questions are:

> Can food fish production be targeted upon the poor through the leasing of private and *khas* ponds? Must the poor be organised and supported either by an NGO or the BRDB for this purpose? How can the assessments of technical, social and economic viability best be organised and combined? Is credit most effectively used by the poor under this heading in the fish culture system, given caution about the distribution of risk in the political economy?

UNIFORM DATABASE ON PATTERNS OF POND USE BETWEEN PROJECT, DOF AND OTHER PARTNERS IN THE AREA

The natural resource study of four villages in the project area, carried out by BCAS and Project personnel, may provide a model format for a pond and aquaculture status study of the whole project area. It is suggested here that a regional database on pond resources could start to be compiled at Parbatipur for use by extension agencies in the area. This natural resource inventory would also provide baseline data from which the progress of aquaculture development could be monitored.

3.3.4 Monitoring of extension activities and action-research

Extension is notoriously difficult to evaluate. The project should perhaps look towards an independent monitoring body to help in assessing the impact that extension (and training) strategies are having and to help guide future work.

Several options can be considered: partners such as NGOs with their own monitoring and evaluation capacity; DOF staff currently on training courses at AIT, Bangkok investigating aspects of the extension programme for their thesis work; staff (possibly with graduates) from the Departments of Aquaculture, BAU, Mymensingh for technical issues and from Anthropology at Jahangirnagar University for more socio-economic issues; consultancy organisations; and of course directly by Project personnel. Different options may apply for different aspects of the extension strategy.

4. Agenda for Action-Research

Introduction

At this stage in the evolution of thinking within the project, when the implications of recently completed studies are only just being absorbed, it is premature to design a comprehensive action-research plan for the coming year. There remain, too, key problems of the project's own extension staff capacity to pursue the various proposals arising from the studies and project papers. The studies contain staffing proposals and the Wood January 1990 report contained suggestions for the appointment of additional staff to create an extension team consisting of: 2 Senior Extension Officers (male and female); 2 Assistant Extension Officers (male and female); and the retention of the 2 Field Assistants. These suggestions would, to some extent, absorb the proposals for staff arising from the studies and project papers. In addition, some consideration will have to be given to On-Farm Applied Research posts to pursue production and technological development issues. For the purposes of concentrating discussion in the final session of the workshop, the various proposals from the studies and the project papers have been culled into the following structured list:

Subsequent to the workshop and a review of the minutes the extension staff of the project met with the TCO team leader and the consultant sociological advisor and suggested the following (bold) priorities for action over the coming year.

PRODUCTION

1. Demand for fingerlings by NGOs with groups leasing ponds for food fish culture.
2. Recognise the concept of 'poor people's fish' and therefore target poor people's consumption by concentrating on the production of species which are suitable for smaller ponds, many of which dry out (which may have less commercial use).
3. Recognise that the sequence of production from FSMF will be via a substantially reduced target of fingerlings, switch of emphasis to hatchlings, and the development of broodstock services (in the context of mini-hatchery development in areas where ponds dry out or may not be available for storing broodstock).
4. Genetic improvement trials to development species more adaptable to conditions in the N.W.
5. **Experimentation with species varieties and combinations through collection of full range of brood stock species.**
6. Flexible use of pond lay-out to reflect switches of emphasis in production objectives.
7. **Targeting sales of hatchlings to the poor through NGOs. Fingerling sales to be directed towards NGOs and small fingerling traders.**
8. Pituitary Gland services for mini-hatcheries in the region.
9. **Pituitary gland collection for the purposes of project site spawning.**

TECHNOLOGICAL DEVELOPMENT

1. Examination of fish disease, and development of methods to control it (not only among cultured fish).
2. Experiment with short-term seed storage for small traders to increase value-added and survival rates.
3. Understand further the interaction between poor people's species and pond types.
4. **Resolution of iron problems in groundwater supplies for hatchery development, and experiment with surface water (filtered and cooled through water hyacinth), to produce hatchlings at the project site and to assist subsequently the development of micro - hatchery technologies.**

5. Investigate through experimentation with the use of RDRS treadle pump at various points in the fish culture system.
6. Research and monitoring of integrated pond use through a 'farms systems' approach, and develop the old site for trials, experimentation and demonstration.
7. Undertake comparisons of rates of return on extensive/intensive options to gain optimal levels of investment and minimise risk (especially for poor producers).
8. Search for ways of improving quality and survival rates of seed during and after transportation (to aid traders and their pondholding customers). Experimentation to take place at site comparing the performance of seed, in *happas* which had travelled long distance to the area and that which had been produced locally.
9. Examine ways of reducing the deterioration of food fish during marketing to relax the spatial constraints in commercial food fish production. Testing fish in the market to determine deterioration and comparing with their immediate pre market history.

EXTENSION

1. Develop the Extension Traders Strategy (ETS) as a way of linking extension constraints to poverty-focused and private sector development objectives. To be achieved through training selected traders and monitoring their performance, through widening the traders' participation in the fish culture system through a bundle of pond culture services, and (perhaps in conjunction with partners) to encourage share returns for traders from their customers.
2. With partners, promote local nursery pond businesses among landless groups of men and women.
3. Focus on smaller pond operators and ponds which dry out to grow appropriate species for poor people's consumption.
4. Experiment with a group lease of a co-ownership pond to test the co-ownership constraint. Note the possibility that women in co-owning families may commercially cooperate with each other more easily than their menfolk.
5. Use the *Gender and Development Forum* to develop strategies with small NGO partners.

6. Attempt extension strategies with women (using homestead ponds or small family *digis*) through: grassroots female workers with Dipshika; and a private female entrepreneur in Dangapara Village, Khansama Upazila.
7. If working RPP, concentrate on Chirirband Upazila to test and strengthen its organisational capacity.
8. Assist the Lamb Hospital in its work with 20 women's groups to excavate ponds and invest in fish culture.
9. Recognising the significance òf RDRS and its plans to expand in to fish culture, to continue to provide assistance in pond selection, reexcavation, stocking policies and analysis of rates of return.
10. Support CARITAS if services requested.
11. Continue to support the Grameen Bank's development of mini and micro hatcheries.
12. Recognise the key points and moments in the fish culture cycle for credit support, advising partners accordingly. Immediately, to recognise the significance of credit to traders in the ETS, and the problems therefore of mobilising non-partner organised fingerling traders. Some thought can be given to a Revolving Loan Fund for traders.
13. Concentrate initial activities with the poor on a few, well organised groups and projects most likely to succeed with low risk--therefore smaller ponds, nursery ponds, mini/micro hatcheries, trading and fishing, rather than leasing of large *khas* ponds and co-ownership ponds (apart from one or two experiments).
14. In addition to working with NGO and RPP partners, negotiations with the Banks in the locality should be started to extend the availability of credit `in a responsible way` to the fish culture system.
15. The project will need to adopt a `facilitating role` in the development of credit support to fish culture in the region.
16. To respond to items 12-15 above it is proposed to hold informal introductory sessions between project extension staff and credit dispersers within NGOs and commercial banks (starting at senior levels for the latter).
17. Using the natural resource database methodology, extend it to partners to develop compatible, interactive data-sets between organisations, and experiment with 'scoring' villages to target extension activity.

18. **A one day workshop with fingerling producers at Champatali in Chirirbander to review current nursery pond management practices.**

TRAINING

The training implications derive logically from much of the preceding agenda, but some specific proposals are:
1. Training of trainers among the partners to develop knowledge of fish culture issues and socio-economic context.
2. Assist in devising 'partner courses' for the groups with which they are working.
3. Training selected fry traders as part of the development of the Extension Trader Strategy.
4. Training staff of credit disbursing organisations (e.g. Bank workers in the assessment of fish culture applications).
5. Specific credit training for different categories of actor in the fish culture system, to develop awareness of the requirements of lending agencies, risk issues in fish culture and minimum rates of return for viable enterprises.

ESTABLISHMENT

Alongside the opening comments about staffing, three specific suggestions arise out of the studies and project papers:
1. The appointment of a women's co-ordinator (Bangladeshi, either DOF or TA) to oversee the proposals for developing women's involvement in fish culture.
2. A senior extension appointment to co-ordinate and negotiate/advise partners on credit and fish culture.
3. The deployment of an extension officer to assess market demand for fish seed (including targeting) for output from the FSMF and to define production targets for the hatchery staff.

MONITORING (Short and Long-term)

The following options (some in combination) are being considered:
1. NGO partners with their own action-research capacity.
2. DOF staff on courses at AIT using project issues as topics for Masters theses..

3. Staff and graduates from BAU, Mymensingh (Department of Aquaculture) and Jahangirnagar University (Department of Anthropology) to undertake studies.
4. Consultancy organisations or individuals.
5. Project personnel.
6. Continuation of ODA appointed advisory services.

Chapter 15

Open Water Bodies and Capture Fishery: The Poverty of Policy

Introduction

As indicated in the introduction to this volume, and the preamble to Part IV, I was invited in early 1990 by the UK-ODA to examine the World Bank Appraisal document of the Third Fisheries Project (TFP) Proposal since the UK-ODA would be a co-funder. Those in ODA responsible for social development wished to understand the poverty implications of this large-scale stocking policy for open-water bodies in Bangladesh. Prior to this exercise, I had already made some comments to ODA on an earlier pre-appraisal paper, and some references are made to this in what follows. Although a separate and exercise from my involvement in the Dinajpur fish culture project, the appraisal document was relevant to that work since to some extent the point of departure for this open water policy was an acceptance of the inefficiency of the public sector fish seed multiplication farm strategy and an assumption that such a large scale stocking policy would stimulate the private sector hatchery and nursery pond system. My terms of reference were rather ambitious for the time made available, but given the significance of the policy for rural development in Bangladesh it seems right to include the paper in this collection. The limited time available has determined the style of the report as a close, textual analysis of many of the propositions made in the Appraisal document.

The TFP has proceeded, with UK-ODA participation in some of the training and monitoring components of the project. The ODA expatriate team, with a particular brief to secure a poverty focus for the programme, were surprised to learn of the existence of this critical commentary in April 1993, some 3 years after I had written it! The team have regarded this commentary as highly relevant to their present agenda and feel that it could have saved much of their early learning time. Now it can enter the public domain as a set of issues to be considered in the context of the expanding fisheries programmes in Bangladesh.

1. Terms of Reference

In the light of the draft appraisal document for the proposed IDA Fisheries III Project, and his knowledge of the social context of the fisheries sector in Bangladesh:

   i. Assess the implications of the various components of the project for the distribution of benefits between fisherfolk and entrepreneurs, between part-time and full-time fisherfolk, and between men and women.
   ii. Assess the likely social impact of the procedures proposed for auctioning fishing rights in the flood plains on the current structure of formal and informal rights governing access to fishing areas, especially by part-time and self-employed fisherfolk.
   iii. On the basis of the experience of similar projects, consider the implications of the institutional arrangements proposed for the duties of fishing rights for both equity and cost recovery.
   iv. Comment on the adequacy of the proposed institutional arrangements for (A) the formation of groups to bid for fishing rights in the flood plains (B) the participation of farmers in coastal shrimp culture.
   v. Comment on the adequacy and feasibility of proposed arrangements and methods for monitoring institutional development and social impact due to the project.
   vi. Suggest what, if any, modification to project design would be necessary to alleviate any undesirable social effects of the project or improve its impact on the poor.

Three days' additional work was suggested in the TOR, which therefore defines the scale of the exercise. I have excluded coastal shrimp culture from my comments, since I have little direct knowledge, while being aware that the conflicts of interest between different classes of farmer, tenant, labourer and entrepreneur are acute, and that local bureaucracies (including police) have been incorporated into these struggles. The stakes are very high, both nationally in terms of policies for generating foreign exchange earnings, and locally in terms of the potential value of securing appropriate water bodies into private entrepreneurial hands. There is a case for someone having a specific look at these issues, starting, perhaps, with some of the information provided by NGOs whose group members are being adversely affected by these developments.

Before reading the Appraisal document, I was led to understand that the previous criticisms of fish (hatchlings and fingerlings) production in the public sector made in the pre-Appraisal paper had been moderated. This is not the case, and the implications for the Parbatipur FSMF of this critical policy stance towards the production activities of the DOF remain as indicated in my October 1989 report. I note that the follow-up to that report included a brief visit and report by an ODA Fisheries expert, which confirms my earlier assessments, and is, if anything, more pessimistic about the adoption of the Parbatipur complex to perform training, applied research and local policy development functions as part of TFP.

It is not my intention to repeat those parts of my earlier report which discuss the pre-appraisal report on TFP strictly in terms of its implications for the Parbatipur FSMF. However, for those new to the TFP proposals, some repetition of discussion will occur on issues related to the TOR above.

2. Summary of Project Components

The main objectives of TFP are to assist GOB in: increasing incomes, particularly of the poor, through fish and shrimp production for domestic consumption and export; supporting fisheries development in the West of the country; accelerating expansion of floodplain fish production; and strengthening sectoral institutions. To achieve this, the main components are (excluding shrimps):

   i. Stocking of about 100,000 ha in PY6 of high potential floodplain with selected fish (i.e. fingerlings, and it is proposed that 70% of this stocking will be silver and bighead carp). A longer 10 year strategy increases this area to 550,000 ha. in 3 pockets: Chalan Bheel, Gopalganj depression and Khulna depression. All these areas are in the SW rather than NW.

   ii. Intensifying extension to about 13,500 ha of other fisheries resources (small floodplains, ponds, rivers) and selected target groups (women, fingerling traders). A longer 10 year strategy increases this area to 50,000 ha. including the Teesta floodplain (Kurigram, Lalmonirhat

Nilphamari Districts) as well as elsewhere. Some of these areas are closer to the Dinajpur Project Area. This component to be achieved through a mixed bundle of 55 finance packages for *Upazila* and District level programmes to develop fresh water fish and shrimp ponds.
iii. Supporting administration and research: financing incremental facilities and operational costs of DOF and BWDB; and providing funds to FRI to carry out project related, applied research.
iv. Providing technical assistance and training in management (experts in management, planning, monitoring and procurement) and support services (studies, NGO services, study tours).

It is proposed that 60% of the project cost will be allocated to the development of resources, with institutional development, TA and training representing 40%. (para 3.03, page 14) Within the development of resources the main thrust is floodplain stocking. The scale of investment is justified on the calculation that planned incremental production represents about 5.5% of present domestic consumption, but this will only be available at the end of 7 years by which time population growth will have increased demand by 18%. Expansion in coastal or pond fishery will not be adequate to satisfy this demand. (para 5.03, page 32).

Attached to this scale of investment are: savings of $3m p.a. by divesting the DOF of loss making, superfluous fish seed farms (at least 50% of those in West Bangladesh) and discouraging the building of new and pipeline ones; and a 50% target for the recovery of annual stocking costs. (para 4.02, page 27-28).

3. Public and Private Sector

Within the stocking strategy, a distinction is made between major floodplains "for which a direct public sector intervention is appropriate", and minor floodplains "which could be developed through private sector initiative" (para 2.12, page 4).

However, it should be recognised that this reference to public sector does not apply to production, but to procurement and stocking (para 2.13). The approach to minor flood plains is really an extension to the model for pond fish culture. "The main difference from ponds is one of scale and of increasing demands on financing, protection and cropping." (para 2.26) "Acquaculture in ponds, lakes and minor floodplains shares a common technological approach, based on stocking and managing a confined water body". (para 2.27). The contribution of TFP to increasing pond productivity would be "tackled via extension of existing technology" where it will be "important to offer pond production models that are well adapted to the flooding or dry season risks and to profitable new possibilities such as fingerling... production. The availability of hatchlings is not, as sometimes assumed, a constraint." (para 2.25). The public sector's role would be confined to providing "advisory services on organisation according to the type of private operator, especially for groups of fishermen (women) or rural communities." (para 2.26).

The report goes on to state that the constraints to expanding the private supply of fingerlings are: effective market demand; access to low cost and appropriately organised finance; advanced hatchery techniques; and credit plus know-how for nurseries. It also refers to the Wood and Gregory proposals for the developmental use of fingerling traders as conveyors of technical messages to fish farmers. It argues that the DOF will have to change its promotion strategies through producing fish seed. "..the main thrust should now be on disseminating more sophisticated technology to the more advanced pond, lake, floodplain, hatchery and nursery operators. DOF funds and personnel presently engaged in fish seed production should be freed for this more important task." (para 2.28, page 8) The critique of public sector DOF production is sustained through remarks like: "...a major DOF activity, the production of hatchlings, has become redundant due to the dynamism of the private sector" (para 2.33, page 9). It is worth quoting para 2.35, page 10 in full as the last word on this point to demonstrate the underlying policy position being pursued by the World Bank, even in the context of an open water, floodplains stocking project where demand for hatchlings and fingerlings will increase dramatically:

> "Public hatcheries and seed farms were developed in the past—there are over 100 in the country and over 60 in the project area—to spur acquaculture development through the provision of fish seed. Meanwhile the private sector has taken on fish seed production and public sector hatcheries now provide a declining hardly sellable proportion (less than 5%) of the national spawn and hatchling production, resulting in substantial annual losses. Nevertheless, DOF still continues to develop new ones. It is time to adapt to the changed situation and to liberate the public budget and to use DOF staff in other, more important activities, such as research, extension and training support to the private sector seed industry." (see also paras 4.06-4.08, page 29).

4. The Target-Group Case for Open Water Stocking

The appraisal estimates that fisheries provide full-time employment to about 2m. people (fishing, fish farming, distribution and processing). Full-time fishermen are seen as mainly landless, with their numbers rising at 3%

p.a. Part-time fishermen and women are estimated at about 11m., peaking in the flood season from June to October. With population increases and pressures on declining fish stocks, per capita consumption of fish is declining. (para 2.01, p 1).

The main recent initiative has been the New Fisheries Management Policy of 1986 with equity and conservation objectives. The focus has been to limit access to public water bodies to genuine fishermen, on the principle endorsed by the President that "Water bodies belong to those who own the nets." This policy was based upon the analysis that over-fishing occurs through the exploitation of genuine fishermen by entrepreneurs who have been successful in the past in monopolising access to the leases. The transfer of fishing rights to such genuine fishermen would lead either to a self-interested voluntary restraint (to protect stocks for future years) or to compensation through stocking with fingerlings. These policies have been pursued especially with the aid of NGOs and the *bittaheen* groups organised by them, but also through DOF *Upazila* officials with other programmes (e.g. BRDB/RPP and the Nordic funded PEP). There have been several reports (see e.g. Wood in chapter 17, Part V) on groups and open-water bodies in Nageshwari, in Kurigram) indicating difficulties in selecting genuine fishermen, collecting fees and promoting a collective sense of responsibility among fishermen. (para 2.41, pp 11-12).

In this sense, TFP could be seen as an extension of the recent direction of MFL policy, with the World Bank drawing positive conclusions from the 79-86 Oxbow Lakes Fisheries Project in Jessore, which increased yields and stimulated the development of private small-scale hatcheries. However, the appraisal recognised weaknesses as being: insufficient devolution of hatchling and fingerling supply, stocking, lake management and fish marketing tasks to the beneficiaries. (para 2.47, page 13).

The quantifiable benefits of the project (fish) would be annual incremental production of about 45,390 t, resulting in increased net incomes to "subsistence and professional fishermen,..., pond owners, women, nursery and hatchery operators and their families." (para 6.01, page 34). Additional benefits (besides institutional, management and research ones) would be: substantial increases in government revenues through improved cost recovery and to budget savings from divestment of loss-making government fish seed farms; further mobilisation of the already strong private enterprise activity, particularly in the area of fish seed production (This would be achieved through procurement of fingerlings for public stocking and through a re-orientation of fisheries services from public sector production to support to the private sector.); increased services to groups of persons needing more attention than previously, such as women and non-institutional extension agents (sic); equity improvement that would go beyond the self-targeting mechanisms of most fisheries programmes (such as benefits to fishermen in floodplains).(para 6.02, p 34). This final point is elaborated further by the following statement, made in the context of the major floodplains:

> "Since access to fisheries is not controlled, there is a high probability that the increased levels of income would be diverted into increased employment in fisheries of lower income groups. Under the assumption that 30% of the income potential would go to existing professional fishermen, while 70% would go to newcomers, increased employment of the order of 12,600 more full-time fishermen jobs at Tk 5000 p.a. could result from the project's floodplains component." (Para 6.04, p 35).

Further categories of beneficiaries are nominated for minor floodplains (36,000 subsistence fisheries dependent persons, and 2,000 persons dependent on professional fishing); 40,000 pondholding families (para 6.06, p 35); and "a small group of some 600 business, entrepreneurs and rural financiers (sic) would reap considerable increases in annual incomes through operating leases in floodplains and shrimp culture thereby controlling possibly 50% (!) of overall production. In floodplains their incomes would be strongly influenced by the cost recovery policies." (para 6.07, pp 35-36).

5. Key Policy Issues

The Appraisal sums these up as: public and private sector roles; size of public agencies; cost recovery; and balance between equity and growth. (para 3.07, pp 14-15). The public/private sector issue, especially with respect to the relation between DOF production and extension activities, has been discussed above. The public agencies issue is a reference to proposed arrangements for public stocking, where service contracts with private firms are favoured. The issues which concern the TOR for this review are cost recovery and equity, and they are highly interrelated.

Equity

The Appraisal's opening remark on equity is essentially of the 'trickle-down' variety. It basically argues that income distribution is improved whenever production technology and the management of respective development activities are well established, and whenever public service concepts are overtaken by the pace of social change. It then argues that since "floodplain fisheries is at the stage of establishing technology and management;

overloading it too soon with ambitious income distribution aims or cost recovery targets is risky." (para 3.07, p 15). For my first observation on this, I shall adapt the remarks made in the Wood October 1989 report on the similar analysis in the pre-appraisal on TFP.

It is depressing that the Appraisal, with all the Bank's contemporary rhetoric, should be expressing these views concerning income distribution in 1989. One cannot divorce production from the social relations of production. If private sector production is to be encouraged in the open water bodies and fishponds (both stocking and fishing), and if this encouragement is to be left to the free interplay of market forces (i.e. those with capital or collateral first able to respond to the procurement-led demand and price increases for fingerlings, and to avail themselves of enhanced mid-water catching opportunities, where the bulk of the new stock will be located—silver and big-head carp), then a commitment to socially regressive production relations is also being expressed. Such relations become institutionalised into vested interests, and cannot be subsequently changed by sleight of hand.

Credit

This problem is further reinforced by the absence of any targeted credit component to TFP (para 3.08, p 15). While recognising the problems of socially targeting credit, it is laughable for the Bank to refer to the availability of funds with the banks, to identify the problem as absence of "well-prepared and eligible clients", and then to argue that it is preferable to concentrate on creating DOF capacity for better collaboration with the banks." This does reveal the emptiness of the Bank rhetoric on giving effect to equity objectives, and there appears to be no provision to study the credit issue in the TOR for studies in Annex 9. The Bank must be aware of the reluctance and obstructionism of the NCBs in Bangladesh to disburse even earmarked funds to the poor, even in the context of close project or NGO supervision. One assumes that the ODA's support for the BRAC Bank is an acknowledgement of this problem. In the Dinajpur Fish Culture Project, we will be proposing some collaboration with Banks (as well as with NGOs) in the project area with the project extension staff providing a 'fish credit advisory service'. But this will initially be on a small-scale location and Bank specific, and monitored very closely. And even then, it is more of a test or exploration of a possible institutional arrangement entered into without a great deal of optimism. It is very unrealistic to imagine progressive collaboration between the DOF and the banks on the scale required to give effect to equity objectives in the TFP.

Cost recovery

However, it is when we come to the Bank's discussion of cost recovery that the priority given to equity is revealed. The Appraisal refers to the two opposing concepts within GOB: one concerned with increasing revenue to develop the domestic resource base by auctioning leases on water bodies to the highest bidder (Ministries of Finance and Land); the other concerned to improve equity and access to public resources by professional fishermen as supported by MFL and NGOs and pursued in the New Fisheries Management Policy from 1986 (para 2.41, pages 11-12, and para 6.09 page 36). The Appraisal observes that "the bid process has been progressively eroded into a process aiming at distributing fishing rights at low rates to groups selected on the basis of low income and landless, viz professional fishermen. Recovery by this means is far less than the true auction value of fishing rights." The Appraisal then appears to abandon its previous optimism about the targeted beneficiaries of TFP by arguing: that selection of target groups poses problems; that dependency ties accompanying poverty cannot be broken even by effective support systems, particularly institutional credit; that professional fishermen, bound by local financial interests, are used for cheap access to fishing rights under the NFMP; and that, therefore, there is negligible revenue to GOB and little improvement in equity. It is unacceptable for the Appraisal to justify the floodplains stocking policy in terms of wider benefits to poor fishermen, and then to present this analysis of social leakage around water-bodies and to note that: "The problems of recovering costs is aggravated when the public sector invests funds in improving the resource e.g. in floodplain stocking." The analysis of social leakage is not incorrect. Furthermore, even the idea that 'professional fishermen' are poor has to be treated with caution, especially in the context of open-water bodies and the auctioning of rights to fish. These 'fishermen' are often themselves businessmen, well-networked to the local administration, managing the access and then employing labour fishermen for the catch. This is not just a labels issue. The Appraisal's designation of 'professional fishermen' is clearly intended to refer to those who may be "bound by local financial interests", in which case they are most likely to belong to the Hindu or *Adhivasi* community (or low 'caste' Muslim) and almost by definition are unlikely to be able to manage the networks to gain direct access to fishing rights. They can therefore only be indirect beneficiaries at best as labour fishermen, employed by those who both manage access to fishing rights and have the capital or collateral to purchase boats and nets—labelled interchangeably as businessmen or 'professional fishermen'. While the Appraisal recognises that discussion of a cost recovery system remains academic until the feasibility of public stocking has been demonstrated (a study on cost recovery options is proposed in Annex 9), it does argue that:

> "The financial models in Annex 14 indicate that stocking costs could be recovered fully from the holders of public leases or professional fishermen whilst maintaining financial attractiveness and allowing part-time fishermen to take half of the incremental fish catch without payment of charge." (para 6.09, p 37).

This represents an extraordinary contradiction with the social leakage analysis earlier in the same paragraph, since here cost recovery is expected from professional fishermen (despite being "bound by local financial interests") or professional fishermen have to be seen as businessmen with the ability to hold public leases, thus undermining the connection between floodplains stocking and poverty targeting. The second part of the statement also contains problems, since part-time fishermen are generally bank wading cast net fishermen who will be unable to reach the enhanced stock of silver and big-head carp who are phytoplankton feeders and therefore located away from bank sides.

6. Implications of TFP Components for Poverty-Focused Objectives

Floodplains Stocking

Leaving aside, for the moment, the value of floodplains stocking, there can be no doubt that the stocking policy will increase the demand for fingerlings, and therefore for expanded nursery and hatchery capacity. Given the parallel policy of favouring the private sector and divesting public sector fish seed farms, there should be a proliferation of private sector hatcheries and nursery pond businesses in the major areas of stocking, which are at present to the south of the Western zone. Some demand will be created in the NW around Teesta floodplain stocking. These private sector units will in effect be sub-contracted to procurement agencies. (See paras 3.42-3.45, p 25 for the description of procurement arrangements.) These agencies will obtain contracts through a tender process and be subsequently supervised by the DOF. Despite attempts to set up cross-checks in these arrangements, the opportunities for corruption will be substantial at both the tendering and supervision stage. However, in principle, some thought should be given to bringing the poor into this set of opportunities—as hatchery, nursery pond and transportation sub-contractors, and even federations of groups (supported perhaps initially by NGOs) acting as primary contractors supplied by their constituent groups.

There is no *a priori* reason why the poor, technically and financially supported by NGOs, should not enter these fingerling supply markets. Models exist for their entry into more risky markets than envisaged here. Indeed the existence of publicly funded procurement policy minimises risk. At present, the Appraisal does not emphasise these opportunities for the poor, an opinion reinforced by the absence of a credit line in the TFP and a reliance on the banks to use their funds on bankable schemes. Thus an opportunity for targeting is being missed, and there is no provision in Annex 9 to research into these possibilities or to provide technical support. The Dinajpur Fish Culture Project intends to experiment with *bittaheen* group participation in hatchery and nursery pond businesses, and is proposing a variety of ways in which the public facility at Parbatipur can be reoriented to this purpose.

Fishing Rights

There has been some discussion of these issues in 5) above. As demonstrated, the targeting of rights to fish cannot easily be separated from cost recovery objectives. But, in principle, three alternative conditions could obtain: no leases, open access; 'peppercorn' leases as an attempt to establish targeted access for the poor (as is attempted under the NFMP), cost recovery leasing (either full or partial) through auctions, qualitative tendering at a fixed price, lottery at a fixed price. Cost recovery may be achieved in other ways, unconnected to the allocation of fishing rights. Although some difference should be acknowledged also between major and minor floodplains, the principle of targeted access can only apply to water-bodies which shrink into confined areas as the floods recede.

Open access would clearly favour those people who owned or managed large boats and nets, and who could mobilise the 'muscle' to prevent competition.

Peppercorn leases may enable rights to be allocated to designated groups, but this cannot guarantee such groups' rights to the productivity of the confined water-body unless they can also mobilise the boats and nets, and police those who are ineligible. Furthermore, in the large water-bodies, rights cannot reasonably be concentrated upon a sub-set of *bittaheen* groups to the exclusion of others. Considerable co-ordination is therefore required either to organise the collective holding of these rights (with many problems in the division of the product) or to allocate specific areas of the water-body to different groups. Even if such groups are genuine, and under the guidance of a single NGO, this can be difficult. However, the experience with group access to water-bodies has revealed the presence of other vested interests in these groups. Sometimes they have been created fictitiously for 'professional fishermen' or businessmen to gain access to fishing rights, sometimes existing groups have been penetrated by the henchmen of such businessmen. The allocation of a peppercorn lease to a group may have been achieved by the successful networking of such businessmen with District or *Upazila* officials. Where the groups have no prospect of acquiring boats or nets, they are then obliged to contract 'professional fishermen'/local businessmen (if they are not already connected to the group) to undertake the actual fishing. As noted above, the labour fishermen may belong to a different communal group anyway and it is not clear to what extent the performance of these activities are socially segmented.

Cost recovery leases through auction (with reserve prices?) reintroduce the principle of competitive bidding, and would represent a move back from the 'equity' to the 'revenue' objective. Given the Bank's view on social leakage, upheld by the remarks above, it is difficult to see how even an organised federation of groups could bid successfully. It is possible that NGOs could do so on their behalf, and then in effect administer the waterbody through 'their' groups. One might imagine the Grameen Bank extending its 'fish factory' approach (Neemgachi and Dinajpur) to seize these opportunities. Cost recovery leases through qualitative tendering at fixed prices (in effect similar to peppercorn lease allocation to 'genuine' target-groups, but at prices which pay for all or a proportion of stocking costs) may stimulate recipient groups to engage in policing and self-employed fishing in order to protect their investment. However, some of the social problems above will apply and credit support will be necessary to purchase leases and equipment.

A central issue which applies to the social targeting of fishing rights and opportunities concerns the process of group formation. One of the weaknesses of the NFMP has been the induced haste to form groups around water bodies in response to the policy of preferential allocation. Most NGOs, seriously working with the poor, realise the long term negative consequences of rapidly forming groups to obtain an instant material asset. Their internal solidarity quickly breaks down. Groups are formed by NGOs as part of a political analysis of poverty, in which the development of group solidarity is the paramount objective to assist individuals in a collective struggle against the classes and state apparatuses which oppress them. There are no shortcuts. There are dangers in the present fashion to incorporate NGOs and 'their' groups into large-scale development schemes. For the TFP, the conclusion should be that any attempt to target access to fishing rights to the poor needs to be sensitive to a sequence of events in which the physical availability of resources follows the development of the social and political capacity of the poor to acquire, retain and fully utilise such assets. That attempt should also be sensitive to the cultural constraints which may be responsible for the division between lease-holders and the activity of fishing. There are gender as well as communal and class issues here.

Fishing Opportunities

These are to be distinguished from rights, in the sense that floodplain stocking is intended to provide incomes for the casual, part-time fishermen as well as full-time, 'professional' ones. Such opportunities are supposed to exist outside the period of major catching, where the incremental productivity from stocking provides the incomes from which cost recovery policies can be sustained. The Appraisal estimate is that casual, part-time fishermen (among whom destitute women and children are predicted) will gain access to 50% of the incremental fish catch through TFP at no charge. (para 6.09, p 3) As suggested above, there are several problems with this prediction.

First, there is a 'management of the commons' issue—the avoidance of early fishing of large, recently stocked, fingerlings. Newly stocked fingerlings do not readily disperse and so are vulnerable to a boat-net catch. Fingerlings, it is now acknowledged, will have to be sufficiently large to survive significant predation. In this condition, they have greater attraction as foodfish. However, the tradition on non-stocked open water-bodies, auctioned out to leaseholders, is that casual fishermen are only allowed to fish after the main catches have been taken—i.e. at the end of the fishing season, when floods have fully receded. It is unlikely, therefore, that there will be significant 'poaching' of large fingerlings by casual fishermen at the beginning of the season, but competing 'professional' fishermen may do so.

Second, as mentioned before, the predominance of silver carp and big-head carp in the stocking strategy will locate the fish out of reach of casual cast net fishermen. The incremental fish catch will, therefore, be mainly restricted on species grounds to those with access to boats and larger nets. These are usually owned by entrepreneurs who have gained rights to fish (directly with a lease, or in some way through a group) and who employ labour fishermen. They may employ several teams working in a peripatetic way, moving from one water body to another. The difference between major and minor bodies is significant here—i.e. more peripatetic between minor bodies, more 'resident' for the season in the larger water bodies. Such labour fishermen are also expected to perform *chaukidar* or guarding functions, and both for this reason and the possible occupational specialisation of their communal group, such fishermen are likely to be migrants with few contacts (and therefore few alliances) with the local community.

Third, if cost recovery policies are being applied (either through the price of leases, or through other suggestions such as the licensing of boats and equipment), then we must expect entrepreneurs and 'professional' fishermen to have an incentive to fish out the end of flood, confined water body as efficiently as possible. No doubt marginal utility calculations will apply, but they are likely to extend beyond 50% of the incremental catch, especially in the more confined areas—i.e. precisely in those areas where casual fishermen might otherwise have had the greater opportunity for significant catches.

Fourth, we should not imagine that casual, part-time fishermen are allowed to fish a leased area free of charge. Usually they have to pay some kind of fee or incur some other obligation (e.g. agricultural labour at below market rates) to gain the right to fish even at the end of the season.

Fifth, if attempts have been made to target access to *bittaheen* groups, then such groups will also have an interest in either preventing or charging casual fishermen from outside their groups.

These comments, based on direct observations elsewhere (e.g. Nageshwari, Kurigram) and on discussions with PROSHIKA who have wide experience to relate, therefore challenge the statement that "Since access to fisheries is not controlled, there is a high probability that the increased levels of income would be diverted into increased employment in fisheries of lower income groups." (para 6.04, p 35).

Financial Packages

These are the 55 packages, allocated thus: ponds/17; minor floodplains/13; rivers/9; women's development/8; freshwater shrimp/5; coastal & estuarine fisheries/3. These packages would finance particularly dynamic Districts and *Upazila* fisheries officers to intensify their activities beyond the routine. Packages are expected to take different forms: studies, surveys, temporary peak needs of personnel in extension and technical services, training and operating costs. With such an open-ended, flexible component, judgements at this stage are premature. Indeed, this 'process' approach is to be welcomed especially where it is supported from the UNDP grant for studies and monitoring.

However, the prognosis cannot be optimistic with the current DOF commitment to production units, and a concept of research in MFL (as well as among donors) which is restricted to technical, biological issues as represented by FRI. The response from DOF and MFL (and ODA!) to the proposals for a revised PP for the Dinajpur Fish Culture Project, including a revised purpose for the Parbatipur site as a poverty-focused applied research, training and extension complex (RTEC), will be a good test of the commitment to: poverty-focussed extension; the switch of DOF efforts from production to extension and the need for applied research combining technical and socio-economic issues to guide extension strategies. The TFP appears very ambitious, but Parbatipur might constitute a pilot model for the approach.

At the same time, it is worth recalling some of the unresolved socio-economic issues which will be examined through an action-research programme in Dinajpur: ability of poor groups to acquire and retain significant water bodies (e.g. *khas* ponds and derelict, large co-owner ponds); a genuine rather than token entry of women into foodfish pond production; the portfolio of commitments for pondholders determining their interest in pond culture; modes of extending fish culture knowledge to widely dispersed pondholders, with possible use of fingerling traders as informal extension agents; securing a share for traders and/or fishermen in the incremental productivity of cultured ponds; the entry of poor groups (men and women) into small scale hatchery and nursery pond businesses (resolving technical issues like iron content in the water and availability of broodstock, as well as credit constraints and the leasing of ponds); the impact of culturing fish as a cash crop upon domestic consumption of fish; the transportation problems associated with moving large fingerlings over long distances (the connection between nursery pond sub-contracting and open water, floodplain stocking).

Switch of DOF from Production to Extension

This is clearly a major element in the overall strategy for TFP, especially for the 55 financial packages, but also for the stimulation of private sector capacity to meet the demand for fingerlings for floodplains stocking. It is connected to the conditionality for divestment of public sector production capacity, which is also supported by a research study (Public Seed Fish Farms 7.3 of Annex 9, p 15) to "assist in preparing a policy for MFL/DOF to divest itself of surplus capacity and avoiding construction of further capacity". However, despite the stated aims to switch the DOF focus towards extension, there is little provision to support this switch in either the ODA funded MTA component or the UNDP STA component. The reference to training under STA mainly concerns short study tours abroad in the region (Annex 9 pp 17-18), and under MTA 24 training awards for management in DOF. Neither of these really amounts to an input into improved, poverty conscious extension. The apparent commitment to a range of poor beneficiaries cannot be achieved through investment alone, especially if attached to a cost recovery strategy. The absence of any provision for a poverty-focused extension appears therefore as a glaring omission in the Appraisal. The commitments to research and monitoring social issues, have little operational meaning unless some thought is given to transferring the knowledge gained to DOF District and *Upazila* staff.

Given the DOF's historical weakness in extension (since it has been promoting fish culture through its own production activity), and given the fish biology background of its District and *Upazila* staff, as well as those in HQ management positions, the TFP should include explicit provision for in-service training and re-orientation of staff at all levels in extension support to the private sector, with special interest in opportunities for the poor. This training cannot be classroom based, but should be undertaken by a roving team which also facilitated introductions and short placements with local NGOs working with small groups in fish culture, or intending to do so.

The Dinajpur Fish Culture Project will try to work in this way, and again it could be regarded as a pilot for such an approach. The Parbatipur complex could well function as a centre for poverty-focused extension training in fish culture for TFP. We must recognise that the statements about beneficiaries and the extension orientation of DOF will remain as platitudes unless some provision along the lines suggested is forthcoming.

Technical Assistance (technical and social support)

It is clear from the applied research in the Dinajpur Fish Culture Project that much needs to be understood about the interrelation between fish culture interventions and their socio-economic context. But the headings for project support and research (funded from UNDP grant but administered by IDA) in Annex 9 of the Appraisal remain very open-ended on fishermen organisation, private sector stocking of minor floodplains, women's involvement (where provision for research seems to be especially inadequate), and of course cost recovery.

A particular worry here, with such open-ended TOR, is the implementation. It is intended (para 4.13, p 30) that the IDA would contract the implementation to a "reputable international consulting firm with long experience in planning, fisheries and rural development; it should have experience in South East Asian countries where experience in a wide range fisheries activities is abundant. ... At the outset the firm would enter in a cooperation agreement with a reputable domestic consultant firm with experience in the same areas and with a reputable NGO for the social support services." The pursuit of poverty-focused social objectives in TFP is going to depend crucially upon these technical and social support services. However, I have not been made aware by fish specialist colleagues of an international firm which has a strong socio-economic and fish culture CV, and certainly not with an extant knowledge of Bangladesh. The domestic partner choice is also narrow. There is a danger that a local NGO may well carry the burden for incompetence elsewhere. And, to the extent that the NGO will be expected to carry out action-research (as well as monitoring and research), it is odd that an IDA document makes no mention anywhere of connection with the BRDB/RPP.

Monitoring

Some of these issues have now been referred elsewhere in these remarks. However it is noted that the provision for monitoring in the MTA concentrates on physical and financial targets (2.4 Annex 8, pp 4-5), supported by provision in the STA for 'production monitoring' (2.2. Annex 9, pp 2-3). Otherwise there is scattered provision on social impact issues under STA in Annex 9: 2.4 on Fishermen Organisation, which is heavily reliant upon an NGO contractor for action-research; 4.2 on Minor Floodplain Development, with a study which "assembles experiences, successful and less successful, of the private sector (private individuals or groups, fishermen and rural poor groups with or without NGO or public sector support) practising stocking of minor floodplains.."; 4.3 on Women in Fisheries where a core study team will spend 3 weeks in each Division to assist project staff to develop plans for feasible 'women in fisheries' programmes. It is disturbing that there is no regular, high status, provision for social impact monitoring within the DOF and that the studies are time-bounded and specific, and do not address the poverty issues in the main investment part of TFP—namely the stocking and fishing of major floodplain areas. Social impact monitoring should at least be added to the TOR for the ODA funded MTA Monitoring function (Annex 8, para 2.4, pp 4-5). The management training should emphasise such issues. The TOR for the International Consultant firm under the UNDP funded STA should include, as an objective, the setting up of a social impact, poverty focused monitoring system for subsequent and routine use by the DOF, with the information generated during the lifetime of the TFP used for evaluation purposes.

PART V

STRATEGIES WITH THE POOR

Preamble

As indicated in the introduction to this volume, the preceding three parts have focused upon three major sectors of rural development activity: agriculture, rural infrastructure and fish. Part V now brings together a collection of papers which have addressed institutional issues in poverty-focused development. The pieces have been written over a period of nine years, but the last three chapters within the more recent 1992-93 period. The first paper (a reduced version presented here as Chapter 16) was prepared in 1984, partly at my own suggestion. With criticisms of the SIDA funded IRWP mounting from the poverty-oriented aid officials and their social science friends, there was concern about the efficacy of trying to work for the interests of the rural poor solely via the cumbersome strategy of rural works employment. As new directions were considered, including hungry looks at the NGO models with thoughts of donor replication (as subsequently implemented in the Production and Employment Programme—PEP—and criticised by me in Chapter 17 here for being too donor-led), I argued that the experience of the Bangladesh Rural Development Board's (BRDB) work with the rural poor should be reviewed in order to see how it could be further supported. This was a forerunner of a later, more explicit argument about state versus NGO responsibility for working with the poor.

The original report on which Chapter 16 is based was rather long, and some of it must now be reckoned as water under the bridge. In editing, therefore, general arguments about mobilisation and the social conditions of poverty have been cut, and the history of BRDB involvement with the landless has been severely pruned. The reason for this is to emphasise the case studies of BRDB rural poor mobilisation and the conclusions to be drawn from them. It has been easier to do this because some of that discussion was summarised for the 'Sirs and Sahibs' report, which appears here as Chapter 17, which again also includes case studies and analysis of mobilisation practice. Thus these two chapters sit interestingly side by side. Two sets of practices (BRDB and a donor-led sequel) can be compared.

Chapters 18 and 19 are much shorter (possibly to the relief of the reader!) and more recently written: 18 is an extract from a paper for a SIDA exercise reviewing its aid stance towards Bangladesh; and 19 an extract from the recently published *Breaking the Chains* (Kramsjo and Wood 1992). Both of these pieces reflect a degree of impatience with mobilisation panaceas. Chapter 18 appears here as a corrective to the fashion for entrepreneurial models of income generation; 19 draws attention to a more mundane definition of struggle and consciousness, arguing that we should use terms like revolution and struggle to refer to more modest, incremental, small-scale and local victories by the poor.

Chapter 20 is an adaptation of an annex written by me for the recent evaluation and appraisal of GSS in Bangladesh. This was an opportunity to bring together some further thinking about the relationship between income-generating activity and mobilisation, and in particular to draw upon collective thinking at CDS concerning the notion of 'resource profiles'.

Chapter 16

Government Approaches towards the Rural Poor in Bangladesh

Introduction

The Rural Poor Programme (RPP) as it was emerging within BRDB from 1977 onwards represented the GOB's main direct efforts to address the problems facing the rural poor in the context of its overall rural development programme. It received a stimulus through the RD-I Project (Rural Development-I, a World Bank funded special programme) in selected *upazilas* in Mymensingh and Bogra districts, especially from 1981 when the 'landless cell' was created in BRDB in response to the joint IBRD/GOB review (1981). A further, more explicit stimulus was given with the RD-II-RPP programme (see below). This development of the rural poor programme within BRDB reflects in principle, then, a set of evolving institutions to which the Nordic Donors might attach an additional component within its broader, revised framework of an 'Employment Sector Programme' as a sequel to IRWP. The BRDB Rural Poor Programme (RPP) therefore had to be examined.

The criteria for assessing the RPP at this stage derive to some extent from the current experience with IRWP. To summarise then: there is a need for a more realistic if slightly more complex understanding of the nature of rural poverty in Bangladesh. Interventions by donors should be made on the basis of that deeper understanding rather than on the simplified notion of delivery via benign institutions to an abstract category of recipients who remain individualised, exposed and vulnerable in the context of a temporary and partial flow of resources, dominated by the recipients' habitual oppressors, into the community. The interrelationships between various elements should be recognised. These are: a conception of poverty as a social process; the need for the existence of real groups capable of united, meaningful social action instead of groups formed by merely drawing circles around convenient numbers of similarly designated people; income generation conceived not just as precarious employment given by some external institution (and therefore always susceptible to withdrawal) but which is secure because it derives from the use of one's own assets; and the prevention of income loss via a direct attack on rural non-institutional moneylending, which is complementary to the process of income generation, and indeed underpins it.

History of Landless Focus in BRDB

IRDP/BRDB Mobilisation of Landless

The IRDP (BRDB since December 1982) started in an ad-hoc way to organise landless groups for men and women from around 1977. Initiative was clearly demonstrated in the RD-I *thanas* of Ghaffargaon, Trishal and Muktagacha in Mymensingh District, and Gabtali, Shariakandi, Kotwali and Sherpur in Bogra District, where the Project Director took a personal initiative. However, the World Bank report for 1982 on 'Selected Issues in Rural Employment' noted that "...of the 303 societies organised by March 1982 under RD-I which is implemented by IRDP, only 138 had been registered..." (*ibid* P 68 para 5.12). The remainder of the paragraph is also revealing and worth quoting in full:

> "As unregistered societies cannot get credit under the program, by March 1982 only Tk. 1.2 million had been loaned, 68% of which was for cattle fattening. Because of the problems associated with the lease of ponds and availability of fingerlings, of the 122 societies organised for pond fisheries only 92 had been registered and only 41 had been able to lease ponds. Furthermore, because of unfamiliarity with the techniques of promotion and development of non-crop programs for the landless, there was a general lack of commitment from within the IRDP." (*ibid*)

There are a number of highly significant themes in this statement. Certainly it conforms to the experiences of the groups who were interviewed by me in the *upazilas* in Mymensingh district. However even with this experience, IRDP was instructed in 1982 to mobilise the landless all over the country following the joint GOB/IDA review of IRDP in 1981, and the subsequent creation of a landless cell within IRDP. An example of the confusion this has caused can be seen in Phulpur where 11 BSS (*Bittaheen Samabay Samity*) were 'formed' in response to this instruction, but only three have been registered and are therefore eligible for loans. The RD-II Appraisal Report (1983) observed that:

> "Major problems faced in the implementation of these programs of BRDB have been delays in the registration of new societies, delays in the flow of credit, training and extension support and a lack of management focus in BRDB to implement a more extensive program." (p. 20).

BRDB Women's Programme—description

In this review of rural poor mobilisation activities in the Rural Development Division, there are programmes which are gender-specific to women. While other programmes do not formally exclude women, in practice the specific circumstances of rural women are not recognised. The BRDB contained a Women's Programme entitled 'Population Planning and Rural Women's Cooperatives'. The formal position on this programme is that it started as a Pilot Project in July 1975, completing its first phase in June 1980. The second phase is 1980-85. As of June 1984, 1477 women cooperatives (*Mahila Samabay Samity*, MSS) had been formed, across 40 *upazilas* of all districts, with a membership enrolment of 58,564. (Islam, S 1984). The main objective of the programme is:

> "To integrate rural women into the development process in a manner which will enable them to have access to the services provided by different organisations" (Development Schemes of R.D. & C Division in LGRDC 1984-85; Planning Cell, Programme Section).

The components of the programme are:
- Organisation and formation of women's cooperative societies
- Orientation and motivation
- Training of the staff
- Production of the training manual
- Adult literacy
- Family planning
- Health and nutrition
- Development of cottage crafts.

The cooperative societies range in size from 15-90 members, with an average size of 41.5. Representatives from each cooperative receive training in the preparation of production plans, management etc. Plans are submitted to the TCCA for approval and the normal two-tier rules apply for registration, savings and shares.

The main issue for the donors in considering any involvement with this programme is whether the claims to be working with *bittaheen* women are really justified. There have been differences of opinion on this issue. The programme's own documents refer to the "mobilisation and organisation of the poor and disadvantaged rural women through MSS" (*ibid*, Page 3). However, the Joint Director in charge of the programme reports that although the programme is meant for the poorest women, only about 50% are from landless families (Interview 10.9.84). The programme was thoroughly evaluated under a CIDA sponsored project in June 1980 ('An Evaluation of the IRDP Women's Programme' Feldman, F. Akter and L Banu CIDA June 1980), which found that richer village women dominated the cooperatives and then monopolised the services and credit, and that this pattern was especially institutionalised in the nomination of the five representatives from each cooperative.

In relation to RPP, some clarification was necessary. There was some danger that the women's groups in the RPP would in effect not be MSS but MBSS—which should mean that the criticisms of Feldman et al have encouraged a greater concentration upon the *bittaheen* women. This 'clarification' should therefore settle the issue of terminology between MSS and the currently informal notion of MBSS. The MSS in the Women's Programme of BRDB did not, have the membership composition which conforms to the objectives of the RPP, RD-II RPP or IRWP.

Lost Inputs to RD-II RPP Design

It is relevant to note that many of the officials with experience from SFDP and other experimental programmes who were involved in the preparation for the RD-II—RPP component, were not convinced of the TCCA two-tier structure as the vehicle for the RPP. They made innovative suggestions on loan practices (e.g. multiple, not single purpose) and the value of mutual aid within the groups for distress credit etc. It was also pointed out that the share deposit—credit entitlement ratio involved group members in taking loans from village moneylenders to raise the share capital to release significant levels of funds for worthwhile productive activity. The value of the credit was thereby instantly reduced because of these other obligations. A third issue, raised especially in the context of women, was the proposal that the female 'inspectors' (assuming TCCA structure) should only work with 4-5 MBSS since it was physically impossible to do the necessary legwork for more if the programme is serious about mobilisation. It would also be difficult to motivate such 'inspectors' on their low salaries. The RD-II RPP ratio of one 'inspector' for 10 groups would be impossible. This was rejected by BRDB on the grounds that neither ERD nor the World Bank would accept it, even though a high proportion of the budget is for transportation and infrastructure. There seems to have been considerable slippage between the thinking arising from the final RD-II RPP proposal.

RD-II RPP—description

The final section of this review consists of a summary description of the RPP component within RD-II which began as a legal entity on 15.3.84. It constituted an extension and intensification of the landless mobilisation activity under RD-I, and in particular relies heavily upon the TCCA (now UCCA) two-tier model despite the doubts expressed at the time of the project's design. While RD-II was intended to apply in 13 old districts, the RPP component of RD-II was originally confined to six districts 'because of BRDB's currently limited implementation capacity' (Project Proforma for RD-II. Planning Division, BRDB March 1984 p6).

Over the four years (in effect) of the project up to June 1988, the number of *upazilas* included in programme was planned to rise to 100, involving the creation of 2,500 new BSS and 400 exclusively MSS. Membership of the BSS would be limited as now to those owning less than 0.5 acres of land and earning their basic livelihood from casual labour. By year 4, BSS membership was expected to reach about 120,000 men and 9,000 women. Leaving aside the estimated numbers of BSS members in these six districts (i.e. mainly in the RD-I areas of Bogra and Mymensingh) of 21,000, this target figure would produce average group membership of 40 for BSS and 23 for exclusively MBSS. These are large groups, adding a new dimension to the facilitator—group ratio.

The major income-generating activities were expected to be in pond fisheries, beef fattening, weaving, rickshaws, cane and bamboo work, pottery, carpentry, oil-milling, and bio-gas units. (No mention of irrigation, despite an experimental programme for the provision of irrigation by landless groups with BRDB-IMP, supported by a grant and loan guarantee deposit from the Ford Foundation.) Fish pond excavation was to be supported by medium term credit (see case studies to follow) and short term credit available for some of the other activities, along with a cattle insurance programme. Other agencies to be involved were the Bangladesh

Handloom Board (BHB), Bangladesh Small and Cottage Industries Corporation (BSCIC), Bangladesh Council for Scientific and Industrial Research (BCSIR) for developing bio-gas units.

The programme depended heavily upon a training input both for BRDB and for *Upazila* Central Cooperative Association (UCCA) staff, and upon the provision of additional staff for the RPP. The 'innovative rural poor' training for BRDB would be conducted at RDA, Bogra, while the UCCA staff training would take place through the *Upazila* Training Units (UTU). The main 'rural poor' training would be for: UCCA promotional staff; BSS (and MBSS) Chair-persons and managers, and skill training where necessary. The training of UCCA staff would focus on "how to motivate the disadvantaged people to form BSS, create an attitude of mutual trust and build up an absorption capacity for project activities" (Project Proforma 1984 Page 16 or World Bank Appraisal Report 1983 p 31). It is in this section that in addition to institutions like BSCIC, an input from 'willing NGOs as appropriate' is made but with reference to skill training only. It is clear from my discussions with the World Bank ('Skills come first in our view') BRDB and RDA, Bogra that no 'training' role outside skill training was envisaged for the NGOs.

The main elements of the programme, therefore, were: targets for group formation; phasing of *upazilas* to be involved; specified income-generating activities; other organisations to be involved; strengthening of BRDB staff and absence of corresponding UCCA expansion; training; and relationship to NGOs. After the project was designed, the Canadian International Development Authority (CIDA) became a major funder of RD-II. As a result, a CIDA consultant produced an Inception Report which referred to the reluctance of BRDB to involve NGOs in a broader training input, not just confined to skills training despite the fact that their strengths lie in basic social orientation and conscientisation which are not what BRDB wish to include.

Case Studies of the BRDB Rural Poor Programme (Mymensingh)

Introduction

In the preceding sections various criticisms have been made of the experience so far of BSS/M(B)SS mobilisation through the BRDB/UCCA system. It is necessary therefore to present in more detail some of the evidence for this criticism. In the time available, discussions with nine groups together with a significant interview relating to a further group were held. Therefore ten summary case studies will follow. These groups are distributed between three *upazilas* in Mymensingh District: Muktagacha (five groups); Nandail (three groups); and Phulpur (two groups). The gender composition of these groups is: seven male, two female and one mixed.

I was assisted in these group discussions by a Research Assistant, a long time associate of mine. We have jointly been engaged in discussions with landless groups in many different parts of Bangladesh. In each *upazila* visited, we interviewed the Rural Development Officer (RDO)-BRDB and obtained a union wide distribution of the BSS and M(B)SS. We selected at random, making certain, except in case 3 (mixed gender), that the village locations were not on the main *upazila* road. Except for Group 3, in no case were the UCCA staff or RDO able to prepare for our group visit in advance. In most cases we 'released' our UCCA guide as soon as we made contact with the group. We used a loosely structured interview proforma to provide a checklist of questions. However in the time available, the extent of systematic detail we sought was over-ambitious. This is why case studies are presented here, rather than any spurious attempt to quantify, aggregate and make comparisons on this basis. Interview sessions lasted for a minimum of two hours, often longer. But no-one can expect accurate detail of landholding indebtedness or institutional loan transactions, only indications and types of relationship. A number of common themes emerge from these cases and these will be discussed in a concluding section at the end of the chapter. The cases are summarised in the order in which we met the groups.

1. Chapuria Dighi S.S, Mankon Union, Muktagacha Upazila

Membership: 20 males and one widow. Formed in September 1976 and registered one month later. The manager has 0.6 acres and leases in 0.4 acres. One other member owns 0.4 acres and leases in 1.36 acres. The remaining members are dependent on wage labour and fishing from the tank. The group was formed on the initiative of ex-Minister who resides in the village, by his friend Omed Ali. It was formed around a derelict tank. The group has an eight-year lease, but is confused over longer term rights despite paying 50% excavation costs. UNO are now reclaiming pond, ordering an open auction. We were informed the group has paid a bribe of Tk. 7,000 to try and stop this order. Fishing is clearly profitable—Tk. 1 *lakh* profit claimed. Each member has cultivated a plot on side of tank for fruit, vegetables. Loans from UCCA have been for rice-husking, duck-raising, beef fattening and the purchase of one LLP. Beef fattening loan in 1981 of Tk. 5,000 still in default from individuals who will not repay, so no more short-term credit to individuals is available. LLP loan is medium-term (over five years) taken in 1982. The group manager was given 20 days mechanical training for LLP; the UCCA inspector advises on irrigation management and maintenance. Payment from farmers in command area of only five acres(!) is unreliable. Group accounts are very confused, despite requests from RDO to UCCA inspectors to sort them out. Previous manager was thrown out of the group in 1980 when it was discovered he had misappropriated group funds for four years. He was the village *chaukidar*, and was made leader because he had access to UCCA and knew how to handle papers. There is a hostile relationship to KSS, since it is waiting to compete in the auction for the fish tank. Omed Ali, one of the founders, now Vice Chairman of the group, and friend of ex-minister, is a labour *sardar* for an earthcutting group. Members extensively in debt to local farmers, business and service-holders. A frequent condition of loans is a promise to work at harvesting/weeding for Tk. 10 a day instead of usual rate at that time of Tk. 15. Effective interest rates seem to range between 7%-15% per month. Although an income exists from sale of fish and fruit, it is distributed among individuals to meet immediate consumption needs. Despite this income, group savings are estimated at Tk. 2,000. But there is some evidence of collective activity in trying to retain control over the tank, and to make the LLP profitable by expanding the five acre command area. Union Chairman is lobbying for auction on pond. The interviewees claim that he only employs group members on earthworks to 'show-off' if visitors come. Group members used to have to try to sell their labour every day, but this pressure has been relieved by their access to the pond. Also they used to outmigrate for work during August/September and April/May.

2. *Chapuria Bhuminin Mahila Samity, Mankon Union, Muktagacha Upazila*

Membership: 20 female. Formed November 1983, registered January 1984. This is in the same location as the male group (1). Fourteen are married to members of group 1, four others are married and two are widows. Nine are closely related including managers. Their main work is agricultural or in houses of their husbands' moneylenders for 1 *seer* rice per day. Group formed after female IRDP officer from District came to look at male group's pond. She said if they formed a group they could get loans. After three months this was followed up by TCCA inspectors and one of the male group members. They promised a large loan for a handicraft centre if women deposited Tk. 20 each. Chickens were sold to raise this deposit. Tk. 400 deposited. Nothing happened for three months again until women went to *upazila* HQ where they were informed that a loan for Tk. 10,000 allocated to them, Tk. 500 each (which is the loan ceiling). The loan was for rice-husking and some bamboo/cane work. Each member only received Tk. 440 (thus Tk. 1200 'lost' somewhere). The inspectors told them that Tk. 300 was for a bribe. The members had to thumbprint sign for Tk. 500. Inspectors also demanded Tk. 5 per head. Repayment of loan is in monthly instalments, but only after first month did women realise the 12% interest plus 5% service charge was also due. Defaulting has occurred, mainly because of confusion over this issue. Manageress of the group studied up to Class 5, and can read and write a little. She is married to someone who has studied up to Class 10. He is not a member of the male *samity*. The rice-husking activity is individual rather than collective. Loan required to buy a *dheki* (Tk. 40), the paddy, fuel for parboiling. Husbands and sons buy the paddy and fuel (wood) and sell the rice, unless a neighbour comes to purchase it. The drying problems this year have forced members to use the loan for immediate consumption needs. Any 'profit' accrues at a rate of two *seers* of rice (Tk. 20) for two days of processing. The loan ceiling is so low that only one *maund* paddy can be purchased at any one time, so sale has to be completed before paddy can be purchased again. The work, therefore, is not continuous. No other M(B)SS in village, but other women would like to join. They are excluded because of ceiling on group numbers. Manageress requested by a woman from another village when she met at Muktagacha Upazila centre to help her form a *samity*. Earlier one of the inspectors had tried to form a *samity* with those women, but had demanded a deposit of Tk. 40 from each and they could not raise this amount. They remained interested, however, and therefore approached this manageress.

(In both these groups (1 and 2) the names of two inspectors appeared with regular frequency. In fairness to them, their identities are not disclosed. The women especially, with whom we spoke in private inside a *bari* with no onlookers, reported that they were totally dependent upon them for getting any access to loans, therefore "we do what they say").

3. *Monirambari Pulpurpara, Bittaheen Telojibi S.S., Bhashati Union, Muktagacha Upazila*

Membership: 16 male, 12 female, total 28. Formed mid 1982. Registered June 1984. Our discussion with the group was hindered by the presence of a Union Member who initiated the group just before the Union elections in 1982. The group's 28 members are distributed between five *baris* in one *para*. It is a traditional oil-crushing village. All 12 female members are involved in oil-crushing, along with eight males. Seven males are rickshaw pullers, and one is a blacksmith. He is the manager. The Union Member is from the village. He is an inter-district truck driver and claimed to be landless. The group was started by him in association with Abdul Halim and Abdus Salam, who are both labour *sardars* contracting labour for loading trucks with oil seed, earthworks and harvesting. Their main contact for earthwork opportunities is the Union Member. The member obtained one machine at first, and over time 12 more were acquired out of savings and loans from the local *mahajan* Abdul Huq, who owns nine acres in the village. He is the group's source of oilseed (as a merchant), charging 5-7% higher than the normal price to be repaid after two days! He employs group members in agricultural work and they receive advances from him. The labour *sardar* members contract labour (from the group and outside) to work for Abdul Huq. Women joined the group a year after it was formed. They are not the wives of male members but often kin-related. They have skills in oil-crushing through this kin-relation whereas those from other families have not yet. Husbands of female members are mainly rickshaw pullers. Oil-crushing has been a traditional family based occupation for these *baris*. The advantage of forming a group was to raise capital to buy oilseeds and machines. Member took initiative to get group registered this year since he learnt that "government was going to help the landless". He went to UCCA office so an inspector came to see the group. No loan has been taken yet from UCCA, but the member is hoping a large enough loan can be obtained to run the business during the year, to purchase oilseeds directly from producers; and therefore to cut out the merchant/*mahajan's* domination of their business. The business essentially consists of the groups acquiring the oilseed from *mahajan*, crushing it; selling it locally; and then repaying the *mahajan*. The women it seems do most of the hard physical work of turning the crushers (normally done by cattle), with the men handling the transactions or working elsewhere. Profit from oil crushing has been distributed directly and equally to the members engaged in the oil-crushing. This amounts to Tk. 300 per member, per month for six months for immediate consumption. Tk. 2 is deducted per week per person for the whole year for savings (Tk. 2,912 per year). Since registration the group has bought 28 x Tk. 10 shares, and deposited Tk. 1,500 savings with UCCA. This group is an example of collective economic activity occurring independently of the UCCA so far, but not independently of the village and Union power structure, labour management and debt dependency. It would clearly be in a stronger position if the links (collective, individual and diverse) with the *mahajan* could be broken. The significance of the Union member connection is difficult to assess in the way that the interview was set up.

4. Sagradi BSS, Batagair Union, Nandail Upazila

Membership: 25 male. Formed April 1983, registered May 1983. Some of the members of this group have recently had land, (such as 2.0 acres, 1.5 and 1.2) but it has been mortgaged out—sometimes to relatives, sometimes to a jute businessman from a nearby village. Others still have small amounts of land (5 to 50 decimals), and six are still sharecropping. The group was formed after the UCCA Vice-Chairman and KSS manager of this village approached the *madrasa* teacher and suggested he form a group. The group Chairman is also a *madrasa* teacher. We learnt that this UCCA Vice-Chairman/KSS manager was also a principal moneylender to the group members as well as their employer. The other main moneylender/employer was the Union Member from this village. The need for advances was highest in September—October when a *maund* of paddy would be repaid by two *maunds* of paddy (or equivalent cash value) after 3-4 months (after the harvest), and the workers would also have to accept wages below the market rate. At least five members were in debt to the KSS manager/UCCA V.C. in this way. One of the group members, Abdul Kadir, is a labour *sardar* especially on Food for Work Schemes. He contracts labour for the 'member's' schemes. They are from the same *gosti*. In September 1983, Tk. 35,000 was sanctioned by the RDO as a loan from UCCA to the group. Not all members took a loan by choice. The average loan per member receiving was Tk. 900 (loan ceiling?), but three members of the Managing Committee including the manager, had loans each of Tk. 1500. No repayment had been made, since "we still have the grace period", and at the interview (18.9.84) it was clear that the manager had little idea about the interest rate. The UCCA inspector had told him "not to bother about it". The loans were taken for beef fattening and rice-husking. All loans are individual. Loans are regarded as too small to accumulate savings for people to reinvest on their own without having to repeat annually the elaborate (and expensive!) documentary procedures to secure a loan from UCCA. We had the details of this procedure described to us (see also Banking Plans both of 'World Bank' September 1983 and BRDB August 1984), and there are many points of access where 'payments' have to be made. The involvement of UCCA staff or BRDB staff is restricted only to advice on income-generating activities, shares and savings discipline as a precondition for loans, and on how to organise a meeting with tables and Chairs(!!).This final sentence could be repeated for many of the cases being described.

5. *Mozzempur M(B)SS Mozzempur Union Nandail Upazila*

Membership: 20 female. Formed June 1983, registered August 1983. The main feature of this group is the relationship between manageress and the group members. She has completed Intermediate (Class 12). Her husband has two acres, and her father who is a neighbour, has four acres and is a service-holder. The manageress was our informant and she was well-dressed. She heard about *samities* at college, so when the manager of the KSS told her to form a society she responded. The KSS manager is the husband of her girlfriend. He took her to the UCCA office and she was told to contact some women and collect their deposits. Twenty women responded to her call, and they deposited Tk. 17 per head. She did not know why this figure was requested. Most of the women are dependent on their husbands. There is no earthwork or agricultural work; some have been involved in basket making. The manageress' aunt is the chairwoman, and lives in neighbouring *bari*. The manageress is the only member who is literate. Loan from UCCA was for goat raising. Allocation was Tk. 6,500 (August 1983), so the 20 members received Tk. 300 each (who buys and sells the goats? Do the women know what those prices were?). The manageress received Tk. 500 on advice of UCCA inspector. She has spent her loan on consumption. "I can repay at anytime. I just have to get the money from my father". She was confused over interest rates and service charges—she thought they were due to pay 12.5% interest and 15% service charge. Tk. 1,000 is outstanding, and all the interest to be repaid. We did not learn much more from this interview since our informant excluded other interested group members from the discussion. When one came into the room, she was sent out for the reason that "she knew nothing". Others were too shy to enter. This is all data of course, since it reveals the extent to which the educated manageress was not of the group but a 'broker' or *sardar* for it. She dominates all communication with outsiders.

6. *Chamta BSS, Mozzempur Union, Nandail Upazila*

Membership: 32. Formed September 1983 and received loan from UCCA in November 1983. Informant was an ordinary member, a rickshaw pulled. His lack of knowledge and that of other ordinary members who joined later is itself evidence of the lack of involvement of the membership. The Chairman of the group has most land, but still less than one acre. Five more have some land but below 0.5 acres; 20 members sharecrop. Twelve have land mortgaged out on the local *girbi* system. The members are otherwise agricultural, and two of them are rickshaw pullers. There was no explanation of '*Bittaheen*' for them. The manager, who is a petty trader (e.g. paddy), called a meeting and said he had been told that if a group was formed there would be credit for them, The manager knows the *upazila* offices as he was often there for business. The manager, chairman and vice-chairman all have some land and can read and write. Two of the group members are also labour *sardars*. The proportion of landless in this area was estimated at between 75-85%. Most of the group members have lost land over the last ten years, so that most can remember when they were small farmers. Now they have to out-migrate for work during April—May, and October—November. Sometimes 12 of them will work as a gang organised by the *sardars* on the Food for Work Programme. The *sardars* receive the payment from Union member or chairman. Most of the group members work for one large landowner and businessman—Ismail. He owns more than 20 acres. He uses wage labour, not tenants (so those who sharecrop do so on other land). He does not pay advances for lean employment periods unless land is offered in mortgage. This is a labour surplus area, so he does not need to offer advances to secure his labour supply at peak periods. A loan of Tk. 550 per person was taken in November 1983, for goat raising, beef fattening and rice husking. A high proportion of this loan, according to our informant, has been spent on immediate consumption. To release the loan a deposit of Tk. 20 per head was paid (share presumably), but just before the receipt of the loan a further Tk. 1,500 had to be deposited. The total loan was Tk. 17,600. (It is difficult to know whether Tk. 1,500 was a straight bribe, or if added to 32 x 20 = 640 + 1,500 = 2,140 which would yield a loan of Tk. 21,400. In this case Tk. 3,800 would have been 'realised' by the officials managing the loan at various levels.) To raise the money for these deposits, the group members had to take loans from local moneylenders. Now they are being continually urged to pay back their loans by the UCCA inspectors but they have little choice but to default, thus preventing access to any further institutional credit.

While we were in this discussion a *bittaheen* man from another village joined. He wanted to merge with this *samity*, as 21 of his fellow villagers collected Tk. 21 each and gave it to a UCCA inspector. They have heard nothing now for over half a year. Some of this Tk. 442 had to be loaned.

7. *Shahapur Majapukur BSS, Pyari Union, Phulpur Upazila*

Membership: 23 male. Formed 1977. Informants: five ordinary members and the vice-chairman. This group was formed out of a struggle over the rights to a derelict tank. It had been used by everyone in the village until a rich group tried to take it for their own exclusive use. The rest of the village was against this, and heard that if a *bhumihin* group was formed they could lease it from the Circle Officer (now UNO). A group of 43 was formed, and filed a case against the rich group. The rich group accused them of not being *bhumihin* so the Sub-Divisional Officer (SDO) and Circle Officer (CO) came, sat by the tank and worked out who was landless! Twenty people had to leave the group, among them five, including an ex-Union member, who had initiated the formation of the group. The tank was excavated at a cost of Tk. 17,000. Some informants assumed they would have to pay the whole amount, others only half (which is the provision in Circular No. 5). The RDO has no trace of the papers referring to this lease. He speculated that at that time, the rich group bribed the CO to prevent the *bhumihin* from obtaining the lease. Some of group worked on the excavation under a labour *sardar* from the group. A shopkeeper on a road about two miles away who is related by kin to some of the group members has been handling some of the money transactions with the UCCA on the group's behalf (perhaps because the literate initiators from the original membership had left). According to him the group had received no income from the tank as they were preoccupied with their obligation to repay the half excavation cost which they considered as a loan. He was now withdrawing his 'services' as an informal broker with the UCCA because the group's accounts were so chaotic. However, there are other good reasons why little income can be derived from this tank. It has a small canal running through it. When it was excavated it was blocked. But after the re-excavation, the farmers led by the excluded rich group insisted that the *khal* be opened up again otherwise their paddy land would be overflooded. Now the tank is useless for fish cultivation. Only wild fish can be caught. Furthermore, the group tried to plant coconut and jack-fruit trees on the pond, but these were repeatedly stolen at night. There are labour *sardars* in the group closely linked to the Union member "on whom we are dependent for the earthwork." About 19 of the group members are in debt mainly to five or six members of the rich group and their outside relatives. Cash loans are repaid at 20% per month. For one *maund* of paddy advanced in September, two *maunds* have to be repaid in December (300-400% per annum!) Finally there are UCCA elections each year, and the BSS manager has a vote. The UCCA Chairman is a large landowner from another union but he reportedly comes offering food to the BSS manager at each election time.

8. Emadpur BSS, Pyari Union, Phulpur Upazila

Membership: ten male. Formed: September 1982. Some confusion in the group about whether it was registered a year later. None of group members has possessed land in last five years. Their land has been sold, not mortgaged. Between 15-20 local families are buying up land from others in distress. No-one in the group has cattle, therefore no plough and no sharecropping. Only agricultural work available, and occasional earthwork. These same land-purchasing families are the moneylenders, and advances in cash or kind are repaid by working for them to the value of 1.5 times the advance. They were encouraged to form a group by the IRDP village accountant who promised loans for rickshaws, shallow tube-wells etc. Even drains were constructed in preparation on a command area. The group pressed the initiator to arrange registration but he did nothing. The group has raised Tk. 130 deposit so far and paid it to the initiator. The group has no idea where that money is now. The vice-chairman, a rich-looking *madrasa* teacher, claimed that the initiator had a reputation for collecting money from people in this way, promising them loans and then keeping their deposits. "Tk. 25 here, Tk. 25 there". A UCCA inspector has advised them to hold weekly meetings, raise share capital and they will get a loan— the familiar, repetitive theme. This group has given up now. (Part of the explanation for this problem is the drying up of funds for RPP groups in *upazilas* like Phulpur which find themselves not included from RD-II RPP after being urged to form BSS and MSS.)

9. Shashara Laxmipur Bhumihin Pukur S.S., Ghoga Union, Muktagacha Upazila

Membership: 37. Formed: 1979, Registered 1980. Informant Dudu Mia, Manager. Group formed by Union Council Chairman, who handled paperwork with UCCA, to obtain the lease on a derelict tank as per Circular 5. Chairman came and sat by the pond and by talking with people constructed a list of the 'poor'. This list was sent to UCCA, which sent an inspector who gave a lecture on shares and savings. The tank cost Tk. 10,500 to excavate; the group was obliged to pay Tk. 5,250 as a condition of the eight-year lease. No papers exist referring to lease. Of the funds for excavation released to the Chairman, only a little was spent. Excavation was incomplete, so the tank dries out in winter. In the first two years, as a result, there was no catch. In the third year Tk. 3,500 earned but it was spent on litigation against a former member who has had to leave the group. Last year (4th) a catch worth Tk. 4,000 was divided between a bank deposit to pay for excavation and individuals for immediate consumption needs. The catch is made in September/October when fish price is high and little work is available. When complaining that the tank is dry in March/April, the Chairman told the group to pay any further re-excavation themselves. (This Union Chairman has a reputation for this kind of behaviour elsewhere in the Union. It would be fair to suggest that he used the group as a way of getting the re-excavation funds released to him). Five group members have no homestead, most of the others have some cultivable land. Six have about one acre and cattle, so they along with four others are also sharecroppers. Land was lost mainly by fathers. But also this was a Hindu area, and when Hindus left many outside merchants and businessmen moved in quickly and acquired the land here. Ten are engaged in rickshaw pulling, hiring them from someone in Muktagacha *Upazila* town. The manager is a shopkeeper, and lives in the *bari* of his richer, landholding brother. Most of the members are in debt to two large landholders (at least 50 acres and 40 acres each) for whom they also work. The Chairman of the group, Panjab Ali, is the richest group member, and is also an *Imam* of the village mosque. One of the Managing Committee members is a labour *sardar*, with close links to the Union Chairman, The group has taken a series of loans from UCCA since 1980, mainly for cattle but also fingerlings and fishfood. A problem has arisen over the 1982 cattle loan since five cattle died. At the time the group paid Tk. 40 per person for cattle insurance (and we examined their papers which verify this). However, this insurance is being ignored by BRDB who insist in the meantime on repayment. With this defaulting no-one in the group can now receive a loan; and yet the low ceiling on loans produces insufficient profit for group members to withdraw from their dependency on those loans. The previous Project Director, Mymensingh, threatened the group with jail if repayment was not made.

10. Palgai BSS, Choga Union, Muktagacha Upazila

Membership: 14 male. Formed in 1979. This case is based on a short interview with the KSS manager. The purpose of including it here will soon become apparent. The Union Chairman (same as for (9) above) instructed the KSS manager to organise a *samity* around a derelict tank in the village. Many in the village wanted to join, but were later informed by UCCA that the *samity* had to be *bhumihin*. This meant that the KSS manager could no longer be the BSS manager. So his brother, a school teacher, was appointed to manage the BSS. Close contact between the two brothers is maintained. Meanwhile the Union Chairman had obtained access to another tranche of re-excavation funds. Perhaps the observation (referred to earlier) of the RDO Nandail is wrong—the connection between BSS and other opportunities as in Circular 5, which IRWP are supporting, seems precisely to provide the Union Chairman with access to additional resources!

Analysis of Themes from the Case Studies

A number of issues emerge from these case-studies, some of them recurring strongly. They are discussed briefly below but not necessarily in order of priority.

Obviously there is the issue of corruption involving the UCCA, the Union Council and the *Upazila* administration. The group members lack any internal coherence or independence to resist these demands. The group members are directly affected when their entitlement to credit is effectively reduced as the transactions take place. In some instances groups are enticed into making deposits, and immediately become net losers. In other instances they lose over time, paying interest and principal on the full value of a loan which they never received in the first place. Finally, the formation of groups especially around fish ponds has provided the opportunity for other classes in the network to have access to public funds under conditions of low accountability.

Related to this is the attitude towards the idea of a 'group'. Throughout these cases (with the exception of the oil-crushing group) the groups appear more as an institution of bureaucratic convenience—the end point in a credit chain, where it is easier to deal with a manager who 'represents' the individuals. There is no input into what the NGOs call 'human development', which might make the manager (and managing committee) more accountable to them, and strengthen their overall capacity to interact with other institutions and classes on more equal terms.

The importance of this is demonstrated by the extent to which the members of these groups are imprisoned within these hierarchical relationships of employment and indebtedness. In some cases, the roles of mobilising or initiating the formation of the groups overlaps with the oppressive/exploitative roles. The UCCA structure is deeply integrated into these relationships. The members are presented with little choice but to rely upon their traditional brokers—and this has always been an ambivalent role. The UCCA presents no alternative which can be trusted.

This has been revealed in the way many of the groups have been formed via the Union Council Chairman, via the KSS manager, both of whom are likely to exercise significant 'private' economic power within the locality.

Patron-client relations seem to be abundant, with clients being approached to form a group and subsequently becoming its leadership. In most cases the ordinary members have little idea of what they have joined, becoming in turn clients of the managers, Chairman etc. In many cases this is reinforced by the labour *sardar* structures with some individuals in the 'group' being well connected with the more powerful actors/patrons in the system. This is another way in which the 'group' is incorporated.

The credit provision through the UCCA is conceived very narrowly. It is individual, jointly timed, restricted to production or income-generation, with tight loan ceilings. There is no recognition of the rhythm and pressures of family survival which can divert such loans through necessity into immediate consumption, or meeting their village debt obligations. The rules on default are apparently so strict that the rest of the group members are denied access to further loans. Yet a loan dependence is built-in to the low ceilings, since in most cases both for beef fattening and for rice-husking (for example) the turnover is so low that insufficient profit/savings can be realised for the activity to continue independently.

It is clear from these cases that the notion of a 'target-group' has to be defined in a looser way while avoiding some of the glaring inequalities in social and economic status. But many of the groups involve people in different states of land loss, and the restructuring of access to land is so dynamic that the landless and the soon-to-be landless have more in common than outsiders with their definitions think. In most cases, although no reference has been made in the case-by-case presentation, the composition of the group is spread across almost as many *baris* as *paribars*. There are a few examples where the membership is more kin/*bari* related and concentrated into a smaller area of the village. It is possible that the more normal widespread distribution of membership is also responsible for the lack of group coherence and accountability.

There is obviously a problem of gender relations: in the mixed group, it appeared as if the women were being 'employed' by the male members, with little involvement in the transactions side of oil crushing business, where the decisions are made. Instead their role seemed to be restricted to the hard physical work of operating manually the oil-crushing turntables. When loans are taken by men for rice-husking, it is of course the women in the family who do the physical work, although the men do have some load-carrying in addition to handling the transactions. When loans are taken for the same purpose in a women's group, similar issues arise. In both cases, how do the women acquire any knowledge of the monetary transactions through which the value of their own labour is expressed?

Finally, it is worth pointing out that the UCCA inspectors and BRDB officials work in the daytime, but group members can only really appear together in the evening after sunset. This further accounts for the exclusion of ordinary members from any effective participation in the group. Usually only the manager meets the officials. It is in this relationship that the formal and informal transactions occur.

Conclusion: Bureaucratic Convenience versus Group Mobilisation

It is with this information on the government institutions currently focused on the rural poor, on the search for alternatives within government to the prevailing system, and on the experience of these mobilisation efforts with rural poor people, that the limitations of the main RPP strategy can be identified. In turn this analysis becomes the basis for advising on the way forward for a donors' supported ESP.

RPP—Bureaucratic Unrealism

It must be clear by now that I regard the Rural Poor Programme as a case of 'bureaucratic unrealism'. The choice of this critical term is quite deliberate. The equation of 'convenience' with 'realism' must be challenged. The RPP may be bureaucratically convenient in that a universal institutional structure (the two-tier co-operative system) is adapted for the purpose. But it is a hollow structure, with much evidence of its domination by classes of landlords and farmers not just at KSS level, but therefore at the UCCA level too. This must certainly rule out any incorporation of the BSS/M(B)SS into the same UCCA institution. The two-tier co-operative system remains attractive to its advocates because it offers, on paper at least, the chance of distributing credit on a wide scale without having to administer it down to the level of the individual recipient. It has become, in effect, a substitute for a commercial banking service. That service, especially to the poorest, was certainly very weak with signs now of marginal improvement.

However, a response to the limitations of commercial banking for the poorest does not have to be in the form of a set of institutions which have been tried and tested for a long time, and found wanting. In this sense the decision to transfer the UCCA—KSS practices into the RPP has been both bureaucratic and unrealistic. I can understand the frustration of policy and programme-designers with socio-economic consultants who present complex pictures. But complexity has to be confronted with more subtlety, if the objectives of working with the rural poor to enhance their material and social position are serious. It is better to do something small-scale and well, rather than become tempted, as IRWP was, to have a broader yet immediate impact. The lessons from IRWP seem clear in this respect. There is no virtue in supporting or implementing a paper programme.

Narrow Understanding of the Social Relations of Poverty

The problem with the RPP as currently conceived is that it reflects a narrow understanding of the social relations of poverty. It assumes essentially that what people lack is capital and skill and that a combination of credit and skill-training is an adequate response which can be delivered through the two-tier co-operative in association with BRDB and other line departments of government. But these institutions are not benign transcenders of the social relations of poverty, they are inextricably part of them. Secondly, there is a narrow conception of credit needs, and little appreciation of the impact of village debt relations upon the efficacy of that credit.

Formalistic Groups

Several issues arise from this criticism. The use of the term 'group' is both formalistic and misleading. The social relations of poverty are about power. The purpose of group activity under the Small Farmers Development Programme (SFDP) or the Action-Research for Destitute Women (both were special programmes of the Rural Development Division), or among the NGOs is to alter the balance of local rural power as part of the process of increasing the material condition of the poor. This is not a separate issue from employment generation and the acquisition of productive assets by the poor—it cannot be relegated to the future, *after* some economic activity has taken place. At the same time, the achievement of group consciousness cannot merely precede economic activity—ideas and hopes do not fill stomachs, so hungry people are not impressed. Therefore, I am not arguing that attempts to expand employment opportunities, generate markets for products, capture the rents from technological innovation by acquiring the appropriate skills, and to provide credit where necessary, are unimportant. But I am arguing that other dimensions are important. Realism is the recognition of the complete rather than partial (though bureaucratically convenient) picture.

Complete the Picture—example of credit

Consider credit for example. As already suggested, the interaction between village indebtedness and income-generating activity stimulated by institutional credit has to be recognised. To make a distinction between 'consumption' credit and 'production' credit, leaving the former (which is the real, ever-present demand for credit) to the local usurious system and concentrating only on the latter for institutional credit constitutes an arbitrary selection of one aspect of survival arrangements. But the 'profitability' or 'viability' of a production plan does not only depend upon its internal composition. It is crucially affected by the unavoidable demands on income, especially debt repayment. Perhaps rural poor credit needs should be described in 'cash-flow' terms. Certainly there are seasonal variations in employment opportunities and therefore income. The whole IWRP is based on this premise. But the village interest rates are so high that the debtor's position structurally changes and declines continuously. It is not simply reproduced.

At the same time we should recognise the advantages of village credit. It can be delivered quickly, and demands are often made urgently when a few days pass without employment. It can be provided for any purpose. With some exceptions, there are no documentary procedures. Access therefore is resolved if the loan-seeker is an employee, sharecropper or some kind of client to a patron. These advantages are of course in contrast to the elaborate, drawn-out procedures for obtaining UCCA or commercial bank loans. Clearly under present arrangements these institutions could not service such immediate but persistent and significant—not least for the efficacy of other categories of credit—needs.

Real Groups and Mutual Loans

This is where the development of real group coherence is so important beyond some general notion of unity in the struggle against other classes. The mutual loan activity in the 'Destitute Women's Programme' in Bogra is highly relevant and there are similar cases among the NGOs. Groups become more than just an administrative collection of individuals when they develop their own capacity out of savings, or from a line of institutional credit specified for the purpose, to offer loans to their own members in emergencies. The members have the best knowledge of a family's need and the 'track record' of the person in distress. The loans can be immediate. The need for them can be perceived before it is too late. A minimum of documentary procedures is necessary. Access is easy. Interest rates can be near the prevailing commercial rate and variable. Therefore much flexibility is possible. Other possibilities exist, such as lending to non-group poor, taking the land of small farmers in mortgage. In this way *real* groups are not only enhancing the value of any 'production' loans which they receive, but also breaking their dependence on the village hierarchy. These possibilities do not exist in the formal UCCA/BSS/M(B)SS approach, because no one in authority (BRDB, UCCA staff, dominant village and Union classes, as well as dominant classes elsewhere in the system) has an interest beyond small-scale loans to individuals, provided they remain unable to monitor effectively the financial transactions in which they are involved. Real group solidarity turns individuals into active claimants.

Bureaucrat-Dependent Clients

In addition therefore to the formalism and unrealism which pervades the practice so far of the RPP, there is almost a 'relief' tendency in its reliance on low ceiling credit to individuals. This loan ceiling is due to be raised under RD-II RPP, but it will still be directed at individuals through the convenience of the BSS/M(B)SS. For example, collective provision of irrigation services by the *bittaheen* is not mentioned in the proposals for RD-II RPP. Yet this activity lends itself simultaneously to a joint economic enterprise and a reinforcement of solidarity in relation to other, richer classes. Instead most of the income-generating activities cited in various reports are oriented to individuals, and revolve around the idea of petty business, livestock-rearing, handicrafts and so on. There seems to be the assumption that all *bittaheen* as individuals can be turned into businessmen (or women), which is very optimistic. But certainly loan dependency for individuals on an annual basis with a repetition of documentary procedures seems designed to produce a class of bureaucrat-dependent clients.

Skills Training—necessary but not in isolation

This approach to income-generation among the rural poor is further reflected in the conceptions of training embodied in the RPP. The emphasis is clearly on 'skills' mainly of the technical/artisan kind, although some attention to co-operative management is also included. The assumption here is that an increase in self-employment must involve either new technology or people learning new skills. Experience elsewhere both among NGOs and the action-research programmes within the Rural Development Division suggests that rural people overflow with skill and talent and that more attention should be given to the abilities which already exist. Actually, I have a more open mind on this, since there is rapid technological development in Bangladesh and the *bittaheen* people should be assisted to capture more levels of this technology. In the minor irrigation sector, for example, there are many possibilities beyond merely providing an irrigation service from a group's asset. Workshops for repair and maintenance, spare parts businesses, peripatetic mechanical services all represent opportunities for which skill training is clearly required.

So again I am not dismissive of skill training, but in isolation it will not produce the desired results. 'Training' as a process is normally conceived in such a formal, classroom way, that it becomes a blunt instrument. This is increasingly true as the 'subject matter' moves from technico-artisan into management into forms of appropriate social organisation into the analysis of political economy and the necessity for collective forms of socio-economic activity.

NGOs' Training Involvement—skills or group conscientisation

There are references to some fashionable words like 'conscientisation' and 'mobilisation' in the training outline for BRDB Assistant Rural Development Officers at RDA Bogra. Central staff from BRDB HQ themselves undertook the 'RPP—component' in the course (five days out of three and a half months), which is evidence of their personal commitment. One cannot re-orient a culture in five days in a classroom, when the 'audience' has so clearly invested in that culture to become the audience in the first place. So despite the presence of the fashionable (not wrong for that, just over-used) terminology, one questions if there is any real intention to include it in course content. The attitude towards NGO involvement in training is itself evidence of this problem. Several agencies, including the World Bank, have been urging their involvement in 'training', but have restricted the purpose of this involvement to technical skills. It is true that the NGOs such as BRAC and Proshika have some of this capacity but they also hire other people for such skill-training on their own courses. Their own expertise lies elsewhere—working to form and sustain groups around income activities, and struggles for land and wages. Work which occurs explicitly in the context of an understanding about power, inequality, dynamic agrarian change—in short the social relations of poverty occurs outside the classroom, and the animators themselves have in effect been apprenticed in the countryside to others with more experience, and then have learnt by doing. Of course they have had their work supported by courses—on the subject of 'Alternative Strategies for Development'; 'Social Analysis'; 'Participatory Research'; and so on. For over a year now since RD-II RPP was approved, negotiations between NGOs and BRDB/GOB have occurred—but the gap between them is a large one. No NGO which is committed to a 'Freire' type strategy can feel comfortable simply being slotted into another institution's highly-formalised training programme to provide a 'skills' input. However, alongside credit and skills, 'training' seems to be regarded as a panacea for orienting both official, UCCA inspector and recipient towards income-generation. Indeed these panaceas constituted the title of the Feb. 1984 World Bank Mission, as a sequel to the 'Selected Issues in Rural Employment' paper. But it is the *style* of institutional involvement with the rural poor which is important. As I understand the situation, poor people are not impressed with agencies which deliberately choose to ignore the main explanations for their poverty. If the action-research programmes in the Rural Development Division or the NGOs are being relatively more successful in their involvement with the rural poor than other government programmes, then it must be because their style is more consistent with the real problems suffered by poor people.

High Staff-Group Ratios—brokers at a price

However, we must beware of inventing a new panacea. These relatively more successful activities have depended on a high staff-group ratio, with quite intimate 'broker' functions being performed between groups (as groups) or group members and other agencies, whether banks, line departments, public sector corporations and privatised sources of inputs and materials. These 'brokers' have sometimes been of high status (instructors at the academies/university for the SFDP and Action-Research on Destitute Women Programme) which has clearly helped in facilitating access to these other institutions. In the case of SFDP some withdrawal of these 'broker' functions has occurred. This of course is a crucial test of a group's ability (or its members) to interact directly with organisations such as commercial banks. The awareness of this staff-group ratio issue has obviously prompted the RD-II RPP to include an increase of staff at ARDO level as part of the intensification of the general RPP under BRDB. If we connect this increase to the August 1984 Banking Plan proposal for the ARDO—RPP to work outside a UCCA—BSS (or MBSS) structure, then together this might constitute the institutional basis for a whole series of SFDP-type activity independent of the two-tier formula. However the other dimensions of style, orientation, and a recognition of the complete picture confronting the poor would also have to be present as part of the content, so long as the 'broker' (in this case the ARDO—RPP) function does not smother the development of poor people organised as groups towards an independent ability to interact with the rest of the system. We should be aware of the dangers of 'official dependency' as a substitute for other forms. But this is not a pure 'extension' model where officials deal with individuals. On the other hand if officials start to deal with groups *only* as a convenient means of reaching larger numbers of individuals, thereby being preoccupied with a more conventional extension agenda, then the virtue of group formation among the poor is lost. Groups become an institution of mere administrative convenience.

Chapter 17

Sirs and Sahibs: Government and Technical Assistance (TA) Relations in Rural Development Projects

Introduction

This chapter is derived from a short study undertaken for BRDB and SIDA to analyse the Production and Employment Programme (PEP) funded by SIDA but located as a project within the BRDB's Rural Poor Programme. For SIDA, PEP appears as a component of a wider programme (the Rural Employment Sector Programme RESP) which retains a hangover from the earlier Intensive Rural Works Programme with the Infrastructure Development Programme (IDP). Later in the chapter, there is some discussion about the relationship between these two wings of the programme. There are also references to RESP2, which is simply the project sequel to RESP1. The PEP focuses upon group mobilisation and income-generation, and employs specialist TA and BRDB project staff *to implement within* government a programme which resembles closely the principles on which many NGOs work. In this sense, it has had an agenda of replicating some of the NGO experience into government practice. In undertaking this study, I was asked specifically by SIDA to focus upon the relationship between the project PEP and BRDB.

The strategy adopted for this paper is to draw on extensive field notes for a summary of issues leading to a proposition for PEP and RESP reorganisation which reflects more closely the realities of donor bilateral assistance to Bangladesh in the 1990s, while seeking to preserve a distinctive quality for Nordic official aid in which the concern for poverty and inequality has been paramount. I was, of course, aware that the project TA staff had contributed to a paper entitled 'Working Paper on RESP II' which draws rather different conclusions from those presented here for the reorganisation of the management structure for RESP and PEP. These conclusions derive from an analysis of problems in the implementation of PEP so far, and reflect a depth of frustration felt by many of the expatriate and Bangladeshi TA staff. However, the central thrust of those conclusions argued that the project should be allowed to operate more independently of its host institution, the BRDB. Such conclusions for reorganisation reflected a deep misconception about the purpose of PEP and indeed Nordic official aid in Bangladesh. They also revealed a misunderstanding of certain institutional histories, since the liberation of Bangladesh, which will be discussed further below. At the same time, the project TA staff can be forgiven for some of these misconceptions since they also reflect an unfortunate dimension of style in the donors' attitudes towards 'policing' their aid to Bangladesh, with expatriate TA staff to some extent masquerading under 'technical' headings and coopting Bangladeshi TA staff into the same culture of mistrust. Thus the experience of PEP has posed a fundamental issue about alternatives—whether to work constructively with the responsible GOB agency or not.

In order to focus upon this central policy question, this revised chapter has been organised into 6 principal headings: position of PEP in long-term GOB/donor policy; relations between BRDB and TA; institutional alternatives for continuation of PEP; group formation and mobilisation within PEP (including the case studies); requirements for working with the rural poor (a manifesto of good practice); the case for PEP remaining within BRDB.

It should be recognised that RESP (and PEP within it) was formulated partially as a departure from IRWP, not just a continuation of it. While various strategies could be adopted to improve the participation of labour in short-term infrastructure employment, such as Labour Contracting Societies, the provision of rural works employment is too partial an intervention in poor people's survival strategies to become the vehicle for access to longer-term, more secure, employment. Furthermore, if improved participation in short-term employment required forms of action and struggle which put other survival linkages in the local political economy at risk, then it was irresponsible to encourage narrow, short-term processes of mobilisation among the poor. PEP was conceived out of this recognition that a target-group approach to rural works employment generation actually required a broader conception of group mobilisation among the poor, into which such employment opportunities could be located as appropriate. When working at its best, the IRWP formula was contradictory in putting vulnerable people at risk in their normal survival structures as a condition of programme success. Clearly the PEP response runs a high risk in agency-client dependency, but at least it has the virtue of trying to conceptualise the totality of poor people's lives.

This report is based upon 16 days of discussions and interviews conducted at all levels in the organisational structure of PEP/RESP, and with representatives of SIDA and NORAD. Outside Dhaka, discussions were held with 15 groups, along with Group Organisers, Assistant Rural Development Officers (ARDO), *Upazila* Socio-Economists, District BRDB and TA staff across the 3 districts of Faridpur, Madaripur and Kurigram, 7 *Upazila* and 12 Unions. In all of the discussions outside Dhaka, as well as for some in Dhaka, I had the invaluable assistance of Mr. Shah Newaz, a senior coordinator with Proshika and a recent Masters graduate in Development Studies at the University of Bath in the UK. We had worked together previously in one of the preparation exercises for RESP 1, and enjoy close conceptual affinity, as well as a common understanding about the limitations of NGOs and the necessity of connecting the poor to the resources disposed of by the state. It should also be noted that in our previous work, examining BSSs and MBSSs (a term invented by us in 1984 to distinguish them from MSSs) as part of the BRDB RPP (see Chapter 16), we were highly critical of the RPP group formation process and the non-target membership of the RPP groups.

The argument of this report is that Nordic bilateral official aid has to contribute to institutional progress in GOB's responsibilities to its rural poor, since in quantitative terms Nordic aid can make very little immediate impact upon the condition of the poor. But in pursuit of this objective, the donors, and SIDA as the lead agency in particular, will have to accept a change in aid-giving style if constructive and positive working relationships with government are to follow. The report which follows seeks to elaborate this argument.

Position of PEP in Long-term GOB/Donor Policy

Many of the problems associated with PEP at this time arise from a lack of clarity about the overall purpose of the programme in the context both of

GOB and donor policy perspectives. As a result PEP is part of the familiar *ad hoc* process of rural development initiatives in which the host government is confronted with a range of separate models and approaches presented by donors eager to show 'value for money' over short time periods in terms of quantitative achievements for the rural poor. The EEC funded project in Rangpur is probably the best and most recent example of this. At the other end, the CIDA funded RD-12 represents an attempt to conform to the BRDB RPP approach, with the principal modification of a separate *Bittaheen* UCCA in the 17 Districts and 129 *Upazila* of its operation. This de-linking of the UCCAs has now become, I believe, general policy and has certainly been one of the conditions for PEP. With a much larger scale of operation than PEP, RD-12 has very little TA—the Canadian Resource Team (CRT) based in Dhaka. This consists principally of a team leader (expatriate), a training advisor (Bangladeshi), a credit advisor (Bangladeshi), and a monitor (Bangladeshi) who is continually touring to check on group formation. There are no TA staff at District and *Upazila* level. The administrative structure for RD-12 is conventional for RPP: Additional Director-General (BRDB/RPP) at HQ, who chairs regular management meetings for RD-12 (this is effectively the Central Coordination Committee), with a line of authority through the District Project Director, the Deputy Project Director (RPP), a male and female ARDO at the *Upazila*, and 3-4 Field Organisers in each *Upazila*, supported by the same Basic Training as for PEP staff at Comilla/BARD and in-service training at the Bogra and Sylhet academies, along with specialised training provided by NGOs. RD-12 has a target of the formation of 16,000 BSSs and MBSSs. The function of the CRT is to strengthen the Research, Evaluation and Monitoring Cell for RD-12, to strengthen the action-planning at District and *Upazila* level, and to carry out directly monitoring and review functions. Clearly the RD-12 approach is much more 'hands-off' than that of PEP, especially in terms of TA, and yet it shares the same components of credit, training (with a strong emphasis on Human Development) and group mobilisation (though using a cooperative formula, parallel to but de-linked from the KSS/UCCA system). RD-12 was initiated as a project in 1987, but actually started in May 1988. It is therefore too early to assess the quality of the programme especially with regard to group formation and mobilisation. It will be very informative to conduct an explicitly comparative and independent review of PEP (in whatever form it continues) and RD-12, at some point in the near future, to test the efficacy of the different approaches. Clearly if one was satisfied with the RD-12 approach, then the requirements in PEP for greater staff intensity and TA support would disappear. It is because one remains highly sceptical of the claims made for this 'hands-off' approach, that the case for some version of PEP is made.

Given these two extremes of donor stance towards rural institutions and target-group formation (the EEC funded RD-9 and the CIDA funded RD-12), the justification for the Nordic funded PEP has to be placed in the context of an awareness of the history of rural development institutions, rural social and economic change, and donor involvement in GOB development programmes. The intention here is not to embark on a long digression, but to point out that times have changed in several crucial respects in Bangladesh since donors in general and SIDA in particular formulated their aid style to Bangladesh.

The first and most fundamental issue concerns the changes in agrarian structure. Reviews of programmes and proposals for organisational reform have to be strongly connected to an understanding of these changes in agrarian structure. There is, of course, a strong consensus now over the extent of rural landlessness and the severe constraints on availability of non-farm employment (especially urban, but also rural). These constraints have been discussed extensively elsewhere (see also Wood 1988, and Wood & Palmer-Jones 1990) and amount to an argument for the necessity to expand employment opportunities for the landless directly or indirectly in agriculture, an agricultural sector which is itself expanding and proliferating new roles in supply, production, processing and marketing. This process of expansion is occurring substantially in the private sector, but considerable opportunities can be envisaged for the poor if their entry into the private sector is prepared for and supported. Thus it is not only landlessness to which the GOB and donors have to respond, but the structural conditions under which this process is occurring and the opportunities which are also released by it. It is no longer enough to see employment opportunities for the poor restricted to the temporary rural works programmes in the locality, or to small-scale, low turnover, self-exploitation activities such as paddy husking and petty trading, or minor livestock rearing. This is a very patronising, indeed ghetto approach to employment and income-generating (EIG) activities for the poor.

Secondly, these structural conditions of agrarian change also imply that the organisation of farming itself will change, with non-land agricultural means of production becoming increasingly significant especially in the context of intensified land fragmentation. Ironically, this fragmentation is itself stimulated by the changing values of land with the expansion of irrigation, as well as by population growth. I am suggesting here a process of disarticulation/re-articulation of farm organisation, in which the holders of non-land means of production like water or ploughing services increasingly hold the consolidated agricultural assets as closely or loosely networks of entrepreneurs, perhaps even resembling 'agricultural companies'. The farmer with land has plots distributed between a number of such 'companies' (as now with plots distributed between different STW command areas), and will increasingly have to surrender any autonomy of production decisions to them, especially if in receipt of credit from them. If this kind of scenario for the agrarian future of Bangladesh is correct (see Wood 1988 and Lewis 1989 forthcoming for further elaboration of these arguments), then the grouping of farmers together as input supply cooperatives (i.e. the KSS) will become increasingly irrelevant to their structural circumstances. I notice, with some alarm, that the UNDP Agricultural Sector Review (December 1988) is proposing that KSSs should be restricted to farmers below 2.5 acres to avoid their capture by the rich—but this is to propose cooperatives for precisely that category of farmers who will lose their land first in the process of disarticulation/re-articulation described above. In the transitional stages of this process, farmers are better organised as consumers of services and inputs rather than as self-providers. In this way, they have a common interest and may indeed hold on to their land more effectively and interact on more equal terms with these 'agricultural companies'. At the very least then, the BRDB would have to reconsider the purpose of its farmers' cooperative strategy towards consumer association models, but if the process described above became quickly pervasive then farmers' cooperatives would become redundant as a set of rural institutions.

Thirdly, the rapidity of these changes has to be appreciated. The Comilla two-tier cooperative system was devised over 25 years ago and to some extent reflected accurately at that time the small farm agrarian structure in the Academy's area of institutional experimentation. There may be criticisms of the model's relevance to other parts of Bangladesh/East Pakistan for that time, but the main issue now is that in the intervening years the agrarian structure in Bangladesh as a whole has changed quite dramatically. This has been reflected in a partial shift of focus in BRDB over the last years with the creation and expansion of the RPP. But the difficulty of making this shift should be acknowledged. Many of those now in the senior positions in BRDB had their formative years in the Academy at Comilla (now BARD), under the strong leadership of Akter Hameed Khan. The spirit of innovation and experimentation at that time must have been exciting, and a strong commitment to the model naturally emerged. In many ways the strength of this attachment, integrated with a respect for its 'author', has acted subsequently as a constraint upon continuing the spirit of experimentation. What was innovative became routine, and the holders of the tradition were understandably reluctant to show disloyalty to it. This attachment has to be respected not scorned by outsiders who are now confronted only with the new reality of pervasive landlessness and deep class division. But certainly now the agrarian structure in Bangladesh has reached a stage when this spirit of experimentation needs to be rekindled within BRDB in order to devise institutional strategies for the 1990s and beyond. The observations about agrarian structure deliberately look to the future—the long term outcomes of an intensive interaction of agricultural capital with a highly fragmented farm structure under conditions of extensive landlessness. This is the scenario for which we have to prepare with and for the poor, and the BRDB is the agency of GOB charged with the responsibility for that preparation. Despite the many criticisms of its practices (connected partly to this problem of applying an outmoded model) its current leadership does appreciate the need for a grounded phase of institutional experimentation in which mobilisation and service strategies with the poor move beyond formalism into a real achievement of quality. The risk must be recognised that the CIDA funded RD-12 'hands-off' approach may not be institutionally sufficient to move beyond that formalism. In this complex institutional context, there is an important role for a small donor with these concerns to work constructively with the BRDB in an experimental way in order to explore how to be effective in a hostile but dynamic rural political economy on issues like: staff intensity; pace of group formation; flexibility of target-group definition; mobilisation of local resources; functions of local staff as facilitators and brokers rather than managers and planners; and so on.

Fourthly, however, if SIDA (as the lead agency) and NORAD are to claim this role, then some rethinking of donor style is also required. Perhaps, first, the limits on Nordic aid to make a significant quantitative impact upon current poverty in Bangladesh should be recognised. This does strengthen the view that the principal justification for official Nordic aid lies in providing additional and appropriate resources to assist GOB in institutional experimentation and policy development in areas of donor concern such as the rural poor. It makes no sense to attempt this experimentation outside the agencies of government charged with this concern even if GOB did permit such a use of official aid. This is the basic weakness of the EEC approach, because models created under artificial institutional conditions cannot simply and subsequently be transplanted to an unprepared host. Rejection will follow. The EEC approach is, at best, an example of the Balkanisation (apologies for the inappropriate metaphor!) of official rural development efforts in Bangladesh. Secondly, SIDA must therefore reconsider the role of Technical Assistance (TA) in PEP, and to some extent RESP as a whole. Taking PEP alone the disposition of professional TA staff is:

	Bangladeshi	Expatriate	Total
Kurigram	17	1	18
Madaripur	15 (+1 Vacant)	1	16
Faridpur	9	1	10
HQ	4	6	10
Total	45 (+1 Vacant)	9	54 (+1 Vacant)

The programme is currently working in 10 *Upazila* with a greater intensity of BRDB staff at *Upazila* level (i.e. ARDOs and Group Organisers) than for RD-12 which will operate across 129 *Upazila* with a small HQ TA unit (the CRT). This contrast in the intensity of TA reflects a profound difference of perception between two donors with similar objectives in their official bilateral aid policy. Again, a little history is in order. In the mid-seventies, when the quantum of aid to Bangladesh started to rise rapidly, there was a genuine problem of appropriate professional capacity among the national population. Many, especially in the social sciences, had been killed during 1971 and others had their training and careers seriously interrupted by the disruption of that period. It was perhaps inevitable that expatriate consultants and TA staff would be heavily involved in donor-funded projects as well as providing inputs directly into line ministries, agencies and research and training institutions. A tradition of intensive TA (both expatriate and Bangladeshi) has been established as a result, in marked contrast to India for example. The situation in the late 80s, and looking into the 90s, is now also dramatically different with significant capacity now available both within and outside GOB. But the tradition of TA (both expatriate and Bangladeshi) has persisted strongly in SIDA funded projects, and the relevance in the future of this extent of TA must at least be questioned.

There are several issues to unravel: how 'technical' is the assistance; does the strong, almost executive, presence of TA undermine the capacity and confidence of those being 'assisted' to take responsibility; is the TA functioning as a source of independent monitoring to the donor; should the TA capacity which lies outside the GOB agency be recruited directly into the agency; is the TA in some sense functioning as the 'conscience' of the project? It is possible that the reliance by SIDA on extensive TA has become self-defeating in that it constitutes a barrier of implied mistrust between the donor and GOB and prevents the development of any kind of constructive dialogue in which honest scepticism is openly discussed with a BRDB leadership which is able to share with the donor the constraints on its own room for manoeuvre.

This returns us to the issue of donor style. If SIDA with bilateral aid accepts that its concerns for the rural poor in Bangladesh can only be expressed through institutional cooperation with GOB/BRDB to move the RPP from formalistic to more progressive content, and if it further argues that this can only be done through experimentation with mobilising informal groups, greater staff intensity located at the Union, and so on, then it has to convince the BRDB by argument and the analysis of existing practices that such experimentation is required as an integral part of its own policy development in the context of a changing agrarian structure. This demands that SIDA and its consultants accept the national sensitivity on the use of expatriate TA, or donor loyal Bangladeshi TA, and redefines the role of TA to support constructive dialogue rather than act as a barrier to it. This also requires that SIDA breaks the connection between a concern for the poor and its lack of trust in any GOB institution to honour that concern. No progress with constructive dialogue and institutional reform can be made as long as that connection remains. If SIDA wishes to make it clear that the state in Bangladesh cannot work constructively with its rural poor, then it should face up to its own analysis and leave the stage. However, it is the view of many both within SIDA and among NGOs working with the poor in Bangladesh, that since the state disposes of such significant resources in a resource scarce economy, the search for the space to create positive institutional connection between the poor and the state has to continue. This is the common ground between those in SIDA who have struggled with the incarnations of this programme and those in the Planning Commission and BRDB who are engaged in a parallel search for room to manoeuvre.

Fifthly, but intimately connected to the discussions above, the donors in Bangladesh concerned with institutional reform in rural poor development must have a longer-term perspective which coincides more closely with the obligations of responsibility upon the government. In the case of SIDA, this is a particularly strong imperative for the efficient use of its aid since it is unable to make significant short-term contributions to the relief of poverty.

In this sense, it should see the value of its aid under RESP/PEP as contributing to the development of institutional capacity within the relevant agency of government, a capacity which is then more able to absorb larger amounts of aid from other sources directed towards the interests of the poor. In many ways, this has been the achievement of both IRWP and IDP in developing the institutional capacity of LGEB to the point where it is able to absorb the infrastructure components under the EEC-Rangpur project. In order to maintain such a long-term perspective, a dimension of faith with the counterpart agency is required. There have been many moments (especially after the 1985 Evaluation of IRWP) when that faith was placed under extreme pressure, but perhaps with the advantage of hindsight SIDA should be congratulated for maintaining that commitment to LGEB. This is not say that problems do not remain, but that a conception of the long term has certainly contributed to the development of technical capacity and some sensitivity to the linking of infrastructure to target group EIG activity. In this way, the EEC and SIDA approaches may almost appear compatible: both share a certain critique of the weakness of BRDB institutional capacity to work constructively with the poor, but the EEC's overall scale of funding to Bangladesh is justified to its member states by producing a tangible immediate impact upon poverty through direct intervention, which in the opinion of its consultant advisors demands an institutional by-pass strategy; whereas SIDA should acknowledge a division of donor labour and see the PEP project TA staff's critique of the BRDB (assuming it is fair) as justification for its efforts to contribute to its reform. It should also be acknowledged that this reform may either take the form of a transformation of BRDB into a *Bittaheen* Rural Development Board if the UCCA/KSS becomes steadily redundant to the structure of the agrarian economy, or the creation of a separate board for the *bittaheen*, as has been discussed for several years now in the *bideshi* salons of Dhaka.

Relations between BRDB and Project TA Staff

This has become the fundamental issue for both sets of staff in the project involving allegations and counter-allegations. During the 'fieldwork' for this paper, sessions have been held separately with both sets of staff at all levels in the project, though this was not possible in all locations, especially towards the end of the period as different sets of people departed for leave. My gratitude for the courtesy shown to me in these sessions should be noted, since potentially very threatening questions were being posed in order to elicit responses, and frequently informants were asked to back up their statements with case study illustration. Informants were certainly patient, and mainly appeared very frank in their responses. I acknowledge, the trust bestowed upon me, and seek here to present a distillation of the issues rather than the graphic details, some of which have to be considered as confidential. The reader can be assured that cases have been presented to support these points. While it is of course difficult to check the veracity of these statements, the real data (in terms of evidence of the breakdown of trust between BRDB and TA staff) is revealed in the belief of the informants that their respective versions are true.

There are several differences in perception about the project which express the tension between both sets of staff: the extent of experimentation intended for the project; whether the project is intended primarily for service delivery or mobilisation; where responsibility for implementation really lies. The significant presence of TA staff from the *upazila* level upwards, operating under unclear guidelines about both the extent of their executive responsibility, has led in effect to a dual command structure through all levels of project administration. It is simply not clear to all concerned whether TA staff should have implementational or advisory roles. Given the balance of staffing, in which TA staff outnumber BRDB staff at all levels until the union, it is hardly surprising that the TA staff have taken much of the initiative. Indeed in Kurigram, the TA staff seemed to have behaved in the past as if PEP was 'their' project, but it is possible to detect similar attitudes prevailing in the other districts. There are many ways in which this is revealed. In each district or *upazila* visit, the TA staff have controlled the data and filing system and have been most efficient and gracious in supplying me with information. They have direct access to secretarial support and to transportation. Since there are three TA socio-economists and often only one ARDO (there should be a male and female) at *upazila* level, the TA staff have a greater frequency of contact with the Group Organisers in the unions which also confuses the line of authority. This 'taking over' of the project is also reflected in ideas about the future. There appears to be a strong consensus among all the TA staff that PEP in RESP2 should be more clearly separated from BRDB in order for the project to be successful and genuinely experimental. Is this a fundamental misconception of the purpose of the project or not? Is there a PEP project outside BRDB, if the project is to contribute to the development of good practice inside the responsible government agency? As discussed in the previous section, SIDA and GOB have to clarify this overall policy position for PEP in RESP2. This plea for separation was clearly expressed by the TA staff in Faridpur who regarded such separation as a precondition for 'getting on with the work', indeed the suggestion was made that the TA staff should have their own cadre of Group Organisers and work in an entirely self-contained way. The project was seen in terms of the achievement of localised short-term objectives in group formation, savings, training and credit extension and was not connected to a broader conception of developing institutional capacity within BRDB. BRDB 'control' over training and upcoming credit through the Banking Plan arrangements was resented as an obstacle. This view was shared completely by the TA staff in Madaripur.

How do the BRDB staff perceive the overall relationship? The issue of experimentation has attracted differing responses. The assumption that BRDB is a resistant host to experimentation is somewhat dispelled in discussion with the D-G, BRDB, who listed the departures from normal operational practice which were accepted by BRDB in the setting up of PEP. These were: the formation of informal groups instead of formal cooperatives (i.e. BSSs and MBSSs); their separation for savings and credit purposes from the UCCA; and the location of TA staff at the *upazila* level. He also suggested that the project had not used its experimental space sufficiently, introducing considerable rigidity of practice in group formation, savings discipline, discretion of judgement allowed to GOs. Indeed he observed that the management of these informal groups resembled too closely the management of cooperatives! While the TA response to these points might well be that GOs were either too inexperienced or could not be trusted to operate outside tightly defined guidelines, the D-G's remarks indicate a willingness to explore further limits to constructive room for manoeuvre for PEP within BRDB. The outer limit is expressed in his observation that "not everything can be regarded as unanticipated, the project activities cannot be left totally vague". But he suggested forms of words in a future Project Proposal which could describe the conditions under which flexibility and discretionary judgement should be exercised—e.g. for size of groups, degree of geographical concentration, frequency and regularity of meetings, and so on—within the spirit of the overall objectives. Finally, although the project TA staff have often complained that when PEP operates in an *upazila* the normal RPP tends to withdraw, the D-G argues that this is a positive step and should be deliberately part of future policy. In this way PEP becomes the RPP in the *upazila* in which it is located—the informal group model with a greater intensity of GOs (though not of course on a fixed ratio of 1:10 groups) covers the *upazila*, and the quality of work can be subsequently compared with other RPP programmes (i.e. the CIDA funded RD-12 *upazila*).

For the PEP Project Director (PD), the position has not been so clear. The tolerance limit for experimentation and departure from normal BRDB practices is a constant source of tension. Different sets of expectations exist for his leadership of the project, which leave him stranded as being obstructive when conforming to BRDB rules on e.g. tendering for training contracts and authorising training budgets or exposed to GOB reprimand if the rules are not followed. Such rules are not necessarily inconsistent with experimentation if experimentation is not simply associated with short-notice, *ad hoc* decisions (i.e. one month instead of 6 months). "Some stability in strategy is required, the modality has to be settled." Without this resolution, the PD is in the unenviable position of having responsibility without much authority: within GOB a dual line of authority exists for him between the BRDB and the Planning Commission; and in the project he is outnumbered in performing the programme development function as well as at the District and *upazila* implementation levels, where the bulk of the project staff are employed by HIFAB (the SIDA hired TA consultants) not by BRDB and not therefore under his guidance or authority. As a result he admits to a strong sense of isolation in the project.

In a discussion with BRDB staff in Madaripur, involving the District PD, the Deputy PD (PEP), the URDO for *sardar upazila*, and the two ARDOs (PEP), the advantages of PEP over the normal RPP were summarised as: the greater intensity of fieldworkers at the union level; the appointment of an additional ARDO at *upazila* level; the better pay received by GOs under PEP because of the union-location living allowance along with some extra facilities supplied by the project like torches, bicycles, umbrellas and 'PEP' bags; the availability of transportation through PEP such as jeeps, motorcycles and speedboats which enable *upazila* and District staff to visit the field; the RPP cooperatives have been under the UCCA, an autonomous body managed by a committee of 12 members who dominate the local political economy and use the cooperative inspectors as a source of informal income; and ARDOs and GOs are provided with additional training by the project which improves their performance. When it was pointed out that the presence of TA staff had not been mentioned in this summary, the response was definite and forthright:

"In Madaripur, we can do without TA staff. This is not a personal criticism, just that if we have the resources (as summarised above) we can work effectively. At the moment, we are confused by the parallel administrative structure. For example, many of the TA nationals come from NGOs, so when we are considering training for Traditional Birth Assistants we think that someone from the Family Planning Ministry is appropriate, whereas they favour someone from GUP (the ex-NGO of the Add. DO) which has a health programme. Furthermore, the salary gap between the GOB and TA staff is about 3 times, which affects the psychology of our staff. We think this is a BRDB programme but the TA staff think this is a TA programme. We are at the mercy of the TA staff. We have to ask for transport. Even to attend this meeting with you (GDW) we were commanded to attend by the TA staff, we had no say over the timing of the appointment. There are 15 programmes in BRDB but this is the only one with two lines of command. We are just sitting here. The responsibility for PEP at the *upazila* is with the ARDO, but all the resources are with the socio-economic counterpart coordinator. He controls the means for the ARDO to exercise his responsibility. [Does the ARDO have autonomous use of a motorbike under the project?] We do not understand the Planning Commission involvement, a Planning Commission man came here, asked us in BRDB for transport and to our embarrassment we could not help him. [The discussion moved onto credit.] We will take projects, based on GO reports, which do not have high technical content. We know the situation, and the two ARDOs and Dep PD can advise the GOs, so TA staff will not be required because of the introduction of the Banking Plan because we are not dealing with sophisticated projects. Anyway, the orientation of group members in this area is more towards getting employment rather than getting credit. PEP does not need to become dominated by the credit dimension." This was said, unprompted, despite the near universal TA view that BRDB is obsessed with credit, though in Madaripur this has not yet replaced the obsession with training! This preference of group members towards employment rather than credit was confirmed by our own discussions with groups.

This series of statements by the BRDB staff in Madaripur has been quoted at length because although the veracity of its tone may be disputed by the TA staff, as it was at a subsequent long meeting, it does to some extent state the position 'as the BRDB see it'. The TA staff counter-criticism was equally strongly worded, and covered the following issues: direct involvement of TA staff in the formation of groups; ARDO's visits to the field too infrequent to supervise adequately the GOs, so the TA staff have to check on the application of target-group criteria in group membership; TA presence also necessary since ARDO is often asked by local elites to form a group; the ARDO will often go to a *haat* and call a teacher or member of the local elite to assist in forming a group; cooperative inspectors of the UCCA have been used by the ARDO in group formation—e.g. 5 groups formed in Mohindrodi Village in Kubirashi Union, Rajoir Upazila during flood relief by cooperative inspectors, even though GOs were present, promising relief goods but at least 2 of these groups subsequently had to be 'cleaned up' and some unmarried women had been included contrary to PEP guidelines; in the same *upazila*, when the socio-economist (Training) was herself absent for training, the ARDO formed a group with many non target-group members in it, and selected an unmarried woman for literacy training; indeed during this period of 21 days, 7 groups were formed by the ARDO in a union which was supposed to be outside the project because it was a GUP area as per PEP agreement. In one of these groups, the Union Parishad member tried to keep his daughters in the group, and when the TA staff objected they were overruled by the ARDO; in another group in the same *upazila*, the secretary is a cattle dealer, the chairman an ex-Ansar and 2 members are owners of a pond to be excavated under the project; GOs would not stay in the union unless the TA staff were monitoring and authorising the union-stay allowance; GOs are constantly looking for other jobs, and this lack of commitment to the group means that TA staff have to be more involved in supervising GOs and group meetings, savings and training schedules. The *Upazila* Resource Centre idea will be impossible to operate because BRDB has control over training budgets, credit and extension; ARDOs trying to use budget to purchase own stationery and hire more loyal typists and *peons* who are at present paid by the TA budget; BRDB could not manage the extent of training, especially Human Development, provided by the TA between May-June 1988 (15 Human Development courses, 25 on each course in each `upazila`) financed out of a TA advance, which the BRDB has yet to reimburse.

This presentation of the Madaripur case on the relations between BRDB and TA captures some of the flavour as well as some of the content of the differing perceptions about each other's roles and performance under the project. While the case of this district is fairly typical for the other districts, obviously with local variations in case examples, the atmosphere in Ulipur Upazila, Kurigram was more constructive, even though the District TA staff were as critical as in Madaripur. There has been a recent and substantial relocation of staff within both BRDB and the TA as a response to earlier problems, some of which reached a highly personal level. Our meeting with the Ulipur BRDB staff (2 ARDOs and 7 Group Organisers) and 2 recently appointed Bank workers lasted for 3 hours. The ARDOs had also been recently appointed to PEP, one with 7 years' experience in BRDB's normal programme and the female ARDO was recently recruited after her MA in Bangla at Rajshahi University. Three of the GOs had the most experience and provided many of the responses. They made similar points about the advantages of PEP over RPP, and considered that the staff intensity and union residence were crucial variables although accommodation especially for women was difficult to find as rental houses were not available. These problems were more acute if GOs came from outside their locality of work, and with the insistence on BA or BSc qualifications more women were likely to be outsiders than men.

The role of Group Organisers arose in the Ulipur discussion among the TA staff as well, and constitutes an interesting example of differing perceptions about the purpose of the project. The issue was raised about GO motivation and the extent to which a promotion ladder would encourage good practice among them. The TA staff responded that "if GOs are promoted to ARDO then they will consider themselves equal to rather than junior to us, and what role would be left for the TA? And, furthermore, it is not right to promote people who are basically village accountants and cooperative inspectors." Such a statement, of course, overlooks the qualifications required to be recruited as a GO to PEP, indeed it is commonly remarked among TA staff throughout the project that the GOs are overqualified. It also overlooks the additional training which GOs receive. This thinking by the TA staff seems to reveal a sense of permanent rather than temporary function for the TA, and indeed would appear to be in contradiction with the objective of experimentation. Could it be that neither BRDB nor the TA staff really consider this project to be experimental, but for different and opposed reasons? The BRDB staff in Ulipur noted a different potential for the GOs. While acknowledging the cooperation between the TA staff and the GOs in survey and identification, group formation, mobilisation and informal training follow up, they also reported that TA staff withdraw from supervision of the more experienced GOs and allocate their time to the newer GOs, that experienced GOs could and should provide the role of assistance to new GOs, and that under these conditions the more experienced GOs could take on the TA role, thus releasing TA staff to work in new *upazila* rather than expanding their number in the project as the geographical area expanded. They pushed the argument further with the observation that in 'mature' *upazila* for PEP, the assistance to BRDB staff needed to be more technical rather than 'socio-economic', that this technical support could be available on deputation from other GOB departments, and that 'socio-economic' support for Human Development training could be available in a planned way on request from the District. In this way, there would be no confusion about responsibility at the *upazila* and union level for group mobilisation and service delivery, as there is now with the dual structure. Furthermore, with presence of PEP trained Bank workers at the *upazila* working alongside GOs, there would be no need for credit TA at that level. The implications from this discussion are that a District Resource Centre might replace the idea of an *upazila* one (which has not yet been attempted), and that if some TA 'socio-economic' staff were at *upazila* level they would be working in an explicitly temporary way in new *upazila*—a sort of mobile group formation advisory unit.

This description of the relations between the BRDB and the TA staff has been presented here to illustrate with case examples the misunderstandings and mutual criticism which have existed. Obviously this description is not a revelation to the staff directly involved, and it is not exhaustive. But I hope it serves to provide some insight to the Evaluation Mission and the donors, as well as to the staff concerned. The examples have been presented to avoid too much direct attribution, although that has been unavoidable in some cases. It would, however, be useful to conclude this section by summarising these and some additional issues of concern. Together they constitute an agenda, which the draft 'Administrative Guidelines' have tried to address within the framework of the project as it is currently formulated. But the Guidelines themselves reveal the depth to which problems have reached, since they give the overriding impression of continued mistrust between the sets of staff by defining functions and tasks so tightly. If these Administrative Guidelines were to be introduced into this climate, everyone would be using them to police everyone else—they represent an institutionalisation of mistrust. The agenda, then, has to be seen as evidence that RESP2 has to be formulated on significantly different lines rather than incremental adjustments to existing practice. This report, the Project TA staff proposal and the promised one from BRDB all agree on that, even if the conclusions vary.

To return to the agenda:

1. There is a duality of command structure, in which the distinction between implementation and advisory responsibility is not clearly specified.
2. The extent of tolerable deviation from normal BRDB administrative practices is unknown for clearing expenditure, recruiting staff, working unsocial hours and living in unsocial conditions.
3. The distribution of compensatory reward for such work within the BRDB is restricted to GOs (the Tk. 500 Union stay allowance) and authorised by the TA staff.
4. The remaining salary differentials between BRDB and TA staff at *upazila*, District and HQ are hugely unequal (average a multiple of 3),

and not justified by references to job security for BRDB staff especially when TA Bangladeshis, many of whom are ex-NGO often BRAC, are experiencing successful job mobility and receiving incomes which are significantly larger than the directors of large NGOs. The motives of TA staff are naturally if unfairly questioned by their BRDB counterparts under these conditions. Nothing contributes more to a lowering of morale than differentials of this kind in what is trying to be a cooperative venture. The expatriate TA staff cannot be entirely excluded from this problem, but if the market is being applied in one sphere then perhaps it should be in the other.

5. The expatriate TA staff constitute a further problem in that a very high turnover is currently envisaged, which dislocates the project, contributes to very short-term perspectives, distorts the ratio between learning and functional effectiveness, and undermines their technical or professional credibility with their TA and BRDB counterparts.

6. The TA staff of all nationalities are mainly generalists operating in areas of *bittaheen* group formation, training, mobilisation, and EIG activities feasibility assessments which are all familiar to Bangladeshi NGOs which have to some extent pioneered them internationally. These NGOs have already been involved contractually in the training of BRDB staff. As generalists, the expatriate TA staff would seem to have less to offer than the Bangladeshi TA staff who have by and large come from that relevant NGO tradition, yet the expatriate staff can be observed in a command role over their TA counterparts trying, as 'socio-economists' (for the most part new to Bangladesh), to wrestle with administrative matters such as planning, scheduling and mobilising budgets for training.

7. The distribution of TA and BRDB personnel within the project at *upazila*, District and HQ is unbalanced to the point where no development of capacity within BRDB personnel is possible.

8. Perceptions differ over the extent to which this is a service delivery or group mobilisation project. This is expressed in the nervousness about the impact of credit upon the project. The control over credit looks likely to be disputed for example over the affiliation and group maturity/savings record procedures.

9. Disputes exist over flexibility in administering the training budget.

10. Disputes exist over the extent to which RPP groups should reorganised or allowed to wither in the project areas.

11. Disputes exist over the flexibility of response to new EIG activities, with the expatriate TA staff wishing to be very adventurous—e.g. on a Tk. 5 lakh bee-keeping programme in which consultants from Scandinavia were apparently used despite the expertise in Bangladesh, while the BRDB are cautiously wedded to 'traditional' activities for the poor such as rice husking and petty trading.

Policy Implications for PEP in RESP2

This is not intended to be a long section, but rather at this stage in the report to outline the various possible conclusions from the preceding account of relations between the project and the BRDB, and to make clear the strategic choices which face the donors in preparing for a continuation of PEP in RESP2. It is certainly my judgement that there should be a continuation. It would make no sense at all to withdraw from the principles associated with PEP in target-group formation and support for such groups. However, the problems which have been recounted in the previous section are susceptible to at least two rather opposite conclusions, both of which imply continuation: that the purpose of the project is to improve the capacity and commitment within BRDB in its work with the rural poor, so that as long as problems of practice exist the project and its TA component is justified; or that the purpose of the project is to provide immediate assistance to the rural poor in the project areas through a directly and separately administered strategy of target-group mobilisation and service delivery (including credit) to develop EIG activities and access to basic rights. There is a further version of the second alternative, which responds to the problems of the BRDB by seeking to relocate the PEP in another part of the Ministry, namely the LGEB. This version is currently the preferred position of the 'Draft Working Paper on RESP2'. This section will discuss each of these alternatives briefly, so that the reader has the options clearly in view when reading the rest of the report.

First, let us examine the 'immediate assistance view'. It has already been implied that Nordic or at least Swedish aid to Bangladesh is too small to make a significant quantitative impact upon contemporary poverty. While this may be true in national terms, it could still be argued that in the project areas (10, potentially 22 *upazila* if all IDP ones are covered by PEP) some quantitative impact is possible so that whatever institutional formula is effective for those areas is justified. This is to some extent the position adopted by the EEC, which has used its control over significant imports of food aid to Bangladesh nationally to insist upon a self-contained mobilisation and service model regionally in Rangpur. This might be regarded at the 'enclave' position. In one sense there is no TA in this model, because there is no counterpart agency to assist except for a token project manager who in effect acts as a conduit for EEC funds to be routed through BRDB directly to the project. However, in another sense it is all TA with expatriates having counterpart, junior, Bangladeshi TA. The problem here is the extent to which such an 'enclave' model fails ever to break out of the enclave and have some broader institutional effect to ensure that its principles of operation are continued in the future. Without the prospect of such a transfer, projects with time limits (even if it is 10 years) could be guilty of irresponsibility in setting groups of the poor up in exposed and vulnerable positions (because other survival mechanisms had been abandoned in favour of project supported struggles and EIG activities) and then withdrawing, leaving nothing behind. It is simply not good enough to pretend that groups will cohere, and become successful, collective entrepreneurs after a few years of intervention. Without wishing to argue for long term dependency, it has been our experience in Proshika that the nature of support changes as old problems disappear and new ones emerge —staff intensity in a locality may drop or it may be restructured to provide, for example, more genuinely technical inputs. But neither Proshika nor RDRS (in the Rangpur Region) have contemplated a total withdrawal of some organisational backstopping to sustain groups in their solidarity and to provide 'institutional protection' for groups or individuals to draw on. The enclave model shares nothing and leaves nothing behind. Furthermore the option is expensive unless it is all contracted out to NGOs, since it becomes in effect a 'donor NGO' trying to learn and replicate what local NGOs already do with longer relevant experience, less staff intensity and much more local knowledge than any expatriate-led project team. At the same time, we should not imply a contradiction between immediate assistance and longer-term institutional development. The objective should be to attempt both simultaneously. PEP is not merely an experimental exercise.

Secondly, this brings us, therefore, to the variant of the 'immediate assistance' position, in which the enclave preferences of the TA staff are modified by the ideas of relocating PEP within LGEB. This position is described in the 'Draft Working Paper on RESP2'. It should be noted, instantly, that part of the motivation for this proposal is to establish more integration between the PEP and IDP components of RESP. It has been a further criticism of the BRDB relationship that the administration of PEP is not sensitive and flexible enough to optimise these potential linkages on behalf of the poor.

In particular, the integration and phasing of project attention on LCSs becoming PEP groups and, where appropriate, vice-versa. Certainly this objective of integration is important, and what is proposed resembles strongly the experiment in the latter days of IRWP with informal group formation in Bhanga *upazila*, Faridpur. PEP emerged from this experiment. However, the arguments and proposals in Chapter 2 of the 'Draft Working Paper' put the clock back in several ways. It seems to object in principle to incorporating PEP into BRDB's RRP through a process of RPP withdrawal and complete PEP occupation of the areas in which it is located (para 3.5). It dismisses the commercial bank option for credit (under Banking Plan conditions) before it has even been tried, with the implication that a self-contained project RLF is preferred (3.4). This attitude to credit implies no wider reform agenda for the banking sector at all, which is something that PEP was set up to test—as part of its experimental purpose. The arguments (para 4.1) concerning the problem of groups capturing some of the benefits of the infrastructure investments is a familiar one: As the author of the LCS concept, I am pleased that it is going reasonably well, but it was never intended as a panacea for group mobilisation, since infrastructural work is such a small though important component of poor people's survival strategies. The proposals in chapter 3, para 7 nevertheless appear to reinstate the principle of infrastructure work as the lead concept in group mobilisation. This is the only way I can interpret the proposal that: "Socio-economic interference must *precede* and *continue* after the construction work of IDP, and this shall be given priority over mobilisation/organisation of RESP 2 groups outside IDP investment areas." This constitutes a major reworking of the principles on which RESP was founded—namely a bifurcation of IDP and PEP, as a reflection of the lack of interest of the Local Government structures and LGEB in the rural poor, and the problem of infrastructure as a limited intervention in people's lives, around which aggressive mobilisation without broad based support across the year was problematic for the poor. This process of redefinition is clarified further in Chapter 3 para 9.3 and should be quoted in full to underline the position taken in the 'Draft Working Paper':

> "The RESP2 socio-economic resources will be concentrated to areas of direct programme investment. Schemes which are not area restricted should to the extent possible be implemented in areas where present PEP groups have been mobilised. Group mobilisation with no linkage to investments/constructions should be discontinued. Exisiting PEP groups in areas where no further investments can be made should however receive continued support until they are self-supporting."

What is proposed here is a 'cleaning up' exercise, with a responsible concern of not leaving PEP groups outside investment locations stranded, in order to achieve an investment-led target-group programme with further support to LCSs and support for two-tier systems (as the NGOs are currently practising). While obviously being driven by a profound concern and commitment to the rural poor, the 'Draft Working Paper' is in effect arguing for a more restrictive approach to group identification and mobilisation, because infrastructure opportunities are so limited to the poor in time and space, especially in the flood-prone areas (to which RESP is historically restricted) with extensive male outmigration in which regular, external patterns of survival have been developed through annual agricultural labour, sharecropping or rickshaw contracts. At the same time, the proposals lose any longer term institutional significance. There is a consistency in seeking to locate the more restricted approach in the more single function part of the ministry. The problem, therefore, with this version of the 'immediate assistance' option is that the enclave approach only appears to be modified. Since the LGEB is not in a position to recruit a permanent (or even temporary cadre) of GOB counterparts (the equivalent of ARDOs and GOs), PEP could only be implemented through the direct executive authority of TA with its own cadre of project-tenured GOs. This was of course the preference of the TA staff as reported to me in Faridpur.

The third option, the development in BRDB of capacity and commitment towards working with the rural poor, certainly needs to contain the dimension of immediate assistance to the rural poor in its areas of operation, otherwise the test or experiment would have clearly failed! The argument here is that for Swedish aid (and perhaps Norwegian) 'immediate assistance' is a necessary but not sufficient condition for supporting the project. The option should also give serious attention to optimising the interaction between the two wings of the project—PEP and IDP. However, the relationship should be the inverse of the funding ratio. The temporary nature of infrastructure-related employment (except for maintenance, pipe-casting and roadside forestry) in time and space suggests that IDP schemes are incorporated into the portfolio of PEP group overall survival and advancement strategies, as appropriate and where other precious relationships are not undermined in consequence. This is the logic of group mobilisation taking precedence in time over the introduction of IDP schemes into an area. With these caveats, the argument here is that the donors should not 'lose their nerve' at this early stage in the project, but should seek to use the information arising from the TA staff, the BRDB staff, this report and the Evaluation Mission to re-assess and renegotiate the 'space' which PEP can occupy in BRDB in order to strengthen GOB's own capacity to assist the poor. To an important extent, a 'green light' has been offered by the D-G, BRDB for this exercise, and constructive dialogue should be maintained. The final sections of this report will seek to clarify the requirements for occupying this space, so those arguments will not be rehearsed here.

Group Formation and Mobilisation within PEP

Discussions were held with 15 groups and several group organisers in Faridpur Sardar and Bhanga Upazila, Madaripur Sardar and Rajoir Upazila, and Kurigram Ulipur, Rajarhat and Nageshwari Upazila. In the last two *upazila*, the focus of attention was on 'waterbody' groups, a label which may carry equivalent problems to organising poor people narrowly around infrastructure opportunities. A number of themes arise from these discussions, some of which are repetitive and others more unique to the circumstances. It should be obvious that in short meetings of about 2 hours, only a limited insight can be gained of the complexity of people's lives and how the project appears in them. Although some standard questions were asked, our method of interviewing was as far as possible designed either elucidate the group members' primary concerns, or to corroborate or refute other propositions being made to us by project staff or previous groups. Selection of groups in this kind of exercise is a combination of random intent, availability of group members, and accessibility of group (both in the sense of achieving a balance between travelling and meeting time, and the need for guidance to the location—especially for evening meetings). The willingness of group members and project staff to be available at short notice is a tribute to the hospitality for which Bangladesh is so famous, but to some extent it perhaps also reflects a command relationship between the staff and the groups, which is expressed in the language of 'sir' and 'sahib' which pervaded most of these occasions. NGOs do not entirely avoid this hierarchical relation, which is so deeply embedded in the society generally and in the relations between government and the people in particular, but they are certainly more alert to the need to overcome this deferential form of interaction. This should be a key objective of Human Development training, since people will never stand up to officials and other classes if the training culture unwittingly reinforces. Brief sketches of each of these group interviews will be presented for the first 13 cases, followed by a discussion of the waterbodies in PEP. These sketches do not amount to a contribution to knowledge for the project staff, who live daily with much of the detail, but rather a selection from it informed by our own antennae (i.e. Shah Newaz and Wood). My fieldnotes contain much more detail than the points summarised here which reflect the selection of issues which seem most to affect good practice.(NB BPD indicates a male group, and BMD a female group, and note also that the style of these sketches is semi-note form, with some use of the first or collective person without quotation marks.)

1. Kabirpur Madhapara BMD, Koijuri Union, Faridpur Sardar

The group has 22 members, 16 with husbands. Three had to drop out after screening, though one of them, Rehenu, had been instrumental in setting up the group. She is the wife of a typist, and was approached by the GO to form the group. She maintains an interest in the group through conducting literacy training and helping the cashier, Amena. She participated prominently in the discussion and must be considered a social if not formal member of the group. Other neighbours in the *para* who are considered by themselves and the group members to be equally poor have been excluded from membership because of the ceiling on size. Both the GO and the TA staff favour a concentration of group formation which overcome this local division, but the ARDO is insisting on a wider geographical spread. There has been a 50% drop out on literacy classes because of the violent objection of husbands, involving the beating of wives. The project is described as a '*bideshi* committee', with an almost 'cargo' attitude to what might be offered. This is partially confirmed by the two women in the group who have regular maintenance contracts—'they go to work, and just come home and cook food'. They no longer participate in group meetings. Savings are not used to provide mutual assistance, since they are nervous about repayment.

2. Mamudpur Paschimpara BMD, Gerda Union, Faridpur Sardar

17 members from an original 24. The 7 were screened out more than one year after formation, although the group members considered them to be equally poor. The group members protested, but the GO was adamant that they were outside the criteria. One of the members has two sisters-in-law who she claims are equally poor, and they and she are resentful and angry about their exclusion. The TA staff would like people who are useful to the group to stay in it (like Rehenu in group 1 above). The husbands are uncooperative about their wives' membership. They will not help with savings, they say the '*samity* is not needed', and they too are violent over

their wives' attendance at literacy classes. They consider the household duties and the children to be neglected, but the women deny this. The literacy teacher, Hashi Begum, is also the cashier as the only literate person in the group. The savings are irregular since the women have to rely on unpredictable earthworks income or taking handfuls of rice from the pot (which is common behaviour for women in households anyway). Some of the members have been engaged in earthworks but not in LCSs. The leader of the group has recently been working for another group member in raising the ground level of the house, but this group member also earthworks.

3. Paschim Para BMD, Koijuri Union, Faridpur Sardar

Although not much detail is being provided for this group, it is significant in that 6 women (not the whole group) have received a loan from the TA fund (i.e. prior to the introduction of the Banking Plan) for Tk. 12,000 to sharecrop in 6 *bighas* of IRRI *boro*. They have also used some of their training allowance money for this purpose, and 4 seedbeds have already been prepared. Two of the women are operating the account on behalf of the other 4, and they will rent in ploughing services. These 6 women took the initiative to approach the owner and lease in the land. This is seen as an action-research activity by the TA, which is why these 6 have received the loan. This is an important innovation not only for PEP but in Bangladesh. The TA staff intend that an action-research fund of this type will remain with them for this sort of use after the introduction of the Banking Plan, although they are worried that they will lose all control over credit. Again, this group reported much bitterness over the exclusion of members after a screening exercise in which the GO and ARDO decisions "had to be overruled by the TA staff". As a result the TA staff from the *Upazila* are now involved directly in the process of group formation in order to avoid subsequent bitterness. (Is this the right reaction to the problem?)

4. Bakunda Pulbopara BPD, Gerda Union, Faridpur Sardar

15 members at present. Had lost land to brickfields and compulsory sales to Railway Board. Main motivation behind membership was savings as a trigger for loans from the project. The rates of savings vary with fluctuations in employment, with the worst time in September. Had considered asking Yusuf Sahib (the ARDO) for a loan to lease land and cultivate IRRI *boro*, but subsequently put off the idea by the cyclone. The ex-GO, Anwar (who is now in Kurigram) used to stay in the locality in a clinic. The group is interested and is under pressure to expand its

membership to 25, and the new GO, Farouq Sahib, is assumed by the group members to have the final decision. However, there is a problem about expanding membership when significant savings have already been accumulated by the group. If group savings are used for joint activities how will the dividends be apportioned? This problem disappears if the EIG activity is loan supported, but even a mix of loans and savings represents a problem. The group asked for Tk. 20,000 loan from the TA fund for cultivation, and the TA staff offered Tk. 15,000 with the balance to come from group savings, but the group have been reluctant to accept this because of these problems of different periods of membership. (Of course if savings are irregular for some group members who are 'in default', which we have observed as quite a regular occurrence then the problem is quite widespread for this reason.) The group leader owns a rickshaw, and there are many rickshaw pullers in the group, but an earlier loan request for the purchase of rickshaws was turned down by the TA staff, even though rickshaw owners are bringing new ones into the area continuously. It is odd that a group dominated numerically by rickshaw pullers should be encouraged to take up cultivation instead of rickshaw ownership. The group often uses the savings collected at meetings as an immediate loan to a group member in need. The group's savings are reported as Tk. 4,600, but its assets are therefore higher at Tk. 5000, since Tk. 400 is currently loaned out. (The use of savings for instant interest free credit to fellow members might be encouraged, since cash flow problems mean high interest loans and a further subsequent loss of assets. However, the practice does raise issues about accounting procedures since this loan activity does not seem to be recorded anywhere.)

5. Muraridha Pulbopara BPD, Koijuri Union, Faridpur Sardar

28 members, and 2 additional members are expected to join. 20 members of the group work in brickfields for 5/6 months. Four *paras* in the locality, comprising about 450 households, of which approx. 100 are organised into PEP groups. The members knew that some groups were to be formed from the Member Sahib (member of the Union Parishad). A big meeting was held in his house (he owns about 15 acres), and the GO announced that groups were to be formed of the poor, i.e. those with less than 1 *bigha* (which is presumably less than the half acre in the guidelines). "We know who the poor are, so we selected from among ourselves." After about 4 months Yusuf Sahib (the ARDO) and Mannan from the TA staff came and described the reasons for being in a group, such as savings, credit, training

and regular meetings. Initially the group meetings were held in the Member's house, but the group was told by the District TA staff that the group should hold them elsewhere, so they are now held in the house of 4 brothers, who live in separate *chulas*. One of these brothers is the group leader. (This might suggest a strong kin-group domination of the PEP group.) Loans from savings are made to individual members, and are recorded on a piece of 'white paper', which is destroyed when the loan is repaid. After the groups were formed in this area, there was some more earthworks for group members and women group members. 15 women from 3 women's groups in the area have routine maintenance contracts. Many more people would like to join the group. The records of its meetings (attendance, minutes and resolutions) are well maintained, but sometimes with the assistance of the TA staff. There is some confusion over whether the group is affiliated to PEP (for credit purposes), but the group is interested in a loan to cultivate 2 *bighas* of IRRI *boro* collectively, by working in the fields in rotation after finishing the day's work in the brickfields at 2.30 pm. The members have clubbed together to buy *saris* for a member's wedding. The Member employs most of the group members at some point in the year for agricultural labour, but he treats them with more respect now and is interested in their votes at the next election. The group felt more confident about their strength. "The rich used to tether us like goats, but now we can control the rich, and we are insisting that the brickfield wage rates should be raised to Tk. 35 a day."

6. Adampur Mirpara BPD, Manikdha Union, Faridpur Bhanga

25 members all from one *para*, where there are 100 households. About 80 are regarded by the group members as poor like themselves, and would like to join the group, but are excluded by the ceiling on membership. The GO approached the Union Parishad Chairman, who held a meeting in his house. 19 attended initially, and a further 6 enrolled later. For 6 months of the year, about 20 of the members are engaged in continuous earthwork, but there is also work with rickshaws, vans and on the campus of the Atroshi Pir. One of the group members is a labour *sardar* who mobilises up to 70 men to work if the contract requires them. He does not work himself, but gets the contracts, supervises the work and takes a commission. He therefore organises most of the work for most of the group members, and takes a commission on their labour. Since the land is quite poor and very flood-prone in this area, the value of half an acre is much less than in other locations, yet there appears to be no flexibility on this single index. (There

is a need for a more situational definition of poverty, and the GO with us agreed.) Savings are deposited at monthly meetings because the bank is too far away to travel there every week. The group members have cooperated more with each other since it was formed, e.g. over getting bamboo to repair flood damaged houses this year. The group leader, who was very quiet during the discussion, was selected because "he has a good way of explaining things", but he is a date and palm tapper and is unable to attend training except during the monsoon—but no training is held at that time. 12 of the members have been on Human Development training and were able to describe its content well.

7. Manikdha BPD, Manikdha Union, Faridpur Bhanga

The discussion with this group also revealed other 'data', in the sense that one of the TA socio-economists from the *Upazila* who was present kept trying to answer the questions, and tended to order the group members around. Of the 25 members, 2 are Hindu. The membership is drawn from a BSS group formed 5 years ago. The BSS group was formed by Motaleb and a cooperative inspector, with the promise of loans at Tk. 4000 per person to buy rickshaws. The promise was not kept and the members lost interest in the BSS. There are 11 rickshaw pullers in the new group who were in the previous one. Motaleb had some earlier connection with the KSS, so the cooperative inspector had used him to form a BSS. The cooperative inspector then organises savings and deposits with the UCCA, though he makes 'additional charges' for 'registration and account books'. No-one of the old group can say what the total savings figure was. The new group has passed a resolution trying to retrieve the savings of those in the old group. However, the cooperative inspector's verification is required and he has been transferred, and only Motaleb even knows what his name is. (Motaleb only turns up at the end of session, and also claims ignorance.) Motaleb is still the 'manager' of this new group and initially described himself as a woodcutter, but on questioning he revealed that he was really a businessman: buying wood from those in need of cash, employing 2 people to cut it, and arranging the sale. Savings in the group are deposited weekly but only paid into the bank monthly because it is too far away. According to the group, savings can only be used with the approval of the project staff and must not be used for immediate consumption purposes. If members need money then they have to go to the moneylenders (often the local shopkeepers) at a rate of 10% per month. Sometimes during the monsoon months, members are borrowing in order to fulfill their savings commitments. This is also the period of outmigration for 4-5 months, with members (along with non-members) often travelling together to Dhaka and to Jessore. Some of the travellers are then 'delegated' to bring back money to the families every couple of weeks. Some of the group members have been engaged in local earthworks, but not in LCSs and they are not sure what the scheme was. The Member of the Union Parishad arrived at the meeting and was shown much deference by the group members. He admitted to owning 15 acres, and employs most of the group members who engage in agricultural labour. The group's meetings have to be held at the group leader's house which is by the roadside, so there is no possibility of the meetings being held in private. (This is a general problem, since non-group members who casually turn up to a meeting in a *bari* cannot be easily excluded. Rich neighbours and employers are obviously curious to know what is being discussed, and this clearly restricts the strategic options of group members. (Yet in my many meetings with Proshika groups, the occasions have always been more private.) The group considers it has been promised loans by the project for fish cultivation in 2 nearby *khas* ponds, and they would still like to buy rickshaw vans either collectively or individually. None of the 'officers' of the group have received any training, though 6 'ordinary' members have, and no-one has received any LCS training.

Before leaving Bhanga, its special problems for PEP should be noted. PEP has been given the mandate to reorganise the 70 BSS/MBSSs which existed in the *Upazila* RPP. The problem of screening out non target-group members, is compounded by this problem of not only extracting the savings of PEP group members, but also disentangling loan commitments. In many of these cases, no independent documentation remained with the group members so the cooperative inspectors have been able to 'hide their tracks' where they are accused of informal charging and mismanagement. Sometimes the PEP project staff are perceived as cooperative inspectors and treated with much caution and suspicion. As elsewhere, Bhanga also illustrates the problem of insisting on regular savings and meetings in areas of significant male outmigration.

8. Chandiborder Pulbopara BMD, Dudkhali Union, Madaripur Sardar

15 members from the same *para* which is usual for women, but not 'necessary' for men. They come from 4 *baris*, out of 7 *baris* and 50 households. These 15 describe themselves as more motivated to join a group although there are others who are poor. 13 of them have husbands, 8

of whom belong to a group outside the *para*. All the women work inside the household, which actually includes cattle raising, poultry rearing and paddy husking, with husbands trading. "Our husbands could not manage if we worked outside." (The concept of 'outside' has to be seen culturally contingent, and in this case really refers to earthworks.) The source of savings is from husbands, selling eggs, and saving rice from the pot: in fact, the usual source of savings by poor rural women. Group savings are not used in loans to individual members, but individuals will loan to each other to enable savings to be deposited. The floods have interrupted the regular savings discipline. They intend to use savings in. EIG activities (unspecified), the education of children and the marriage of daughters. While one can understand these last two objectives, the use of savings for dowry purposes could constitute an important source of gender leakage in the use of female group savings. Th group members were very positive about the Human Development training (8 of them had attended) and recited the themes of the course. (The comment of the Training Specialist at HQ PEP should however be noted: "group members are parrots of human development training, they do not come with their own analysis, and they cannot react to questions because there is no appropriate follow-up" Interview 18/12/88.) Half of the group meetings are held without the GO present, but both the GOs (male and female) come frequently at other times.

A. *Interview with Group Organiser, Bahadur Union, Madaripur Sardar*

This session took place between the discussions with groups 8 and 9, and helps us to understand the process of group formation, and background of the organiser. (A further GO interview follows later. Other interviews were taken but are not reported here.) He is a BA in Political Science and Language from Rajshahi; then a teacher; then 3 years in BRDB *Upazila* Accounts; joined BRDB after seeing a newspaper advertisement; transferred to Madaripur before PEP started and worked on loan realisation (relief of interest after 87 flood) before transferring to PEP; wanted to work near Madaripur town, but TA HQ argued that government does not normally work effectively in the 'interior' (interior to whom?). Initial work was tough. We had to take bedding and sleep in village houses. Went from door to door talking to people. Started to form groups, initially 18 male and 2 female. Two male groups were 'dropped out' because they were greedy, obstructive and always causing trouble over savings. Regular work follows monthly work plan: attends meetings, takes care of account books because most group members too illiterate to cope (How accountable are GOs under these circumstances? Is the temptation to behave like cooperative inspectors offset by additional allowances?) 5 days in a month are spent in Madaripur, submitting work plan, drawing salary and attending meetings. If the female GO is away e.g. on training, then also looks after women's groups. After the flood, heavily involved mother/child immunisation and cattle vaccination. Supervises action-research on vegetable cultivation with women's groups. Selects members from groups for training. Facilitates informal training on issues such as: need for savings and regular meetings, group organisation, and explanations of poverty. Responds to emergency requests to handle visitors (like me!), and to provide extra information. Organised people from groups to work on rehabilitation of a bridge—not for proper wages, but tea and snacks. Helped a group to get a pond for fish cultivation on lease from an owner for 15% of the catch. When PEP started, Union Chairmen were invited to familiarise them with project and introduce them to GOs, and they were invited to describe the social condition of their Unions. This gave us some idea where to go. Also met the Member of this village area in the *haat* who helped to direct us to some villages. In the different groups, only 4 were dropped out after screening: the decisions were made by the group but after prompting by the GO. This screening was done after a regular meeting about 5 months after the group was formed. This screening has to occur, because initially the GOs are quite ignorant about different households, and it is impolite to ask so many questions of personal detail when meeting people first. Later we have more information, which is also necessary when we start the credit service. The work plan is designed by the *Upazila* office, and supplied by the ARDO. The form has recently been changed. It is a guideline. (However, on examination, it does look very specific.) We work to a 20 day cycle of visits around in my case 16 groups —which should mean 16 group meetings with informal visits in addition. Requests come to me from various sources: ARDO, TA, BRDB, other GOs, but our responsibility is to the ARDO. There have been some embarrassing conflicts of instruction and advice. Most meetings take about 3 hours, with about half taken up with savings business.

9. Mirikdia BMD, Bahadur Union, Madaripur Sardar

25 members, of whom 5 are widows, and 4 divorced. Originally 27 members 16 months ago, but 2 have died, one in delivery. The members come from 2 *paras* and 11 *baris*. There are about 45 households in the 2 *paras* across 13 *baris*, all of whom are regarded as equally poor. This configuration means that about 7 of the *baris* appear arbitrarily split between members and non-members. This has caused much bad feeling,

and there is pressure by the non-members because they see the members getting relief, training, status, many visits and obviously suspect the group of receiving much more. (This is a clear demonstration of a pervasive problem about concentration of group formation to avoid local conflict, and of the problem about a rigid limit to group size.) The rich families in the area, holding e.g. 50-100 *bighas*, are a little hostile. They say the women will be converted to Christianity when they go for training. (This theme has been deployed in several of the project areas.) They object to the women working in the field, and it is difficult to rear cattle, goats and poultry because if they stray on to a rich man's field then he confiscates the animal. "We are trying to avoid confrontation with the rich until we are stronger." About 15 of the members have husbands who both work as agricultural labourers for the rich, so we have to be careful, and outmigrate for about 5 months a year to Jessore and Sylhet between November-March. In their absence, along with the widows and divorcees, the women have to cope on their own. The men are away for about 20 days at a time, coming back to give us money. If there is no money, we ourselves go directly to the moneylender, we also do domestic labour for the rich families if we can. It is because of these connections that we cannot confront the rich to get land, ponds, better wages, or to reduce interest rates. But we try to manage savings through selling coconuts, eggs and saving rice from the pot. We are free to go for training. We are saving to build up a big fund, but hopefully the project will help us with credit for poultry, livestock rearing, pisciculture, paddy husking and small trading. The group is also interested in trying to sharecrop in IRRI *boro*. (This group is to be contrasted to Group 8, where there are more husbands present, doing more work locally. The women of Group 8 see themselves as doing more 'household' work, even though the functional reality may be otherwise. But the women of Group 9 certainly give the impression that they have to fend more for themselves, and they were much more confident and forceful in their statements. The long absences of husbands through outmigration, combined with the condition of widows and divorcees, has given these women no choice but to challenge further the cultural expectations of their sex.)

10. Shutagandi Paschimpara BPD, Bajitpur Union, Madaripur Rajoir

There is a sense in which this case should not be presented. It is a group reorganised from a BSS, but the story of its disentangling was so confused, constantly changing with different opinions of group members and group organiser/TA staff present that it became impossible to derive a consistent description of the process. I am not going to attempt a phoney reconstruction here. But the confusion is itself data about the process of reorganising BSSs. The elements of this story involved conflicting ideas about membership of the respective groups, about the role of a KSS member who had formed the BSS and then initially the PEP group (a sort of 'career' or 'groupie'!), about the status of the 2 male groups in the village, about outstanding loan obligations from the BSS days, about the procedures and information on how to extract savings for the members of the new PEP group. However, the group did have a clearer idea about the differences between an 'IRDP *Sarkar*' (government) group and a PEP group: PEP groups pay their money to the Sonali Bank not the Krishi bank (actually there was some confusion over whether it was Sonali, but Sonali was the final agreed position of the members); some of the PEP people are in LCSs but BSS cannot belong (is this true?); it provides employment; PEP staff visit more often. However, PEP was described as a '*bideshi sangsta*' (a foreign organisation), but only 'sirs' can tell which *bideshi*. PEP is a branch of IRDP *sarkar*, but that may be foreign. The GO, Nasrul, is a frequent visitor and he asked us to call him '*bhai*' (brother). Many of the members regularly outmigrate, but their family members come to the group meetings to deposit savings on their behalf, in this way the savings are quite regular. (The corresponding women's group from the same *para* had 24 members, 20 of whom were ex-MBSS, 6 married to male group members. The chairwoman was the literacy teacher, owns 4 cows and has a large house with good walls and a tin roof. Her husband is a rickshaw puller in Dhaka. 7 of the other women in the group live in families which own cattle. Does ownership of cattle before group formation constitute disqualifying 'assets', whereas it is 'EIG activity' if acquired through the project after the group has been formed?)

11. Charbozra Madhapara BPD, Bozra Union, Kurigram Ulipur

As the name suggests, this group lives in the flood prone part of the *upazila*, though it does enjoy some embankment protection. There are 25 members from 24 *baris*. Of 132 households, more than 80 are regarded by the group members as equally poor and have tried to gain entry to the group. Membership has been restricted by the 'officer', and we would need permission from him if we wanted ourselves to organise a neighbouring group. (Would they have an interest in doing this? The group reports that there is strong competition among all the poor for scarce local employment opportunities, in which the old and the weak do not get hired. There are only 3 major employers in the *para* with 95 *baris* competing individually for the work. A *bari* here is more approximate to a *chula*/household.) From this discussion, considerable variation in family size was revealed which led into the problem of using fixed indices like landholding in the context of variable need as indicated by family size. The group was strongly against using savings as short term loans to group members, this would break down the discipline and sabotage the savings strategy. If people are desperate for cash, they have to go to the moneylender. One old member is currently 18 weeks behind in his savings—another reason why paper group assets might exceed the actual amount deposited, though the reality of these assets is something different if defaults cannot be made up. Now that the group holds regular meetings and has a savings discipline, there is less need for the GO to attend our meetings. He can be used elsewhere to form more groups like us. It also emerged that this was essentially a group of petty traders, so that their interest in PEP credit was to support these existing EIG activities as individuals. Some of them had been involved in a food for works scheme February-March. It was a 'Union Chairman and Member scheme' in which we were paid 40 *seers* of wheat per 1000 cft, when it should have been 50 *seers* plus 2.5 *seers* for the *sardar*.

12. Ramkhana Paschimpara BMD, Hatia Union, Kurigram Ulipur

The village lies outside the protection of the embankment, the settlements are prone to extensive flooding and sand rather than silt is everywhere. 25 members from 9 *baris* in one *para*. There are 60 households in the *para* only 3 of which are seen as marginally better off. Of the 25 members, 18 are widows or divorced, 6 remaining in their husband's house, 6 in neighbours houses of their husband's village, and for 6 this represents a return to their natal home after abandonment by husband or husband's family. All of the 18 have 2 or 3 children with them. The males in their homes are unemployed for 6 months of the year ('idle'), and for the other 6 months they have work inside and outside the village. The men are constantly searching for work in these lean periods but are reluctant now to raise a loan to keep their families while they travel outside to search during the lean period. The women do not know during the day what their men may bring home in the evening for them to cook, and many times they come empty-handed. Last year, Hashi Appa, the GO provided 34 days of earthwork for all the members, but there is no work this year. Two women from each of the 2 women's groups in the area have road maintenance contracts, but actually the work is rotated among us to share it. A male group was formed here first, so we wanted one. We asked the GO and staff who visited the male group, but they did nothing, so we went to the Ulipur office and finally got a response. The old and the pre-marriage young have been excluded from group membership. The savings have to be very irregular. Hashi deposits them for us at the bank in Ulipur (Janata) as it is too far away for us. We do not want loans from the project, we want earthworks.

13. Ramkhana Pulbopara BMD, Hatia Union, Kurigram Ulipur

We visited this neighbouring group quickly, because some of the members had come over to join the previous discussion, and were insistent that we should visit them also. This itself was data, since they forcefully said that they were always being neglected by the GO in favour of the previous group. We got no (LCS) earthwork last year when they did. However, they did receive a Tk. 11,000 TA fund loan for rice husking last year, and the final balance of Tk. 300 was paid to the GO during our visit. It was also evident from the group, that family fortunes depended very heavily on the presence of men, and that in this group the membership was heterogeneous with respect to this variable with half of them being without men. (What impact might this have on their common interest?)

B. Short Interview with Women GO, Hatia Union

This discussion was conducted walking back to the vehicle. She appeared to be very lively and committed. It is also clear that she wields potentially significant patronage over 'her' groups, e.g. on the selection of group members for earthwork and maintenance contracts. Even if she does not actually have this power, she is perceived as having it. Her father's house is in Ulipur, but her husband is an Associate Professor at Jamalpur College in old Mymensingh district. She has 3 children, the eldest is 17, the youngest stays with her in the Union. When recruited, she claims she was promised a transfer to the TA staff. Since there is no prospect of promotion as a GO, what is the point of staying, especially when my husband is pressurising me to come to Jamalpur. She is not clear who makes the decisions in the project, but she knows they keep changing their minds, and is always getting new instructions. Some things get done in a rush, like being asked to form 8 PEP groups in one week in order to fit with an IDP scheme. We should be asked to do things in a planned way, with enough advanced notice. I have problems getting around. How should a woman like me with a son of 17 be asked to go round on a cycle? There are plenty of motorbikes sitting at the district and *Upazila* (that's true), and the women TA staff use them, so why don't we get them? I have applied for one but don't get any response. She has 17 groups to look after. There are no social problems for her in moving around, partly because of age, but also because my father was a headmaster in this area. I have no social problem in walking around at night, but the problem is transportation. (This is a very revealing account of a number of problems faced by GOs in general and women in particular: place in local power structure; promotion linked to motivation; comparisons with TA staff; inconsistent and *ad hoc* policy and operational directives, coming from different sources in the administrative structure; family pressures on women, but different kinds of problem according to whether single or married, young or old, with or without children, and the extent of freedom in upbringing. This last point was also revealed strongly to us by a young, newly recruited ARDO in Madaripur Sardar.)

Waterbodies in Rajarhat and Nageshwari Upazilas, Kurigram

This is a complex picture to summarise. Since there are 16 'Waterbody' groups in Rajarhat (in which PEP has only just started) and 56 in Nageshwari, the Nageshwari experience will dominate the discussion. There appear to be several problems. First, there is a labelling problem. The groups in Rajarhat were not aware that the project was referring to them in this way, which to some extent removes the criticism that these groups are performing as rentiers to actual fisherman. If the groups do not see themselves as being primarily interested in the neighbouring *bheel*, then the provision of a loan to purchase the lease under new government arrangements has to be seen as some form of 'cargo'. The 'waterbody' label is akin to seeing a group only as an earthworks group, but at least earthworks labour requires little skill and no capital assets to enter it. This strategy runs against much of the other ideology in the project about group mobilisation, self-awareness, and empowerment, since the rationale for their formation here is to capture a specific resource. This issue is further reinforced by the experience in Nageshwari in which 48 of the groups were formed in 2 months of March/April 1987, in fact the staff were asked to form the group in one or two weeks. Secondly, then, is the rapidity and phasing of group formation. The membership could not be scrutinised, strong local elite connections were represented in the groups, making their subsequent exclusion almost impossible. Indeed if it were tried in some cases, the project staff would not be allowed back into the area. In one area, the local MP, Union Chairman and their networks have strong interests in the waterbodies, so their representatives cannot be displaced from the groups. Third, the waterbodies are perceived locally as such a valuable resource that the leases will not be surrendered to the *bittaheen* without a struggle, with the main rearguard strategy being the penetration of these groups. This undermines the other dimensions of group mobilisation and human development training. The value of these waterbodies is embodied in the large profits possible over the lease price to the government; the power which can be exercised in other relationships by controlling local people's access to fishing through sub-leases; long lease rights have to some extent been regarded as *de facto* ownership; within the *bheels* during the low water season, there is *khas* land which can cultivated or informally leased out by those with the main lease over the area. Fourth, there are about 15 below-20 acre *bheels* left to allocate, which are therefore in the control of the *Upazila* Parishad (i.e. Chairman and UNO). This category of lease allocation is, therefore, penetrated by their local landed and business networks more easily, although there are factions and alliances among them through which the Chairman has to steer an optimising course in terms of votes and contributions to elections funds, and so on. Fifth, as others have pointed out in tour reports, the fishermen who took sub-leases from the leaseholders are now transferring their requests to the Managing Committees of these 'waterbody' groups, who are located within one km. of the *bheel*. The groups have little incentive to invest themselves in closed waterbodies, since they can gain from this sub-leasing. In this way, they avoid investing in capital (nets, boats), and they avoid risk of theft and policing costs. As always, 'sharecropping' functions as a form of risk sharing in an unpredictable and undercapitalised production cycle. In this case the share is now 10:6 to the groups. Sixth, with fixed share ratios, fishermen competing for rights have to be distinguished by their willingness to fish longer in the season (with what longer term ecological effects?), and the amount of *salaami* deposited. These contributions are non-refundable, and in one case reached Tk. 10,000. Seventh, with an elite-penetrated 'federation' of groups around the *bheels* receiving credit to purchase these leases, the extent of leakage of targeted project funds is really unknown. At present, leakage has yet to be activated as income has been deposited in the bank rather than distributed to the groups and their members. Indeed, the groups never see any of the money in these transactions, it is all being managed for them by the 'office', with TA and GO staff present to supervise the catches, the sale to invited traders, and the distribution of the shares between the groups' managing committee bank account and the professional fishermen. Any relaxation of this close supervision is regarded by the project staff as an open door for the locally dominant classes to re-enter the process and re-assert their claims to income, albeit through these rhetorically *bittaheen* routes. Eight, caution needs to be exercised over the definition of 'fisherman'. The image of the groups on-leasing to poor fishermen who ought to be enjoying the lease rights directly is not strictly true. Some of these 'fishermen' are large-scale businessmen who certainly engage in fishing (as buyers, sellers, employers of labour, holders of relevant capital), but they are also well connected, indeed part of the local elite structure. Some of these have taken care to penetrate strategically groups in 'federations' to give them easier access to different fishing rights: a father here, a son there, and so on. In this way they can rise to prominence in the Managing Committees and influence the sub-lease decisions. Most of the professional fishing in the area has been carried out historically by this class of fisherman. The second category of fisherman are 'locals', who are part-time and traditionally use the *bheels* on cheaper, more casual leases from February to April after the large professional catches have been completed in the peak season (i.e. mid-November to January). It is estimated by the project staff that about 90% of this second category are now members of the groups, and are indeed employed as per PEP regulation for these leases by the professional fish businessmen. Nine, it should also be noted that 25% of these 'waterbody' groups in Nageshwari are ex-BSS, reorganised groups with all the attendant problems which have been noted in previous discussions together with those referred to in this section.

Summary of Issues Arising from Case-Studies

Presented below is a list of 33 issues which arise from the material presented above. The items are listed without comment since their implications have been discussed above.

1. Use of relevant and committed non-target friends in the groups.
2. Exclusion of non-selected poor in a locality, and implications for concentration of groups and size of group rule.
3. Splitting of *baris* in group selection.
4. Male hostility to female literacy classes.
5. 'cargo' dependency of groups on GOs and project.
6. Lack of interest in group participation once maintenance contract is acquired.
7. Social problem of screening groups justifying presence of UZ TA staff in group formation.

8. Self-exploitation as a source of savings especially for women.
9. Innovation of women preparing to sharecrop IRRI *boro*.
10. Rates of savings varying with fluctuations in income flows and employment, stimulated by outmigration and sometimes prompting expensive borrowing.
11. Irregular, or less frequent, pattern of collecting and depositing savings because of distance to the bank.
12. Problem of changing composition of groups once savings have started.
13. Appropriateness of TA/GO EIG suggestions—e.g. cultivation when rickshaw pullers want rickshaws.
14. Use of savings as instant interest free credit to members.
15. Is project permission required for group use of savings? If so, too directive.
16. Strong involvement of Union Chairman and local Member in group/individual identification.
17. Presence of labour *sardars* in earthwork dependent groups
18. Deferential relations between group members and project staff-'Sahibs and Sirs'.
19. Difficulty of extracting the savings of former BSS members now in PEP groups.
20. Inability of groups to hold private meetings.
21. Impact of home based husbands/outmigrating husbands on women's behaviour.
22. Heterogeneity of women's group members by marriage status.
23. Use of savings for dowry—gender leakage or purchase of security for daughter?.
24. Degree of flexibility of GO Work Plans—they closely resemble T&V, and can be very formalistic as a performance indicator.
25. Balance of time used in meetings between savings business and informal human development training.
26. High dependency of group male and female labour on local landlords and political elites constrains struggles for wages, low interest rates, and rents.
27. Variable family size undermines standard asset thresholds as indicator of target-group status.
28. Perceived and actual patronage power of GOs over 'their' groups.

29. Lack of promotion prospects, and lack of access to transport undermining morale and motivation of GOs (similar problems, including salary differentials exist for other levels of BRDB staff).
30. Inconsistency and short notice of instructions to GOs.
31. Overrapid group formation in IDP and Waterbody areas leading to subsequent problems.
32. Labelling of groups around earthworks or waterbody opportunities contradictory to purpose of the project.
33. Membership of groups as cynical instrument of access to targeted resources.

Requirements for Working with the Rural Poor

The purpose of this section is to move from the description and often critical analysis of the administrative and target group formation problems which exist in PEP towards a set of arguments for the future shape of the project. As a step towards this, and partly as an implied set of criticisms of current practices, it may be useful to present some ideas about the institutional goals to which a future PEP could contribute. To some extent, then, this pretends to be a manifesto of good practice for government, TA and NGO staff working with the rural poor. But unless some exercise of this nature is conducted, the word 'experimental' has very little content. It has to be defined in some way, and by definition constitutes a departure from existing practice which is seen to be problematic, as noted in the previous sections.

The first issue is the hierarchical structure of most administrations, but especially those in Bangladesh. Perhaps as part of the colonial inheritance, the structure of the administration seems to be built upon a principle of mistrust in the competence and honesty of one's juniors, especially when decisions have budgetary or financial management implications. As a result, elaborate rules have developed over time, setting very severe limits to the discretion which junior officials can exercise. This structure is anathema to the requirements of rural development administration, especially in a complex and dynamic agrarian structure where local variation makes nonsense of attempts to standardise too closely rules of implementation. Central administrations of course find universal and standardised frameworks comforting and convenient especially on the grounds of equity both for client populations and for staff management. But this assumes that client populations will achieve equity through applying standard criteria of group membership size, or defined assets.

Second, as pointed out above, equity is not actually achieved in this way. Some variation in the application of the objectives of a programme is always required to be sensitive to the meaning of local conditions. Local specificity has to be recognised, and universal indicators avoided. Situational definitions of poverty are required, in order to make programmes compatible with that poverty. In a Matara District in Southern Sri Lanka, during another evaluation for SIDA, I identified at least 8 quite distinct detailed explanations of poverty, but the programme was really only based on one of them. Many of the 'target' were being missed as a result.

Third, therefore, there is a strong need for local socio-economic analysis. There is undoubtedly much tacit knowledge among the fieldworkers of this project, some of which is lost when personnel are transferred, or leave altogether. What I have missed in each of the districts and *upazila* is any coherent sense of the local political economy in the which the project is working. I have been told many valuable anecdotes which have informed some of the preceding analysis, but has there been a conscious devising of locally specific strategy in the context of this knowledge? Clearly it is convenient to an outside visitor in desperate need of a 'quick fix' to be able to read such an analysis. But in many ways it needs to be verbal and oral, conducted in association with the groups in a 'continuous seminar' approach. There has not been time to look at the Human Development training in detail, but discussions with trainers reveal that some of this occurs in the classroom and of course it is meant to be followed up informally at group meetings. However, our impressions concur with those of the HQ Training Specialist: that platitudes of a very general kind are being repeated almost like a litany. I have not been made aware of any systematic *upazila* or district analysis of this kind, and yet it is indispensable to advice on EIG activities, arenas of struggle on wages and rents, legal rights issues, identification of local resources, and so on. This should be the work of the *upazila* or possibly District Resource Centre. Although this aspect of PEP has not really be pursued yet (which leaves room for manoeuvre in relocating it), the purpose of such a Centre needs further clarification. Is it primarily a mobilisation forum in the sense of identifying local resources and assets to which the poor can gain access with appropriate support, or is it conceived as a top-down service delivery function—connecting training, extension, and capital to EIG activities in the area? These options are not necessarily mutually exclusive, but their joint potential would only be realised by an explicit analysis of the local political economy.

Fourth, this implies a substantial change in the patterns of authority in rural development administration. PEP and RPP have an ideology of empowering the poor through their mobilisation into groups (otherwise why do it at all?), but where is the ideology of empowering the fieldworker? If local specificity is to be recognised, then flexibility of response is required of the fieldworker. If the fieldworker is part of the system of local analysis, then the fieldworker is empowered by this knowledge. Their judgements are important, because they are the most informed of local detail. But now they exist in a top-down command structure. This can be observed throughout both the BRDB and the TA structures. Indeed the TA HQ staff have used terminology like 'rebellious districts' and the need for the field to conform to universal guidelines. The relationships between the different levels within both administrations needs to be more symmetrical. And staff in the 'field' will have to be trusted much more to use their discretion. Initiative goes with responsibility. Ironically, this was the principle of colonial field administration: socialise and train the officer well, and then stick him in the district and let him get on with it, he can be trusted to act within the spirit of the organisation's objectives without having his every action supervised and scrutinised. No wonder there is little job satisfaction throughout the different levels of the project. So the hierarchy to some extent has to be adjusted, and the investment in good fieldworkers has to be matched by trusting them. It has been frequently argued by more senior staff in the project that GOs are overqualified (and some have even said: "overpaid"), and, therefore, they are constantly regretting their 'imprisonment' in the field and looking elsewhere. The solution has frequently been presented that less qualified, local people should be found. This sentiment is very revealing. It seems like a plea for a docile labour force, who are used to the local living conditions but will carry out instructions. But these qualifications can be put to good effect in the scenario outlined here, especially if they are relevant. (The label 'socio-economist' is widely used in the project, but is presumably intended more as a description of function and not of the qualifications and abilities of the person concerned. This might be part of the explanation for the absence of serious local 'socio-economic' analysis.).

Fifth, the issue of investing in and then trusting in good fieldworkers must be connected to motivation. Dynamic, innovative, and risk-taking (but not adventurist) performance cannot be sustained indefinitely without some prospect of recognition. We should not expect fieldworkers to be saints, especially when those above them have rarely worked under similar conditions of deprivation. Those not engaged in fieldwork have no right to romanticise about it. (I have similar criticisms of culture of poverty arguments.) Fieldworkers(i.e., in the case of PEP, GOs) have very limited prospects of promotion, although there is possibly a rule that 50% of ARDOs can be recruited in this way. ARDOs have few prospects of promotion. If these field level workers are to succeed in the field, the prospect of advancement through the system must be available. Furthermore, over time, it will mean that increasingly senior cadres of staff will have had progressive field experience. This will enrich future rural development administration both with their experience, and with a revised relationship between themselves and those in the field at that time. This, of course, is also the way to recruit well qualified and enthusiastic staff, and to keep them. The current system, together with the stated inclinations of some of my informants, is not a recipe for good field practice.

Sixth, as strongly implied in the preceding paragraphs, senior staff in rural development administration should then be seen much more as facilitators and monitors, than as commanders. This means that they have to be prepared to be 'led' by those in the field, responding to their agendas, debriefing them to put their ideas under scrutiny, offering comparative experience from elsewhere, ensuring that their proposals conform to the spirit of the programme, but not necessarily to the letter of it if deviation appears justified by the explanations presented. Of course, senior staff will always have to apply certain kinds of constraints when scarce resources and money are required, since these have to be allocated. But these allocations have to be informed from the field. If we expect GOs to behave to groups in this way (i.e. removing the 'sirs' and the 'sahibs') then the same has to apply throughout the system.

Seven, working with the rural poor anywhere, and especially in Bangladesh, requires a balance between mobilisation and delivery objectives. Part of the analysis of poverty is the individuation and consequent vulnerability of those without control over the means of production. The whole basis of the strategy is to enable those who are poor in resources to become rich in people. Forming groups has been adopted in Bangladesh as the first step in achieving this. But there is always a danger in a society and rural administration, which has a cooperative tradition, to see such groups as a convenient unit for the delivery of services (credit, skill training, physical inputs). On the other hand, the preoccupation with 'rich in people' should not encourage us to ghetto the poor into some perpetual backwater of mutual support among the poor as a substitute for connecting them to the major resources and opportunities in the society. This is why governments have important roles to play as the only legitimate authority to set the terms of exchange and distort the market to some extent in favour of the poor. No-one is suggesting that the state has complete room to manoeuvre either domestically or internationally, but those working for the poor should be doing their utmost to connect them to what the state can offer. Delivery, then, is an essential part of the equation, and the poor's access is made more realistic by the organisational innovations in their mobilisation.

Finally, in this short 'manifesto' of institutional objectives for working with the rural poor through government in Bangladesh, the issue of staff intensity should be noted. The objection to PEP and much of what has been proposed here is that, as a system and style of intervention, it is very staff-intensive both in terms of quantity of staff but also in terms of the scale of investment in quality. There are many ways in which this could be answered, some of which lie outside the remit of this report and the project, though not necessarily outside that of Nordic aid. With policies of privatisation and corresponding structural shifts in the economy, several areas of state monopoly are in decline, with consequent room for manoeuvre in overall establishment deployment (NB not redeployment!). Cadres in some sectors of government will wither, and indeed if my opening remarks about the structure of agrarian change and the declining relevance of the KSS/UCCA structures are correct, then room for manoeuvre will exist within BRDB. Secondly, it must clearly be a remit of a future PEP to experiment with a scaling down of staff intensity. As groups become more mature and secure, the need for intensive supervision can be relaxed. In Ulipur, for example, Proshika have 5 staff working with 350 groups—a ratio of 1:70, instead of PEP's 1:10. Field staff intensity should increasingly be rotated, withdrawing from older groups and concentrating on newer ones. Third, however, the concept of withdrawal is subtle. Staff functions will change as requirements in different areas alter from initial mobilisation into more routine service and technical advisory roles. Generalists can be replaced by specialists. But occasions will demand that the generalist reappears if new issues of broad strategy arise—e.g. new struggles, or new earthworks opportunities for which LCSs can be mobilised. All this demands a much more flexible attitude to the way that staff are deployed in a locality. Fourth, this issue of staff intensity is also connected to the now familiar processes among NGOs of groups forming federations in villages, Unions and Districts in which increasingly their own committees steer the local agenda of group activities. This process should not be rushed in PEP, since strong groups are a precondition to ensure effective accountability of these committees. Move too fast, and the risks of repeating the UCCA experience become stronger. Fifth, while one of PEP's objectives for BRDB should be to test the level and type of staff intensity required over the period of group maturity, the donors should certainly continue with and seek ways to strengthen the investment in the quality of staff through supporting proper schemes of service in which length of service, quality of performance, in-service training at several career points including overseas if necessary, and promotion are all interlinked. The motivation, morale and long term quality of staff at all levels of a government rural poor programme (in what ever institutional form) would be enormously improved as a result. The donors and the GOB are long term institutions, and have the responsibility to prepare in the long term for the future. In 10-15 years' time, a whole new cadre of rural workers with field experience, commitment and appropriate qualifications could be leading GOB's efforts with the rural poor. In my discussions with various BRDB staff at different levels in the project, I have encountered many candidates for this process. It is for this reason that the experimental strategy within BRDB should be adopted.

The Case for PEP Remaining in BRDB

The argument here draws together and summarises the preceding elements of the report. The purpose of the opening discussion was to remind the reader of the processes of agrarian/rural change in Bangladesh on the one hand and of institutional history and change on the other. It is clearly the duty of anyone in this public policy area to be aware of both processes and to be sensitive to the constraints and opportunities which exist at this time for both the rural poor and the relevant officials in government. To move their interests and abilities closer together over time is a worthy objective, and a window exists to do so. It can also be restated that the Nordic donors have shown their commitment to contribute to this process, and Swedish and Norwegian aid has a size and quality to reinforce the efforts of other donors (e.g. CIDA and ODA) through sponsoring with BRDB further action-research to develop a more effective style and appropriate capacity in assisting the rural poor.

The 'manifesto' in the preceding section has been informed to some extent by NGO experience in Bangladesh and other parts of the world. But at the same time, NGOs have limitations and do not constitute a sole panacea for assisting the rural poor. For the forseeable future, they will not seriously compete with government or the market private sector in the scale of resources at their disposal in agriculture and rural works. Furthermore, their own expansion will also set limits to their capacity to maintain their present styles of operation. They are not therefore in a position to expand their staff in the field indefinitely, although they are certainly in a position to redeploy experienced staff in new areas of group mobilisation. With that caveat, we can conclude that there are limits to NGO practice and geographical coverage. If the government claims the lion's share of foreign aid to Bangladesh, it therefore has the responsibility to develop its ability to address the needs of the majority of its population.

The process of rural social transformation over the next decade or so will consist of a simultaneous rise of landlessness and *bittaheen* status for men and women, and the development of new forms of entrepreneurialism directly and indirectly in agriculture. It will entail the disarticulation and rearticulation of farm organisation to resolve the tension between the rise of capital and extreme land fragmentation. Marginal farmers and small farmers are initially the most exposed to this process, and are unlikely to be 'held in' landowning status by existing cooperative strategies. For larger farmers, consumer associations for making the suppliers of services accountable to them are much more relevant than persisting with self-supply cooperatives. This is the only way that larger farmers will be able to reconcile their own scattered holdings with the new opportunities for increasing land productivity arising from the intrusion of capital. This process is well underway in the more advanced IRRI *boro* areas of Bangladesh, and various events are conspiring to push this along even faster. As the economic and social institutions change in the countryside, so must government institutions be reformulated to stay in touch with the new reality.

The rise in significance of the RPP, partially in response to these processes, has occurred over a very short time period of 5 years, and has been undertaken by personnel who had been educated and trained with a small farmer rural development orientation, and have had extensive experience of working with farmers. Most of them have spent more time in the field than their critics. The fact that this change to RPP has come about at all from this background should be respected even if the content of field practice with the rural poor leaves much to be desired. It would be very ahistorical to abandon support for that process now. It needs improving instead. The final institutional form of government support to the rural poor has yet to evolve. It might continue its present form as an increasingly significant programme within BRDB. It might displace the UCCA/KSS origins of BRDB as a reflection of the scenario outlined above. Or, as some are arguing, a separate *Bittaheen* Development Board might emerge.

474 Bangladesh: Whose Ideas, Whose Interests?

However, for PEP it does not really matter which form evolves as long as it is working directly with the personnel concerned, and experience and learning is gained to provide future leadership of whatever institutional form emerges.

There is, then, an agenda of improving BRDB's practices through continuing to support its own programme of experimentation: with situational definitions of poverty as the basis for local flexibility of practice; more symmetry of interaction between staff levels; more discretion to field staff; schemes of service to stimulate motivation; rotating staff intensity; optimising positive linkages between group mobilisation and infrastructure employment opportunities (where RESP has a strong comparative advantage over other BRDB projects); stimulating initiative and responsibility among BRDB staff with project-related compensation for unsocial hours and duties, and by the relocation of TA staff from the *upazila* level to the District level in order to emphasise their advisory, research and monitoring functions rather than implementational ones; to substitute for TA at the *upazila* level by the insertion of a promotional grade for experienced GOs to guide new recruits and to work in new areas of group formation, and by the greater use of technical staff from the line ministries in the dominant EIG activities of the *upazila*; for this activity at *upazila* level to be supported by District Service Centres which respond to *upazila* requests, initiate applied and action-research, feasibility studies, assist in the analysis of local political economy, coordinate training across *upazila*, coordinate inventories of local resources, monitor performance both quantitatively but most importantly qualitatively; to extract lessons from this experience and apply it to the in-service and entry training schemes (i.e. informal as well as formal) as well as presenting such findings (in association with HQ staff) to regular policy workshops in BRDB, in which the different RD projects are compared as the basis for recommendations about organisational reform.

Chapter 18

Philosophies of Economic Change: Three EIG Models

Introduction

This short chapter was originally drafted in late 1992 as part of a review for SIDA *'Making a Little Go a Long Way'*. This was a background paper for an exercise in which SIDA was reviewing its aid policy to Bangladesh. This explains the various references to SIDA and Swedish aid in the following discussion. I was in the fortunate position of having no terms of reference beyond contributing relevant observations and ideas. Some further extracts from the same paper appear in Part VI of this volume. In this extract, I was concerned to elaborate a comment which I have recently been making on various public occasions in Bangladesh and Europe. Among the large number of poverty-focused rural development initiatives, projects and programmes, there seems to be a convergence upon target-group formation and individual or group based entrepreneurial models of income generation. Furthermore this convergence is occurring without any reference to the agrarian social and economic reality in Bangladesh. Despite all the various studies commissioned and all the social science PhDs completed, we have policy and project ideas plucked out of thin, indeed vacuous, air. Fashions, panaceas and sheer intellectual laziness prevail. There is also a dishonesty on the part of donors in advocating convenient solutions like a small-business strategy not only because other structural inequalities in the political economy can thereby be overlooked, but also because they know that such solutions have not solved structural unemployment back home, where the economic and educational conditions might be regarded as more favourable. This chapter therefore points out the need to reconsider the natural resource endowments in Bangladesh and to connect appropriate strategic employment generation options to them. The discussion which follows is admittedly incomplete, but may provoke a more systematic examination of options in future planning and aid strategies.

Two Myths: Land Reform and Trickle Down with High-Tech Agriculture

The dimensions of the landless problem have been dealt with in previous chapters in terms of: structural change in agrarian relations; the constraints to wage employment opportunities outside agriculture (to be distinguished from non-farm, but within the agriculture system); the constraints to wage labour absorption in conventional high-tech, farmer-based, trickle down agricultural strategies; and, of course, absolute numbers. In my view, the intellectual response to the scale of these problems has been entirely inadequate, though Swedish aid can be congratulated for perceiving the opportunities within rural works programmes for labourers to acquire longer term employment rights along with the possibilities of integrating such labour opportunities into the agricultural annual labour cycle, thus enhancing the bargaining power of agricultural labour. The implementation of these hopes has, as we know, been restricted by a series of vested interests (officials, contractors, local patrons and labour gang leaders).

We can easily agree that there are no panaceas, no single set of responses to the landless problem. However, dangerous myths abound. One concerns land reform, which has been discussed above and dismissed as a credible response to the problem under Bangladesh conditions of a rapid decline in landholding size for existing farmers. A further myth concerns the capacity of high-tech, HYV agriculture to absorb labour directly in field operations beyond the rate of incremental additions to the rural labour force. Even World Bank documentation has acknowledged the paucity of this assumption. It is also the case that any prospect of increased wage rates in high-tech localities (with additional crop seasons and therefore locally intensified demand for labour) is off-set by the rural-rural migration of increasingly desperate, though informed, labour. It is within the context of these myths that we have to consider contrasting philosophies of economic change for Bangladesh. What is the vision for providing incomes for the rural, not yet significantly urbanised, landless over the next 10-20 years?

Three EIG Models

Given the conditions discussed above, there would seem to be a basic choice between: a small-scale entrepreneurial model; off-farm but agrarian wage employment; and a 'Keynesian' employment generation model. The wage employment industrial model of advanced capitalist societies would not apparently apply under conditions of weak effective domestic demand for manufacture of wage goods, and the absence of clear comparative advantage in internationally tradeable goods (often the result of rich country protectionism). It is these options, together with the possibility of other more 'radical' models arising from organic, low-tech production, which require further dissection.

Without wishing to incur the criticism of proposing a panacea, the key idea or vision has to be based upon the development of a package of non farm, income-generating opportunities within agriculture, horticulture, forestry, pisciculture, open water capture fisheries, and other categories of natural resources. This is not to preclude continuing to search for employment generation outside agriculture and other rural natural resources, within the context of: urbanisation; an increased demand for services; and industrial options. However the rural, and specifically agrarian, vision leads as the more realistic response for the foreseeable future. The critical question, however, is whether these opportunities can consist of waged employment in addition to the current fashion for small business activity through entrepreneurial initiative. Group based, income generation initiatives have been demonstrated within the context of HYV high-tech agriculture (e.g. irrigation services by the landless) and may also be extended to a low-tech, organic production system though with different implications for structure and activity. From this we have a matrix of existing policy practice to structure the following discussion:

Natural Resource Sector	Entrepreneur Model	Wage Employment	'Keynesian' Model
Agriculture (off-farm) (Incl. livestock)	+	(-)	-
Horticulture	+	(-)	-
Forestry	+	(-)	(+)
Pisciculture	+	-	-
Open Water	+	(-)	(+)
Rural Works	(+)	+	+

where: + = existing policy practice
- = absence of potential
(+) = high potential
(-) = lower potential

This matrix draws our attention to the dominance of the entrepreneurial model as the favoured form of response to employment and income generation needs within these natural resource sectors. The question is whether the employment and income generation vision for the rural poor in Bangladesh can be adequately served by the '+' in the first column, or whether that vision requires converting the second column into '+', and the extent to which the third column should contain more '+' in rows where '+' are less likely to be achieved in the second column.

If this approach is helpful, then we can proceed by discussing the entrepreneurial model in the context of 'off-farm agriculture' because of its structural significance for the society as a whole. The other natural resource sectors are important subsidiary contributors to the theme, and are subsequently discussed in sequence.

Entrepreneur Model for Off-Farm Agriculture

Much of the focus on the landless by NGOs and GOB, supported by elements of foreign assistance, consists of forming groups and developing small scale business activity among them. Programmes consist of mobilisation, training, and credit. In essence, these are programmes of group entrepreneurialism. One of the paradoxes behind these programmes, especially for donors, is that the small business recipe is advocated as a universal solution by agencies who would not dream of applying the same hopes to their own countries, which have a strong, industrial wage employment tradition. The poor, working class in such societies would be regarded as incapable of such a development (through lack of capital, skills, education, market awareness, and other socio-psychological factors referring to risk taking propensity). Such propaganda in the UK during the 80s has been revealed as a hollow sham, with the rate of bankruptcies exceeding the rate of newly created businesses. One of the most common criticisms of these efforts in Western societies is over-optimism or ignorance of the market. This is a criticism which is increasingly made of small scale, business activities in Bangladesh.

This simple comparison is not sufficient, though polemically useful to gain the reader's attention. The analysis of the limits of the entrepreneurial panacea have to go further into the nature of markets in societies like Bangladesh. Suffice to observe here that they are far from perfect, but interlinked and interlocked, requiring the utilisation of kin and other networks, involving all kinds of hidden prices and exchanges. Indeed a formal analysis of the financial viability of such businesses is highly complicated by this reality, as was shown in Wood & Palmer-Jones 1991. It would make an interesting study to discover the rate of failure among group based, small scale activity in Bangladesh for a sample across a range of activities. The *Water Sellers* is probably the only rigorous attempt to show this for one sector of activity (the landless irrigation service). It seems as if there is a conspiracy of fantasy among donors and their clients in Bangladesh concerning these strategies. Is it fiddling while Rome is burning? A small sub-set of the poor are identified, mobilised, trained, and made functionally literate. They are offered credit and staff intensive agency back-up through monitoring and evaluation. This is fine as a welfare project for those involved, but does it constitute a realistic strategy for development for the society as a whole, for those millions among the unselected majority, and for those constantly joining the ranks of the needy either through dispossession or by simply becoming adults with no claim on other adults in their family? Is it possible for rural, still agrarian, Bangladesh to become a nation of small businesspersons; i.e. can the targeted model be replicated as a universal model of the structure of economic activity for the society as a whole? Furthermore, does the 'group formation' dimension spread the benefits of such a strategy more widely than otherwise, or does it actually function to undermine aspects of group solidarity as some members become more skilled, adept and dominant than others?

These criticisms of the entrepreneurial model do not amount to a total rejection. That would be silly—and inconsistent with earlier propositions concerning the potential for non-landholding classes to participate in agricultural growth through the provision of agricultural services. It is evident that the simple and expanded reproduction of the agrarian system depends very heavily upon private sector activity, a kind of peasant informal sector. However there remains, firstly, the problem that this activity is characterised more by trading and rent-seeking than it is by investment in production (leading to employment). This even applies to much of the private investment in high-tech, HYV agriculture, where even for the larger individual farmer, the mechanical (as opposed to biological and chemical) technologies are lumpy, requiring either the renting out of surplus capacity or under-utilisation matched by defaulting on credit (a form of rent-seeking on public capital). Furthermore, as noted earlier, direct private individual farmer investment in such high-tech agriculture has limited employment implications since it tends to intensify the use of family labour first.

Secondly, the entrepreneurial model is over-optimistic concerning the unsupported entry of non-landholding classes into the market for agricultural services. Although we might envisage an agrarian structure in which the fragmented farm disappears as an integral unit of production to be re-articulated through the provision of consolidated services upon coterminous plots of different farmers (now as passive rentiers on such land), can the poor enter this market for services only as capital borrowing and investing entrepreneurs with all the attendant risks, or can they enter as wage labour of larger, 'agricultural service companies'? Can the second column of row 1 gain a '+'?

The same question applies when we take the agricultural service model and consider the backward and forward linkages of such strategy. Are there more opportunities for wage employment rather than entrepreneurial incomes in the 'servicing' of agricultural services? This question refers to: the production of agricultural machinery (including processing) and rural transportation; distribution outlets; the maintenance of equipment; distribution of spare parts; and rural workshops. Larger scale production and distribution units require larger scale investment for capital goods as well as carrying stock. How realistic is it to see these activities as representing opportunities for direct entry by the poor (either as individuals or collectively)? Is it more realistic to see these linkages as the basis for expanded wage employment, provided the society as a whole internalised more of these production and servicing activities, though probably it could only do so through partnership deals of some kind with MNCs? At present, too much of this value-added is being enjoyed by, for example, Japan and Korea at the expense of Bangladesh. It is of course in this sense that Swedish aid policy on Bangladesh cannot restrict itself to the micro issues.

Horticulture

The horticultural case differs from agriculture even in Bangladesh by remaining more completely subsistence in orientation at this stage. With commercial forms of horticulture still confined to small, family holding activity, it is unrealistic to expect '+' in the second column. Two points, however, should be considered: the implications of urbanisation (especially rurbanisation) and the organic, integrated horticulture and agriculture strategy. For Bengal, the current emphasis upon cereal and jute crops derives from a combination of colonial imperatives, Pakistan's needs for foreign exchange, and more recently population pressure and the quest for food security in terms of cereals. Although the increase in cereal production has partly been achieved through high-tech tactics, much of the increase is from expanded acreage at the expense not only of jute, but crucially of pulses and legumes. This component of poor people's diet has declined to dangerous nutritional levels. The homestead has been the vital source of this production for middle and richer peasant families.

If this process continues alongside an increasing urbanised proportion of the population, then the objective need for horticultural products will rise significantly and the effective demand also, but to a lesser degree. In the hinterlands of existing and rising urban centres, we may therefore expect a substantial rise in horticultural activity, which is increasingly commercialised. With the population structure of Bangladesh, these hinterlands will be widespread. What will be the form of this horticultural production? Will the landless again miss out? Can homesteads be made more efficient to produce a commercial surplus? Will poorer producers sacrifice their own dietary needs? Will cereal land be reconverted for the first time in living memory to horticulture? Will there be a positive impact upon rotations and soil fertility? What are the employment consequences of converting arable land to horticulture, which is more labour intensive? What are the employment and entrepreneurial implications of small scale production units (i.e. homestead) and perishable goods in terms of storage, marketing and transportation?

The organic, integrated agriculture and horticulture strategy for the agrarian system raises slightly different possibilities, as discussed in Chapter 5, Part II of this volume. First, such a strategy will take a family form, indeed part of its rationale is to protect the family against dissolution as a production unit. In this sense, it is a peasant solution, in which the natural resources available even to small cereal, pulse and legume producers, together with the waste products and inputs to essential livestock, are harnessed together for mutual productivity. To some extent it involves turning back on the market, towards, instead, a model (maybe over-romantic) of family self sufficiency for efficient, sustainable family production. The labour intensive character of this approach is more likely to rely upon family labour rather than wage employment.

Forestry

Perceptions about the significance of this natural resource are growing daily in Bangladesh across a range of social actors. At the same time, the pressure on these resources is immense: from rural families in search of fuel and village housebuilding materials, as well as boats and agricultural implements, to more organised commercial interests serving urban construction markets. With steady deforestation over many years, these resources are increasingly regionally concentrated, often in the habitats of ethnic minority populations whose socio-cultural livelihoods are threatened. Some NGOs, especially the more environmentally conscious ones such as Proshika, are already substantially committed to family, group and community-based responses to sustainable forest areas. In their programmes, they encounter the interrelated vested interests of commerce and official bureaucracy.

Forests function in three ways: as an immediate family resource, as part of subsistence reproduction; also as a legitimate commercial product with widespread utility, conditional upon satisfying sustainability criteria; and as a common good in both national and international terms. This accounts for the complex representation in the matrix above: small scale, group based entrepreneurial behaviour in planting, protection and exploitation; some possibilities for larger scale wage employment (especially in forward linkages); and public large scale investment in forest labour, especially in *khas* land and coastal areas, generating local incomes and effective demand against a long term, common good objective. The latter category is a clear case for aid, with Swedish aid well placed after its IRWP and RESP/IDP experience.

Pisciculture

Pisciculture has been conceived in Bangladesh primarily as a family or small scale, group activity. Where more capital intensive activity has been envisaged (e.g. Fish Seed Multiplication Farms-FSMF), it has not at the same time been labour intensive except at the construction stage. Furthermore large scale, and certainly public sector, FSMFs have now been discredited as efficient suppliers of cultured hatchlings to regions deficit in wild sources of hatchlings. The employment opportunities in pisciculture do not lie in food fish ponds (rearing for ultimate consumption) but are derived from the rapid growth of stock which requires an intensive sequence of transactions from a central point in early stages of growth to increasingly scattered nursery and destination ponds. Further transactions and employment opportunities exist at the catching, transportation, storage and marketing stage. However, the opportunities for large scale employment generation seem remote. Single traders can cover a substantial number of pond holding customers.

Open water bodies do offer considerable scope across all three columns. SIDA is aware of some of these issues from the work of Maal in Kurigram. I have also commented on the income generation and distributional issues of the Fisheries 3 programme. CDS at Bath is currently undertaking a major research programme in Bangladesh, Thailand and Indonesia on some aspects of poverty and sustainability around such bodies. The various activities of fishing and stocking have offered small scale, group based entrepreneurial opportunities; there are also examples of wage employment, where 'fishermen' turn out to be contractors of other (sometimes indebted) labour, especially during intensive 'harvest' periods. There are backward and forward linkages in terms of boat building and maintenance, net making, transportation and marketing. However, it would seem that further opportunities exist for public sector employment in terms of investment in the water bodies themselves: dredging, weed clearance (especially water hyacinth), embankments, dams, integrated canal irrigation systems, excavation.

Rural Works

Given the extensive discussion in Part III of this volume, there seems little to add to the extensive knowledge within Bangladesh on rural works and the various opportunities for different organisational forms of employment generation: short and long term; seasonal to coincide with troughs in agricultural labour demand; wages only or longer term control over assets created; regular maintenance contracts; local government project committees and private contractors; variations in forms of payment; and more recently, labour contracting societies to create the possibilities of entrepreneurial models in a situation when only various forms of public and private sector wage employment had been considered. This is a natural resource sector where the potentiality for '+' in all three columns has been recognised by SIDA.

Chapter 19

NGOs and the Theory of Struggle

Conditions for Emergence of Development NGOs

The emergence of the early development NGOs in Bangladesh was not just a reflection of a gap in analysis and policy, or even primarily of the growing awareness among concerned Bangladeshi observers about the dimensions of poverty. The political conditions in the aftermath of liberation were crucial. The Awami League, as the nationalist party of Bengali independence, inherited political office but rapidly abused its mandate in an orgy of incompetence and corruption. From the beginning of 1975, it engaged in the wholesale repression of political opposition, especially that of the Marxist left, organised in various regional groupings, as it attempted to establish a one-party state under presidential rule. The coup in August 1975, followed by the sequence of coups in November 1975, brought the army under Ziaur Rahman to power, and continued the repression of the Left, both within the military and outside it. With the domination of the army continuing more or less unbroken until December 1990, the opportunity for political parties openly to espouse the interests of the poor against the vested interests of those dominating the state was severely restricted. At best, a benevolent dictatorship could be hoped for which would, of course, be more accountable to foreign donors than to progressive internal opinion. However, the more imaginative development NGOs played an increasingly crucial role in this delicate political equation.

Their origins, from the mid-seventies, arose from a combination of the repression of formal, radical and progressive politics; a generation of young activists with political values and courage shaped by their experiences during the struggle for liberation from Pakistani domination; and the availability of donor support. It is also true that some were set up by expatriate Bangladeshis returning in the later seventies from successful professional activities abroad (for example, accountancy and academia). The main impetus, however, reflected a concerned generation of students, active on the University campuses in the late sixties against Pakistan's domination, active during the liberation war itself both within refugee camps in India as well as underground within East Bengal, and active in the immediate aftermath of the war in various emergency relief campaigns, often involving foreign charities and voluntary organisations. Without the opportunity to translate these immediate concerns into a longer-term political commitment, alongside their witnessing of the criminal neglect of the poor by the state, this generation of young Bangladeshis turned to the voluntary, and eventually funded, NGO sector as the outlet for their frustrated ambitions on behalf of the poor in their newly liberated country.

Proshika's Agrarian Analysis

I was privileged to come into early contact with these groups, and one in particular, Proshika. My assistant during the research at BARD, Comilla, was Md. Yahiya. After my departure at the end of 1975, he joined Proshika there. This organisation had started in 1975 as a project of the Canadian University Service Overseas (CUSO), and became a separate Bangladeshi organisation in the following year with two branches, one in Dhaka and one in Comilla supported by core funding from the Canadian International Development Agency (CIDA), and continuing inputs from CUSO until the early eighties. CIDA continues to support the two, now separate, organisations of Proshika. Md. Yahiya introduced me to the two branches in 1978 on a return visit, and eventually I worked most closely with the Dhaka wing, now Proshika Manobik Unnayan Kendra (henceforth referred to here as Proshika). The remarks here are derived from our shared view of the problems which the poor face. The basis of our initial association was the analysis contained in Wood (1978), and the recognition that few development agencies, (academic, donor and government) really understood the scale of landlessness as the root cause of pervasive poverty and remained locked into trickle-down strategies by concentrating upon the economic development of farmers through enhanced agricultural technologies embodied in the Green Revolution. Few also understood or appreciated the social processes through which landlessness occurred and poverty was reproduced in the rural political economy or that the state was not benign in these circumstances. There was a strong sense that to work with the poor through ideas of empowerment and conscientisation was subversive; some of our early discussions were distinctly clandestine in a way which now appears unnecessary.

The basic ideology as it emerged over time among the initial leadership in Proshika was that class subordination for landless men and women was experienced more as individual rather than collective exploitation. The rural political economy consisted of class relations expressed through patron-client-structured hierarchies, with poor landowners, sharecropping tenants and landless labourers constituting a class of clients tied individually to patrons who might be landlords, moneylenders or employers, usually in combination. The poor rural peasantry in Bangladesh thus exemplified the familiar Marxian problem of a class in itself needing to become a class for itself as a precondition for its own effective action. Original Marxists, up to and including Trotsky, dismissed the capacity of the peasantry to move itself from one condition to the other; the jump from individual perceptions of injustice to collective action to rectify injustice was considered too great, and certainly required leadership external to the peasantry.

The Bolshevik Revolution was premised upon this assumption. Eric Wolf (1969) also pointed out the difficulty for the peasant in moving from a passive recognition of wrongs to implementing the means for setting them right. He listed a number of issues: the isolation of rural work - working as scattered individuals in distant fields; the tyranny of work - long hours dictated by seasonal imperatives; the competition among the poor for scarce land resources to rent, or for security of employment by accepting tied conditions; the inability to attend meetings and rallies because of work obligations and financial constraints; the problem of communication in a non-literate world; and the absence of control over or access to the means of communication.

To these we may add a lack of empathy for other poor outside localised experiences of exploitation and oppression; a loyalty to a familiar individualised survival option involving the acceptance of subordinate status; the fragmentation or disarticulation of potential class solidarity on horizontal lines through incorporation into vertically aligned, factional groups led by leaders of one's lineage (*bangsho*); and under such structural conditions, the untested unity of one's own class to deliver a substitute survival package, sustained over a long time period. I have always been impressed by an encounter in my Comilla fieldwork with a poor, marginal farmer who was obliged to sell his labour. He was about 35 years old. He recounted in great detail how he and others like him in his *para* (a distinct grouping of households in a village, often tracing common genealogical descent) were being dominated and exploited through low wages, high rents, exorbitant interest rates and the manipulation of opportunities for scarce employment by Abdul Ghafor, the largest landholder, businessman, dealer and tout (a Bangla expression to denote a corrupt broker with official connections, especially with government) in the *para*. When I asked him what was the furthest distance he had travelled in his life, he replied with a review of other villages visited. When we identified them on my map, the answer was eight miles.

Converting Affliction into Action

This underlies a distinction made in academic circles between protest and revolution. The term 'revolution', especially in South Asia, conjures up an image of Marxist-inspired armed movements fighting official armies over sustained periods of time with the support, active or passive, of the peasantry. The term therefore induces fear of immediate threat to the interests of the middle classes (commercial, professional and official) who then endorse repressive state action and a widespread application of sedition laws. NGOs in Bangladesh have, therefore, been at pains to describe themselves and their objectives in other terms. If, however, we adopt a more generic use of the term revolution to refer to processes of structural transformation in the society, without assuming violent upheaval as a prerequisite, then the distinction between that and protest remains highly significant to the whole question of extending the horizons of the poor beyond eight miles, together with a corresponding sense of opportunity.

The south Asian region is replete with examples of protest activity among the rural poor. Colonial officers were always dealing with 'agrarian discontent', particularly at harvest times with struggles over crop shares. Protest activity reflects an acute awareness of local, immediate injustices involving intense hostility to those held immediately responsible, whether landlords or officials. The personalities are known, the hostility is not abstract. Typically, such protest was shortlived with single-issue objectives which, if achieved, led no further. Often, the fear of the ruling classes that such protest would spread resulted in quick and comprehensive repression deploying the forces of the local state or their own retainers. Indeed, the anticipation of such immediate counter-repression played a major part in maintaining the quiescence of the poor; apathy is less a state of mind than a rational calculation of the odds.

Thus a large jump in perception and calculation is required to translate discontent into a sense of class and a programme of action leading to structural transformation. The problem, then, is not the awareness among the poor of their exploited condition, but the degree of sophistication in that awareness together with the objective capacity to act on it. I have always been worried by the inference that the poor need to be taught either how poor they are or the immediate explanations for their plight. Poor peasants who have recently lost land through their inability to repay usurious loans basically know the score. We must also accept that such people have kin with similar experiences in their own and nearby villages. And we must also accept that rural-rural migration is long established in Bangladesh and has provided an opportunity for broadening class perceptions. More recently, urban and external migration has added to this.

But the pressures which fragment and individualise the experience of exploitation persist. The family remains the unit of primary reference, in which ultimate responsibility for ensuring survival or enhancement resides. There has to be a strong sense in which Banfield's thesis of 'amoral familism' applies (Banfield 1958): behaviour within the family is bound by moral ties; behaviour outside the family can be structured within concentric circles through which action is increasingly instrumental. But since the poor family as a unit is so vulnerable when acting in isolation from others, there is a strategic imperative to extend the sense of moral community outwards through some of those concentric circles, to include others in similar positions, for example. Conscientisation reminds people of their vulnerability as individuals or families, and encourages the notion that strength lies in unity with other families similarly placed. It is the gaining of this strength which constitutes the revolution for the poor in Bangladesh; how precisely this strength is to be used is a secondary question. Revolution in the rhetorical sense of 'armed mass movements advancing upon the edifices of the state and those classes protected by such edifices' is only one option for the use of that strength. The first step towards gaining strength is the one described in Wood and Kramsjo (1992) namely 'breaking the chains'.

Several issues arise at this point. What are the chains to be broken? Are they the same ones for all the poor? In other words, are the poor homogeneous? Can poor people break their own chains or do they require external assistance? What are the consequences of relying on external assistance? What prevents chains being reimposed, or new ones binding? To remove chains implies the gaining of freedom, the establishment of meaningful, universal human rights of citizenship, that is, the right to participate in the political process and equality before the law, accountable government and unlocked prices in market transactions. In Bangladesh, the acquisition of freedom, so defined, would be the revolution, to be distinguished from locally resolved protest (whether through concession or repression) implying no structural transformation in the relation between rich and poor, and to be distinguished from guns on the parapet.

Chapter 20

Resource Profiles and Social Mobilisation: The NGO Case of Gono Shahajo Sangstha

Introduction

In February 1993 I led a small international team in an evaluation and appraisal of Gono Shahajo Sangstha (GSS), an NGO in Bangladesh which has placed a strong emphasis upon organising poor people in struggles to claim entitlements in both state and market structures. The reflection on NGOs and struggle in the previous chapter are pursued here through a review of GSS. As indicated in the overall introduction to this volume, this chapter is an edited version of Annex 1 (prepared by me) of the team's report. It should be emphasised here that many sensitive issues had to be discussed in the nature of the subject matter, and that GSS personnel were wonderfully frank in sharing their dilemmas with me and the team. In editing this paper for this chapter, I am removing some detail and specificities in order to focus upon the key issues of wider interest. Nevertheless, some issues remain sensitive at the level of political process in the society. These issues should be openly analysed in the public domain, and we should be grateful to GSS for sharing its general experience with us.

1. Principles and Theory

GSS and the theory behind its social mobilisation strategy was bred out of local activists, unconnected to the international community and development assistance, during the turbulent years of the late sixties and early seventies. These activists combined a sense of Bengali nationalism with an empathy for the human rights of the poor. Some were imprisoned by the state during the seventies and early eighties for these commitments. These roots are significant to understanding the evolution of GSS and its approach to mobilisation work among the poor. It derives from a basic lack of faith in other classes and their institutions to work for the poor in a sustainable way, i.e. in a manner which changes permanently the structure of socio-economic and political relations through which poverty is reproduced. There has been a distrust of anything less than this objective as essentially charity and relief. Only the poor have an ultimate interest in helping the poor.

This point of departure fits in well to a 'resource profile' approach to an understanding of poverty and developing appropriate and progressive responses to it. This approach involves assessing the resources which a household has at its disposal. The categories of resources which must be taken into account in this system, however, go beyond those normally perceived by economics. In this approach four categories of resources can be identified:

material: those resources normally accounted for by economics, including wealth and capital assets such as land, machinery.

addressing these four resource dimensions, though in different measure. In this sense, GSS is much closer to 'state of the art' Human Resource Development approaches to poverty.

Within this approach, there is a strategic complication which requires acknowledgement, though not necessarily resolution. Where does the ultimate guarantor of rights reside in a society? Although regimes come and go with good or bad records, the state can be seen as conceptually that dimension of society which codifies rights and guarantees them (to the extent that regimes can be impeached for failure). The alternative view rejects the state's positive role for the poor on the grounds that it is indistinguishable from a succession of regimes representing richer interests (either coherently or haphazardly via corruption).

This alternative view looks to a series of institutions and agencies in the society, as opposed to the single, universal institution of the state, as the most effective guarantor. No doubt the respective merits of these contrasting positions can be debated (statutory versus voluntaristic, the place of legislation etc.), but the alternative view continues the theme of rejection and distrust of the state by the local activist founders of GSS. It nevertheless raises a question of whether the state is targeted for change (through making it more accountable to majority interests) or being bypassed as always an ineffective guarantor—in short, a hopeless case? Clarification of thinking on this fundamental strategic issue is necessary in order to justify especially legal and advocacy aspects of an NGO programme. There are dangers of disenfranchising people, removing citizenship from them, in the alternative view.

Care has been taken not to describe this contrast of view as a contradiction, since over time we might hope that institutions in the civil society, albeit voluntaristic and statutorily unaccountable to their target clients, may themselves become a force for raising the accountability of the state on behalf of all the poor and other classes.

2. The Process

The GSS strategy towards social mobilisation has changed over time. Initially the entry into the minds of the village poor was via adult literacy, mainly for men. Adult literacy classes for women followed, and continue as a strong element within GSS. From the content of the literacy books which are used in teaching, it is clear that there is, intentionally, a high 'social' content in terms of raising gender and class issues in the teaching material. These course were initially for 8 months, with up to 6 nights (days for women) of teaching per week, in 2 hour sessions. The strategy has evolved from this in the light of experience. The main difficulty has been to attract males into such an extended programme of literacy classes. Now, the strategy of 2 month 'functional' or 'night' schools has been adopted for men. Alongside this, a recent revision of the literacy curriculum reduces the time to 6 months, taken mainly by women although the programme remains available to men where demand is sufficiently strong. To summarise this evolution: the strategy of mobilisation for women remains through 6 month literacy schools; the strategy for men is via 2 month 'functional' schools.

These 'functional' schools for men are taught by the GSS field workers (*mat kurmi*) of the Adult Literacy and Social Mobilisation Programme (ALSM), but are supported by 'volunteers' with some education from the local community. These volunteers take about 3 classes per week, with the GSS field worker taking the other three classes. The division of labour is that the 'volunteer' will concentrate upon some literacy skills (without attempting complete literacy), while the field worker deals with the 'social' agenda.

After these 2 months (for men) or 6 months (for women), the group, together with other groups involved in a similar process, are encouraged to form a village preparation committee. At this point, cadre are identified by GSS Field Workers from the constituent groups for more intensive training in committee management and ideas at Regional GSS Training Centres. These selected cadre are considered to be the more dynamic and committed members, showing leadership potential. After this stage of both cadre training and discussions among constituent groups about the prospects for their internal cohesiveness and common issues for their action, a village committee is formed. At this point, the village committee technically separates from GSS and becomes part of the People's Organisation (Gono Sangathan-GS), though continuing to receive services from GSS such as Legal Aid and Legal Education (LALE), involvement in Popular Theatre (PTE), in some places the Primary Health Education Programme (PHEP) and the Primary Education Programme (PEP) for their children. Uncertainty exists at present about income-generating programmes and the issue of credit (see below), although the organised quest for local unused resources like *khas* land and ponds occurs from the outset.

These village committees may subsequently combine with others within the local GS movement to form Ward level or Union level committees. These combinations may be permanent, with established 'executive committees', or temporary in order to secure a particular objective, such as an area of *khas* land lying between 2 villages.

3. The Message

The social agenda in the 'functional schools' consists of an analysis of the causes of poverty within the local, national and international political economy (taught of course in simple terms, using analogies and incidents from the group's own experience). Inevitably, the local issues take precedence although the attempt is made to connect them to critiques of the state (bureaucratic corruption etc.), an analysis of class and the rationale for global patterns of investment (e.g. the location of garments factories in Bangladesh by international companies to take advantage of cheap, female, Bangladeshi labour). The local issues focus upon: wages, conditions of share-tenancies, moneylending, bonded labour relations, availability of *khas* land/ponds, dowry, rights of women in marriage and divorce, the regressive influence of rising Fundamentalism in Islam. This analysis leads into ideas about action and the imperatives of solidarity, organised unity, to achieve improvement on any of the dimensions noted above. A strong theme concerns the necessity for the poor to be prepared to act without support from external agencies. This has included, for example, a hostile attitude towards receiving institutional credit, though changes in this message are on the way. Certainly the whole thrust is towards making demands upon publicly allocated resources and acting more aggressively in the marketplace.

Once started, these schools appear to be well attended, though unlike the primary schools or the adult literacy classes, no records of attendance are maintained. Although members may have had long days in the fields, on roadworks, carrying loads, petty trading or just searching for work, these evenings in the 'functional schools' are a novel contrast from normal life after dark in the *bari*. They represent a chance to meet, talk together, encounter new people (the GSS Fieldworker and occasionally other GSS personnel such as the Area-in-Charge [AIC]), and some excitement was reported in the elementary literacy (being able to sign names, read signboards and signposts). Furthermore, these 2 month schools can coincide with lean employment seasons.

4. Concept of Empowerment and GSS

'Empowerment' is clearly a central concept in GSS social mobilisation activities, as it is in other NGOs. It may be useful to reflect upon the concept in order to understand and assess the process in GSS. I am indebted to a working paper by Judith Turbyne, a research student at Bath who is studying parallel issues in Guatemala, for some of the ideas in this section.

'Empowerment' describes both a process and the learned competences which accompany that process. These ideas are clearly illustrated in the work of Kieffer (1984), who worked with grassroots leaders in the USA and approached the issue from an individual, psychological standpoint. He describes a series of stages through which individuals pass as they evolve from a state of powerlessness. To begin the process of empowerment, individuals must first experience a 'mobilisation event' which will force them to re-evaluate their relationship with respect to traditional authority figures and structures. This event will not be fostered by rational conscious analysis, rather an emotional response to an event of personal significance. From this follows four stages:

- the 'era of entry', in which individuals tentatively make enquiries about the social world in which they move;
- the 'era of advancement', with the appropriate guidance of a mentor and suitable peer support, individuals reflect further on the changes the process is bringing to them and deepening their understanding of social and political relations;
- the 'era of incorporation', when individuals begin to internalise the feelings of competence and mastery which have come out of the process, leading to a re-evaluation of the individual's position in relation to the socio-political structures in which they are participating;
- the 'era of commitment', when the individuals have developed specific competences and applies these to other areas of their lives.

Kieffer rejects the idea of the quick fix in this process, and thinks in terms of 4 years to move through these stages. The acquisition of participatory competences has 3 components: an increasingly positive self-concept; greater analytical abilities in understanding the social and political world; and increasing resources for action in these spheres.

The image of the 'empowered' individual, then, is one who has moved from a situation of being a powerless pawn in a system that is alien to them, to one of being aware of their position in the social and political structure of which they are part, alongside having the skills to question that system.

Kieffer's approach can be contrasted to the more familiar Freire (1970) who worked with the disenfranchised of Brazil and Chile and has been the main inspiration for 'mobilisation' NGOs in Bangladesh. Empowerment for Freire was linked to freedom both for oppressed and oppressing classes (a true Marxian approach to alienation). It would have to be the oppressed classes who act as an agent of change. From becoming objects on which social and political forces acted, the oppressed would become subjects who would take an active role in questioning and moulding these forces. In Freire's model of 'education for critical consciousness', education is not the imparting of knowledge by experts but a partnership. The 'illiterate' person is treated as a subject who will dictate the context and the content of the educational experience, not an object on whom the educationalist works.

The critical issue for Freire was the lack of value of theory without action, and the inadequacy of education as generally practised (i.e. expert to object). For him, also, the mobilisation event could be engineered rather than left to fortunate accident. Freire also emphasises the value of group dynamics over individual behavioural and attitudinal change. Freire stressed the social nature of knowledge and thus the communal activity that must occur to work towards knowledge (and therefore empowerment) entails conscientisation as a group process.

A further contrast can be drawn between a notion of empowerment as the perception that external forces have less 'power over' oneself (more associated with Kieffer), and a process whereby individuals and groups actually gain 'power to' change the structures of oppression (Freire). Of course, we are entitled to conclude that the former will lead to the latter condition.

Certainly GSS is not concerned to limit its view of empowerment to the level whereby individuals perceive themselves to have greater power and control, often expressed in other organisations for example through greater autonomy in economic behaviour, through income-generating activity. GSS argues strongly for a view of empowerment which goes beyond the self and interpersonal spheres into the collective sphere, which critically includes not just the local community but also regional, national and international levels.

While it could be argued that GSS shares this perspective with other organisations less explicit on this point, such as Proshika and perhaps BRAC, GSS is deliberately polemical in these broader ambitions. Isolated individuals, groups or communities may acquire more positive self-images through collective behaviour, but it is the wider economic, political and social realities which have to be challenged and changed. The preoccupation with achieving, for example, greater representation in local community affairs (i.e. Union Parishads, or the *shalish*) would be considered only as a preliminary step. The occupation of *khas* land and ponds, while valuable in itself to an extent, is justified more for its symbolic importance in revealing the oppressive alliance of the state with propertied classes and that successes are possible. Other NGOs would settle for the pragmatic objective of acquiring a redistribution of productive resources locally leading to greater, immediate income-generating activities for poor groups. This is the subtle, but significant, difference between GSS and other NGOs.

It is these more structurally challenging polemics about empowerment which brings GSS into greater conflict with the state, removing the protective labels of service, welfare, and poverty alleviation through income enhancement. However, as discussed below, GSS has begun to encounter the dilemma of empowerment, where groups start to demand the kind of services which GSS would prefer not to deliver directly itself.

By understanding this conceptual point of departure for GSS, some criteria have been established for the following discussion concerning: long term strategic objectives and consistency with more immediate goals and activities; the implications of 'empowered' group demands for more immediate, tangible services from GSS; leadership and participatory processes within GSS in terms of theoretical assumptions and formulation of policy.

5. Problems with 'Activity-input' Indicators of the Empowerment Process in GSS Practice

A useful distinction can be made between performance, effect and impact in attempting to measure empowerment. As an analysis moves from performance to impact, so it has to become less quantitative and more qualitative. Performance refers to the activities or responsibilities which the organisation (GSS) and the target people (Village Committees of Gono Sangathan) are expected to perform: the 'activity-input' indicators discussed below. Effect refers to the evidence of group formation, perhaps to some welfare indicators, and to perceptions of exploitation and the reasons behind it. Impact refers to permanent changes in the way resources, property and power are owned and worked.

> "Impact therefore may be measured from observations of the collective resource position of the target people, in increased wages, in improved sharecropping terms, in the enforcement of legal rights, in a reduction in the incidence of the oppression of women, in the level of literacy [perhaps should be an 'effect'] and indeed in the effective representation of the landless in institutional and political processes such as those in the Union and Upazila levels."

(An extract from a GSS paper, prepared for our evaluation mission.)

Clearly these distinctions relate to the remarks on empowerment in the previous section, especially the distinction between 'power over' and 'power to'. For the purposes of expressing a quantitative-qualitative continuum of measurement and indicators, a two-fold distinction is used here of activity-input and activity-output indicators, recognising that the latter probably encompasses both effect and impact.

There are obviously problems in establishing indicators by which to measure the success of an empowerment strategy directly with target people (in contrast to broader coalition building in the wider society with other organisations). These problems are compounded by continuous refinements of the process over the last few years, particularly the shift from adult literacy for men (as well as women) towards shorter 'functional' schools for men, while retaining the adult literacy for women. However a policy shift is occurring here also, partly by intent and partly by default, towards more adolescent literacy training. This evolving picture reflects a process of internal re-assessment of strategy in the light of experience and should not be criticised for that.

Turning to the formation of Village Committees as an indicator, there is some ambiguity in the figures from different sources. While it is difficult to reconcile these figures, the approximate picture is: 212 formed up to June 1992; 174 additional up to the date of the PP; 180 more planned by June 1993. This gives the figure of 566 as the position in mid-1993. From this pattern we might conclude that past inputs in literacy and recent inputs in functional education are producing a sudden surge in the growth curve.

Other input indicators analysed in the evaluation were: the disposition of GSS employees in the Adult Literacy and Social Mobilisation programme (ALSM) as an indicator of activity and perhaps some notion of cost effectiveness; the number of trained Village Committee members (cadre training; ad hoc committee training; village committee training; accounts training; paralegal training; theatre training; and income-generating skills training.

Although the input indicators are important in terms of understanding cost effectiveness in reaching physical targets, questions can be asked about the level of appropriate staff inputs in order to achieve the intermediate objectives of forming Village Committees (VCs), and thereafter sustaining them. At the same time, the earlier discussion on empowerment cautions us not to look for low staff-VC ratios and not to expect rapid results in VC formation and sustained forms of collective social action. Other NGOs, adopting less ambitious agendas for 'their' groups or defining successful action in more immediate tangible forms such as income-generation activities, may show lower staff-group formation ratios and may show a faster rate of tangible new behaviour in such groups. To the extent that GSS moves in some of these directions, i.e. towards income-generating activities among its VCs, then the tougher judgement on these 'input-output' ratios is invited.

There were various internal issues about the ambiguity of definition in the use of these indicators which are not of general concern. However, these ambiguities are not unique to GSS, and highlight the wider problem of trying to establish universally accepted input indicators to measure performance in social mobilisation. It is worth noting that some NGO donors are becoming increasingly preoccupied with finding measures and are devising various quantitative procedures to measure what is essentially qualitative data. While further comment on the problems with these 'activity-input' figures for ALSM is not appropriate here other inputs into the GSS social mobilisation programme can be noted through the Legal Aid and Legal Education (LALE) and Popular Theatre Education (PTE) programmes (which were discussed in separate annexes of the evaluation report).

6. Social Action Output Indicators

If the conclusion from a consolidated set of unambiguous figures for inputs against the social mobilisation objectives (which probably should include LALE, PTE and the continuing Adult Literacy inputs, despite the 'separation') is that the staff input ratios are significantly over-generous compared to other acknowledged successful programmes, then what remains is a qualitative judgement on the long term value of the GSS strategy, which might claim to be achieving a more thoroughgoing, penetrating and therefore sustainable and structurally significant process of empowerment.

In this context, what indicators of social action are appropriate? By GSS's own agenda (91/92 Annual Report, page 17), we must consider: wages; share tenancies; access to *khas* land and water; access to other leased land and water (e.g. from public sector agencies); dowry; women's rights in the family (divorce, maintenance, rights to children, rape, polygamy, non-registration of marriage, and laws of inheritance); resistance to harassment by a coalition of local dominant classes and the administration; protests against corruption (e.g. concerning short measures of relief wheat). Interest conditions on personal debt should perhaps also be included.

Some of these indicators can be presented more precisely in the LALE programme. The 91/92 Annual Report refers to 500 'social actions' of this kind in 90/91 and 710 in 91/92. Of course it is impossible to standardise these 'social actions' into some kind of equivalent unit. Acknowledging the problems of measurement on these dimensions, GSS needs to generate case material to demonstrate further the content and process of these social actions. Although some stories have been collected during field visits for this evaluation, it would not be appropriate to engage in extended case analysis here. Some quantitative issues can nevertheless be raised.

Systematic evidence on success in increasing daily wage rates could be collected and compared to the general increases in the local, equivalent season wage rates. Some of the VCs interviewed by us reported increases in wage rates, but from further discussion it is by no means clear whether rates for GS members (or other workers in a local labour market influenced by GS action) have risen appreciably higher than the general wage rate for the area. Success in achieving this objective can be empirically examined using control groups. Similar points can be made concerning conditions of share-tenancies and interest on personal debt (if GSS accepted this as an objective).

Information has been provided on *khas* land and waterbodies in the 91/92 Annual Report (p.16): by 1991, GSS VCs had gained control of approx 776 acres of *khas* land and waterbodies; and during 91/92, a further 412 acres of *khas* land was gained. The report acknowledges the difficulties of securing legal possession and therefore long term occupation, and the implications for setting up income generating projects in these *khas* areas. There are various dilemmas here, which will be discussed in the section on income generation. A GSS report shows that of the 776 acres, the GS VCs lost all but 249 acres during the long process of making this land legal.

Other social action indicators are more difficult to quantify in any sense: such as issues of dowry and women's rights in the family. GSS has laudable gender targets which could be quantified in the future: 30% of all marriages in landless households without dowry (presumably in GSS areas); 60% of all marriages of GS members' families without dowry; 80% of all marriages to be registered (100% for GS member families); rate of divorce reduced by 50%; 80% of divorces through state not religious law. Clearly GSS would need to conduct some baseline surveys in its areas and produce monitoring data as the basis for measuring achievement in these targets subsequently. These issues are only now being placed on the agenda of VCs, as we have observed in interesting sessions between GSS and Cell lawyers with both male and female VCs.

The next indicator of impact in the sequence is representation in the local institutional decision-making body: the Union *Parishad*. The concept of political participation is, of course, highly contested in terms of significance and meaning. What is the value of electoral success at this level? GSS papers also raise some of these questions. Certainly, the Union *Parishad* has certain powers, controlling: FFW; Test Relief; Vulnerable Group Feeding cards. It is also involved in the allocation of *khas* land and waterbodies; and has a role in investigations and mediating disputes through the *shalish*.

In the 1992 Union Parishad elections, 90 out of 202 Gono Sangathan candidates were elected as members; and 7 out of 18 candidates for Union *Parishad* Chairman were elected. 2 of the successful member candidates in Nilphamari were women. This undoubtedly represents a certain level of success. However this should be in perspective, since the people's organisations attached to other NGOs have also reported similar successes in their areas of concentrated presence.

During the visits to 3 GSS areas, lengthy discussions were held with 9 Union members from Gono Sangathan. Two dominant themes emerged: the struggle to seek election and the aftermath of success; and the advantages of having GS Union *Parishad* members.

The struggles and victimisation of GS people, including the burning down of 2 schools in Nilphamari, are probably familiar in general terms to the reader of this report and need not be repeated here. Detailed case studies have been collected during these lengthy discussions which reveal the strong solidarity and determination of GS people to resist harassment both before, during and after these elections, including colluding harassment from the local administration. But perhaps even more important was the evidence of preparation and forward thinking once it became clear from election results in neighbouring Unions, earlier in the election cycle, that a counter-reaction was likely. Of course there were accusations and charges of being terrorists, engaged in underground armed struggle. (We attended another meeting of 2 village committees having a session with a GSS Cell lawyer in the Khulna region in which advice was sought to counteract similar charges.) The experience of these struggles, this first major encounter with the alliance between state and dominant rural classes, has been itself a dimension of the empowering process. Solidarity has been tested and it feels good. When the schools were burnt down, GS people erected them again: their collective property to invest in the human capital of their children had been attacked. We can sense a buoyancy, a sense of confidence not just among the GS Union *Parishad* members but among the VCs as well in Nilphamari and Jholdakar. These are important qualitative indicators of empowerment but perhaps at the level of effect rather than impact, what about the 'power to'?

This brings us to the advantages of having GS Union *Parishad* members, even when they are in a minority. The evidence here is unambiguous. Even a sole GS Union member in Mollahat Cluster, Khulna reported the benefits of gaining a place on the *Parishad*. Information is gained: about the FFW programmes and correct payment levels; about the VGF wheat allocation; about projects to be initiated and therefore employment options. He has been successful in putting pressure on other members and the Union Chairman to respond to GS people's interests, even threatening to mobilise a vote of no confidence in the community which brought the Chairman to a negotiation meeting with VCs.

Another lone member in Nilphamari Cluster had successes and failures during the last year. He pointed out that the agenda for Union *Parishad* meetings is set by the Chairman and Members together, so a majority is important to get a controversial item taken. This GS member successfully proposed that the Chairman should be excluded from the 'judicial' functions of the UP (organising *shalish* etc.) since he was too busy and his non-availability would delay settling issues. His presence at *shalish* enabled him to intervene on behalf of GS people. Also he has better access to the Land Records office as a Union *Parishad* member when GS is searching for details of *khas* land. Before, GS people had been ignored by this office. However, there has been a failure too, when the GS member wanted the Union *Parishad* to appoint GS people as day guards to a project (roadside tree planting), and the Chairman refused claiming his right to hire and fire.

Other GS Union *Parishad* members from Jholdakar Cluster, after telling the story of harassment during and immediately after the election, reported greater strength in numbers (including a successful Chairman candidate) though not quite achieving majorities. Again, their presence was ensuring proper entitlement for example to destitute women (and reducing traditional Union *Parishad* Chairman and Members' corruption); greater use of *shalish* in the Union rather than litigation going to the courts where the poor are weaker; process of identifying and acquiring *khas* land is easier (though early days yet); they could insist on timely implementation of projects (crucial in rural works schemes which should be implemented during the lean agricultural employment seasons).

Obviously there is a need for caution in assessing the longer term implications of this participation in local democracy. Many caveats exist. If the poor were to gain significant control in future elections would the Union *Parishads* have even their limited powers subsequently removed? Is there a danger that even GS Union *Parishad* members become incorporated into the power structure, especially if those elected come not from the poorest but the more educated and marginal or even small farmer classes? (see section on leadership) Will there be a counter-reaction when the next Union *Parishad* elections are held, now that other classes and parties realise that they cannot take their traditional pockets of support for granted? Will Union *Parishad* elections be abolished in the future if it looks as if the poor might manage a landslide victory, at least in certain localities? 'Critical mass' strategies for GSS and other NGOs which must be appropriate for local political participation objectives as well as influencing aspects of local labour markets and gaining control over resources, may actually provoke this counter-reaction.

These caveats, however, must be placed in the context of other wider objectives, shared by many of the donors to Bangladesh, namely progress towards good governance, government accountability, more open democracy, and so on. The successes of GS clearly conform to these objectives, and should be applauded for that. At the same time, recalling the discussions above concerning dimensions of empowerment and the notion of 'impact', participation in local democratic institutions may be of very little, long term structural significance. It is an early plateau of success and will contribute to a sense of achievement, proof of the value of solidarity. In itself, however, this is unlikely to be the route to the wider impact: certainly not a sufficient condition, nor probably a necessary one.

7. Income-Generation and Credit: Policy Dilemma

Income-generation and credit have become a central policy dilemma now for GSS, and at this time remains unresolved. At the same time, there is demand from the VCs for GSS support for income generation activities (IGA), accompanied by confusion over the credit issue.

The problem for GSS is that its original development stance consisted of a sharp critique of the strategies of other NGOs as undermining the process of empowerment by engaging in service provision and credit, thus reinforcing the dependence of the poor upon external agencies, potentially dominated by other class interests. In many ways, GSS began as anti-NGO, deliberately adopted a more political strategy in the sense of the poor needing to seize rights for themselves rather than have a limited set of rights granted to them through charitable instincts, rights which in principle could then be withdrawn if the conditions changed. As one GS informant reported (from Hakula Union, Nilphamari):

"When GSS first came with literacy and conscious raising intentions, we said 'are you just going to make us literate and conscious, or are you going to give us credit and other assistance?' In reply, GSS said 'we are not going to give you anything and we are not going to take anything from you either'."

There is an obvious question within the participatory development framework about who defines problems and issues. As Chambers has pointed out, most definitions of poverty have been made by external observers. Perhaps GSS workers can claim that their backgrounds in some of these communities make them insiders rather than outsiders. But as GSS has become, and may become further, institutionalised as an agency of professional 'development' outsiders then the hitherto natural assumption that GSS leaders know best what the interests of the poor are, because they are among them, has to be constantly checked against the opinions of the mobilised and conscientised GS people. What is the point of this conscientisation if the mobilising agency then concludes that it does not like the opinions thereby produced? The question has been put to some of the senior GSS personnel in the field: what if you produce a 'monster', i.e. a movement that rejects your long term empowerment premises in favour of more immediate and pressing concerns for income.

Chambers has espoused the cause of 'participatory appraisal' in the understanding and identification of needs. Why should agency outsiders, using significant amounts of international taxpayers' money and other voluntary donations, presume to override these 'indigenous' definitions of need? People have immediate lives to lead, poor people have very short term survival horizons. From his work in India, Chambers has described the priorities of the poor as identified by themselves:

- Incomes and Consumption.
- Net assets and security.
- Independence and self-respect.

Clearly much of the GSS agenda, as discussed above, is responding to the second and third items here. So its analysis and strategy can be applauded for that. But the first item cannot be overlooked. Other NGOs like Proshika and BRAC included all three from the outset, seeing them as mutually compatible and reinforcing. Proshika was very explicit, for example, in theorising the gaining of material autonomy through IGAs (supported where necessary by credit) as a precondition for wider social and political independence in the society; it also recognised the need for conscientisation to create appropriate IGA forms which would not just lead to individualism.

The point about GSS, and the origins of its current policy dilemma, is that it explicitly rejected this association between IGA and the gaining of resource, employment and other social rights. In this way, it claimed a distinction of analysis and strategy from other NGOs, with the exception, perhaps, of Nijera Kori in its early days. Of course the danger is why an agency with this position should seek funds as an NGO? The perfectly respectable answer for GSS was that it had an analysis of poverty more exclusively in human, social and cultural resource profile terms, and that agency resources (e.g. for education, literacy, legal aid, primary health, social mobilisation around issues) were required to intervene along these dimensions. But it is now difficult conceptually for GSS to resist the demands for services along the IGA dimensions from the very social forces which its previous activities have helped to create. During the field visits, support from GSS for IGAs was a universal demand from the GS VCs. Of course, an aspect of this demand is the awareness of other NGO activities with their client groups in the locality. If them, why not us?

This dilemma, which represents a serious conceptual challenge to GSS, is reflected in some recent GSS documentation. The concern here is not to 'trap' GSS with a contextual analysis of its past positions. Perhaps GSS has always planned to undertake collective income-generating projects. The real point is that this approach has not been a central and explicit part of its practice. To date, after 10 years since formation, there are very few examples of IGAs. If there is a strategy of capturing *khas* land and waterbodies, and acquiring through lease other similar resources from public agencies (Railway Board etc.), then it logically follows that an appropriate IGA should be established. A 'collective v individual' issue arises immediately in the use of *khas* land: evidence from a cluster in Khulna suggests that a large amount of *khas* land/water might be acquired soon, but that it will be distributed to individual VC members in 2 acre plots. This instantly raises a question about the collective strategy and collective or individual demands for credit to occupy the resource successfully through cultivation.

The present activity and plans consist mainly of re-excavating fishponds with WFP wheat as payment for the labour. One such project was visited in Mymensingh. The intention is to save from the wheat payment in order to create the funds to prepare the pond, stock it and provide feed, equipment and so on. Our information from GSS staff is that there are currently 21 such fishery projects to date. GSS has appointed a Fisheries Coordinator to manage this programme and provide technical advice. If there are presently 386 VCs, then this could be regarded as 5.4% of VC activity to date, and this only since 1991. Again, the point is demonstrate that IGA is a recent initiative, reflecting its previous low priority in GSS thinking. When the organisational structure for these IGAs is considered then the proportion of GS people involved is even less. These IGAs will be undertaken by production units or teams (*Utpadak Dal*) within a VC consisting of 5-6 members. Through their labour, these 5-6 people will be the principal beneficiaries.

However, the organisational form is still evolving, since a production team in Mymensingh (Trishal cluster) consists of 25 members (selected by the general committee of the VC from 350 members as the ones most in need), although only 6 of these (again selected on the basis of being the poorest, rather than the most entrepreneurial) will take the management responsibility for the pond once it has been stocked. The definition of 'collective' clearly has to be worked out through experience, with some policy on how profit is distributed to reflect the broader interest of the VC (and of course its role in providing overall protection for the asset, as well, perhaps financial contributions from savings).

Similar plans for social forestry are being considered. A GSS Forestry Supervisor was appointed in November 1992 to coordinate this initiative of roadside tree planting using WFP sources as part of the FFW. WFP has emphasised its priority for women's production teams to undertake this work. Both of these programmes (fish and forestry) have, of course, been operation with other NGOs for some time, thus highlighting the pressures on GSS to move in their strategic direction.

However these IGA approaches, which reflect the understandable desire within GSS to try and retain a collective form consistent with its broad philosophy and contain the possibility of occurring without credit support, do not encompass the range of demands from the VCs for IGAs. These are familiar and conventional: ducks and chickens, livestock rearing, handicrafts (e.g. mat making), small nursery plantations, as well as support for agriculture, fruit and pulses production, small fish ponds, shrimps and so on. The implications of these IGA proposals are: request for technical services and credit; and often individual, household based activities rather than production team or larger scale collective. There are further implications for the issue of group or VC savings which have been avoided to date. Across the three regions visited, the clamour for these services is universal among the GS people and confirmed by the GSS field staff: this can be supported by numerous case examples.

Credit

Apart from the collective-household dilemma, the main issue is becoming credit. Universally demanded, but often very unrealistically: both GS people and GSS staff are confused. A recent GSS paper seeks to establish a policy, but leaves open several unresolved issues. The decision to promote collective IGAs for each VC remains ambiguous given the observation about *khas* land above and the formation of production teams within VCs, and does not deal with the pressure for more household based, small-scale, low risk, low turnover activity. The statement on credit reflects the further policy dilemma:

> "This decision necessitates a new strategy whereby GSS's stated principles regarding credit can be maintained whilst setting up as large a number of projects as is possible. Credit will, therefore, in future be extended to Village Committees. GSS has never been against credit per se. Its position is that:
>
> - Groups must develop their own organisational structures, independent of any economic enterprise, before credit is introduced.
>
> - All projects must be collectively run so that its activities are consistent with members' solidarity. Each project will be run by a work unit, of which there will be four to six in Each Village Committee.
>
> - Projects will be productive as opposed to speculative.
>
> - Credit should not be introduced as starting point to social mobilisation.
>
> GSS's objective in supporting Village Committees is to help them increase their capacity, strengthen their organisational base and reduce their dependence on GSS."

These principles are very similar to those of Proshika. The sequence of organisational forms is different but ends up at a similar point. Proshika form groups of approx. 20-30 people and as far as possible these groups engage in collective projects or manage funds to individual household members of the group. These groups increasingly belong to the wider forum of People's Organisations at village and other levels, approximating GS VCs, joint VCs and Union Committees. GSS historically converges onto this point from establishing the larger village committees first, then breaking them down for productive purposes into smaller production teams (though their standard or average size remains unclear).

However, the omission in the GSS statement is the source of this credit, and any discussion of the conditions for loans. There were no plans within the GSS to establish a line of credit (e.g. a revolving loan fund) or to recruit and train the personnel to run it both centrally and in the field. Given its historical position, and given its small size and the unpredictability of donor support, GSS is understandably reluctant to embark upon the commitment of operating a credit fund. This is a dilemma faced by many smaller target-group forming NGOs with IGAs in their portfolio.

What are the options? Banking for the poor remains problematic in Bangladesh. The record of commercial banks is abysmal. The links between GOB programmes and projects, nominated banks and the poor target groups has yet to produce a viable model (though SIDA funded PEP/RESP within BRDB might be close). Large NGOs (e.g. BRAC and Proshika) or Grameen Bank with substantial credit operations only lend to groups/individuals organised by themselves.

A few years ago there were discussions about setting up an NGO bank run jointly by NGOs to lend to any groups registered with an approved NGO, with the respective NGO acting in effect as a guarantor of good practice. This strategy would have overcome the problem organisations only lending to dedicated groups, and enabled the client groups of smaller NGOs to enter a poor user-friendly segment of the credit market, without those smaller NGOs having to replicate a credit administration and individually seek the funds. The move by BRAC to set up its own bank, supported by a consortium of donors, sabotaged those discussions.

Although not an immediate solution to GSS, such an initiative if renewed, would constitute a longer term solution for organisations like GSS and for example the Bangladesh Khetmojur Samity. There is a bank which lends to NGOs to on-lend to target groups, but this would still involve GSS in developing its own credit service.

These credit problems for GSS are further complicated by its own past critiques of credit in its conscientisation work with literacy and, later, functional schools leading to the formation of Village Committees. When credit is raised in discussions with GS people, it generates heated debate and some internal disagreement. The main response involves referring in very negative, even hostile, terms to the Grameen Bank model 'which insists on instant repayment in weekly installments, and charges interest'. It would appear that in the training with target groups, GSS has set up the GB approach as the only model of credit to the poor, has presented its approach in crude, simplistic terms (since GB does have variations of approach, and retains certain principles of solidarity and human rights with the groups with which it works), and has insisted on some kind of political right to receive credit without interest.

The GB weekly installment system for most of its credit is a problem for seasonal or annual length productive projects, and probably does function to encourage small scale, low risk, high turnover often speculative economic activity. However, other models of credit extension, such as Proshika's, have been ignored in these discussions even when operating in neighbouring areas (e.g. Domar next to Nilphamari). The hostility to paying interest is, of course, highly unrealistic and very unsustainable. The recognition of this hostility to interest (and perhaps by implication resistance to the repayment of principal) by other potential credit collaborators with GSS, such as RD-9 and even RESP in the Nilphamari area, has cut off that option for the time being. As some VCs members have explained: 'they do not trust us to pay back.'

There is a final issue on the credit dilemma which perhaps reflects the over-dogmatic, purist content of past conscientisation messages from GSS. We questioned GS people (men and women) relentlessly on the contrast between the interest which they pay on personal debt (often dowry induced, incidentally) and the rate offered for example by Grameen Bank, Proshika, BRAC and so on. There are well known explanations concerning preference for private over institutional credit up to a certain level of contrast between their respective interest rates in terms of transaction costs (which includes many institutional and cultural features of the local social structure, as well as long period games). But a contrast between the prospect of user-friendly institutional credit at approx. 20% compared to private village credit at between 120-240% cannot be explained in these terms. There was much uncertainty and confusion when this contrast was made. All institutional credit with interest was bad, personal indebtedness to private moneylenders was unavoidable. No connection could be made between the two.

This universality of private indebtedness among the rural poor raises the question of whether organisations like GSS should start to confront the issue politically through widespread mobilisation as in West Bengal. At present the strategy among progressive developmentalists in Bangladesh seems to be to provide IG opportunities supported by subsidised or at least targetted credit as the route to free their target clients of private debt. But why should such efforts be allowed to confirm the validity of the exploitative element of private rural indebtedness with new sources of income being diverted into paying off usury?

8. Re-assessment of Objectives: GSS and GS

The preceding discussion, together with GSS's proposals for expansion over the next 5 years, raises the question of GSS's operational objectives. At the general level of themes and restructuring of the society towards good governance, accountability and improved prospects for the poor to take control of their own destiny and alleviate their poverty in the process, GSS's ambitions for the society as a whole remain intact. And of course many other agencies and donors share this broad agenda--indeed it is difficult to disagree with it. But the question remains: where does GSS as an organisation fit into these general aspirations for the society? A certain ambiguity of objectives may be functional in a fledgling democracy verging on the edge of repression at any time, especially with fundamentalist and secular forces in competition. However, GSS for the foreseeable future will only flourish with donor support, and donors as well as staff and target people associated with GSS will need a clearer understanding of realistic intent.

Three sets of questions arise in the context of social mobilisation, and they apply to other NGOs too:

- whether GSS is attempting, through belief in the superiority of its approach, eventually to provide a total geographical coverage with its philosophy?
- or does it see itself as consolidating in present areas, with some expansion during the next 5 years in the expectation that other NGOs with their People's Organisations are growing in a similar fashion and that some eventual unity between them will be achieved?
- or does it see its approach as model which needs to be tested on a limited scale in order to influence other agencies to adopt similar practices with target client groups?

To some extent, the answer to these questions does not lie solely within the power of the GSS leadership to decide. Much depends, anyway, on the availability of funding, if funding is the key variable. It is already apparent that other NGOs have realistically perceived the constraints placed upon the progressive donors, facing severe public expenditure cuts within their own societies in the context of global recession. Whatever the morality of this, this is the reality. Other NGOs have reduced expansionary ambitions, produced standstill budgetary proposals, even offered cuts on advice from donors. As a result, donors perceive the rationalism of their clients and a framework of trust on a revised, truncated programme is created. In this process, objectives have to be re-assessed and revised. Visions are having to make way for realism.

The conclusion from this process tends to be: greater coverage on tangible services (education, IGAs, primary health); reduced geographical aspirations with a greater emphasis upon learning from models and projecting them into the public domain; at this time, caution in relation to the political process with non-donor funded national political parties legitimately competing for office (though the funding of fundamentalist parties may in part be external in various ways); and for some organisations (like the Khetmojur Samity which recently became a registered NGO) an examination of the minimum possible ratio between funded staff posts and volunteers to sustain and extend present social mobilisation activity.

GSS is now having to confront these issues. Over the past 10 years, other NGOs have expanded and the donor funds have been available. Under these conditions, donors and their NGO clients were not under the same pressure to resolve the questions noted above. GSS is now faced with the awkward situation of seeking dramatic expansion of its field presence on all aspects of its programme precisely at a conjuncture when funds are contracting and when responsible donor behaviour obliges continuity of commitment to a large proportion of the recurrent costs of NGOs, whose expansion was supported by them in their last project cycle. It should be noted that such expansion was an incremental percentage from a higher level of existing activity, with secure management structures in position.

9. GS and other NGO People's Organisations

The second of the three questions posed in the previous section concerns the prospects for collaboration between the client people's organisations of different NGOs where there are overlapping regional concentrations. Perhaps the first observation is GSS's intention to saturate areas of present *thana* coverage. We must presume from the projection of VCs to be formed that the intention is to saturate each union as well. This implies that there is no intent to cooperate with other people's organisations in these areas. It may also imply some direct competition where other NGO people's organisations already have a presence in these unions. (To the reader less familiar with Bangladesh, unions may have approx. 10-15 villages and it is not uncommon for there to be a multi-NGO presence in a single union.).

This implied strategy may be a realistic assessment on GSS's part. Interviews with other major NGOs do not reveal an enthusiasm to merge their client people's organisations with those of GSS in areas of common activity. There is some fear of a repetition of the Nijera Kori attempt a few years ago to form a national Bhumihin Samity. This attempt was considered pre-emptive and, anyway, premature. There is also a distrust at the field level based upon the past criticisms by GSS staff of other NGO activities, and to some extent residues of this distrust remains at the respective centres. However with GSS's shift towards IGA, its interest in credit, its participation in ADAB, and its important lobbying networking on other national issues, more cooperation is evident between GSS and other NGOs at HQ level. But the position of some other NGO leaders is that firm, trusting relationships between NGOs at the centre have to be forged and sustained as the basis for any encouragement to client organisations in the countryside to join forces on various issues.

If GSS is unable to carry out present intentions of saturation to achieve critical mass presence in particular *thana*, then it will have to consider and be more proactive about cooperation. A precondition of this will be to redirect field staff away from present hostile positions, and to start delivering different messages to the VCs. To reproduce in the field, the greater cooperation between NGOs at HQ level, will require a deliberate reworking of basic ideas and strategy throughout the field staff structure. Certainly, over the longer term, with expansion into new geographical areas (i.e. new *thana*) unlikely in the foreseeable future, cooperation between different people's organisations at higher levels of aggregation (district, constituency etc.) will have to be accepted. Without such acceptance, it is difficult to see the overall purpose of these efforts in 'impact' terms, i.e. addressing wider national and international issues. Without seriously considering a route to those objectives, visions remain rhetorical.

The predictions of available funding support on the one hand, and the wary attitude of other NGOs, anyway committed to the validity of their own models of social mobilisation, would seem to rule out positive answers to questions one and three in the previous section.

10. GSS Supporting Programmes to GS and other People's Organisations

One way of overcoming some of the barriers to collaboration noted above is to move back from the idea of GS and other people's organisations joining forces permanently or temporarily towards the deliberate sharing of common support programmes. GSS's LALE for example might also work with other target clients in the locality. Popular Theatre performances can be deliberately taken into the areas of other NGO presence (and, of course, vice versa since Proshika have been developing the popular theatre approach for several years). We might imagine client groups of other organisations seeking to access GSS's PHEP and indeed PEP. Likewise, we might imagine, as an exchange, GS VCs or production teams accessing the credit lines of other NGOs (if there is a genuine desire to accept credit discipline) and receiving specialist extension services. Obviously this would involve considerable cooperation between specific NGOs in specific areas, and it might also involve some financial transfers to reflect the respective use of services. But such a strategy would draw upon the principle of comparative advantage, prevent unnecessary replication of institutional structures, assist organisations like GSS caught in this recessionary conjuncture, and demonstrate the utility of the good GSS practice on the ground in respect of these supporting programmes (Literacy, LALE and PTE), and related programmes (PEP and PHEP).

11. Horizontal Integration: GS, Supporting and Related GSS Programmes

This possibility of NGOs exchanging services between themselves prompts the issue of integration between the various elements of the total GSS package in the field, especially in relation to the central social mobilisation and empowerment strategy. This is, of course, difficult to assess in such a quick, cursory look at the programme in the field. However early scepticism on my part has mainly been overcome. There are no doubts about the integration and value of LALE and PTE, as *directly supporting programmes* to ALSM. Both of these elements are very impressive. LALE is an indispensable element of the maturing of GS, quite fundamental to the empowerment strategy in the direct way of providing information on legal rights and active support in fighting cases or resolving them through pre-emptive action. Several legal education occasions have been witnessed between lawyers and VCs which demonstrate the value of this approach.

The PTE performances, seen during the field visits, were attended by large numbers, women as well as men, despite being late at night at some distance from the homestead. The messages were clear (dowry, debt, sharecropping, opportunist fundamentalism etc.) and the acting held the attention of the audience. It is not just that these performances deliver a message, but that it is received collectively and therefore amenable to subsequent follow up by ALSM fieldworkers, literacy village workers and LALE lawyers, as well as continuing debate among the audience. It has to be worthwhile for men and women together to watch themes of dowry and family law, knowing that each one has been exposed to the issues.

The integration of *related programmes* (PEP and PHEP) with ALSM is harder to assess just from field visits. The relationship between GS and their respective primary education school was vividly demonstrated during the Union *Parishad* elections in Nilphamari. The GS VCs rebuilt the burnt out schools. More routinely, the children of GS members are attending the schools; GS members are maintaining the fabric, replacing bamboo etc. as necessary; repairing after floods, especially rebuilding the hard mud base to the *pucca* construction. Over time, considering long term sustainability, we could imagine GS members contributing fees to the schools to underwrite their recurrent costs. Without extending this discussion into the PEP *per se*, we can reasonably expect GS members to develop a strong sense of ownership of these schools, and to begin to place pressure upon the public system or other private schools in the locality to secure the same standards and quality of teaching in the later educational careers of their children after grade 3. Indeed, we could conclude that such action by GS members might be a necessary precondition for a continuity of experience and quality for the children.

The PHEP has yet to be extended out of a few well-focussed areas. Out of all the programmes it has a particular gender focus within the community and for the women GS members. The test of its integration lies not just in changes in personal health practices among the immediate recipients of educational packages, but also in the translation of primary health awareness into effective claims for services from the public health system. For this translation to occur, the relationship to GS is essential. GS becomes the vehicle by which some of the PHEP objectives have to be realised.

12. GSS Strategy and Resource Profiles among the Poor

From various elements of the preceding discussion, when GSS has its full package of programmes (social mobilisation, directly supporting programmes, related programmes and IGA) in a locality it claim to be operating within a 'state of the art' approach to poverty alleviation. With the full portfolio, it is engaging with all four resource profiles affecting people's life chances: material; human; social; cultural. The logical conclusion from this observation is that GSS has comparative advantage in this package, when the package is operating in full. In the context of both funding constraints and current underdeveloped management capacity within the organisation, it would make sense for GSS to reformulate its strategy on the basis of this principal observation.

A lean version of a GSS programme would consist of identifying *thana*/clusters where it would make sense to concentrate all its programmes in a critical mass, so that a total integrated model could be tested through the stages of performance, effect and impact.

If this strategy was accepted, then it would involve a close examination of existing area/geographical programme strengths as well as the potential for introducing as yet new programmes to complete a full portfolio (presumably a similar exercise was conducted for identifying clusters for the Intensive Literacy Programme). In this way both GSS and the donors could see a clear application of principle to the design of a project proposal. It would be consistent with GSS current objectives to consolidate in existing areas, but the implication of this total portfolio strategy in any one location might be (under funding and management constraints) to withdraw from some existing areas in order to concentrate in more appropriate ones. This is not a suggestion to withdraw from particular regions, but it may lead to a retrenching of clusters within a region as the price to pay for applying the principle of the full portfolio.

PART VI

GOOD GOVERNANCE AND THE FRANCHISE STATE

Chapter 21

Parallel Rationalities in Service Provision: The General Case of Corruption in Rural Development

Introduction

This paper was written in response to an invitation to offer for a seminar at CDR, Copenhagen on State and Non-State Provision of Services in Eastern Africa and South Asia 'an overview of some of the more important theoretical questions raised by different forms of service provision as well as the practical problems and issues involved'. What follows are some thoughts about how to conceptualise state-society relations around service provision through an examination of the problematic of corruption.

Initially, there was a very pragmatic reason for this. The invitation to give an introductory presentation to this seminar competed with a request to provide a note for SIDA on how to approach the whole question of corruption (with special reference to Bangladesh). As I thought about both exercises, it became increasingly difficult to separate them as I contemplated the paradox of corruption as an increasingly acknowledged generic feature of the relationship between official and society and yet seriously overlooked theoretically.

I should also add that I doubt if any of us with rural fieldwork experience in India and Bangladesh have not encountered and recorded almost daily accounts of what we have termed corruption, perhaps failing to realise by doing so the generic rather than specific character of these accounts. Having spent most of the 90-91 academic year in North Bihar, and much of that in one village, I can no longer dismiss 'corruption' as pathological or deviant behaviour. There has been, however, an intellectual route to this position which I will attempt to list (rather than review).

Development Administration has had a long career as an applied discipline, complicated by having to cope with evidence as well as ideology. The central concern throughout this career has been the appropriateness of bureaucracy to make and implement allocative decisions under prevailing conditions of extreme scarcity and in the context of weak general markets in which the poor are considered to be especially handicapped and therefore to have a special claim upon public expenditure. A persistent analytic weakness in this career has been to judge the performance and behaviour of officials and institutional formulae by criteria which lie outside the social structure and culture of the society in question, derived from some model (implicit/explicit) of Weberian rationality. It has rarely been acknowledged that any achievement of instrumental, objective behaviour on the part of officials in the West (and then only post-Trevelyan and post-Trollope) has been based upon the granting of privilege, status and personal resources to the point where their separation from prevailing but imperfect market cultures has in effect been purchased on their behalf by the state, removing the necessity, as it were, for individuals to purchase such security directly. Colonialism and the development of colonial bureaucracy (in the Indian sub-continent, with the transfer of governmental responsibility from the East India Company to the British government, though again we should beware of confusing normative principles with practice) was an extreme example of this process. Indeed the principles of liberal democratic pluralism and the separation of powers owe as much to the legitimation of the state via purchase as they do philosophy. But even if embedded now and rooted in Western democratic culture (though perhaps more tenuously in the practices of local government), such notions of rationality do not transport easily to other, poorer societies where the necessity not just the opportunity for rent seeking by post-colonial indigenous officials is more widespread, and where the expectations to do otherwise are lower (Krueger 1974).

Much of the more interesting elements of the career of development administration has been devoted to wrestling with this problem, but in a non-judgemental way. That is to say, it does not proceed from an *a priori* assumption that such rationality is or can be the norm in society, but examines situations in which recourse to such rationality forms part of the cultural equipment of legitimation in the allocation of public revenues. That is to say, it appears as a contingent rationality. With this in mind, let us consider the evolving development administration agenda, using this principle as a criterion of judgement.

Themes in Development Administration

A list of familiar themes has been: preparation; training and localisation; overdeveloped bureaucracy (2 traditions—Riggs 1960 and Alavi 1972); administrative reform; central planning; parastatals; management of projects—from blueprint to process; access (including clashes of rationality between bureaucracy and community); labelling; room for manoeuvre; community development; cooperatives (including the analysis of village level politics and factionalism); integrated rural development; decentralisation; participation; structural adjustment; market and state; privatisation; NGOs—delivery and empowerment; institution-building and replicability; good governance and accountability; and, though only occasionally, corruption. There has been an undercurrent of populism to parts of this list embodied in such catchy phrases as 'putting people first' (Korten 1983) and 'putting the last first' (Chambers 1983), connected, of course, to the whole participation, empowerment, conscientisation, NGO nexus.

There is much to be disentangled in this list. There are practitioner concerns, ideological actor ones and academic or analytic ones (sometimes interwoven). There are contrasting equally firmly held normative and ethical positions, such as: for and against the market; poverty targetting versus growth and trickle-down; technical versus participatory routes to project efficiency and moral outcomes. Uphoff suggested these paternalistic and populist themes could be transcended by a strategy of 'assisted self-reliance' (1988). There are contradictions between notions of citizenship and accountability on the one hand, and by-pass models of service delivery on the other. There are rhetoric and reality problems in comparisons between for example state and NGO, and state and market. The reality of the former often being compared to some idealised notion of the latter, usually for polemical effect.

The Clash of Rationality Problematic

However, for my purpose, it is important to focus upon those aspects in the broader agenda which deal with, or include, state-society relations in service provision. Some of these themes, above, imply a rationality clash perspective and are concerned to bring about a greater degree of fit between them, usually by trying 'to bring government closer to the people' (hence administrative reform, open recruitment, decentralisation, participation), but sometimes by bringing people closer to government through functional literacy, wider representation on local bodies, mobilisation via NGOs for effective claiming of state administered resources, participation again (though conceived of differently). Although worthy in the sense of recognising a terrain of contested culture, they are not analysing the terrain as such but the circumstances of the protagonists on it. The implied criticism here is that such approaches may take us to the brink of understanding how service provision occurs, but it is only an interactional analysis which can really deliver that understanding, since it focuses upon a structure of relations, the resolution of contested culture through daily practice, rather than a 'relations between' analysis where if a party to a transaction is influenced by the other they are seen as acting out of character (pathologically so for the official, conformity to a modernising rationality by the non-official), entailing negative or positive judgements.

The 'relations between' analysis is static in not allowing for the dynamic evolution of coherent structure and culture through the social action of service provision, since what is emphasised is the lack of understanding and communication between or across discrete cultures. Such forms of analysis necessarily entail deliberate strategies to bring government closer to the people, or vice-versa, through administrative reform, decentralisation or participation without recognising that much accommodation has already occurred, and that actual distribution is determined by the nature of this accommodation. The agenda of development administration has been dominated overtly or covertly by 'what ought to be', reflecting the needs of the practitioner actor (staff colleges, administration policy analysts, TA, donor etc.) with less attention to 'what is'; and the 'what is' has tended to take the form of the limitations of one 'side' in terms of the other.

Interactional Analysis

So which aspects am I going to select from the broad development administration agenda which get close to the interactional analysis required to understand this process of accommodation? Perhaps unsurprisingly it is the anthropologically oriented forms of analysis which have been the most useful, but up to a point. For example, Bailey's 'Peasant View of the Bad Life' (1966) remains a path breaking contribution in describing 'home and away' games between peasants and officials. The analysis was based upon a sharp contrast between two distinct cultures, with respective sets of preferred or familiar interactions, with respective sets of actors able to play at home or forced to play away depending upon other contingencies. Basically these consisted of who needed whom most, i.e. a notion of power that was not structural and universal in the sense of class theory, but itself contingent on circumstance though Bailey accepted too inevitably the ultimate strength of encapsulating structures. (Bailey 1968 and Wood 1974) Bailey's notion of people (for him, especially peasants) acting differently within a moral community from their relations with people outside it (echoing Banfield's instrumentalist view of amoral familism, 1958) set up the problem of trust across these boundaries and how people coped either with brokers (which kept the worlds apart) or by short term anticipatory cheating, which also reinforced the sense of boundary and difference (for Bailey occurring heuristically at the boundary of the village). This is why Bailey takes us so far, but no further. He does not consider the outcomes, the structural implications, of these patterns of interaction between the cultures except insofar that he seemed to accept teleologically the absorption of small communities into the wider, modernised, encapsulating structure.

Intervention and Village-Level Politics

The work of Hart (1971) and Nicholas (1973) on the relationship between development initiatives and village level politics has also been useful, to a point. Both took a then growing literature on village factionalism, combined with their own fieldwork (Nicholas in West Bengal), and considered the impact of factionalism on development initiatives (like cooperatives) and vice-versa. Nicholas' point was that no local level development project could be understood except by reference to the structures of power and conflict in the local society. Hart went a step further to claim that the arrival of external resources in the form of credit, inputs, location of infrastructure projects etc. exacerbated factional forms of conflict and in effect strengthened the networks of patron-client ties in the community as the competition to capture scarce new resources intensified. Although dated now, these were refreshing contributions to the career of development administration in highlighting the necessity for an ethnographic understanding of the context of local level development. However, along with similar forms of argument occurring in *Economic Development and Cultural Change*, particularly in the context of Latin America (ideas about the limited good, and zero-sum games), Hart and Nicholas accepted a clash of rationalities between modern bureaucracy and the traditional community as being responsible for failure of intended project outcomes and did not consider the possibility of unique structural outcomes beyond the hiatus.

Access

Possibly the most important opening contribution to the 'relations between' problematic was Schaffer's theory of access (Schaffer and Wen-hsien 1975, and Lamb 1975). An ethnography of allocative decision-making and implementation was presented by focusing on the experience of encounters between official and applicant (see also Wood 1986a on applicants and clients) creating a conceptual language of queues, gates and counters. Arguments were developed about how bureaucratic devices (such as compartmentalisation and prioritisation) would manage and ration access under conditions of scarcity, and Lamb located such devices within a broader Marxian view of the state and its capacity to disorganise subordinated classes. Attention was then focused on how members of the public handled the problem of making successful organisational connection in terms of presenting appropriate evidence about their qualifications for service, and how these requirements would systematically constitute a bias against classes and groups less culturally in tune with the principles of bureaucratic rationality.

But these arguments, for all their significance, were still couched in terms of clashes of rationality, with the assumption that the official culture was dominant obliging others to alter their behaviour to interact successfully with it. This stance was in effect reinforced by the adaptation of Hirschman's *Exit, Voice and Loyalty* (1970) to the problem of public distribution and access by recognising that when voice and loyalty failed, applicants could always exit (Schaffer and Lamb 1974). Wood (1977a) continued the theme by emphasising clashes of rationality and pointing out the conditions under which North Bihar peasants would 'exit back' into the informal credit system when the rules of access to formal, even though subsidised, credit was at odds with the rhythm of their own rationality (even if initial organisational connection could be managed). Analytically, in this review, the theory of access should have been more about fit and structural outcomes than manipulation and distributional outcomes (whether for state class or state institutional ends).

Power in Policy Making

The next stage in this particular intellectual route through development administration is: room for manoeuvre (Clay and Schaffer eds. 1984) and labelling (Wood ed. 1985). Both sets of ideas were concerned about power in policy making, especially in the context of specific development intervention whether via programme support for sectors or via project support for target groups.

The 'room for manoeuvre' arguments drew attention to the question of extending responsibility for policy as a precondition for its improvement and relevance. 'Such thinking about policy would touch each of the zones of practice involved in the commitment of important resources: the construction of agendas, the establishment and handling of institutions and their processes, and institutional allocations and distributions.' (Clay and Schaffer 1984 p191) The review of rural and agricultural policy via various case studies led to the conclusion that policy dealt with people as imposed categories, which together with sectoralisation constrained the room for manoeuvre and certainly restricted opportunities for meaningful participation and a genuine sharing of responsibility. And the resort by elites to technique increased the opportunities for the powerful to avoid responsibility and therefore limit the capacity to fine tune policy to its circumstances.

Labelling arguments shared some of these concerns, but in particular focused upon the political necessity to manage the allocation of scarce resources without undermining the legitimacy of the state via the designation of universally valid 'target-groups'. In the process, the state manages to exercise considerable power over civil society by organising its categories and ranking them for service. More specifically it has to subvert and substitute for structural explanations of poverty in which the state is a conspirator with 'culture of poverty' explanations in which responsibility for their condition lies with the victims, the poor themselves. In this process, the state separates people from their story and transforms them into convenient cases or labels. It thus creates a definition of target clients (with varying degrees of loyalty/exit options), thereby resolving clashes of rationality in peasant societies by insisting on compliance with the state's view of what constitutes the development problem. This gives rise to the conundrum that 'the good participant is the bad participant'.

Again, neither room for manoeuvre nor labelling quite deal with the structural outcomes of exchanges between state and civil society. The former emphasises the limitations of state and bureaucratic activity in achieving fit with people's rationality; the latter emphasises an overlooked dimension of state power in establishing a discourse about allocation and setting agendas. Labelling does share with access some focus on the way people have to behave in order to handle state power, and opens up the possibility of counter (self)-labelling which is a step closer to meeting the criterion of dealing with structural outcomes. Certainly both entail a discussion about participation, but hopefully without some of the rhetorical cant which characterises much of the eighties literature on the subject.

Participation

What was participation supposed to overcome? We have seen from room for manoeuvre that it would refer to wider 'ownership' therefore responsibility therefore accountability therefore quality and relevance. From labelling we can see that it would refer to people repossessing their own stories, establishing their own agendas and thereby ownership. There are certainly resonances here with other, more dogmatically populist, strands in the literature: an assertion that development interventions, projects in particular, do not succeed without consultation with intended beneficiaries (Bryant and White 1982; Gow and Vansant 1983). This is strongly connected to the top-down themes about planning, integrated rural development and projects in general which have been criticised for their blueprint character (planning by technique with project cycle principles of sequentialism and sectoralisation) when they should be more 'process' in form involving 'trial and error' based on a full interactive relationship with the target population (Rondinelli 1982). Participatory action-research, favoured by so many NGOs, has reflected this criticism (Marsden and Oakley eds. 1990).

Project Cycle and Participation

Certainly reference to the project cycle reflects strongly the notion of state-society relations consisting of the capacity of people to interact with the rationality of formal state intervention. As soon as some external agency intervenes in a situation with funding, other material resources, training, technical advice and general awareness raising initiatives to improve poor people's material and social prospects, then standard procedures of project development are also introduced. With variations, a sequence of institutional behaviour occurs like: identification of goals and selection of means; appraisal; implementation; monitoring and review; evaluation; adaptation. This is the project cycle of agency activity. Conventional projects, particularly those mobilising large-scale resources requiring professional expertise and technical management, will complete such a cycle with paid employees and even more specialist consultants in conjunction with appropriate government officials who might be responsible for maintaining standards of work and policy conformity. Large engineering projects typically fall into this pattern. The stage of identifying goals and selecting means might include a process of consultation with representatives of the affected population (e.g. dams for hydro-electric and/or irrigation projects, large roads, river embankment schemes or major bridges), particularly when resettlement might be involved. However these processes are likely to be part of the government policy-making responsibility rather than embodied into the conduct of the project.

Projects are now, however, expected also to be participatory. They are expected to be 'process' rather than 'blueprint' projects, with continuous dialogue between 'beneficiary' and agency throughout. But what does this mean in reality? How is a project initiated in a specific locality in the first place? Has the major decision in effect been taken before the people affected can ever be involved? Are people, the potential 'beneficiaries', only then involved at the earliest in the second stage of the project cycle, namely appraisal? What power do they have at this stage: veto, rejection, radical modification, redefinition of objectives, redefinition of target groups, design of institutional or physical content, or just minor points of detail? Is this participation or legitimating consultation? How has a needs assessment been carried out, by whom, at what level of technique? What is the level of discourse in the project (in the sense of language, technical specialism, scientific culture etc.) and who is excluded by it? These sorts of questions continue throughout the different stages of intervention between agency and people.

Such questions are most frequently asked about the evaluation stage, but they apply also at the inception. However the discussion about participatory evaluation in Marsden and Oakley (Eds 1990) *Evaluating Social Development Projects* highlights the problems elsewhere in the project cycle. The means-ends conundrum reappears in the set of principles which should guide the evaluation of social development outlined by Oakley on page 32: "4. Evaluation must be *participative*: in the entire evaluation process, the people involved in the project have a part to play. It is not a question of an external evaluator solely determining the project outcome; the local people themselves will also have a voice." The point here is that evidence of participation in the evaluative phase of a project is required for the project as whole to be regarded as participatory, as an example of good social development. But the quality of that participation is the paramount issue. What is the significance of 'also having a voice' in project management terms? Are the local people bit-part actors or central players? Are the personnel of the agency directorial in style or facilitators of other people's involvement and control? To pursue the metaphor, can the potential beneficiaries improvise throughout the project or must they follow the script, even if their parts are large? In other words, when is participation organised co-option and when is it exercising power in the sense that agency personnel are compelled to behave in ways other than preferred by them? And in situations of no conflict, whose agenda actually prevails?

Now while these are pertinent questions to ask about the reality behind the rhetoric of participation in the continuing presence of the bureaucratic exercise of power, albeit legitimated by a judicious mix of technique and people's rights via the notion of 'process', the discourse remains limited in exactly the same way as the Bailey analysis--relations between contrasting cultures rather than structural outcomes.

This discourse point, or class of rationality theme, also applies to another important aspect of the participation literature concerning the contradiction when hierarchical organisations claim participatory objectives. I am grateful to Turbyne (1992) for drawing my attention to Bryant (1980).

'An hierarchical organization with its pyramidal structure of authority has clearly demarcated levels of influence and status which themselves inhibit participation within the organization. Such an organization will not suddenly become participatory when interacting with its external environment, and indeed, there is something of a bias against participatory features of its own program since these would be dissonant with its own operation'. (p.8)

I raise some of these issues in the context of NGOs in the next chapter (22). The argument of this paper is that the process of development consists of *sui generis* structural forms, consisting of parallel rationalities and cultures, which are more widely prevalent when GOs are actively present in the society. NGOs may have prompted this kind of analysis simply because their own claims to a superior morality in working with the poor inevitably invites examination of their practice, to the extent that they appear as powerful, outsider organisations, partly bureaucratic in form but partly a reflection of indigenous authoritative culture expressed through formal organisations as well as within the societal institutions of kin and patron-clientage (see also McGregor 1989 on 'Credit and the Patron State in Bangladesh').

Corruption: A Special Case of Rent-Seeking

This is how I come to a consideration of 'corruption' for its wider utility in understanding these dimensions of the development process, recognising of course that corruption is a widespread international problem involving donors as well; a problem compounded by the difficulties of defining the concept prior to applying it in a particular context.

Any discussion of corruption has to include the problematic of state and market. For some, corruption can only refer to the misuse of public funds. Others, however, find corruption in amoral markets where transactions are not transparent, where formal contracts only represent part of the transaction, where competition is not perfect and interlocking occurs, giving actors advantages outside the price mechanism. Is corruption therefore about the state or about the market? The further dimension exists when the state interacts with the market, rendering neither of them perfect nor transparent, with the possibilities of subversion all round.

Corruption is associated with crime, therefore a legal definition of behaviour unacceptable to society. In this sense it refers to the diversion of public funds into unauthorised private use and accumulation. However there is ambiguity whether it only refers to official behaviour or whether it refers to behaviour of those private individuals who interact with officials. But then there is a question of the intent of such private individuals, in the context of power exercised in such relationships. Are they seeking to capture public funds without moral rights, with the connivance of public officials for private gain and accumulation? Are they seeking to obtain what is theirs by right, but obliged to enter the informal rules of officials in order to secure such rights? Is the universal problem one of monopoly whether public or private; rent seeking under conditions of monopoly power where clients have no options apart from total exit. Is corruption sustained therefore by collusion from civil society, but bred out of necessity because of the weaknesses of other structures of good governance (accountability, regulation, meaningful participation)?

State Intervention as a Market

In the development conditions of societies like Bangladesh, the scale of intervention by the state (or quasi-state) in the economy and society is extensive. The weakness not only of the private sector but of any morality for an imputed transparent market involves the state not only as a substitute for market allocations but as a formal regulator of moral transactions. However, the state itself is not immune from transactional immorality. Indeed, the state increasingly performs as a market with prices for services becoming established not as an outcome of principled rationing but of demand. Corruption in the state is, as it were, paralleled by the hidden operations of the market, and in most societies they interact. Thus, although conceptually state and market might be distinguished by policy rationing and price allocation respectively, actually the two principles of operation are interdependent and indeed prompt each other.

My recent work in North Bihar, India, prompts a similar analysis to the experience from Bangladesh. In the attempt to understand through the single village case study the generic process of resource allocation and distribution of opportunities it seems clear that the central problem is the familiar one of the relation between state and markets, and the strategic options for different sets of actors. The great dilemma for post-independence India has been the attempt to use the state to control the inegalitarian excesses of markets, while being unable to separate the operation of the state from these markets. On the one hand too much has been asked of the state apparatus, and the limits of its competence and integrity have been long overreached. On the other hand, it has been impossible to repress the ability of the stronger classes to turn the state into a market for their own accumulation, but of course nowhere near a perfect market. The ability to pay going rates for service is a necessary but not sufficient condition. Connections, direct or indirect, through caste, kin or batch (in school and college) are required. As I have noted already, even markets outside the state operate in a similar way.

In terms of social relations, there is then little to distinguish interactions between people and the state, and the interaction between people in markets. Furthermore, empirically the conceptual arenas of state and market are almost completely interpenetrated.

The evidence from the village reveals a defaulting state, unable to control markets and itself thereby marketised. The space has been created, room for manoeuvre as it were, within which both public and private predators flourish with interchangeable roles. Public officials and private brokers colluding to defraud consenting clients who rely upon a combination of monitoring incompetence, imperfect auditing, the lottery of state harassment and populist policy concessions to avoid being any more than the nominal victims of such fraud. In this game, the state purports to redistribute resources, assets and opportunities but in practice produces the distributive outcomes which might have occurred in the marketplace but with officials joining in the profit taking as a privileged class of commercial entrepreneurs using monopoly positions to receive rents and avoid risks. All institutional credit in the village can be described in this way, along with any of the targetted programmes. Corruption is the norm not the exception. It is endemic and structural. It stimulates the networks, and in turn is sustained by them. Who loses? The budget suffers, public debt rises, interest rates either remain high or are subsidised causing further public debt, the currency falls in value, balance of payments decline, austerity measures follow (in effect a structural adjustment package), subsidies are cut, salaries and wages pegged, official incomes are squeezed more through taxation thus increasing the incentive to earn more unofficial income, tax avoidance increases on other incomes and wealth. In such a fiscal crisis, the poverty targets suffer. There is less to allocate through programmes and fewer effective controls over prices of essential goods. It is this analysis which partly prompts the title 'Weak States, Strong Markets, Poor Losers' for my Bihar analysis.

This preamble guides the following observations. In societies like Bangladesh, a principal mode of intervention by the state is via projects. Projects are an institutional and social phenomenon. They are devices, legitimated in various ways in the public domain of the state, to steer resources in mode x rather than y...n. In this sense, projects are additions to or substitutes for market forms of producing goods and services (including infrastructure). The assumptions behind such interventions are that the market will allocate goods in socially non-desirable ways and create, sequentially, other problems of dissatisfaction and voice which will involve greater subsequent burdens via tax upon those who would otherwise benefit from market allocations. It is in this sense that no society can reproduce itself solely via market activity, that sustainability (social and political) can only be achieved by interfering in and limiting the scope of the market, which includes the maintenance of codes of conduct on which the market depends as well as deliberate reallocation of resources in favour of x when y would have otherwise occurred. But this is not to say that market behaviour is excluded from the achievement of x. Rather a form of price fixing has occurred, legitimated by moral reference instead of effective demand. But moral legitimation does not occur independently of the expression of effective interest.

How do we apply this analytic framework to an understanding of corruption as a generic process? The point about projects and interventions is that opportunities before, during and after implementation are not acquired through formal market transactions. The attempt to achieve x rather than y involves direction, targetting and therefore exclusion. The rationing of scarcity is occurring by ignoring price signals, that is an activity of defining needs instead of responding to wants where the definition of needs contains political calculation. Thus needs become wants via political rather than economic demand. The state deals in a market of political goods rather than economic ones. But the state is not just a site either of coherent policy or conspiracy, but also consists of actors and sets of actors making calculations in this political goods market, seeking rent seeking opportunities by charging for what they must anyway deliver. The state is therefore also the sum of these actors.

Conditions for Rent-Seeking

With the rationing of scarcity comes access, voice and loyalty. Demands have to be managed in ways which maintain the integrity/legitimacy of the state without over delivery. This is an objective function, a feature of all state intervention in all societies with states. But within the performance of this objective function lies the opportunity for official actors and their allies to maximise rent seeking. The question is: why does this rent seeking happen in some societies more than others? Why does it apparently happen more in India and Bangladesh than e.g. Scandinavia? Perhaps the key variable here is exit. Are societies distinguished from each other by the opportunities which their citizens have for exit into non-state provision of goods and services? This is not to presume that immoral behaviour is only confined to the circumstances when the state intervenes in society, but that the concept of corruption has usually been restricted to a description of such acts rather than applied to immoral behaviour in the market.

The point here is that monopoly and exit go together. If monopoly refers to absolute rent seeking, with a continuum of relaxed oligopolistic forms shading into crowded competitive markets and wide consumer choice, then the only alternative to paying such 'rents' is exit into no service at all. If the monopoly conditions are progressively relaxed, then of course rent seekers have to compete and reduce price thereby giving consumers/clients more options before considering total exit. Societies can be arranged along a monopoly-choice continuum. In this way, Bangladesh, perhaps even more than India, can be seen as a society in which many goods and services (political as well as economic) are in the control of the monopolistic state, despite attempts to privatise certain aspects of delivery (e.g. the dismantling of the BADC monopoly over agricultural inputs or the rise of development NGOs as a competitor to the BRDB). More precisely, the state retains area/sectoral monopolies via projects. Almost by definition, projects have no direct competitors in their locations (otherwise they would not be needed) unless the project has been misconceived (like the ODA NW Fisheries Project). It is the predominance of projects in the overall allocation of goods and services in societies like Bangladesh which ensure that Bangladesh lies at the 'absolute rent-seeking' end of the continuum.

With this in mind, we can consider actual behaviour. Before doing so, it is important again to reflect on the motives for this focus on corruption. Usually the concern about corruption is judgemental: moral and ethical. For those concerned with public revenue and poverty, the judgements are prompted more by the diversion of rights and entitlements away from the poor towards to the already wealthy and well connected, a confirmation of the significance of the state in the accumulation of certain classes. These are structural and moral concerns beyond the mere indictment of criminal activity in terms of misappropriation and theft. However corruption is of developmental interest beyond the judgemental since it refers to a process of public expenditure allocation, and concerns an important interface between state and society, between state and market, or more accurately a process of marketisation of the state in the provision of public services.

So the concern is not just the moral and ethical issues of misuse, leakage and misappropriation of scarce resources (though these are justifiable concerns), but the patterns of behaviour together with the structures, cultures and codes deriving from them, which determine to a considerable extent how all public services are managed in absolute rent seeking conditions. That is to say, 'corruption' is a heuristic way in to the examination of more pervasive developmental aspects of state-society relations under absolute rent seeking conditions.

Parallel Rationalities: Markets and Bureaucracy

At one level, we can say that corruption is a phenomenon of parallel rationalities. It has been common in the past for sociological explanations of corruption to rely upon the idea of a clash of rationalities as part of the broader tradition in development administration, as outlined in earlier in this paper. For example, a patron-client, multi-transactional culture interacting with the principles of bureaucratic rationality and subverting them. But the analytic problem with this formulation is that the comparison is being made between reality and rhetoric respectively, or between actual behavioural patterns on the one hand and ideal type propositions on the other. The post-colonial inheritance in the Indian sub-continent, and in other societies, of formal rational bureaucracy has, to borrow terminology from another discourse, constituted a formal rather than real subordination of culture. Thus explanations have relied upon a false behavioural or empirical contrast between peasant and bureaucratic rationality, as discussed earlier. To put the point another way, public officials in absolute or near absolute (i.e. strong) rent seeking conditions continue to act more as patrons or clients (depending on the circumstances of the encounter, including whether internal or external transactions are at issue) than in an ideal image as bureaucrats, though this has not ruled out the utilisation of bureaucratic rationality to legitimate non-bureaucratic decision making and resource allocation. This is why I introduce the idea of parallel rationalities to denote the two interlaced worlds which the official (and the non-official) inhabits: the world of markets, society, and multi-transactional cultures on the one hand; the world of case law, standardisation, equitable treatment, sequentialism, sectoralisation and appeal on the other.

Each world has rules, but their utilisation is contingent upon the needs of the moment and may be deployed by either official or non-official actors. Thus officials may rely upon the mobilisation of networks in the society for both official achievement as well as personal gain; the non-official may refer to the rules of the bureaucratic world to secure a favourable decision, playing the game and participating for example in the rules of qualification to legitimate what might anyway be delivered via the use of networks. A further analogy to understand this point may be that each set of official and public actors play either at home or away.

Does this formulation extend the notion of corruption too far? Some would like to restrict the concept to the use of public resources for immediate financial gain (Gould and Amaro-Reyes 1983, and see their bibliography), but I think we should be concerned with a wider interpretation of the concept as representing non-transparent, rule bypassing allocative decisions and implementation which reflects longer more disguised interests of power and institutional influence as well as more immediate, obvious and personal ones. The proposition is that under strong rent seeking conditions, officials have more scope to participate in the parallel, non-official rationality sometimes building long term, stable constituencies of allies within both state and societal domains (i.e. within their own official institutions as well as among the 'public'), and sometimes just acting out the parallel culture in one-off encounters, where short term financial gain (e.g. through bribing) is more self-evident. In this way, corruption is better regarded as symptomatic of wider issues concerning state-society relations in service provision.

Summary of Concepts

So now we are armed with a battery of concepts: a continuum of strong to weak rent seeking conditions (determined by the quality of public exit options from state service), distinguishing the degree to which corruption represents the general character of state-society relations or only a deviant version of them (if strong then general, if weak then deviant); parallel rationalities which refer to the 'cross-dressing' activities of officials and public clients; and strategic corruption options which range from long term alliance building (cooptation) by officials among constituencies of public collaborators (who may indeed be the initiators of such relationships rather than passive or unwilling participants) to short term, immediate gain from one-off encounters by acting out the expectations which the public have of official action.

Sui Generis Structural Outcomes

What structures of social action evolve from this process? What practices emerge and become institutionalised over time? In effect, how do these parallel rationalities come to occupy the same cultural space, how do the interlaced parts become an integrated whole—to produce a *sui generis* structural outcome? And what evidence can we accept that this integration has occurred, that norms and expectations have become established rather than identified as pathological and deviant, which is the normal, Western observer's view of corruption?

We must remember that there are no easy or Utopian escapes for people from being caught up in this process. This is why the judgemental position taken by 'holier than thou' foreigners is so untenable. To the extent that we accept that 'corruption' is merely a special case of general behaviour circumscribing state-society relations around service provision (articulated to a considerable extent through 'projects') then we can see that the notion of escape is meaningless since it would involve people opting out of their own society and ceasing to be social animals (i.e. human). Village people in the Indian sub-continent might talk endlessly about 'rip-offs' in which they or friends have been victims; but they will also recount tales of success when the game has been played sufficiently well for a share in the benefits. People also talk about class relations in which they are victims, but we do not expect them to be able to escape from such relations merely because they observe them.

In this sense, corruption as a strategy is not an option, only the tactics to be deployed and the degree of caution required, especially when playing away from home when uncertainty about the codes, rules and culture is at its highest. And since all corruption involves some element of 'home and away', these aspects of state-society relations are particularly distinguished by brokerage and intermediaries. The more 'away' performance is required, the less the actor can go it alone, the greater the need for brokerage. McGregor's analysis of brokerage and rural credit in Bangladesh illustrates this theme (1989).

The Problem of Trust

The key element in this process is trust. If we consider the range of human resource allocation behaviour to be conceptually framed by the 'moral community' at one end and the developed, perfect market at the other, then most real allocative behaviour occurs at some point between these two extremes in which trust is a problem. In the complete moral community, perfect knowledge of transacting partners equals perfect trust, at least in the sense of predictability. In the developed, perfect market, trust is not required just a belief in stable currency and the value of competition to overcome rent seeking and achieve acceptable norms about rates of profit, although contract remains a problematic. But where neither sets of conditions prevail, and this is certainly the case in strong rent seeking societies, then we witness imperfect parallel rationalities in which trust within them as well as between them is a problem, but the latter is the greater problem—hence brokerage and intermediaries.

This issue of trust has been addressed recently in the policy context of establishing free markets as part of the expansion of modern, 'efficient' economic activity (Platteau 1991). If one accepts the proposition that the required degree of trust in transactions can only be guaranteed within a narrow framework of highly personalised relationships (Bailey's moral community and also Polanyi 1977), then it is difficult to see how fraud and deceit can ever be overcome in more open society conditions in which the state (let alone other economic agents) have come to visit the local community. Platteau points out that Granovetter's (1985) extension of the principle of personalised trust to the conduct of modern market economies can only control opportunism by subverting the other (neo-classical) condition of market activity, namely the independent, freely acting agent. At the same time, Platteau acknowledges that Granovetter has a point in rejecting this neo-classical principle of economic behaviour since the activity of economic agents is embedded within 'concrete, ongoing systems of social relations' which apply to modern contexts as well. This exchange returns us to the problem of non-transparent resource allocation behaviour when the state intervenes in the locality. The one-period transaction analogy of cheating opportunists unconstrained either by generalised morality in the society (Durkheim on moral order as a precondition for economic relations) or by prisoner's dilemma options has to give way, empirically, to a multi-period game in which the principal players have a vested interest in the continuation of the game.

This leads to a fission-fusion model of trust and deceit in these quasi-market state/society transactions, whereby the principal players (local officials in contingent command of strategic resources—whether e.g. credit or road building contracts—and local patrons) collaborate or fuse into a new moral community for the purpose of cheating others, both locally and nationally (the locally poor with policy entitlement or the anonymous taxpayers). This is clearly a multi-period game in which present behaviour (too high a price, too low a kickback) is modified by the future and the need to retain goodwill. Of course, these kinds of moral communities are regularly threatened by the arrival of new, official, players who have to be quickly incorporated—not necessarily a difficult task since they are likely to have come from equivalent moral communities elsewhere.

This 'fusion' idea entails, in effect, a merging of parallel rationalities, that is to say the discourse has moved away from the 'relations between' form towards *sui generis* structural outcomes. The weak or soft state is unable to resist its own marketisation, subverting any other operation of markets in the process. Indeed the idea of the soft state as an explanation of corruption is well established (Myrdal 1970).

Conclusion

Are NGOs the Solution?

The purpose of this paper has been to demonstrate the necessity of reformulating the discourse in development administration in this way in order to address more accurately the ethnography of service provision in poor societies. The issue has been approached via 'corruption' on the grounds that a more socially rooted and less judgemental view of corruption draws our attention sharply to pervasive structural characteristics of service provision which are shared in some important respects by non-state service providers (typically NGOs). NGOs only differ insofar as their rent seeking opportunities are not so absolute and many operate under conditions of external monitoring (though this can be more superficial than donors may realise, or care to realise). The point has often been made over the last decade by nationalists in the region that NGOs in Bangladesh, for example, must be seen as artificial constructs to the extent that their insulation from local morality (i.e. the absence of a generalised morality to regulate economic behaviour or institutional resource allocation) has been purchased in a non-sustainable way, and that to the extent that they return to the culture (e.g. via revenue earning activity on target group banking activity or business enterprises), so they will be absorbed in the relationships that envelope everyone else.

With donor policy emphasising, for different parts of the world, a strategy of privatisation extending beyond 'returning activity to markets' (where in the ideal sense it never was) to a greater reliance on NGOs as substitutes for the inefficient or corrupt state in the delivery of basic services, this nationalist criticism reminds us that NGOs do not represent some kind of magic solution to the state-market dilemma. In countries like Bangladesh, the strategy could produce an early version of a more generic future: namely the Franchise State (Wood 1992b). This might take the form of NGOs in various combinations undertaking contracts with the state to provide key sets of services in the absence of a trusted private sector culture, with the state acting as auditor, regulator and evaluator on behalf of civil society. At present, of course, such NGOs have resisted accountability within their own societies on the grounds that their *raison d'etre* is to challenge prevailing power configurations to which they might otherwise be accountable! The rise of democratic politics in Bangladesh, for example, does not ensure that accountability of either NGOs or state to the poor will be improved. There is, of course, the further problem that NGO-oriented strategies which by-pass the state undermine other sets of donor and even NGO objectives such as good governance, meaningful citizenship, as well as popular accountability (or 'participation'). To the extent that increases in NGO responsibility re-absorb them into the *sui generis* structural outcomes outlined above, then their claims to 'alternative' status begin to look very hollow indeed.

So, in conclusion, we have to return the analysis to the world as it is rather than how we would like it to be, though we should never give up on changing it. This paper has not spelt out the 'ethnography' of these structural outcomes in terms of: brokerage; preferential voice; queue jumping; surrogacy; overlaying of informal criteria upon formal rules; entrepreneurs in differentiated but parallel factor price markets; delayed transfer payments; purchase of office; official sponsorship of candidates for elected office (cooperatives, *panchayats* etc.) entailing subsequent management of public goods; continuous patron-client structures connecting official and non-official worlds; simultaneous applicant behaviour as clients and cases; responsive, legitimating, layers of cross-checking and auditing which extend opportunities for rent seeking and reaffirm configurations of power; and so on. All of these headings can be elaborated both theoretically and with empirical support from Bangladesh and North Indian material, and no doubt from elsewhere; and they can apply to state as well as non-state (NGO) practices of service provision.

Strengthening Resource Profiles for the Poor

The final question to be asked, then, in this overview of approaches to service provision is how the poor fare in these structural outcomes? At this stage in the paper, there is no intention to answer this question extensively but simply to use it to raise a further set of key issues, concerning the resources which people need in order to participate in these *sui generis* structural outcomes. It is in this way that the development administration discourse has to connect with the entitlement and capacity ideas, deriving more from the famine and hunger literature such as Sen (1981) and Swift (1989).

In a recent CDS-Bath paper for the UK-ODA (1992) on female-headed households (principally drafted by David Lewis, but with extensive participation(!) by Wood, Glaser, McGregor and White), we have suggested a resource profile approach to the classification of such households. While the broader context is not relevant to this discussion, four categories of resources were identified in this approach: material; human; social; and cultural. Material and human referred approximately to assets and labour power respectively. Social refers to significance of one's place in networks (via kin, caste, patron-clientage etc.); while cultural refers to the status and honour in which one is held. Adapting Swift to earlier ideas by White (1988 —published 1992) that people could be 'poor in people', we have argued that both social and cultural resources are crucial factors in determining a household's future entitlements.

To some extent, these approaches continue Schaffer's problem of how people manage organisational connection, but they are now applied to this complex structure of parallel rationalities in which imperfect markets coexist in the same institutional space with imperfect official service provision. Thus we have to acknowledge that people, even poor people, do obtain service in this complex structure. The degree to which they do so, and the quality and sustainability of it will depend upon their particular resource profile. The shape of this resource profile will also indicate the degree of dependency or loss of personal freedom entailed in participating in the structure service provision. Receipt of goods and services may be welcome in the short-term but may reflect and indeed reinforce dependency in the long term, thereby undermining the value of the social and cultural elements of the resource profile. Social and cultural resources may be ingredients in understanding how trust is maintained in multi-period games, but the acceptance of degraded forms of delivery may also reflect fear concerning the precariousness of such resources.

To the credit of many NGOs, their efforts have focused upon the strengthening of these social and cultural resources through various strategies of group and wider organisation. But there is a huge difference in developing such resources via a highly segmented and targetted system of dedicated service delivery by NGOs themselves in which the value of such resources is not really tested, and the greater challenge (in sustainability terms) of deploying these 'created' resources in the wider, more generic structure of service provision and markets. Nevertheless the strengthening of such social and cultural resources among the poor (including via support for material and human resources) must remain high on the rural development agenda to enable people to cope with the real world of service provision rather than an imaginary one of transparent bureaucratic allocation curbing the unequal distributional outcomes of a general equilibrium market. This real world is a complex and opaque place and the discourse of development administration does the poor no favours when it fails to recognise this.

Chapter 22

States without Citizens: The Problem of the Franchise State

Introduction

To remind the reader who has reached this part of the book, the main focus of this chapter is the problems raised for governance when the state, for whatever reason including international pressure, devolves some of its functions to other bodies. This is becoming a universal issue, especially in societies which had a large public sector such as post-war UK. In the context of Bangladesh, this 'franchise' strategy refers more to NGOs (often significantly funded by external donors) rather than the private sector per se. The central question posed is: to what extent do citizens of the state lose basic political rights if the delivery of universal services and entitlements is entrusted to other bodies which would at best be accountable to the state rather than directly to those with service entitlements? Secondly, can the state devolve responsibility for implementation without losing control over policy (since practice is policy) and therefore losing responsibility for upholding the rights of its citizens? If the answer to the first question is 'yes' and to the second question 'no', then we have states without citizens. Some of these points were included in the paper for SIDA 'Making a little go a long way', since donors have to be particularly alert to the contradiction that 'structural adjustment' or 'good governance' conditionality for their aid might actually undermine governmental accountability through encouraging this franchise strategy.

Given that part of the argument for the franchise option is the supposed superiority of NGOs over the state along dimensions of competence and integrity, the question of NGO distinctiveness must also be raised. These have been alluded to in the previous chapter, but are raised here in the context of indigenous organisational culture. Can NGOs insulate themselves form prevailing cultural norms and expectations about hierarchy, power and recruitment? These issues were first raised by me in a longer draft contribution to OXFAM's Field Directors' Handbook, but have also been discussed very openly in these terms with the leadership of Proshika.

States without Citizens

The argument about good government is both a conventional preoccupation with political scientists and theorists or apologists of democracy as well as a sign of the times. Although we may understand the recent history of ideas about state and society relatively easily in the context of Western societies, the picture is not so clear for other societies. It is complicated by differential histories of insertion into international economic structures, by particular post-colonial legacies, and by indigenous organisational culture set in wider frameworks of status systems, property relations and deference. A key element of this complexity is the problem of the nation-state and legitimation for any kind of national level, universal institutions whether in the state or the society. The market itself has limits to its universality in the absence of generalised morality about economic transactions, about exchange. But the state has even greater problems of legitimacy as essentially a foreign institution, regarded with suspicion and hostility, dealt with instrumentally and reluctantly when in monopoly control of certain sets of resources but best avoided whenever possible. It is in this sense that we observe states without citizens.

Some of the concern with good government rings hollow under these conditions, especially when it is translated into conditionality. It is not so much that conditionality around some notion of a human rights agenda is intrinsically unethical, but that the setting of impossible and unrealistic conditions is unethical. This is why careful analysis is so important: ethics are contingent on realpolitik.

The Problem of Accountability

The problem for states without citizens is accountability—the culture and structures for it are missing. It is impossible to address the issue of 'good government' without tackling the issue of relativism both in the sense of cultural expectations about performance as well as in the sense of distributional outcomes (arranged by class, ethnicity, gender, age, household composition etc.): good for whom in whose eyes? Probably, therefore, the question of what is 'good' cannot and should not be resolved in a universal sense, out of context: such an exercise is too normative and likely any way to be strongly ethnocentric since the relativism applies also to any author on the subject. What can be addressed is: how 'good in context' can be obtained. This is a question of the *process* of good government where the problem of accountability seems to return us inevitably to Hirschman's (1970) propositions concerning exit, voice and loyalty. The conclusion from that discussion seems to be that the process of accountability is contingent upon several factors: the monopoly character of the service or provision and the opportunities for non-degraded exit; the significance to the agency of the voice being exercised (the voice of some can be more discounted than that of others); the capacity of different groups in the society to exercise voice effectively, and over what aspects of agency performance (since governments have many products and many discrete sets of clients); the degree of exposure or responsiveness of governments to criticism in terms of legitimation (especially when it is explicitly reliant upon popular support); and the extent to which the creation of specialist client constituencies has ensured loyalty and dependency. As in committees, are we supposed to take silence for assent, indicative of satisfied customers? Apathy has long been regarded as an ingredient of democracy. Are good governments identified by the extent of voiced opposition and criticism, or by the absence of such voice? Is loyalty just co-option?

Paradox of Neo-Liberal Views on Good Governance

There are some paradoxes in the current pre-occupation with good government, where in a sense the thrust of policy is to undermine the monopoly of the state in service provision and the allocation of resources, thereby to create more opportunity for exit choices thus reducing, as it were, the necessity for governments to be good. More exit equals less voice and loyalty. The preoccupations with privatisation and markets on the one hand and good governance on the other do not easily sit side by side. Furthermore adherence to neo-liberal views about the efficacy and responsiveness of the market as an allocator of public goods crucially slides over the issue of responsibility. If the theme of good government means anything it has to concern rights. In the market one only has exchange entitlements.

Rights and responsibility go together. The state as guarantor and regulator implies a different view of responsibility to the state as implementer. When my local government sub-contracts the allocation of tennis courts to a private company to whom do I complain about the principles of allocation and to whom do I complain about the operation of those principles? One is never consulted about these principles in the first place, let alone the principle of sub-contracting which disperses both responsibility and the process of accountability.

Along with accountability, responsibility and rights, we then have to reflect upon 'participation'. There are further ironies here since those concerned now with 'good government' have not been so interested in the more established theme of participation. The preoccupation with good government has been more obviously political in the sense of international political actors from rich countries deploying the theme as a way of rationing aid and to achieve some policy consistency now that the CIS countries feature as recipients with voice. It also helps that certain hitherto well supported military regimes have now sufficiently collapsed to allow this semblance of consistency. One wonders whether we should regard 'human rights' as the penultimate ideology (i.e. before the end of..); as the refuge of tired development pundits and politicians exhausted by trying to make structural adjustment work according to neo-liberal predictions. Participation has attracted different supporters, having almost a bottom up career by contrast to the top down 'good government' agenda.

To a considerable extent participation did start with grassroots radicals preoccupied with class and gender struggles, dealing directly with the unsophisticated consciousness of poor, isolated, alienated people well aware of their own conditions but with limited empathy for others and a consequent inability to translate immediate senses of grievance into sustained political and armed action. Such radicals were informed by classic revolutionary theory which had brought them to an understanding that cadre leaders eventually betray their followers. For activists with a longer vision and a conscience, this problem had to be overcome: hence conscientisation leading eventually to participation.

So far so good, a decent career despite the contested nature of the concept especially when the moment of truth arrives and activists are expected by their rhetoric to relinquish their leadership roles in favour of the mobilised and animated. But then the trouble starts as mobilisation takes the form of projects, and as the official aid community adopts the dream formula for their own implementation. Participation enters the project lexicon and becomes rapidly subverted by tokenism, routinisation and overuse. Again, at this point in the concept's applied career, there is a fundamental assumption to be examined: does participation ensure improved project performance. In effect, is participation a precondition for good development; or is reference to it merely a sign of quality, a hallmark? And of course as it moved from means to ends status in the lexicon, so it finally merged into the broader more contemporary concerns with human rights and good government, and in the process was made technical and stripped of its erstwhile radical implications.

Fragmenting Responsibility for Delivery

The concern with 'good government' would seem then to be less a concern with improving the capacity of the state and more with restructuring it with a sharper distinction between the functions of defining, guaranteeing and regulating entitlements on the one hand, and delivery or implementation on the other. While the government remains responsible for the former as part of the maintenance of political and human rights, it is seen in most societies (especially poor ones) to reach quickly the limits of its own competence when it comes to the delivery of such rights/entitlements.

As noted in the previous section, this conclusion has led to a widening of the organisational landscape to legitimate the public activity of alternative agencies to government: the private sector (both banking on the one hand and profit making services on the other), NGOs, management companies, consultants, education and other service institutions. Such agencies are similar to each other in sharing contracts with the state, but they may differ in other important respects: for example, their degree of dependence upon 'government' work or the composition of their public/private portfolio; profit-taking philosophy; market versus service motives determining relations to clients; assets held or owned; tax liabilities; internal management styles; the existence of consumer bodies specific to the service.

However, this fragmentation of the delivery function entails a corresponding fragmentation of voice, with political parties and unions sidelined in the process, their respective voices denied primacy and legitimacy in the specific, sectoral contexts of service provision. In this way accountability is diluted at both ends: source and destination. The function of resource allocation and public expenditure is performed by disaggregated non-public, non-transparent agencies, insulated from a universal system of accountability. The state has discarded responsibility along with implementation by extending and diluting the definition of what constitutes the state.

The theme of 'Good government' therefore risks self-contradiction by diluting these dimensions of responsibility and accountability through the creation of the franchise state as a solution to delivery incompetence. It is as if the basis of citizenship has been systematically removed: the right to attribute performance to the state alongside the existence of formal mechanisms to bring that bureaucratic performance to account.

This suggests that the 'good government' formula is a trick that cannot work since it is based upon convincing us that this distinction between defining entitlements and implementing them is tenable, and that the former function is accountable through the disaggregated, quasi-state directed monitoring of the latter. This separation enables responsibility for policy to be side-stepped by focussing accountability upon the 'product' in action. In this way, the meaning of participation is restricted to the management of outcomes. Citizens become consumers (clients, passengers, even patients become customers) though often without meaningful access to a choice of suppliers. The loss of rights in the state is not adequately compensated for by acquiring them in the market.

The proposition, therefore, that 'good' can be achieved either through improving accountability, or through less government, or through greater reliance upon consumer rather than citizen models looks rather fragile when examined. Certainly these variants of the proposition amount to a substitution of the corporate state model by a franchise state model.

Franchise Model: Alternative to State and Market

In the West, while we continue to deal with the consequences of such policy (especially in health and education, but also in key public utilities such as transport), the philosophy itself is under attack. The market is increasingly revealed as an inefficient allocator of resources in the real as opposed to economic theorist's world, with the market both as a concept and as actual state-market relations being re-assessed. Interestingly, it has been the experience of East Asian NICs, so often held up by the monetarists as their theory in action, which has also prompted this re-assessment along with the evidence of crisis in the North Atlantic economies.

The problem for poor country economies, dependent on key international agencies such as the IMF and network of Development banks, is that while this re-assessment is underway in the North Atlantic economies (and then not widespread and not yet populist) they remain trapped within this monetarist perspective in terms of advice received and conditions set for financial aid. Of course the first point to establish is whether the critique of monetarism as applied in the North Atlantic economies can be transferred to a poor country context. Perhaps the big distinction here is that the poor country state has been palpably more incompetent and indeed corrupt (for structural rather than pathological reasons), so that even though the market may be shown as an inefficient and unfair allocator of resources because of its interlocked and non-transparent character, a return to any form of direct state option for the provision of services remains unacceptable for both theorists and the populace alike.

Under such conditions, the Franchise model has more going for it: the logical outcome of simultaneous critiques of state and market for their inability to achieve either productive efficiency or equity redistribution. However, it would be wrong to assume that somehow the application of this model simply arose out of theory, embodied in the practice of structural adjustment policy. Abstract expressions require concrete institutional forms and it would be a mistake to assume that the latter merely derive from the former.

We have to consider the significance of political process in producing institutional forms which might be adapted to, indeed captured by, the theory. Bangladesh has been a good example of this process. This is best illustrated by considering the early career of the NGO phenomenon: relief, political repression of political parties and unions, military regimes, advocacy, conscientisation, small-scale income generation, credit, pilot, innovation, experimentation, foreign sponsorship moving increasingly from the voluntary to the official sector, domestic government control over flows of funds and monitoring of objectives, organisational expansion, routinisation of programmes, contracts with government (sometimes as part of aid projects) both for training and project implementation in particular localities or among particular target-groups. Alongside this career has been the use of conditionality to break state monopolies in both service and goods delivery and to remove regulations and licensing to allow the market to breathe. But since the market cannot breathe normally (as demanded by theory), then NGOs in all their shapes and forms have been available to fill the institutional space arising from the critiques.

NGOs, Parastatals and Co-operatives

At this point it is important to reflect first upon the claims of candidates other than NGOs to occupy this institutional space: such as parastatals and cooperatives. Both sets of institutions have involved the notion of distance from the state, though to varying degrees. In this sense, the claim of the parastatals can be dismissed quite easily. Despite the principle of management autonomy, they have always operated within a framework of open-ended government support. The impossibility of bankruptcy and absence of personal managerial liability underpinned incompetence, inefficiency and non-accountability. In short, both in theory and practice they have been indistinguishable, in the eyes of critics, from the state. The pretence of separation from the state merely enabled responsibility to be avoided. Poor countries in the Indian sub-continent and Sub-Saharan Africa had this record in common, which constituted the empirical basis for structural adjustment initiatives.

Co-operative institutions, however, represent a more complicated story. Although the *KSS* and *BSS* system has been substantially discredited among observers concerned with the long term interests of the rural poor, it raises several different sets of issues: a process of descent from the high ground of quasi-socialist principles, through a brief ideological spell as banking groups towards the low ground of incorporation into the local patronage networks run by alliances of officials and patrons. Throughout this sad history has been the problem of organisational culture. At the same time, they have represented GOB's successive attempts to maintain a universal system of servicing farmers (initially) and the *bittaheen* (later, after much 'encouragement'!). The cooperative career has moved, as it were, from the quasi-state to the quasi-market. This has happened partly because the quasi-socialist principles could not be reconciled with the inability to insulate the institutions from the prevailing organisational culture of hierarchy, command, authority and patronage (and much more). The obvious question to ask at this point is whether this history for cooperatives will be repeated among the NGOs?

NGOs and Indigenous Organisational Culture

With such a strong role envisaged for NGOs, especially by the donors, in this 'franchise state' model, let us consider to what extent NGOs might conform to negative rather than positive features of the model. Although assumed to have comparative advantages over large-scale government organisations in social development, NGOs cannot easily insulate themselves from the prevailing organisational culture. This is especially the case as they become larger, more significant, receive larger funds and have to account for those funds through various review and reporting procedures.

Many countries in the poor world contain strong hierarchical and authoritarian social forms, often stemming from the authority structures from within peasant households and wider landlord-tenant estate structures. These are reproduced throughout the political economy in all forms of social interaction (employment, credit, exchange of services), sustained through poverty and pervasive inequality, and reinforced by deeply rooted religious philosophy and associated social practices (e.g. caste and gender differentiation). Within this hierarchical and authoritarian structure, relationships of dependence and deference are widespread. In the villages, these are often referred to as patron-client relations in which strong families dominate weaker ones through multi-stranded, paternalistic ties rather than single-stranded ties more associated with the idea of contracts and market transactions. These patron-client relations are transferred into official organisations, despite any appearance of rational, bureaucratic and objective practices.

Most NGOs in Asia (and Bangladesh is certainly no exception) are started by one or a few individuals who retain charismatic control over the organisation subsequently as it grows. As these organisations increasingly interact with other large organisations (i.e. government and donors), these contacts are monopolised by a narrow NGO leadership. Indeed, donors (whether official or NGO like OXFAM) are often very guilty of reinforcing the position of central leaders at the expense of other staff by insisting upon dealing with the 'executive director' only.

But these pressures are secondary explanations to the expectations which both staff and family will have of their leader. He/she (though mainly 'he') will have widespread obligations to staff which will spread beyond the workplace into their personal, domestic lives. Staff will be recruited using a leader's personal networks, sometimes including immediate and extended family, but certainly including friends as favours for friends. This is not just nepotism, since the principle of loyalty among new staff has to be ensured especially if the NGO operates in a hostile institutional environment. Promotions and advancement within the organisation will follow the same pattern, other things being equal. Many trivial decisions will be pushed up to the top for resolution, through fear of giving offence. These centralising tendencies can only be overcome by establishing more standardised and less arbitrary patterns of decision-making, thus encouraging a re-assertion of bureaucratic practices which nevertheless remain hierarchical for the resolution of non-routine issues. Participatory decision-making within such structures is virtually impossible, yet the same organisation may be promoting participatory styles of collective development activity among its target group clients. There is a contradiction here, where the means are not consistent with the ends, with social development objectives undermined as a result.

Perhaps one of the most obvious areas of contradiction concerns the work of large-scale NGOs operating in a variety of regional contexts, with different ecologies, resource endowments, educational levels, social structures, cultural practices and political options. In these circumstances, the agency should be promoting location-specific analysis and problem-solving between local staff and target groups. This in turn requires local staff with considerable autonomy from the centre to commit resources and innovate as appropriate to local conditions. Indeed a principle of good rural and social development practice is that hierarchical centre-field relations are inverted, with field officers using central officers as facilitators, advisers and consultants.

However, with the increasing preoccupation among donors and governments for accountability among NGOs for their use of funds, and especially with the increasing involvement of NGOs in credit programmes, the management and control functions of a large organisation overcome the participatory ones. Reporting, for example, becomes a reaffirmation of hierarchy rather than a sharing of information and experience. Staff cease to be generalists as sections and divisions are created to cope with the increasing specialisms and complexity of the programme. Under such conditions, vertical lines of communication replace horizontal ones with only central leaders having a sense of the whole picture, which is what they in turn report to outsiders.

While this kind of analysis is salutary, it actually arises from a critique of the proposition that NGOs represent an insertion between state-society by successfully brokering the respective cultures and combining the positive application of technique with an appreciation of people's wants and cognitive maps. That is to say, it notes the reversion of NGOs into powerful, outsider organisations—partly bureaucratic in form, but partly a reflection of indigenous authoritative culture expressed through formal organisations as well as within the societal institutions of kin and patron-clientage. NGOs may have prompted this kind of analysis simply because their own claims to a superior morality in working with the poor inevitably invites examination of their practice. Some observers might argue that the preceding account of NGO organisational culture amounts to a charge of corruption. By some standards of moral judgement that may be so, but the question really is whether such 'standards' are realistic in cultural context, and whether anyway the application of such standards gets in the way of understanding structural or institutional outcomes in the public resource allocation dimension of the development process. As noted at the beginning of this section, ethics have to be applied realistically to be ethical.

Conclusion: Strategic Dilemmas

The analysis offered in the preceding chapters of this volume brings us to a series of strategic dilemmas for poverty alleviation policy in Bangladesh. In various ways, the different chapters have been wrestling with the institutional triangle of state, market and community (or non-state). Community in this context refers not only to people's own spontaneous efforts based upon a range of social solidarities, but also what has become known as the NGO sector. Each element of this triangle has been shown to have weaknesses. There can be no realistic dogma which permits opting for one set of institutions (within the state, market or community) over the others.

When I first arrived in Bangladesh almost 20 years ago, fresh from the experience of revolutionary Marxian discourse in Bihar, there was a certain comfort in finding like-minded comrades, teasing out the agrarian class dimensions of this apparently homogenous deltaic region of Bengal and adding Bengal's revolutionary tradition to it. Such enthusiasm, and perhaps optimism, was driven by moral reactions to the experience of the 1974-5 floods and famine, the clear evidence of persistent and pervasive structural poverty in the countryside and the cities, and innocence about power.

This innocence had failed to take account of, *inter alia*, the alliance between domestic class forces and the growing donor presence in the society. It had also overestimated the prospects for establishing an autarchic economy. Bangladesh is not only an aberration in terms of historical Bengal, but greater Bengal as a whole cannot survive without interacting with other external economies. The room for manoeuvre is determined by export potential and terms of trade on export commodities. With population growth and other reasons to shed rural labour, together with weak effective domestic demand for consumption goods, the absorption of labour depends to a considerable extent upon a buoyant export sector. Among all the debates about appropriate rural development, it is difficult to escape the fundamental proposition that people need to work to survive. I hope this message shows through clearly in various chapters.

To some extent, therefore, my 20 years of association with Bangladesh have seen a steady loss of innocence, an erosion of dreams. Although I have been critical of the short time horizons of others, perhaps a desire to see positive change in one's own lifetime makes one guilty of a similar offence. Perhaps one should not compromise one's ideals in order to see some changes, however feeble, occur. At the same time, on my return to North Bihar in 90-91, I discovered that unreconstructed Marxism is alive and well, and still failing to engage with the reality of poor people's lives. Surely those actually living out poverty (in contrast to those of us merely observing it) have a right to expect some positive outcome in their lifetime, a right to die with some sense of optimism for their surviving children? The erosion of innocence may be a cause of personal regret for me as I come to terms with my Marxist past, but to cling to purist principles, to a vision of total revolution is a disaster for the poor. They may wait forever for the big heave. This is why I wanted to include the extract from *Breaking the Chains* as Chapter 19 in this volume. We have to celebrate the small, tangible victories.

Despite a general sense of regret that more has not been achieved, nevertheless I feel an enormous sense of respect when I witness the enthusiasm, the confidence, the optimism, the strength on display in the village meetings of the poor, and more so among women with their double chains. I am grateful to Proshika and GSS for the many occasions on which I have been such a witness. In such a situation, who am I to regret the passing of my revolutionary hopes, who am I to disparage a retreat into revisionism, who am I to dismiss the significance of small victories, small steps, scraps of additional income, a reduction rather than the abolition of dowry? However, if I am prepared to accept that there are no revolutionary panaceas, then I expect others to abandon their respective dogma too. We have to live with the proposition that none of the three elements of the triangle can deliver on their own, and that all will perform imperfectly in conjunction. We have to look for satisfying solutions, in the knowledge that those with power in any of these three arenas will not totally relinquish it and that no-one else in the society is sufficiently strong to take it all from them.

It is in the context of this personal voyage (my mother would call it 'growing up') that I offer the following working principles as a guide through the maze of poverty alleviation in Bangladesh. I cannot, at this point, resist recalling a comment from my daughter when she was much younger. Watching me packing for yet another trip to Bangladesh and resenting another period of absence, she observed that I must be very bad at my job since I had to keep going there to get it right!

Guiding Principles

- The significance of the market has to be acknowledged. The Marxian analysis of the market under general capitalist relations in which sellers of labour operate in buyers' markets has long dominated the discourse about how poverty is reproduced. At the same time, we have witnessed in many parts of the world the various experiments to suppress the workings of the market, and to create an alternative institutional culture of transactions and exchange instead, managed by sets of officials. It has not worked. The social relations of production and service provision entail exchange which cannot be policed at every moment of transaction. The market, however imperfect and for whatever reason, is the arena of people's natural, daily behaviour. That is to say, it is part of culture, not just a series of technical interactions, adjusting supply to demand via fluctuating prices. The mistake of anti-market dogma, based on the valid assumption that poor people cannot meet their needs in the marketplace, was the belief that people could be separated from this culture and operate in an officially managed substitute culture. Thus given the proven inability of the state to operate this substitute culture on behalf of the poor, we had better search for what the market can do for the poor, given appropriate support for them to compete in these markets.

- Other strategies are then only justified to the extent that the market does not adequately reflect and deliver poor people's needs, and that under some conditions effective demand has to be created or assisted, especially when the character of the market is interlocked against the interests of the poor (through being poor in human, social and cultural resources as well as material ones). Households need to be understood in terms of their capacity to perform in the different arenas of state, market and community. Rather than simply targeting people on the basis of material and human criteria alone (which prompts asset and wage generation responses often through narrowly defined projects), a broader consideration of social and cultural resources can lead to more integrated thinking, perhaps prompting programmatic responses which address wider issues of longer term capacity-building for households while not necessarily yielding immediate material outcomes. Thus functional literacy, education, primary health care and education, provision for extending legal rights and so on may be much more developmental than employment and income generation projects.

- These other strategies in effect add the dimension of rights to that of exchange entitlements (citizenship and poverty cannot be separated), and establish that 'rights' are fundamentally a political phenomenon to be legitimated and guaranteed at the level of the nation-state, not handed over to the voluntarism of agencies which are not charged with this ultimate responsibility.

- These other strategies present the dilemma of choosing between the state or non-state as implementors of policy in two senses: how are rights to be guaranteed and performance held to account; and does the choice of a non-state 'franchise' option entail the self-fulfilling prophecy of state incompetence through the state being by-passed? With this second question, we have to recognise that the key actual and potential political actors in Bangladesh are constrained in the degree of choice they can exercise over these issues. Let us assume a desire on their part to be more democratic and accountable to a socially wider class of constituents. Nevertheless, they cannot easily translate this into performance if 'structural adjustment' and 'good governance' conditionality is urging them to divest themselves of responsibility through a 'franchise' strategy.

- There are further issues about the franchise model. Can we envisage the non-state, 'franchise model' as a universal strategy, covering more than a minority fraction of the population? Without some kind of coordinated universality the model cannot be a substitute for citizenship as a route to the removal of poverty. In this context does the 'franchise model' actually harm the majority of the poor in the long run development sense by concentrating achievement in the minority, non-state sector; or does it contribute to the improvement of parallel state performance by offering both experiment and competition? If the latter, then we are back to the triangle, with the state learning from the experience of its innovative contractors and replicating best practice on a wider scale. In this volume, some 'partnership' projects have been analysed in various sectors, but their enclave character has been criticised. An exception has been the development of the LGEB through the SIDA-funded infrastructure project. Overall, there is a huge information loss to the system by failing purposively to consolidate experience or to set up deliberate experimentation within mainline ministries.

- In the context of this information loss, to what extent can we expect that the new political conditions of 'democracy' in Bangladesh will ensure that this learning takes place and that citizenship becomes sufficiently effective to demand widespread service (whether direct or franchise) as of right rather than through the voluntaristic chance of being selected as a target by an NGO or a technical-assistance dominated project? Anxiety about a successful process of learning transfer has prompted the creation of an Institute for Development Policy Analysis and Advocacy in Proshika. It has also prompted the creation of a rural management institute within BRAC, and ideas from GSS about a centre for advocacy. These NGOs have essentially lost faith in the conventional academic institutions and private consultancy agencies (there is some overlap) to perform this task of influencing national policy. These conventional institutions have tended to become incorporated into GOB and major donor agendas (witness the FAP), and have lost a critical edge. Furthermore they have little grassroots experience where the real innovation is occurring, as this volume shows, whether in social development, institutional sustainability or environmental management.

- However, there are dangers for NGOs in this process. Can they bear the burden of expectation placed upon them by donors (and the reluctant state) within a franchise model without becoming absorbed into the prevailing organisational culture from which they sought initially to escape? Chapter 22 reveals some scepticism on this. Some observers are attributing the possibility of NGOs, as delivery and mobilisation agencies, losing relative advantage over government to the problem of 'scaling-up'. My view is that NGOs have never been able to separate themselves easily from the prevailing culture, and that rhetoric has always belied practice to some extent. This is not to be judgemental. I do not think any less of NGOs for operating within their own culture, I just want us to be realistic about their room for manoeuvre, to have reasonable not unreasonable expectations. If the culture is dysfunctional to achieving social objectives, then NGOs as well as governments will have to continue to search for ways to reconcile practice with objectives.

- In this sense, can best development practice be assisted within state and non-state through support for a more deliberate and sustained challenge to the negative features of organisational culture, achieved not via training but through guided and supervised programmes of personal action-research in which hierarchical inversions are tested, underpinned by programmes of support for basic research within Bangladesh on these issues? Are such challenges, which amount to an agenda of administrative reform requiring much dialogue with sympathetic allies within GOB and NGOs, as well as coordination with other donors, best achieved by using experience from enclaved, experimental projects like

PEP, through demonstration and learning, or through strategic programme support (e.g. the Rural Development Division on the one hand, and ADAB on the other)? Increasingly I favour the latter.

- Connected to this line of thought about organisational cultures and administrative reform within the state and non-state is the rather blunt question: do projects with target populations address long term development goals for the society (i.e. fundamental changes in economic, social and cultural behaviour) or merely provide welfare for the selected target for the duration of the project's life? This question should be asked since the criterion for donors' involvement in development projects (in contrast to emergency relief) must surely be the internalisation of new behavioural patterns in the society which are regarded as more conducive to the long term, sustainable alleviation of poverty. This question applies to donors and government alike, since they have joint responsibility for defining the ultimate purpose of aid.
- A further dimension to this question concerns the portfolio character of people's lives, which means that great effort in one sector of a target's life may make little difference across their portfolio of activities, and indeed may be offset by them. In this sense, selecting crucial sectors of target's life is as important as selecting the target itself. The implication of this must be to apply a deeper understanding of poor people's resource profiles (material, human, social and cultural) and to use this understanding as the basis for strategic intervention to improve people's capacities to function effectively within the three arenas of state, market, and community. Some actors within GOB and among donors and NGOs have been very astute in identifying key activities in poor people's portfolios, the improvement of which would affect the quality of life across the whole of their activities. Rural works has been one example, but also irrigation, homestead gardening, functional literacy, legal rights, abolition of dowry and so on. 'Spotting the key activity' should be extended more purposively to agriculture (in both high-tech and organic paradigms) to reflect the agrarian analysis made in Part II of this volume.
- Such resource profiles have to be understood in the context of a basic appreciation of change, and the determinants of it, in Bangladesh society: agrarian reformation (sic); processes of urbanisation; employment opportunities within the domestic and international economy; evolving natural resource endowments (given various processes of environmental deterioration); and the range of available technological responses. People only have resources in context.

Conclusion: Strategic Dilemmas

The analysis offered in the preceding chapters of this volume brings us to a series of strategic dilemmas for poverty alleviation policy in Bangladesh. In various ways, the different chapters have been wrestling with the institutional triangle of state, market and community (or non-state). Community in this context refers not only to people's own spontaneous efforts based upon a range of social solidarities, but also what has become known as the NGO sector. Each element of this triangle has been shown to have weaknesses. There can be no realistic dogma which permits opting for one set of institutions (within the state, market or community) over the others.

When I first arrived in Bangladesh almost 20 years ago, fresh from the experience of revolutionary Marxian discourse in Bihar, there was a certain comfort in finding like-minded comrades, teasing out the agrarian class dimensions of this apparently homogenous deltaic region of Bengal and adding Bengal's revolutionary tradition to it. Such enthusiasm, and perhaps optimism, was driven by moral reactions to the experience of the 1974-5 floods and famine, the clear evidence of persistent and pervasive structural poverty in the countryside and the cities, and innocence about power.

This innocence had failed to take account of, *inter alia*, the alliance between domestic class forces and the growing donor presence in the society. It had also overestimated the prospects for establishing an autarchic economy. Bangladesh is not only an aberration in terms of historical Bengal, but greater Bengal as a whole cannot survive without interacting with other external economies. The room for manoeuvre is determined by export potential and terms of trade on export commodities. With population growth and other reasons to shed rural labour, together with weak effective domestic demand for consumption goods, the absorption of labour depends to a considerable extent upon a buoyant export sector. Among all the debates about appropriate rural development, it is difficult to escape the fundamental proposition that people need to work to survive. I hope this message shows through clearly in various chapters.

To some extent, therefore, my 20 years of association with Bangladesh have seen a steady loss of innocence, an erosion of dreams. Although I have been critical of the short time horizons of others, perhaps a desire to see positive change in one's own lifetime makes one guilty of a similar offence. Perhaps one should not compromise one's ideals in order to see some changes, however feeble, occur. At the same time, on my return to North Bihar in 90-91, I discovered that unreconstructed Marxism is alive and well, and still failing to engage with the reality of poor people's lives. Surely those actually living out poverty (in contrast to those of us merely observing it) have a right to expect some positive outcome in their lifetime, a right to die with some sense of optimism for their surviving children? The erosion of innocence may be a cause of personal regret for me as I come to terms with my Marxist past, but to cling to purist principles, to a vision of total revolution is a disaster for the poor. They may wait forever for the big heave. This is why I wanted to include the extract from *Breaking the Chains* as Chapter 19 in this volume. We have to celebrate the small, tangible victories.

Despite a general sense of regret that more has not been achieved, nevertheless I feel an enormous sense of respect when I witness the enthusiasm, the confidence, the optimism, the strength on display in the village meetings of the poor, and more so among women with their double chains. I am grateful to Proshika and GSS for the many occasions on which I have been such a witness. In such a situation, who am I to regret the passing of my revolutionary hopes, who am I to disparage a retreat into revisionism, who am I to dismiss the significance of small victories, small steps, scraps of additional income, a reduction rather than the abolition of dowry? However, if I am prepared to accept that there are no revolutionary panaceas, then I expect others to abandon their respective dogma too. We have to live with the proposition that none of the three elements of the triangle can deliver on their own, and that all will perform imperfectly in conjunction. We have to look for satisfying solutions, in the knowledge that those with power in any of these three arenas will not totally relinquish it and that no-one else in the society is sufficiently strong to take it all from them.

It is in the context of this personal voyage (my mother would call it 'growing up') that I offer the following working principles as a guide through the maze of poverty alleviation in Bangladesh. I cannot, at this point, resist recalling a comment from my daughter when she was much younger. Watching me packing for yet another trip to Bangladesh and resenting another period of absence, she observed that I must be very bad at my job since I had to keep going there to get it right!

Guiding Principles

- The significance of the market has to be acknowledged. The Marxian analysis of the market under general capitalist relations in which sellers of labour operate in buyers' markets has long dominated the discourse about how poverty is reproduced. At the same time, we have witnessed in many parts of the world the various experiments to suppress the workings of the market, and to create an alternative institutional culture of transactions and exchange instead, managed by sets of officials. It has not worked. The social relations of production and service provision entail exchange which cannot be policed at every moment of transaction. The market, however imperfect and for whatever reason, is the arena of people's natural, daily behaviour. That is to say, it is part of culture, not just a series of technical interactions, adjusting supply to demand via fluctuating prices. The mistake of anti-market dogma, based on the valid assumption that poor people cannot meet their needs in the marketplace, was the belief that people could be separated from this culture and operate in an officially managed substitute culture. Thus given the proven inability of the state to operate this substitute culture on behalf of the poor, we had better search for what the market can do for the poor, given appropriate support for them to compete in these markets.

- Other strategies are then only justified to the extent that the market does not adequately reflect and deliver poor people's needs, and that under some conditions effective demand has to be created or assisted, especially when the character of the market is interlocked against the interests of the poor (through being poor in human, social and cultural resources as well as material ones). Households need to be understood in terms of their capacity to perform in the different arenas of state, market and community. Rather than simply targeting people on the basis of material and human criteria alone (which prompts asset and wage generation responses often through narrowly defined projects), a broader consideration of social and cultural resources can lead to more integrated thinking, perhaps prompting programmatic responses which address wider issues of longer term capacity-building for households while not necessarily yielding immediate material outcomes. Thus functional literacy, education, primary health care and education, provision for extending legal rights and so on may be much more developmental than employment and income generation projects.

- These other strategies in effect add the dimension of rights to that of exchange entitlements (citizenship and poverty cannot be separated), and establish that 'rights' are fundamentally a political phenomenon to be legitimated and guaranteed at the level of the nation-state, not handed over to the voluntarism of agencies which are not charged with this ultimate responsibility.

- These other strategies present the dilemma of choosing between the state or non-state as implementors of policy in two senses: how are rights to be guaranteed and performance held to account; and does the choice of a non-state 'franchise' option entail the self-fulfilling prophecy of state incompetence through the state being by-passed? With this second question, we have to recognise that the key actual and potential political actors in Bangladesh are constrained in the degree of choice they can exercise over these issues. Let us assume a desire on their part to be more democratic and accountable to a socially wider class of constituents. Nevertheless, they cannot easily translate this into performance if 'structural adjustment' and 'good governance' conditionality is urging them to divest themselves of responsibility through a 'franchise' strategy.

- There are further issues about the franchise model. Can we envisage the non-state, 'franchise model' as a universal strategy, covering more than a minority fraction of the population? Without some kind of coordinated universality the model cannot be a substitute for citizenship as a route to the removal of poverty. In this context does the 'franchise model' actually harm the majority of the poor in the long run development sense by concentrating achievement in the minority, non-state sector; or does it contribute to the improvement of parallel state performance by offering both experiment and competition? If the latter, then we are back to the triangle, with the state learning from the experience of its innovative contractors and replicating best practice on a wider scale. In this volume, some 'partnership' projects have been analysed in various sectors, but their enclave character has been criticised. An exception has been the development of the LGEB through the SIDA-funded infrastructure project. Overall, there is a huge information loss to the system by failing purposively to consolidate experience or to set up deliberate experimentation within mainline ministries.

- In the context of this information loss, to what extent can we expect that the new political conditions of 'democracy' in Bangladesh will ensure that this learning takes place and that citizenship becomes sufficiently effective to demand widespread service (whether direct or franchise) as of right rather than through the voluntaristic chance of being selected as a target by an NGO or a technical-assistance dominated project? Anxiety about a successful process of learning transfer has prompted the creation of an Institute for Development Policy Analysis and Advocacy in Proshika. It has also prompted the creation of a rural management institute within BRAC, and ideas from GSS about a centre for advocacy. These NGOs have essentially lost faith in the conventional academic institutions and private consultancy agencies (there is some overlap) to perform this task of influencing national policy. These conventional institutions have tended to become incorporated into GOB and major donor agendas (witness the FAP), and have lost a critical edge. Furthermore they have little grassroots experience where the real innovation is occurring, as this volume shows, whether in social development, institutional sustainability or environmental management.

- However, there are dangers for NGOs in this process. Can they bear the burden of expectation placed upon them by donors (and the reluctant state) within a franchise model without becoming absorbed into the prevailing organisational culture from which they sought initially to escape? Chapter 22 reveals some scepticism on this. Some observers are attributing the possibility of NGOs, as delivery and mobilisation agencies, losing relative advantage over government to the problem of 'scaling-up'. My view is that NGOs have never been able to separate themselves easily from the prevailing culture, and that rhetoric has always belied practice to some extent. This is not to be judgemental. I do not think any less of NGOs for operating within their own culture, I just want us to be realistic about their room for manoeuvre, to have reasonable not unreasonable expectations. If the culture is dysfunctional to achieving social objectives, then NGOs as well as governments will have to continue to search for ways to reconcile practice with objectives.

- In this sense, can best development practice be assisted within state and non-state through support for a more deliberate and sustained challenge to the negative features of organisational culture, achieved not via training but through guided and supervised programmes of personal action-research in which hierarchical inversions are tested, underpinned by programmes of support for basic research within Bangladesh on these issues? Are such challenges, which amount to an agenda of administrative reform requiring much dialogue with sympathetic allies within GOB and NGOs, as well as coordination with other donors, best achieved by using experience from enclaved, experimental projects like PEP, through demonstration and learning, or through strategic programme support (e.g. the Rural Development Division on the one hand, and ADAB on the other)? Increasingly I favour the latter.

- Connected to this line of thought about organisational cultures and administrative reform within the state and non-state is the rather blunt question: do projects with target populations address long term development goals for the society (i.e. fundamental changes in economic, social and cultural behaviour) or merely provide welfare for the selected target for the duration of the project's life? This question should be asked since the criterion for donors' involvement in development projects (in contrast to emergency relief) must surely be the internalisation of new behavioural patterns in the society which are regarded as more conducive to the long term, sustainable alleviation of poverty. This question applies to donors and government alike, since they have joint responsibility for defining the ultimate purpose of aid.

- A further dimension to this question concerns the portfolio character of people's lives, which means that great effort in one sector of a target's life may make little difference across their portfolio of activities, and indeed may be offset by them. In this sense, selecting crucial sectors of target's life is as important as selecting the target itself. The implication of this must be to apply a deeper understanding of poor people's resource profiles (material, human, social and cultural) and to use this understanding as the basis for strategic intervention to improve people's capacities to function effectively within the three arenas of state, market, and community. Some actors within GOB and among donors and NGOs have been very astute in identifying key activities in poor people's portfolios, the improvement of which would affect the quality of life across the whole of their activities. Rural works has been one example, but also irrigation, homestead gardening, functional literacy, legal rights, abolition of dowry and so on. 'Spotting the key activity' should be extended more purposively to agriculture (in both high-tech and organic paradigms) to reflect the agrarian analysis made in Part II of this volume.

- Such resource profiles have to be understood in the context of a basic appreciation of change, and the determinants of it, in Bangladesh society: agrarian reformation (sic); processes of urbanisation; employment opportunities within the domestic and international economy; evolving natural resource endowments (given various processes of environmental deterioration); and the range of available technological responses. People only have resources in context.

Bibliography

Abdullah, A. and R. Nations 1974 *Agrarian Development and IRDP in Bangladesh*, BIDS May 1974, *mimeo*

ADAB 1984 NGOs in Bangladesh—*A Map Association of Development Agencies in Bangladesh*, Dhaka

Adnan, S. & H.Z. Rahman 1978 *Peasant classes and mobility, structural reproduction and change in rural Bangladesh* (Working Paper 9) Dhaka: Village Study Group, Univ. of Dhaka & BIDS

Adnan, S. & R. Islam 1975 *Social Structure and resource allocation in a Chittagong village* (Working Paper 3) Dhaka: Village Study Group, Univ. of Dhaka & BIDS

Adnan, S. & Z. Ahmed 1975 *Integrated approach to village studies* (Working Paper 1) Dhaka: Village Study Group, Univ. of Dhaka & BIDS

Adnan, S. 1976 *Land, power and violence in Barisal villages* (Working Paper 6) Dhaka: Village Study Group, Univ. of Dhaka & BIDS

Adnan, S. 1984 *Peasant Production and Capitalist Development: A Model with Reference to Bangladesh*, Cambridge: Unpublished Ph.D.

Adnan, S. et al. 1974 *The preliminary findings of a social and economic study of four Bangladesh villages* Dhaka: Village Study Group, Univ. of Dhaka & BIDS

Adnan, S. et al. 1976 *An approach to study of village history in Bangladesh* (Working Paper 5) Dhaka: Village Study Group, Univ. of Dhaka & BIDS

Adnan, S. et al. 1977 *Differentiation and other class structure in village Shamraj* (Working Paper 8) Dhaka: Village Study Group, Univ. of Dhaka & BIDS

Adnan, S. et al. 1978 "A Review of Landlessness in Rural Bengal: 1877-1977" Univ. of Chittagong Workshop on the Research Project "Socio-Economics of Landlessness in Bangladesh", Aug. 21

Adnan, S. et al. 1992 *People's Participation NGOs and the Flood Action Plan: An Independent Review*, Dhaka, Research and Advisory Services

Agro-Economic Research, MOA 1979 *Studies on the Small Farmers Development Activities of the Non-Governmental Organisation*

Ahmed, F. 1973 "The Structural Matrix of the Struggle in Bangladesh" K. Gough & H. Sharma (Eds) *Imperialism and Revolution in South Asia*, New York: Monthly Review Press

Alam M.M. 1976 *Leadership Pattern, Problems and Prospects of Local Governments in Bangladesh*, Comilla BARD

Alamgir, M. 1974 "Some analysis of distribution of income, consumption, savings and poverty in Bangladesh" *Bangladesh Development Studies* II (4) pp 737-818

Alamgir, M. 1977 *The experience of Rural Works Programme (RWP) in Bangladesh* UN ESCAP Bangkok

Alavi, H. 1972 "The State in Post-Colonial Societies: Pakistan and Bangladesh", *New Left Review* No. 74 July-August

Alavi, H. 1973 "Peasant Classes and Primordial Loyalties" *Journal of Peasant Studies* 1 (1), October

Alavi, H. 1975 "India and the Colonial Mode of Production" *Socialist Register*, Merlin Press

Alavi, H. 1982 "Structure of Peripheral Capitalism" in T. Shanin and H. Alavi (Eds) *Introduction to Sociology of Developing Societies* Macmillan

Ali, T. 1975 "Pakistan and Bangladesh: Results and Prospects" R. Blackburn (Ed) *Explosion in a Sub-Continent* Pelican

Almond, G.A. & J. Coleman (Eds) 1960 *The Politics of Developing Areas* Princeton U.P.

Arens, J. & J. Van Beurden 1977 *Jhagrapur: Poor Peasants and Women in a Village in Bangladesh* Third World Publications

Arn, A-L. 1978 "A Case-Study from Bangladesh: An Integrated Village in Mymensingh District" *Centre for Development Research*, Copenhagen, Dec.

Arthur, W.B. & G. McNicholl 1978 "An Analytical Survey of Population and Development in Bangladesh" *Population and Development Review* 4 (1), March pp 23-80

Ayub, Md. 1971 "From Martial Law to Bangladesh" Pran Chopra (Ed.) *The Challenge of Bangladesh*, Popular Prakashan, Bombay, pp 40-59

BADC 1977 *Proceedings of National Workshop on Fertiliser Marketing at Small Farmer Level*

Bailey F.G. 1966 "The Peasant View of the Bad Life" reprinted in T. Shanin (Ed) *Peasants and Peasant Societies* Penguin 1971

Bailey F.G. 1968 *Stratagems and Spoils* OUP

Bailey, F.G. 1959 *Politics and Social Change: Orissa* University of California Press, Berkeley

Bailey, P.G. 1963 "Closed Social Stratification in India" *Archives of European Sociology*, 4, pp 107-124

Banerji, J. 1972 "For a Theory of Colonial Modes of Production" *Economic and Political Weekly* VIII 13,23/3

Banfield, E.C. 1958 *The Moral Basis of a Backward Society* Free Press

BARC 1978a "A Select Bibliography on Agricultural Economics and Rural Development with Special Reference to Bangladesh: Supplement and Author Index" E.J. & M.N. Clay (Eds) *BARC* No 6

BARC 1978b "Incidence of Landless and Major Landholding Cultivation Groups in Rural Bangladesh" *Agricultural Economics and Rural Social Science Papers BARC* No 5, May 1978

BARD 1978 *Small Farmers and Landless Labourers* Development Project, Workshop Proceedings, Comilla, May 1978

Barrington-Moore, J. 1969 *The Social Origins of Dictatorship and Democracy*, Peregrine

Bell, F.O. 1942 *Final report on the Survey and Settlement Operations in the District of Dinajpur*, 1934-1940 Allipore: Bengal Government Press

Bertocci, P.J. 1970 "Elusive Village: Social Structure and Community Organization in Rural East Pakistan", Unpub. Ph.D. thesis, East Lansing, Michigan State University

Bertocci, P.J. 1972 "Community Structure and Social Rank in Two Villages in Bangladesh" *Contributions to Indian Sociology*, New Series VI, December 1972, pp 28-52

Bertocci, P.J. 1979 "Structural Fragmentation and Peasant Classes in Bangladesh" *The Journal of Social Studies* 5, October 1979

Bhaduri, A. 1973 "A Study of Agricultural Backwardness under Semi-Feudalism" *Economic Journal* 83 (1)

Bhuiyan, S.I., Groundwater use for irrigation in Bangladesh: the prospect and some emerging issues, *Agricultural Administration*, 16 (4) (1984) pp. 181-207.

Blaikie, P.M., J. Cameron & J.D. Seddon (1979) *Peasants and Workers in Nepal*, Warminster, Aris and Phillips

Blair, H. 1974 "The Elusiveness of Equity: Institutional Approaches to Rural Development in Bangladesh" *Rural Development Committee*, Centre for International Studies

Blair, H. 1978 "Rural Development, Class Structure and Bureaucracy in Bangladesh" *World Development* 6 (1) Jan. pp 65-82

Bose, S. 1973 "The Strategy of Agricultural Development in Bangladesh" *Development Dialogue* 1

Boyce, J.K. 1987a *Agrarian Impasse in Bengal: Institutional Constraints to Technological Change* Oxford University Press

Boyce, K. 1987b "Trends and Projections for Bangladeshi Food Production" *Food Policy* November.

Brenner, R. 1977 "The Origins of Capitalist Development: A Critique of Neo-Smithian Marxism" *New Left Review* 104 July-August pp 24-92

Bryant, C. 1980 "Organisational Impediments to making participation a reality: 'Swimming Upstream' in AID Rural Development" *Participation Review* 1 (3) pp 8-10

Bryant, C. and L.G. White 1982 *Managing Development in the Third World*, Boulder Westview

CDS-Bath 1992 *Going It Alone: Rural Female-Headed Households and their Dependents in Bangladesh* Report to UK-ODA, Centre for Development Studies, University of Bath

Chambers, R. 1983 *Rural Development: Putting the Last First*, Longman

Chashi, M.A. 1976 "Self-Reliant Rural Bangladesh", Vol. 2, *Journal of Bangladesh Economic Association*, Dhaka

Chattopadhyay, P. 1972 "Mode of Production in Indian Agriculture: an Anti-Knitik" *Economic and Political Weekly* VII, Dec.

Chawdhuri, S.D. and D. Gisselquist 1984 *Command Area Development for Irrigation: Integration of Organisational, Economic, Agronomic, Engineering and Institutional Components*, Dhaka, Bangladesh Agricultural Research Council, Dec. 1984

Chayanov, A.V. 1966 *The Theory of the Peasant Economy* (Translated & edited by B. Thorner *et al.*), Irwin

Clay, E. and B.B. Schaffer Eds 1984 *Room for Manoeuvre: An Exploration of Public Policy in Agriculture and Rural Development*, Heinemann

Clay, E.J. & M.S. Khan 1977 "Agricultural Employment and Under-employment in Bangladesh: the Next Decade", *Agricultural Economics and Social Science Paper*, BARC No 4, Dhaka 1977

Clay, E.J. 1978 "Employment Effects of the HYV Strategy in Bangladesh: a rejoinder" *A.D.C.* Dhaka, *mimeo* de Vylder, S. & D. Asplund 1979 *Contradictions and Distortions in a Rural Economy: The Case of Bangladesh* SIDA, May

Edward, C. et al. 1978 "Irrigation in Bangladesh: On contradictions and under-utilization" *Dev. Studies Discussion Paper No 22*, Feb., UEA, Norwich

Faidley, L & M.L. Esmay 1976 "Introduction and use of improved rice varieties: who benefits?" Stevens, Bertocci and Alavi (Eds) *Rural Development in Bangladesh and Pakistan*

Feldman, S., F. Akter and L. Banu 1981 "The IRDP Women's Programme in Population Planning and Rural Women's Cooperative" *Women for Women Research Seminar*, Dhaka University

Foster, G.M. 1967 "Peasant Society and the Image of the Limited Good" Potter, Diaz and Foster (Eds) *Peasant Society* Little Brown

Freire, P. 1970: *Pedagogy of the Oppressed*, New York: Herder and Herder

Ghose, A.K. 1979 "Institutional Structure, Technological Change and Growth in Poor Agrarian Economies: An Analysis with Reference to Bengal and Punjab" *World Development* 7 (4/5), April, pp 385-396

Gill, J. 1981 *Farm Power in Bangladesh*, monograph, University of Reading

Gisselquist, D. 1979 "Land Reform, Productivity and Equity" *Chittagong University Economics Seminar*, April 1979

Glaser, M. 1989 *Water to the Swamp: Patterns of Accumulation from Irrigation in Rajshahi villages* University of Bath PhD thesis

GOB 1960 *Agricultural Census of Pakistan* Vol. 1 & 2, Table 3, p 29

GOB 1973 *The First Five Year Plan* Dhaka: Planning Commission

GOB 1976 *Bangladesh Areas Liable to Famine* Ministry of Relief and Rehabilitation

GOB and World Bank, *Bangladesh: Minor Irrigation Sector*, Dhaka, April 1982.

GOB/US AID 1977 *Land Occupancy Survey* Bureau of Statistics, Government of Bangladesh

Gould, D.J. and J.A. Amaro-Reyes 1983 *The Effects of Corruption on Administrative Performance: Illustrations from Developing Countries*, World Bank Staff Working Papers No. 580 Washington

Government of Bengal 1940 *Report of the Land Revenue Commission* (6 vols.) Allipore: Bengal Government Press

Gow, D.D. and J. Vansant 1983 "Beyond the Rhetoric of Rural Development Participation: How can it be Done?" *World Development* 11 (5), pp 427-446

GPRB 1977 *Land Occupancy Survey* Bureau of Statistics & US AID

GPRB 1978a *Summary of the Land Occupancy Survey of 1978* Bureau of Statistics

GPRB 1978b *The Two Year Plan 1978-1980*

GPRB/FAO 1978 "Agrarian Structure and Change: Rural Development Experience and Policies in Bangladesh" *Min. of Agriculture & Forests* May 1978 (Background papers to Bangladesh Country Report for World Conference in Agrarian Reform and Rural Development, Rome 1979)

Granovetter, M. 1985 "Economic Action and Social Structure: The Problem of Embeddedness" *American Journal of Sociology* 91 (3) pp 481-510

Greely, M. 1979 *Rural Technology, Rural Institutions and the Rural Poorest*, mimeo, Dhaka, Dec. 1979

Griffin, K. 1979 "Growth and Impoverishment in the Rural Areas of Asia" *World Development* 7 (4/5) pp 351-83

Hart, H. 1971 "The Village and Development Administration" J.J. Heaphey (Ed.) *Spatial Dimensions of Development Administration* Duke University Press, Durham, N.C.

Hartmann, B. and J. Boyce 1983 *A Quiet Violence: View from a Bangladesh Village* London Zed Press

Hilton, R. 1976 *The Transition from Feudalism to Capitalism* London: New Left Books

Hirsch, J 1978 "The State Apparatus and Social Reproduction: Elements of a Theory of the Bourgeois State" Holloway & Picciotto (Eds) *State and Capital* Edward Arnold

Hirschman, A.O. 1970 *Exit, Voice and Loyalty* Harvard University Press, Cambridge Mass.

Hossain, M. 1980 "Desirability and Feasibility of Land Reform in Bangladesh" *Journal of Social Studies* (Dhaka) 8, pp 70-93

Hossein, M. 1987 *Green Revolution in Bangladesh: Its Nature and Impact on Income Distribution*, BIDS, Dhaka Oct.

Hussein, M. 1979 "The Nature of State Power in Bangladesh" *The Journal of Social Studies*, No 5.

IBRD 1979 *Bangladesh: food policy issues*, International Bank for Reconstruction and Development, December

ILO 1977 *Rural Poverty and Landlessness in Asia*, International Labour Office, Geneva

Institute of Nutrition 1977a "Nutrition Survey of Bangladesh" *Institute of Nutrition and Food Science*, Dhaka University

Institute of Nutrition 1977b "Economic and Nutritional Effects of Bangladesh Food for Relief Work Program" *Institute of Nutrition and Food Science*, Dhaka University

Islam, N. 1978 *Development Strategy of Bangladesh* Pergamon Press, Oxford

Ito, S. 1990 *The Female Heads of Household in Rural Bangladesh* Japan Overseas Cooperation Volunteers/ BARD, Comilla

Jabbar, M.A. 1977 "Farm Structure and Resource Productivity in Selected Areas of Bangladesh" *BARC* No. 3, December

Jahan, R. 1973 *Pakistan: Failure in National Integration* New York: Columbia University Press

Jahangir B.K. 1979 *Differentiation, Polarisation and Confrontation in Rural Bangladesh*, Dhaka: Centre for Social Studies

Kerblay, B. 1971 in Shanin, T. (Ed.) *Peasants and Peasant Societies*, Penguin

Khan, A.F. 1977 "Poverty and Inequality in Bangladesh" *Poverty and Landlessness in rural Asia* ILO Geneva

Khan, A.H. 1971 *Tour of Twenty Thanas* BARD Comilla

Khan, A.H. 1979 "Framework for Rural Development in Bangladesh and a Plan for 250 Co-operatively Organised Thanas" *BARD* Bogra

Kieffer, C.H. 1984 Citizen Empowerment: A Development Perspective

Korten D.C. 1983 "Social Development: Putting People First" in Korten and F.B. Alfonso (Eds) *Bureaucracy and the Poor: Closing the Gap* Kumarian Press, West Hartford

Kramsjo, B. & G.D. Wood 1992 *Breaking the Chains* IT Publications

Krueger A.O. 1974 "The Political Economy of the Rent-Seeking Society" *The American Economic Review* 64 (3) (June) pp 291-303

Kumari, R. 1989 *Women-Headed Households in Rural India* London: Sangham Books

Laclau, E. 1971 "Feudalism and Capitalism in Latin America" New Left Review 67, May-June

Lamb G.B. 1975 "Marxism, Access and the State" *Development and Change* VI (2) (April) pp 119-35

Lewis, D.J. 1991 *Technologies and Transactions: A Study of the Interaction of 'Green Revolution' and Agrarian Structure in Bangladesh*, Centre for Social Studies, Dhaka

Lifschultz, L. 1979 *Bangladesh: The Unfinished Revolution* Zed Press

Maniruzzaman 1975 ?(Ch.2)

Marsden D. and P. Oakley (Eds.) 1990 *Evaluating Social Development Projects* OXFAM

Marx, K. 1964 *Pre-Capitalist Economic Formations* Lawrence and Wishart

Marx, K. 1970 *Capital*, Lawrence and Wishart

McGregor, J.A. et al. 1992 *PEP Evaluation Studies* SIDA: Dhaka

McGregor, J.A. 1989 "Towards a better understanding of Credit in Rural Development. The case of Bangladesh: The Patron State" *Journal of International Development* 1 (4) (October) pp 467-486

McGregor, J.A. 1992 *Credit and Development in Rural Bangladesh*, Bath, Ph.D.

Messaros, I. 1972 "Contingent and Necessary Class Consciousness" Messaros (Ed) *Aspects of History and Class Consciousness* New York, Herder and Herder, pp 85-127

Myrdal G. 1970 *The Challenge of World Poverty*, Penguin

Nations, R. 1971 "Pakistan: Class and Colony" *New Left Review* 68, July-August, pp 3-26

Nicholas, R. 1963 "Village Factions and Political Parties in Rural West Bengal" *Journal of Commonwealth Studies*, Vol. 2

Nicholas, R. 1973 "Social Service Research in Bangladesh", *mimeo, The Ford Foundation*, Dhaka Bangladesh

Patnaik, U. 1972 "On the Mode of Production in Indian Agriculture: A Reply" *Economic and Political Weekly* VII (40), Aug. 30

Platteau, J. 1991 "The Free Market is not readily transferable: Reflections on the links between Market, Social Relations and Moral Norms" Paper presented to *25th Jubilee of IDS, Sussex*.

Polanyi, K. 1977 *The Livelihood of Man* New York and London, Academic Press

Prasad, H.P. 1974 "Reactionary Role of Usurer's Capital in Rural India" *Economic and Political Weekly* IX (32-34), pp 1305-1308

Pray, C. & Chitrita Abdullah 1980 "A Select Bibliography on Agricultural Economics and Rural Development with Special Reference to Bangladesh" *Supplement II, BARC* No. 8, March

Pye, L.W. 1966 *Aspects of Political Development* Little Brown, Boston

Rahman, A. 1980 "Usury Capital and Credit Relations in Bangladesh Agriculture: Some Implications for Capital Formation and Capitalist Growth" *BIDS*, Dhaka, *mimeo*, March

Rahman, A. 1981 *Rural Power Structure: A Study of Local Level Leaders in Bangladesh*, Dhaka: Bangladesh Books International

Rahman, T. 1985 *Social and Political Implications of Changing Land and Labour Relations in Rural Bangladesh*, Cambridge, Unpublished Ph.D.

Ranton, M. (Ed.) 1965 *Political Systems and the Distribution of Power*, London

Rashiduzzaman, M. 1968 *Politics and Administration in the Local Councils: A Study of Union and District Councils in east Pakistan*, Dhaka: OUP

Rashiduzzaman, M. 1982 *Rural Leadership and Population Control in Bangladesh*, Washington: University Press of America.

Redfield, R. 1966 *Peasant Society and Culture*, University of Chicago Press

Riggs, F. 1960 *Administration in Developing Countries: The Theory of Prismatic Society* Boston, Houghton Mifflin

Rondinelli, D.A. 1982 "The Dilemma of Development Administration: Complexity and Uncertainty in Control-oriented Bureaucracies" *World Politics* Vol. 35, No. 1, pp 43-72

Rudra, A. 1978 "Class Relations in Indian Agriculture" in *Economic and Political Weekly*, June 3, 10 & 17

Salimullah & A.B.M. Shamsul Islam "A Note on the Condition of the Rural Poor in Bangladesh" *Bangladesh Development Studies* IV (2) pp 267-274

Schaffer, B.B. and G.B. Lamb 1974 "Exit, Voice and Access" *Social Science Information* 13 (6) pp 73-90

Schaffer, B.B. and H. Wen-hsien 1975 "Distribution and the Theory of Access" *Development and Change* Vol. VI, No. 2 (April) pp 13-36

Sen, A. 1981 *Poverty and Famines: An Essay on Entitlement and Deprivation* Oxford OUP

Sengupta, A. 1971 "Regional Disparity and Economic Development of Pakistan" *Economic and Political Weekly* VI (45), Sept. 6, pp 2279-2286 & VI (46), Sept. 13, pp 2315-2322

Shanin, T. & H. Alavi (eds) 1981 *Introduction to Sociology of Developing Societies*

SIDA 1979 *Contradictions and Distortions in a Rural Economy: The Case of Bangladesh* by Stefan de Vylder and Daniel Asplund

Singer, M. & Cohn (Eds.) 1968 "Structure of Politics in Villages of Southern Asia" *Structure and Change in Indian Society*, New York

Sobhan, R. 1968 *Basic Democracies, Works Programme and Rural Development in East Pakistan*, Dhaka: Bureau of Economic Research

Solaiman, M. and M.M. Alam 1977 *Characteristics of Candidates for Election in Three Union Parishads in Comilla Kotwala Thana*, Comilla BARD

Stroberg, P. 1977 *Water and Development: Organisational Aspects of a tubewell project in Bangladesh* SIDA, Dhaka

Swartz, M. et al. (Eds.) 1966 "Segmentary Factional Political Systems" *Political Anthropology*, Chicago

Swift, J. 1989 "Why are Rural People so vulnerable to Famine?" *IDS Bulletin*

Turbyne, J. 1992 "Participation and Development: Review of Literature", mimeo, Ph.D. Working Paper CDS, University of Bath

Uphoff, N. 1988 "Assisted Self-Reliance: Working with, rather than for, the Poor" in J.P. Lewis et al. *Strengthening the Poor: What Have We Learned?* New Brunswick: Transaction Books for Overseas Development Council pp 47-59

van Schendel, W. 1976 *Bangladesh: A bibliography with Special Reference to the Peasantry* Univ. of Amsterdam

Westergaard, K. 1978 "Mode of Production in Bangladesh" *Journal of Social Studies* (Dhaka) 2, pp 1-26

Westergaard, K. 1979 "The State and Rural Society in Bangladesh", Ph.D. Submission, University of Copenhagen

White, S. 1992 *Arguing with the Crocodile* Zed Press

Wolf, E. 1969 *Peasant Wars of the Twentieth Century* Faber, London

Wood, G.D. & R. Palmer-Jones 1991 *The Water-Sellers* IT Publications

Wood, G.D. 1973 "From Raiyat to Rich Peasant" *South Asian Review*, 7 (1) (October) pp 1-16

Wood, G.D. 1974 "Peasantry and the State: the Historical Analysis of Underdevelopment and Change" *Development and Change* 4 (2) (February) pp 45-74

Wood, G.D. 1976 "Class Differentiation and Power in Bandakgram: The Minifundist Case" in A. Huq (ed.) *Exploitation and the Rural Poor: A Working Paper on the Rural Power Structure in Bangladesh*, Comilla BARD *(Chapter 1, this volume)*

Wood, G.D. 1977a "Rural Development and the Post-Colonial State: Administration and the Peasantry in the Kosi Region of NE Bihar" *Development and Change* 8 (3) (July) pp 307-323

Wood, G.D. 1977b "The Nature of Rural Class Differentiation in Bangladesh" *Centre for International and Area Studies* University of London *(later redrafted as Wood 1981)*

Wood, G.D. 1980a "Rural Development in Bangladesh: Whose Framework?" *Journal of Social Studies*, 8, Dhaka, pp 1-31 *(Chapter 3, this volume)*

Wood, G.D. 1980b "The Rural Poor in Bangladesh: A New Framework?" *Journal of Social Studies*, 10, Dhaka, pp 22-46 *(Chapter 6, this volume)*

Wood, G.D. 1981 "Rural Class Formation in Bangladesh 1940-1980" *Bulletin of Concerned Asian Scholars,* 13 (4) (1981), pp 2-15. *(Chapter 2, this volume)*

Wood, G.D. 1982 *Socialisation of Minor Irrigation in Bangladesh,* monograph, PROSHIKA, Dhaka

Wood, G.D. 1983 *Landless Labour Participation and Mobilisation in the Intensive Rural Works Programme in Bangladesh,* Swedish International Development Authority, Dhaka *(Chapter 10, this volume)*

Wood, G.D. 1984a "Provision of Irrigation Services by the Landless", *Agricultural Administration,* 17 (2) pp 55-80 *(Chapter 7, this volume)*

Wood, G.D. 1984b "State Intervention and Bureaucratic Reproduction: Comparative Thoughts" *Development and Change* 15 (1) January, pp 23-41

Wood, G.D. 1984c *A Study of Government Approaches Towards the Rural Poor in Bangladesh* GOB/ Swedish International Development Authority. Dhaka *(Chapter 16, this volume)*

Wood, G.D. 1985a "Politics of Development Policy Labelling" *Labelling in Development Policy* (ed G. Wood)

Wood, G.D. 1985b "Targets Strike Back—Rural Works Claimants in Bangladesh" in Wood (Ed) *Labelling in Development Policy* Sage Publications pp 109-131 *(Chapter 11 this volume)*

Wood, G.D. 1985a Editorial in G.D. Wood (Ed.) *Labelling in Development Policy,* Sage, London

Wood, G.D. 1986a "Don't Give Them my Telephone Number—Applicants and Clients: Limits to Public Responsibility" *Public Administration and Development* 6 (4) pp 475-484

Wood, G.D. 1986b "Mobilisation or Delivery: A comparison of Institutional Approaches to the Rural Poor in Bangladesh", *(mimeo,* conference paper)

Wood, G.D. 1988 "Plunder Without Danger—Avoiding Responsibility in Rural Works Administration in Bangladesh" *IDS Bulletin* (October) pp 57-63 *(Chapter 13, this volume)*

Wood, G.D. 1991 "Agrarian Entrepreneurialism in Bangladesh" *Indian Journal of Labour Economics* 34 (1) pp 13-27 *(Chapter 9, this volume)*

Wood, G.D. 1992 *Proshika: Theory and Practice for 90s and Beyond,* Report for Proshika, Dhaka, Bangladesh

World Bank 1982 *World Development Report* OUP

World Bank 1983 *World Development Report* OUP

World Bank 1983a *Bangladesh: Selected Issues in Rural Employment* Washington: The World Bank. Report No. 49-BD

World Bank 1983b *Rural Development—II: Appraisal Report* The World Bank, Washington

World Bank 1989 *Appraisal of Third Fisheries Project*

World Bank/IRDP 1981 *Bangladesh: Integrated Rural Development Programme -A Joint Review* GOB and the World Bank, Washington/ Dhaka

Yunus, M. 1976 *Planning in Bangladesh: Format, Technique and Priority,* Rural Studies Project, Economics Dept. University of Chittagong (quoted in SIDA 1979)

Glossary of Non-English Words

aman	autumn (late summer) rice; autumn season
arotdar	wholesaler
ashol	original/real
aus	early summer rice; early summer season
bandok	mortgaging
baor	lake
bargadars	sharecroppers
bari	homestead, usually consisting of an inner courtyard with huts around (kin groups)
bheel	open water bodies
bhumihin	landless
Bhumihin Samity	Landless Party
bideshi	foreigner
bigha	unit of land measurement, officially 0.33 acre, but with many local variations
bittaheen	assetless
boro	winter rice
braus	hybrid rice season combining winter rice and *aus*
caste-tola	caste hamlet
char	agricultural land formed from silt deposits along the edges of rivers
chaukidar	guarding (e.g. fish stocks from poachers)
chula	oven, hearth, household
dak	so-called
dheki	traditional paddy-husking device
dhani	week-old fish seed (hatchlings)
dhone	traditional low-lift irrigation using dug-out canoe-shaped vessel which is plunged into the river, canal, etc., and raised to shoot the water into a field channel. This can irrigate approximately 2.43 hectares.
digi	small pond
gherao	non-violent occupation (of government office for example)
girbi	a type of mortgage system

godown	warehouse
gosti	patrilineage
Gram Sarkar	village government
haat	afternoon village market held twice a week
haor	low-lying area in N. Bangladesh, perennially flooded
happa	homestead pond
jajmani	caste-based system of patron-client exchange
jotedar	petty landlords/rich *raiyats*
katcha	opposite of pucca - un-made-up, impermanent
khal	canal
khas	government-owned land
khata	measurement
lakh	100,000
madhya-bangsho	middle status lineage
madrasa	village Islamic school
mahajan	moneylender
mahila	women
maund	a measurement equivalent to about .37 kilos
mel-darbar	public court for settling disputes
nicho-bangsho	low status-lineage
para	part of a village, neighbourhood
paribar	cultivating household unit
patil	earthenware (now metal) container carried on shoulder yoke for water and/or small fish
pikar	market retail trader
prantik	marginal peasant
pucca	permanent, solid construction, cemented (of building)
pukur	large pond
rabi	late autumn/winter crop of pulses, oil seeds & vegetables
raiyat	permanent occupancy tenants (pre 1950)
rakhal	bonded labour
reyai	ceremonial group
samaj	Muslim congregation; village association.
samity	society, organised group
sardar	leader, boss
sardari	having traditional high status
seer	a measurement equivalent to about 0.9 kilos.

Glossary of Non-English Words

shalish	village
Swanirvar	name given to government-sponsored NGO created in the aftermath of the 1974-75 floods & famine
taka (tk)	the monetary unit of Bangladesh
thana	administrative unit of Bangladesh (roughly a county)
ucho-bangsho	high status lineage
ulashi	voluntary, self-help canal-digging programme launched by an enthusiastic District Commissioner in the late 70s.
upazila	sub-district
zamindar	superior landlord/revenue collector (pre 1950).
zilla	district

Acronyms

ACF	Agricultural Co-operatives Federation
ADAB	Association of Development Agencies in Bangladesh
ADC	Area Development Centre
BADC	Bangladesh Agricultural Development Corporation.
BARD	Bangladesh Academy for Rural Development
BCAS	Bangladesh Centre for Advanced Studies
BKB	Bangladesh Krishi Bank, the agricultural development bank.
BRAC	Bangladesh Rural Advancement Committee
BRDB	Bangladesh Rural Development Board
BSS	Bittaheen Samabaya Samitis (Assetless Co-operatives)
CARE	Co-operative American Relief Everywhere
CIDA	Canadian International Development Agency
CO	Circle Officer
DMLA	District Military Law Administrator
DOF	Directorate of Fisheries
DTW	Deep tubewell
EIG	Employment and Income Generating
ERD	External Resources Division
FAP	Flood Action Plan
FCDI	Flood Control Drainage and Irrigation
FSMF	Fish Seed Multiplication Farm
FWP/FFWP	Food for Work Programme
FYP	Five Year Plan
GOB	Government of Bangladesh
HBB	Herring-bone brick
HYV	High yielding varieties (rice etc.)

IDP	Infrastructure Development Programme
IGA	Income Generating Activity
IRDP	Integrated Rural Development Programme
IRRI	International Rice Research Institute (High yield varieties of rice)
IRWP	Intensive Rural Works Programme
KSS	*Krishi Sambaya Samity* (farmers' co-operative)
LGEB	Local Government Engineering Bureau
LGRD	Ministry of Local Government & Rural Development
LLP	Low-lift pump.
MFL	Ministry of Fisheries and Livestock
MLGRDC	Ministry of Local Government & Rural Development & Co-operatives
MPO	Water Master Plan Organisation
MTA	Monitoring Technical Assistance
MTFPP	Medium-Term Food Production Plan
NCB	National Commercialised Banks
NFMP	New Fisheries Management Policy
NR	Natural resources
PEP	Primary Education Programme (in Chapter 20)
PEP	Production and Employment Programme (elsewhere)
PFDS	Public Food Distribution System
RESP	Rural Employment Sector Programme
RPP	Rural Poor Programme
RRA	Rapid Rural Appraisal
RWP	Rural Works Programme
SDO	Sub-divisional Officer
SIDA	Swedish International Development Agency
STW	Shallow tubewell.
TCCA	*Thana* Control Co-operative Assistant
TCO	Technical Co-operative Officer (ODA)
TFP	Third Fisheries Project
TTDC	*Thana* Training and Development Centre

TVD	Total Village Development
UCCA	*Upazila* Control Co-operative Assistant
UNO	*Upazila Nirbahi* Officer, administrative head of *upazila*
UP	Union *Parishad* (Council)
VEA	Village Extension Assistant
VGF	Vulnerable Group Feeding
WAPDA	Water and Power Development Authority
WFP	World Fisheries Programme
XEN	Executive Engineer

Index

access, economic & political, 83, 216
access, problems of, 143, 182, 242, 266
accountability, 7, 14, 20, 21, 131, 151, 173, 190, 263, 270, 282, 303, 306, 316, 326, 331, 332, 338, 419, 420, 471, 492, 503, 521, 529, 541-8
administrative reform, 335, 521, 555-6, 562
adult literacy, 22, 405, 491-93, 498
agrarian change, 3, 36, 157, 164, 169, 216, 233, 252, 299, 308, 425, 432, 471
agrarian entrepreneurialism, 5, 163, 233, 251, 254
agrarian reform, 4, 171-2, 181, 185-6, 195, 213, 556, 563
agrarian structure, 25, 101, 108, 133, 163-4, 169, 181, 193, 236, 239, 250, 432-3, 436, 467, 480
agrarian/ rural management, 243
agricultural growth, 3, 6, 92, 166-7, 177, 294-52, 479
agricultural modernisation, 92
agricultural service model, 480
agricultural trends, 166
aid, 1, 8, 14, 17, 20, 24, 100, 108, 129-30, 143, 159, 165, 180, 191, 238, 243, 293
aid agency *see also* external agency, 8, 108, 188, 194, 239, 244
antediluvian capital *see also* Usurious capital, 106-7, 124, 125, 180

aquaculture, 9, 11, 346, 365-68, 371-73, 378, 383
area development, 130, 149, 230, 284, 285
area implementation, 145
area planning, 145
assetless, 159, 290, 294

basic needs, 130, 143, 144, 147, 238
benign state, 129, 151
bhumihin, 290, 297, 303, 308, 416, 418
bittaheen, 290, 298, 299, 308, 377, 389, 394, 397, 406, 411, 415, 424, 431, 445, 464, 473, 548
Bondokgram, 50, 78, 122
broker(s), 37-9, 65, 136, 188, 190, 249, 254, 330, 414, 416, 419, 434, 487, 523, 530
bureaucracy, 34, 129, 143, 148, 175, 179, 190, 242, 278-9, 326, 482, 520-21, 523, 533
bureaucratic interests, 174

capital, productive & unproductive, 101, 106-7, 123, 164, 183
capitalisation (of agriculture), 167, 233
capture fishery, 9, 13, 384
civil society, 20, 338, 492, 525, 529, 538
class consciousness, 49, 88, 94, 96, 97, 231
class differentiation, 38, 48, 50, 78, 85, 97, 98, 99, 128, 133

class formation, 3, 29, 35, 40, 48, 4, 87, 100, 101, 102, 106, 117, 120, 122, 123, 143, 151, 176

class interest, 68, 97, 293, 503

class structure, 51, 80, 81, 93, 144, 214, 316, 342

co-operative(s), 32, 33, 35, 39, 41, 42, 43, 44, 50, 68, 86, 91, 92, 93-5, 99, 102, 105, 131, 133, 134, 136-7, 140-2, 146, 147, 148, 151, 174, 177, 179, 180, 186, 187, 188-9, 195, 260, 287-8, 308, 421-22, 424, 547, 548

Comilla, model, style, 2, 139, 140, 145-47

command area, ref. irrigation, 52, 90, 168, 197-206, 208-12, 220, 221

commercial agriculture, 165

commercialisation strategy, 175, 176, 177, 181, 183

conditionality, 20, 22, 398, 541, 542, 547, 554, 560

conscientisation, 295, 297, 308, 320, 408, 425, 486, 488, 496, 504, 508, 509, 521, 544, 547

consumption credit, 147, 189, 268, 304

corruption, 2, 7, 13, 21, 22, 32, 77, 135, 143, 165, 190, 239, 244, 247, 311, 320, 327, 338, 393, 419, 484, 492, 494, 499, 502, 519, 520, 521, 528, 529, 530-5, 537, 550

credit, 81, 140, 147, 187, 189, 221, 229, 242, 267, 344, 349, 361, 377, 391, 406, 442, 506

credit and fish culture, 382

crop diversification 166, 241, 242

crop trends, 245

cyclical kulakism, 48, 78, 103, 122, 138

cyclical mobility, 80, 81, 103, 138

decentralisation, 150, 243, 244, 313, 521, 522

democracy, 23, 96, 244, 278, 502, 503, 510, 542, 543, 554, 561

dependence, 19, 27, 38, 43, 63, 65, 72, 78, 79, 88, 89, 94, 136, 143, 147, 157, 166, 170, 177, 190, 207, 231, 263, 266, 321, 336, 420, 424, 503, 507, 545, 549

destitute, 136, 158, 265, 266, 298, 301, 303, 316, 355, 396, 422, 423, 426, 502

development administration, 14, 21, 22, 38, 467, 469, 470, 520, 521, 522, 523, 524, 533, 537, 539, 540

development policy, 1, 2, 3, 6, 8, 14, 92, 102, 127, 131, 257, 555, 561

development process, 9, 405, 528, 551

diversification, of economic activity, 78, 82

dowry, 156, 227, 237, 457, 466, 494, 499, 500, 509, 513, 552, 556, 558, 563

draught power, 248, 253

effective demand, generation of, problem of, 125, 164, 179, 16, 194, 195-6, 238

elites, 3, 34, 35, 116, 136, 137, 182, 264, 315, 441, 466, 525

employment & income generation projects, 4, 553, 560

employment rights, 14, 257, 301, 308, 337, 476

empowerment, 20, 169, 171, 463, 486, 494, 495, 46, 497, 498, 499, 501, 503, 504, 513, 521

entrepreneur model, 478

entrepreneurialism, agrarian, 5, 163, 164, 233, 251, 254

Index 585

environment/ecology, 41, 83, 115, 149, 236, 323, 348, 361, 549
exploitation, 2, 3, 11, 31, 40, 43, 48, 66, 67, 78, 79, 82, 84, 87, 95, 104, 10, 106, 111, 117, 119, 128, 155, 156, 157, 176, 190, 234, 237, 253, 280, 328, 389, 432, 466, 482, 486, 488, 497

factional politics, 40
factionalism, 39, 60, 91, 521, 523
factions, 25, 39, 40, 41, 42, 43, 62, 67, 90, 123, 124, 133, 175, 464
FAP (Flood Action Plan), 7, 8, 9, 258, 309, 316
female labour participation, 157
female power and cultural norms, 152
female-headed households, 19, 154, 156, 158, 353, 355, 539
fertiliser consumption, 241, 247
fertiliser distribution, 72, 247
fish culture, 10, 12, 13, 19, 339, 340, 341, 342, 343, 344, 345, 346, 348, 349, 350, 351, 352, 353, 354, 355, 356, 357, 358, 359, 360, 361-65, 367, 372, 373, 374, 376, 377, 379, 380, 382, 384, 387, 391, 394, 398, 399, 400
fish farming, 351, 388
fish pond culture, 2, 9, 341
fish production, 12, 343, 345, 350, 351, 352, 356, 359, 362, 364, 367, 368, 373, 374, 376, 377, 380, 386
fish projects, 9
fishseed trading network, 346
flood control, 7, 184, 241, 245, 285, 310, 318, 320
flooding, 52, 62, 63, 64, 65-88, 108, 109, 110, 166, 191, 222, 235, 240, 253, 310, 322, 327, 368, 38, 461*

forestry, 22, 318, 323, 329, 449, 477, 481, 506
fragmentation, of farm, 1, 26, 164, 209, 220
fragmentation, of landholding
franchise model, 546, 547, 554, 555, 560, 561
franchise state, 14, 20, 23, 517, 538, 541, 546, 548
functional schools, 494, 508
fundamentalism, 25, 494, 513

gender, 24, 152, 153, 154, 155, 156, 157, 158, 59, 235, 310, 321, 356, 376, 380, 396, 405, 409, 420, 457, 466, 492, 500, 514, 543, 544, 549
gender, discrimination, 157
girbi, 121, 415
good governance, 14, 20, 21, 22, 503, 509, 517, 521, 529, 538, 541, 543, 554, 560
good government, 542, 542-43, 543, 544, 545, 546
Gram Sarkar, 174, 178, 179, 294
Grameen Bank, 17, 178, 179, 196, 255, 296, 356, 377, 381, 395, 508, 509
green revolution, 41, 163, 485
group consciousness, 422
group formation, 4, 43, 85, 187, 28, 298, 395, 408, 427, 429, 430, 431, 432, 434, 439, 441, 443, 444, 445, 446, 448, 450, 451, 452, 457, 459, 460, 463, 465, 467, 474, 475, 479, 497, 498
group mobilisation, 21, 286, 296, 421, 428, 430
group solidarity, 4, 218-9, 261, 307, 395, 424, 479
group-internal composition of, 187

high-tech agriculture, 6, 26, 165, 476-7, 479
horizontal integration, 233, 513
horticulture, 22, 170-1, 477, 480-1
human resources, 491, 540

income-generating projects, 198, 219, 291, 295, 500, 505
indebtedness, 95, 114, 125, 138, 140, 147, 174, 219, 228, 267, 311, 409, 419, 422, 509
indicators, activity-output, 498
indicators activity-input, 20, 479
infrastructure, rural, 178, 248, 264, 283, 300, 309, 312, 327
inputs, to agriculture, 27, 92, 176, 312, 323
institutions, rural, 3, 33, 135, 141, 432-3
interlocked markets, 4, 15, 234, 248, 251, 253
intimidation, 265, 270, 360
investment, co-operative, 187-8
irrigation, services, 2, 4, 193, 195, 198, 200, 216, 218, 222, 226-8, 251, 255, 424, 477, 479
irrigation management groups, 190

jajmani, 41

labelling, in development policy, 2
labour contracting societies, 2, 316, 429, 483
labour participation, 7, 157, 259
labour relations, 109, 133, 146, 231, 237, 313, 494
labour sarder system, 261, 267, 272, 274, 275, 277 281, 304, 307, 314, 410, 412, 415, 416, 418, 420, 454, 466

land reform, 2, 4, 25, 95, 130, 170, 174, 175, 182, 184, 476
land-competition for, 85, 486
land-ownership of, 47, 55, 79, 318
landholding size, 45, 47, 54, 55, 58, 120, 476
landless, 2, 4, 26, 45, 56, 57-8, 62-5, 87-9, 97, 110, 125, 135, 159, 178, 193, 195-8, 203, 207, 211, 216, 259, 279
landless irrigation programme, 16, 196, 211, 216, 288
landless labour, 178, 223, 259
landless labour participation, 259
LGEB (Local government Engineering Bureau), 327
local democracy, 278, 502-3

market behaviour, 153, 531
market principles, 70, 84
marxist (ism), 31, 104-5, 129, 48, 486-7, 552
mechanised irrigation, 18, 165-6, 191, 206, 239, 245, 251, 323
migrant labour (ers), 62-4, 84, 88
migrant labour and class consciousness, 88
migrant labour impact on the landless, 87
migrant workers, 78, 98
minifundist farming systems, 4
model of agrarian change, Bangladesh, Punjab, 164
money-lending, 70, 79
mortgage/-ing, 46, 70, 79, 95

negotiation, 215
new agricultural technology, 44, 98

new technology, 11, 43, 82, 92, 135, 245, 312, 424
NFMP (New Fisheries Management Policy), 392, 395
NGO, Non Governmental Organisation, 4, 8, 10, 14, 17-9, 22-3, 279, 280, 287, 289, 295, 298, 355, 374
non-agricultural economy, 88, 145

open market strategies, 182
organic agriculture, 5
organic farming, 5, 170

parallel rationalities, 22, 528, 533-7
participatory appraisal, 504
participatory development, 504
patron-client, 23, 41, 70, 85, 89, 104, 109, 133, 140, 193, 197, 215, 231, 237, 292, 326, 486, 523, 528, 550
peasant economy, 92, 101-2, 110, 137
peasant mode of production, 45
pisciculture, 22, 459, 477, 482
polarisation, 26, 100, 107, 186, 238, 292
political anthropology, 35, 39
population growth, 107-8, 117, 128, 132, 133, 193-4, 235-6, 250, 291, 387
populism, 129, 521
poverty alleviation, 2, 6, 497, 551, 557
poverty focus, target, 348
productive assets, control of, 10, 18, 118, 227, 284, 312
project cycle, 324, 511, 526-7
Proshika, 4, 5, 157, 171, 196, 198-9, 207, 213, 219, 223-4, 226, 229, 261, 270-1, 279, 289, 293, 471, 482, 485
public works programmes, 135, 327

rapid rural appraisal, 149, 264
rationing, 175, 192, 531-2
recruitment, of labour, 266, 301, 303, 315, 328
recruitment, to political groups, 41
regional variations, 2, 45-6, 87, 94-6, 101, 108
regionality, 138-9
rent seeking, 343, 520, 529, 533-4, 536
replication, of projects, e.g. Comilla, 136, 139
resource profile, 19, 490-1, 505, 514, 539-40, 556, 563
resources profile approach, 539
revolution, 17, 41, 96, 105, 485-6, 488, 552
role of the state, 180
rural class formation, 100, 102, 117, 143
rural credit market, 123
rural development policy, 3, 14, 92, 102, 127, 131
rural development strategy, 52, 91-2, 128
rural employment, 57, 130, 163, 194, 291, 298, 404, 426
rural employment generation, 291
rural institutions, 3, 33, 135, 179, 432-3
rural mobilisation, 96, 131, 151, 178, 294, 308
rural poor, 2-4, 6, 15, 31, 89, 94, 143, 172, 259, 295, 311, 403, 408
rural power structure, 31, 44, 277
rural property rights, 184-6, 197
rural workers' federations, 281-7

service provision, 21, 503, 519, 521, 534, 537, 539-40, 545
sharecropping, in and out, 79
skills training, 408, 424, 498
social action, 20, 152, 298, 404, 498-9, 522, 535
social forestry, 318, 329, 506
social mobilisation, 320, 490-4, 498-9, 505, 507, 510-14
social relations, 25, 36, 38, 51, 66, 87, 101, 103-4, 111, 115, 117, 137, 142, 145
social relations of production & exchange, 66, 87, 104, 150, 230, 296
social science/scientists, 1, 7-8, 39, 263, 309, 324, 435, 475
solidarity, vertical, 49
solidarity, horizontal, 40, 49, 216
state, the, 14, 17, 20-25, 36, 66, 86, 93, 111, 124, 150, 183, 491, 494, 497, 530, 536
state intervention, 110, 118, 182, 293, 308, 526, 529, 532
state-society relations, 21-2, 519, 521, 526, 533-5
strategies for rural development, 14, 93-4
structural adjustment, 521, 531, 541, 544, 547-8, 554, 560
struggle, 17-20, 35, 84, 101, 122, 129, 152, 169, 180, 213, 215, 280, 289, 293, 301, 308, 395, 484
subsidies, 22, 95, 130, 134-5, 165
sustainable development, 5
swanirvar, 178, 272

target group, 11, 92, 95, 130, 143-4, 147, 159, 221, 318, 348, 392, 437, 467, 508, 525, 527, 538, 550

technical assistance, 15, 129, 141, 310, 329, 387, 399, 428, 435

ulashi, 178, 294
under-employment, 109, 117, 172, 237, 268, 292, 304
unemployment, 21, 108, 125, 176, 194, 237, 292, 476
urbanisation/rurbanisation, 6, 26-7, 480

victimisation of workers *see also* intimidation, 264-5, 270
voice, 7, 21, 223-4, 527, 538, 543-5

water, allocation of, 68, 184, 210
water, rights to, 189-91, 209
water, sellers, 4, 16, 201, 210, 222, 224, 226, 230
waterlords, 191, 197, 209
women, 152, 155
women, double day, 155-7
women, double exploitation of, 155-7
women's involvement in fish culture, 342, 352-5, 376
women's rights, 499, 500

zoning, 145, 210, 224

www.ingramcontent.com/pod-product-compliance
Lightning Source LLC
Chambersburg PA
CBHW020237030426
42336CB00010B/515